Université Joseph Fourier

Les Houches

Session LXXXVI

2006

Particle Physics and Cosmology:

the Fabric of Spacetime

# Lecturers who contributed to this volume

Alessandra BUONANNO
James CLINE
Walter GOLDBERGER
Thomas HERTOG
Lev KOFMAN
Sabino MATARRESE
Hitoshi MURAYAMA
Sergio PASTOR
Simon PRUNET
Romain TEYSSIER
Peter TINYAKOV

# ÉCOLE D'ÉTÉ DE PHYSIQUE DES HOUCHES
## SESSION LXXXVI, 31 JULY – 25 AUGUST 2006

### ÉCOLE THÉMATIQUE DU CNRS

# PARTICLE PHYSICS AND COSMOLOGY:
# THE FABRIC OF SPACETIME

Edited by

Francis Bernardeau, Christophe Grojean and Jean Dalibard

ELSEVIER

Amsterdam – Boston – Heidelberg – London – New York – Oxford
Paris – San Diego – San Francisco – Singapore – Sydney – Tokyo

Elsevier
Radarweg 29, PO Box 211, 1000 AE Amsterdam, The Netherlands
Linacre House, Jordan Hill, Oxford OX2 8DP, UK

First edition 2007

**Library of Congress Cataloging-in-Publication Data**
A catalog record for this book is available from the Library of Congress

**British Library Cataloguing in Publication Data**
A catalogue record for this book is available from the British Library

ISBN: 978-0-444-53007-3
ISSN: 0924-8099

For information on all Elsevier publications
visit our website at books.elsevier.com

10 11

Working together to grow
libraries in developing countries
www.elsevier.com | www.bookaid.org | www.sabre.org

ELSEVIER    BOOK AID
            International    Sabre Foundation

Transferred to Digital Printing 2009

# ÉCOLE DE PHYSIQUE DES HOUCHES

Service inter-universitaire commun
à l'Université Joseph Fourier de Grenoble
et à l'Institut National Polytechnique de Grenoble

Subventionné par le Ministère de l'Éducation Nationale,
de l'Enseignement Supérieur et de la Recherche,
le Centre National de la Recherche Scientifique,
le Commissariat à l'Énergie Atomique

# Previous sessions

**Publishers:**

- Session VIII: Dunod, Wiley, Methuen
- Sessions IX and X: Herman, Wiley
- Session XI: Gordon and Breach, Presses Universitaires
- Sessions XII–XXV: Gordon and Breach
- Sessions XXVI–LXVIII: North Holland
- Session LXIX–LXXVIII: EDP Sciences, Springer
- Session LXXIX–LXXXV: Elsevier

# Organizers

BERNARDEAU Francis, Service de Physique Théorique, CEA Saclay, Orme des Merisiers, 91191 Gif sur Yvette, France

GROJEAN Christophe, Service de Physique Théorique, CEA Saclay, Orme des Merisiers, 91191 Gif sur Yvette, France and Physics Department, Theory Unit, CERN, 1211 Geneva 23, Switzerland

DALIBARD Jean, Laboratoire Kastler Brossel, École normale supérieure, 75231 Paris cedex 05, France

x

# Lecturers

BUONANNO Alessandra, Department of Physics, University of Maryland, College Park MD 4207-4111, USA

CLINE James, Physics Department, Mc Gill University, E. Rutherford Physics Building, 3600 University Avenue, Montreal, Quebec, Canada H3A 2T8

GOLDBERGER Walter, Department of Physics, Yale University, 217 Prospect street, New Haven, CT 06511, USA

HERTOG Thomas, Physics Department, Theory Unit, CERN, 1211 Geneva 23, Switzerland

KOFMAN Lev, Canadian Institute of Theoretical Astrophysics, University of Toronto, 60 St George street, Toronto, Ontario, Canada, M5S 3H8

MARCH RUSSELL John, R. Peierls Center for Theoretical Physics, Univ. Oxford, 1 Keble road, Oxford OX1 3NP, UK

MATARRESE Sabino, Dipartimento di Fisica *Galileo Galilei*, Universita di Padova, via Marzolo 8, 35131 Padova, Italy

MELLIER Yannick, Institut d'Astrophysique de Paris, 98 bis Boulevard Arago, 75014 Paris, France

MURAYAMA Hitoshi, University of California, Department of Physics, 366 Le Conte Hall, # 7300, Berkeley, CA 94720-7300, USA

PASTOR Sergio, IFIC, CSIC Univ. Valencia, Apdo 220885, 46071 Valencia, Spain

PRUNET Simon, Institut d'Astrophysique de Paris, 98 bis Boulevard Arago, 75014 Paris, France

SERVANT Géraldine, Service de Physique Théorique, CEA Saclay, Orme des Merisiers, 91191 Gif sur Yvette cedex, France

TEYSSIER Romain, Service d'Astrophysique, CEA Saclay, Orme des Merisiers, 91191 Gif sur Yvette, France

TINYAKOV Peter, Université Libre de Bruxelles, Service de Physique Théorique, CP 225, Boulevard du Triomphe, 1050 Bruxelles, Belgique

TKACHEV Igor, Physics Department, Theory Unit, CERN, 1211 Geneva 23, Switzerland

xii

# Participants

AMIN Mustafa, Stanford University, Varian Physics Building, 382 Via Pueblo Mall, Stanford, CA 94305, USA

ANSARI Rizwan Ul Haq, H No 12-2-82, Muradnagar, PO Humayunagar, Hyderabad 500028, India

BAUSHEV Anton, Space Research Institute IKI, 117997, 84/32 Profsoyuznaya str., Moscow, Russia

BEBRONNE Michael, Université Libre de Bruxelles, Département de Physique Théorique CP 225, Boulevard du Triomphe, 1050 Bruxelles, Belgium

BERCHTOLD Julian, I.I.S Zürich Fortunagasse 18, 8001 Zürich, Switzerland

BROWN Adam, Department of Physics, 538 West 120th Street, Columbia University, New York City, NY 10027, USA

CHAMBERS Alex, Theoretical Physics Group, Blackett Laboratory, Imperial College, London SW7 2AZ, UK

D'AMICO Guido, Scuola Normale Superiore, Piazza dei Cavalieri 7, Pisa 56100, Italy

DAS Subinoy, Meyer Hall, Physics Department, New York University, room #424, 4 Washington place, New York, NY 10003, USA

DELAUNAY Cédric, SPhT CEA Saclay, Orme des Merisiers, 91191 Gif sur Yvette cedex, France

DI NAPOLI Edoardo, University of North Carolina, Department of Physics and Astronomy, CB 3255 Phillips hall, Chapel Hill 27599 3255, North Carolina, USA

DVORNIKOV Maxim, Institute of Terrestrial Magnetism, Ionosphere and Radiowave Propagation, IZMIRAN, 142190, Troitsk, Moscow Region, Russia

ERICKCEK Adrienne, Theoretical Astrophysics, CALTECH, 1200 E California Boulevard, Mail code 103 33, Pasadena CA 91125, USA

FU Liping, Room 111, Building 3, Joint Center of Astrophysics, Shanghai Normal University, 100 Guilin road, Mail Box 247, 200234 Shanghai, China

GALAVERNI Matteo, INAF Istituto di Astrofisica Spazial e Fisica Cosmica di Bologna, Via Gobetti 101, 40129 Bologna, Italy

GEORGE Damien, The School of Physics, University of Melbourne, Victoria, 3010, Australia

GONG Jinn-Ouk, International Center for Astrophysics, Korea Astronomy and Space Science Institute, Hwaam-dong, Yuseong-gu, Daejeon, Republic of Korea

GRIN Daniel, CALTECH, 1200E California Boulevard, Mail Code 130-33, Pasadena Ca 91125, USA

GWYN Rhiannon, Centre for High Energy Physics, Mc Gill University, Rutherford Physics Building, 3600 Rue de l'Université, Montréal, Québec, Canada H3A 2T8

HERTZBERG Mark, MIT, Center for Theoretical Physics, 77 Massachusetts Avenue, Cambridge MA 02139, USA

JÄRV Laur, Laboratory of Theoretical Physics, Institute of Physics, University of Tartu, Riia 142, 51014, Estonia

JORAS Sergio, Federal University of Rio de Janeiro Ű UFRJ- Institute of Physics, P.O Box 68528, 21941-972 Rio de Janeiro, RJ, Brazil

KNAUF Anke, Mc Gill University, Rutherford Physics Building, 3600 Rue de l'Université, Montréal, QC, Canada, H3A 2T8

KUMAR Abhishek, New York University, Physics Department, 4 Washington place, New York, NY 10003, USA

LI Nan, Group of Theoretical High Energy Physics, Bielefeld University, Universitaetsstrasse, 33615 Bielefeld, Germany

LOPEZ-HONOREZ Laura, Université Libre de Bruxelles, Département de Physique Théorique CP 225, Boulevard du Triomphe, 1050 Bruxelles, Belgium

LUO Zhijian, Department of Physics, Jiangxi Normal University, Nanchang 330022, China

MAROZZI Giovanni, Department Physics, University Bologna, Via Irnerio 46, 40126 Bologna, Italy

METZLER Falk, Institut für Theoretische Teilchenphysik, Universität Karlsruhe, 76128 Karlsruhe, Germany

MOVAHED Sadegh, Department of Physics, Sharif University of Technology, Azadi Street, Tehran, Iran

NGUYEN Thi Thu Huong, Faculty of Physics, Vietnam National Univrsity, 334 Nguyen Trai Street, Hanoi, Vietnam

PALORINI Federica, Université C. Bernard, Bâtiment P. Dirac, 43 Boulevard du 11 Novembre 1918, 69662 Villeurbanne cedex, France

PORTO Rafael, Physics Department, Carnegie Mellon University, 5000 Forbes Avenue, Pittsburgh, PA 15213, USA

PRITCHARD Jonathan, CALTECH, 1200E. California Boulevard, M/C 103-33, Pasadena CA 91125, USA

SARKAR Anjishnu, Physics Department, Indian Institute of Technology, Bombay, Powai, Mumbai, Maharashtra, 400076, India

SAWICKI Ignacy, E. Fermi Institute, University of Chicago, 5640 S Ellis Ave., Chicago, IL 60637, USA

SCHULZE Markus, Institute for Theoretical Particle Physics, Universität Karlsruhe, 76128 Karlsruhe, Germany

SELLERHOLM Alexander, CoPS Group, Department of Physics, Stockholm University, 10691 Stockholm, Sweden

SHAH Nausheen, E. Fermi Institute, University Chicago, 5640 S. Ellis ave., Chicago, IL 60637, USA

SIEGEL Ethan, University of Florida, Department f Physics, PO Box 118440, Gainesville, FL 32611-8440, USA

SIVANANDAM Navin, Stanford University, Varian Building, 382 Via Pueblo Mall, Stanford CA 94305, USA

SMITH Tristan, CALTECH, MC 103-33, 1200E California Boulevard, Pasadena CA 91125, USA

STURANI Riccardo, Physics Department, Univ. Geneva, 24 Quai E. Ansermet, 1211 Genève, Switzerland

SUNHEDE Daniel, Department of Physics, University of Jyväskylä, Po Box 35, 40014 Jyväskylä, Finland

TULIN Sean, CALTECH, MSC 103-33, 1200 E. California Boulevard, Pasadena, CA 91106, USA

URBAN Federico, University of Ferrara, Department of Physics and INFN, Sezione di Ferrara, via Saragat 1, 44100 Ferrara, Italy

VERTONGEN Gilles, Université Libre de Bruxelles, Département de Physique Théorique CP 225, Boulevard du Triomphe, 1050 Bruxelles, Belgium

VIKMAN Alexander, Universität München, A. Sommerfeld Center for Theoretical Physics, Lehrstuhl Mukanov, Theresienstr. 37, 80333 München, Germany

WELTMAN Amanda, Columbia University, 538 West 120th Street, New York, NY 10027, USA

# Preface

The Les Houches summer school of August 2006 devoted to "*The fabric of space-time*" is the latest of a series of schools on theoretical cosmology. The last two took place in 1993 with "Cosmology and large scale structure" and in 1999 with "The primordial universe". Over the last decade, a fast tightening of the links between observational cosmology, the physics of the early universe and high energy physics theories has taken place. With a new generation of experimental and observational data due to arrive in the coming years, thanks to projects such as the Large Hadron Collider and the Planck satellite, the interplay between these different domains of research is expected reinforced even more. In this respect, this session LXXXVI nicely complements the session LXXXIV of August 2005 devoted to "*Particle Physics beyond the Standard Model*".

One of the aims of the lecture notes presented here is to bridge the gap between particle physics and observational cosmology. And there are many domains where experiments or observations might shed light on the theories aiming at describing the fundamental laws of Nature. For example, the observations of gravity waves or of high energy cosmic rays may provide us with new windows to fundamental physics, especially at energies that will be difficult to probe directly in laboratories. The observations of the large-scale structure of the universe, whether it is through Cosmic Microwave Background anisotropies or cosmic shear mass maps, offer a precious window to the physics of the early universe, and possibly a glimpse into quantum gravity effects. The mechanisms at play for the origin of the observed density or metric fluctuations are still subject to passionate discussions. Inflation may provide us with the correct paradigm but it leaves many questions unanswered: what is the inflaton field? How did inflation end? How did the matter form afterwards? Concerning matter genesis, the mechanism for baryogenesis still eludes investigations, and so does the nature of dark matter and dark energy.

This volume then puts particular emphasis on our current understanding of all these issues, introducing the theoretical tools required for undertaking investigations in any of these fields. The lectures span from various aspects of observational cosmology to notions of string theory.

The lectures given by *Sabino Materrese* cover the physics of the cosmic microwave background. Here the resolution of the Boltzmann equation is the key for the understanding of how the initial metric perturbations can lead to observable temperature anisotropies and polarization. Unlike textbooks on this subject, the presentation includes how to treat not only linear but also second order effects. This might reveal of crucial importance for a full scientific exploitation of incoming observational projects, that is specifically of the Planck satellite due to be launched in year 2008. The contribution of *Simon Prunet* complements these lectures by giving a critical presentation of the current observational situation. He shows in particular how challenging the measurement of the CMB polarization can be. Cosmic shear observations is the other modern mean of investigation of the large-scale structure of the universe. They correspond to the gravitational deformation of deep field galaxies induced by intervening large scale structure of the universe along the lines of sight. *Yannick Mellier* gave a comprehensive review of the theoretical basis of such observations, as well as the difficulties that their practical implementation convey.*

The linear theory of the formation of structure out of dust particles has been developed till the mid-eighties and is now a well established subject. However once dissipative or non-linear physics have to be included, one has to rely on dedicated numerical simulations. *Romain Teyssier* gave an insightful introduction of the art of doing numerical simulations in cosmology. Lessons have certainly to be learned from his presentation of the intricate processes that lead to galaxy formation.

However, a presentation of a theory of the large scale structure of the universe would not be complete without a description of its origin. Inflation is nowadays the only paradigm that can provide us with a genuine theory. The lectures by *Lev Kofman* contain many original and insightful comments on the inflationary physics. His notes provide us with probably the most original presentation of what the physics behind inflation is. But there is one question which is left unanswered here, that is, what is the inflaton field itself. Needless is to say that the identification of the inflaton field in terms of a fundamental field would be a major theoretical breakthrough! For good or bad reasons, string theory has become the most sought after playground for designing inflationary models. *James Cline* presented the logic and motivation of current models of string inspired inflation unveiling the strength and weaknesses of such constructions. His lecture notes are a nice introduction to the subject fot non string-experts. A complementary theoretical construction is provided by the lecture of *Thomas Hertog* who addresses the question of a big-bang singularity in the context of quantum gravity, using the modern tools of the AdS/CFT correspondence. *John March-Russell* detailed the cosmological implications of current developments in string theory that concluded in the existence of a landscape consisting of many metastable de Sitter vacua.*

So far the photons of the CMB have been our primary messengers from the early Universe. In a near future we expect another important source of information in the form of gravitational waves (GWs) that could have been produced during inflation or during a cosmological first-order phase transition. Cosmic and fundamental strings could also be an important source of gravitational waves in the early Universe. *Alessandra Buonanno* gave a pedagogical account of the physics of GWs. Her lecture notes describe in details the interaction of GWs with free test particles and present the key ideas underlying the functioning of GW detectors. Then she concentrates on the production of GWs from binary systems and other astrophysical sources like pulsars and supernovae. The numerous major theoretical advances coincide with an ambitious experimental program soon fully operational at its design sensitivity. It should give full access to the nonlinear dynamics of general relativity 100 years after its discovery. *Walter Goldberger* presented another powerful approach to describe gravitational radiation sources. While traditional technics rely on post-Newtonian computations, he adapted from particle physics the methods of effective field theories and applied them to gravitational problems. His lecture notes first review some of the standard ideas behind effective field theory and then set up the calculation of GW emission from non-relativistic binary stars.

*Igor Tkachev* gave an introduction to the physics of ultra-high energy cosmic rays which is still a subject facing many open questions, most notably concerning the sources of the most energetic ones. The presence or the absence in experimental data of the theoretically predicted GZK cutoff in the energy spectrum is still under discussion.*

*Sergio Pastor*'s lectures focus on neutrino cosmology, one of the best examples of the very close ties that have developed between nuclear physics, particle physics, astrophysics ad cosmology. We know since the impressive experimental results of the late 1990's that neutrino are massive particles. These masses have impact on cosmological observables like the matter power spectrum or the CMB anisotropy spectrum, which can used to placed tight bounds on neutrino properties.

The discovery of the present acceleration phase in the expansion of the Universe puts us in front of what is perhaps the most challenging problem high energy physics has to face. This acceleration is usually accounted for the enigmatic Dark Energy that would weight not less than 70% of the energy budget of the Universe today. On a theoretical side, it motivates the search for possible consistent modifications of the dynamics of general relativity at very large distances. *Peter Tinyakov*'s lectures address the question of whether the graviton could acquire a mass. This question has a long history dating back to the 1930's with the pioneering work of Fierz and Pauli. The lectures concentrate on a novel approach to the modification of gravity which involves the spontaneous breaking of the Lorentz invariance.

The Big Bang connects microscopic physics to macroscopic physics. And studying the physics of the early Universe requires to know what are the relevant degrees of freedom of matter and their fundamental interactions. Up to energies of the order of 100 GeV, the Standard Model is certainly a good description but it is believed to fail to capture the full picture at higher energies. *Hitoshi Murayama* exposed the reasons to believe that the TeV scale is a threshold where new physics should be revealed and he presented various extensions of the Standard Model. Any of these extensions provide extra stable particles that will account for the relic abundance of Dark Matter and these new particles are waiting to be discovered at the LHC. One exciting possibility of new physics that could be discovered at the TeV scale is the existence of large extra dimensions in our Universe. It has been realized in particular that these new dimensions could provide a nice explanation to the question of why gravitational interactions are so weak compared to the three other fundamental interactions. The extra dimension proposal was reviewed briefly by *Murayama* and it was the main topic of the lectures given by *Géraldine Servant* who discussed at length collider signatures as well as cosmological implications of the existence of these extra dimensions.*

Another major question that can be addressed with the Physics beyond the Standard Model concerns the abundance the abundance of matter over antimatter in our Universe. In the early days of Big Bang cosmology, the baryon-antibaryon asymmetry was considered to be an initial condition, but in the context of inflation this idea is no longer tenable. Any baryon asymmetry existing before inflation would be diluted to a negligible value during inflation, due to the production of entropy during reheating. Therefore, a dynamical creation of this asymmetry is needed. *James Cline* reviewed the possible scenarios of baryogenesis.

As noticed by one participant of the school, in particle physics, we face a *hierarchy problem* when two physical scales are widely separated, while in cosmology it is a *coincidence problem* when two quantities are close to each other. The physics of the early Universe has to face both types of problems and this is why it is an exciting area at the forefront of theoretical and experimental research.

The summer school and the present volume have been made possible by the financial support of the following institutions, whose contribution is gratefully acknowledged:

– ILIAS (Integrated Large Infrastructures for Astroparticle Science) supported by the Commissariat à l'Energie Atomique and the Centre National de la Recherche Scientifique in France

– The French National Program for Cosmology supported by the Centre National de la Recherche Scientifique (France).

– The "Lifelong learning" program of the Centre National de la Recherche Scientifique (France).

– The French Ministry of Research through an Action Concertée Incitative jeunes chercheurs.

– The Université Joseph Fourier, the French Ministry of Research and the Commissariat à l'Energie Atomique, through their constant support to the Physics School.

The staff of the School, especially Brigitte Rousset and Isabelle Lelièvre, have been of great help for the preparation of the school and we would like to thank them.


F. Bernardeau, C. Grojean and J. Dalibard

*The lectures notes were not available at the time this volume was printed.

# CONTENTS

*Course 4.    Physics of the early universe and inflation,
by Lev Kofman*                                          *165*

Contents

Contents

Contents

# LONG LECTURES

## Graviationnal Waves

*A. Buonanno*

## Baryogenesis

*J.M. Cline*

## String Cosmology

*J.M. Cline*

## Physics of the Early Universe and Inflation

*L. Kofman*

## Cosmic Microwace Background Anisotropies up to Second Order

*N. Bartolo, S. Matarresse and A. Riotto*

## Physics Beyond the Standard Model and Dark Matter

*H. Murayma*

Course 1

# GRAVITATIONAL WAVES

## Alessandra Buonanno

*Department of Physics, University of Maryland*
*College Park MD 20742, USA*
*and*
*AstroParticle and Cosmology, University Paris VII, France*

*F. Bernardeau, C. Grojean and J. Dalibard, eds.*
*Les Houches, Session LXXXVI, 2006*
*Particle Physcis and Cosmology: The Fabric of Spacetime*
© *2007 Published by Elsevier B.V.*

# Contents

# 1. Introduction

Gravitational-wave (GW) science has entered a new era. Experimentally,[1] several ground-based laser-interferometer GW detectors (10–1 kHz) have been built in the United States (LIGO) [1], Europe (VIRGO and GEO) [2, 3] and Japan (TAMA) [4], and are now taking data at design sensitivity. Advanced optical configurations capable of reaching sensitivities slightly above and even below the so-called standard-quantum-limit for a *free* test-particle, have been designed for second [5] and third generation [6] GW detectors (∼2011–2020). A laser interferometer space antenna (LISA) [7] ($10^{-4}$–$10^{-2}$ Hz) might fly within the next decade. Resonant-bar detectors (∼1 kHz) [8] are improving more and more their sensitivity, broadening their frequency band. At much lower frequencies, ∼$10^{-17}$ Hz, future cosmic microwave background (CMB) probes might *detect* GWs by measuring the CMB polarization [9]. Millisecond pulsar timing can set interesting upper limits [10] in the frequency range $10^{-9}$–$10^{-8}$ Hz. At such frequencies, the large number of millisecond pulsars which will be detectable with the square kilometer array [11], would provide an ensemble of clocks that can be used as multiple arms of a GW detector.

Theoretically, the last years have been characterized by numerous major advances. For what concerns the most promising GW sources for ground-based and space-based detectors, notably, binary systems composed of neutron stars (NS) and black holes (BHs), our understanding of the two-body problem and the GW-generation problem has improved significantly. The best-developed *analytic* approximation method in general relativity is undoubtingly the post-Newtonian method. The errors and ambiguities that characterized the very early literature on the PN problem of motion (for a review see, e.g., Ref. [12]), have been overcome. Robust predictions are currently available through 3.5PN order [13] ($v^7/c^7$), if the compact objects do not carry spin, and 2.5PN order [15] ($v^5/c^5$) if they carry spin. Resummation of the PN expansion aimed at pushing analytic calculations until the final stage of evolution, including the transition inspiral–merger–ringdown, have also been proposed, for the conservative two-body dynamics [16] and the radiation-reaction effects [17]. Quite interestingly, the effective-field-theory approach, commonly used in particle physics, has been extended to gravity, no-

---

[1]GW experiments started with the pioneering work of J. Weber at Maryland in the 60s.

tably to the two-body problem of motion [18]. The recent dazzling breakthrough in *numerical relativity* [19], with different independent groups being able to successfully evolve a comparable-mass BH binary throughout inspiral, merger and ring-down and extract the GW signal, is allowing to dig out details of the nonlinear dynamics which could not be *fully* predicted with other means.

Our knowledge has also progressed on the problem of motion of a point particle in curved spacetime when the emission of GWs is taken into account (non-geodesic motion) [20, 21]. To solve this problem is of considerable significance for predicting very accurate waveforms emitted by extreme mass-ratio binaries, which are among the most promising sources for LISA [22].

The GW community working at the interface between the theory and the experiment has provided faithful *templates* [16, 17, 23] for binaries and developed robust algorithms [24, 25] for pulsars and other GW sources observable with ground-based and space-based interferometers. The joined work of data analysts and experimentalists has established astrophysically significant upper limits for several GW sources [26–28] and is now eagerly waiting for the first detection.

These lectures were envisioned to be an introductory, basic course in GW theory. Many of the topics that we shall address are thoroughly discussed in several books [29–35] and proceedings or reviews [36–42]. The lectures focused more on binary systems, probably because biased towards' the author own background and expertise. The lectures are organized as follows. In Sec. 2 we start by deriving the wave equation in linearized gravity and discuss the main properties of GWs. In Sec. 3 we describe the interaction of GWs with free test particles and the key ideas underlying the functioning of GW detectors. Section 4 reviews the effective stress-energy tensor of GWs. Section 5 is devoted to the generation problem. We explicitly derive the gravitational field at leading order assuming slow-motion, weak-gravity and negligible self-gravity. We then discuss how those results can be extended to non-negligible self-gravity sources. As a first application, in Sec. 6 we compute the GW signal from binary systems. We discuss briefly the state-of-the-art of PN calculations and NR results. As an example of data-analysis issues, we compute the GW templates in the stationary-phase-approximation (SPA). In Sec. 7 we apply the results of Sec. 6 to other astrophysical sources, notably pulsars and supernovae. Section 8 focus on cosmological sources at much higher red-shift $z \gg 1$. We review the main physical mechanims that could have produced GWs in the early Universe, notably first-order phase transitions, cosmic and fundamental strings, and inflation.

## 2. Linearization of Einstein equations

In 1916 Einstein realized the propagation effects at finite velocity in the gravitational equations and predicted the existence of wave-like solutions of the linearized vacuum field equations [43]. In this section we expand Einstein equations around the flat Minkowski metric derive the wave equation and put the solution in a simple form using an appropriate gauge.

## 2.1. Einstein equations and gauge symmetry

The Einstein action reads

$$S_g = \frac{c^3}{16\pi G} \int d^4x \sqrt{-g}\, R, \tag{2.1}$$

where $c$ denotes the speed of light, $G$ the Newton constant, $g_{\mu\nu}$ is the four dimensional metric and $g = \det(g_{\mu\nu})$. Henceforth, we use the following conventions. The flat Minkowski metric is $\eta_{\mu\nu} = (-, +, +, +)$, Greek indices denote spacetime coordinates $\mu, \nu = 0, 1, 2, 3$, whereas Latin indices denote spacelike coordinates $i, j = 1, 2, 3$. Moreover, $x^\mu = (x^0, \mathbf{x}) = (ct, \mathbf{x})$, thus $d^4x = c\, dt\, d^3x$. Partial derivatives $\partial_\mu$ will be denoted with a comma, while covariant derivatives with a semicolon. The scalar tensor $R$ in Eq. (2.1) is obtained from the curvature tensor as

$$R = g^{\mu\nu} R_{\mu\nu}, \qquad R_{\mu\nu} = g^{\rho\sigma} R_{\rho\mu\sigma\nu}, \tag{2.2}$$

$$R^\nu{}_{\mu\rho\sigma} = \frac{\partial \Gamma^\nu_{\mu\sigma}}{\partial x^\rho} - \frac{\partial \Gamma^\nu_{\mu\rho}}{\partial x^\sigma} + \Gamma^\nu_{\lambda\rho}\Gamma^\lambda_{\mu\sigma} - \Gamma^\nu_{\lambda\sigma}\Gamma^\lambda_{\mu\rho}, \tag{2.3}$$

where $\Gamma^\nu_{\mu\sigma}$ are the affine connections

$$\Gamma^\mu_{\nu\rho} = \frac{1}{2} g^{\mu\lambda}\left(\frac{\partial g_{\lambda\nu}}{\partial x^\rho} + \frac{\partial g_{\lambda\rho}}{\partial x^\nu} - \frac{\partial g_{\nu\rho}}{\partial x^\lambda}\right). \tag{2.4}$$

The curvature tensor satisfies the following properties

$$R_{\mu\nu\rho\sigma} = -R_{\nu\mu\rho\sigma} = -R_{\mu\nu\sigma\rho}, \qquad R_{\mu\nu\rho\sigma} = R_{\rho\sigma\mu\nu}, \tag{2.5}$$

$$R_{\mu\nu\rho\sigma} + R_{\mu\sigma\nu\rho} + R_{\mu\rho\sigma\nu} = 0, \qquad R^\lambda{}_{\mu\nu\rho;\sigma} + R^\lambda{}_{\mu\sigma\nu;\rho} + R^\lambda{}_{\mu\rho\sigma;\nu} = 0. \tag{2.6}$$

The latter equation is known as the Bianchi identity. We define the matter energy-momentum tensor $T_{\mu\nu}$ from the variation of the matter action $S_m$ under a change of the metric $g_{\mu\nu} \rightarrow g_{\mu\nu} + \delta g_{\mu\nu}$, that is

$$\delta S_m = \frac{1}{2c} \int d^4x \sqrt{-g}\, T^{\mu\nu} \delta g_{\mu\nu}. \tag{2.7}$$

The variation of the total action $S = S_g + S_m$ with respect to $g_{\mu\nu}$ gives the Einstein equations

$$G_{\mu\nu} = R_{\mu\nu} - \frac{1}{2} g_{\mu\nu} R = \frac{8\pi G}{c^4} T_{\mu\nu}. \tag{2.8}$$

The above equations are nonlinear equations with well posed initial value structure, i.e. they determine future values of $g_{\mu\nu}$ from given initial values. Since $\mu = 0, \ldots 3$, $\nu = 0, \ldots 3$, Eq. (2.8) contains sixteen differential equations, which reduce to ten differential equations if the symmetry of the tensors $G_{\mu\nu}$ and $T_{\mu\nu}$ is used. Finally, because of the Bianchi identity we have $G_{\mu\nu}{}^{;\nu} = 0$, thus the ten differential equations reduce to six.

General relativity is invariant under the group of all possible coordinate transformations

$$x^{\mu} \rightarrow x'^{\mu}(x), \tag{2.9}$$

where $x'^{\mu}$ is invertible, differentiable and with a differentiable inverse. Under the above transformation, the metric transforms as

$$g_{\mu\nu}(x) \rightarrow g'_{\mu\nu}(x') = \frac{\partial x^{\rho}}{\partial x'^{\mu}} \frac{\partial x^{\sigma}}{\partial x'^{\nu}} g_{\rho\sigma}(x). \tag{2.10}$$

We assume that there exists a reference frame in which, on a sufficiently large spacetime region, we can write

$$g_{\mu\nu} = \eta_{\mu\nu} + h_{\mu\nu}, \qquad |h_{\mu\nu}| \ll 1. \tag{2.11}$$

By choosing this particular reference frame, we break the invariance of general relativity under coordinate transformations. However, a residual gauge symmetry remains. Let us consider the following coordinate transformation

$$x^{\mu} \rightarrow x'^{\mu} = x^{\mu} + \xi^{\mu}(x), \qquad |\partial_{\mu}\xi_{\nu}| \leq |h_{\mu\nu}|. \tag{2.12}$$

The metric transforms as

$$g'_{\mu\nu} = \eta_{\mu\nu} - \partial_{\nu}\xi_{\mu} - \partial_{\mu}\xi_{\nu} + h_{\mu\nu} + \mathcal{O}(\partial\xi^2), \tag{2.13}$$

thus, introducing

$$h'_{\mu\nu} = h_{\mu\nu} - \xi_{\mu,\nu} - \xi_{\nu,\mu}, \tag{2.14}$$

we have

$$g'_{\mu\nu} = \eta_{\mu\nu} + h'_{\mu\nu}, \qquad |h'_{\mu\nu}| \ll 1. \tag{2.15}$$

In conclusion, the slowly varying coordinate transformations (2.12) are a symmetry of the linearized theory. Under a finite, global ($x$-independent) Lorentz transformation

$$x^{\mu} \rightarrow \Lambda^{\mu}_{\nu} x^{\nu}, \qquad \Lambda^{\rho}_{\mu} \Lambda^{\sigma}_{\nu} \eta_{\rho\sigma} = \eta_{\mu\nu}, \tag{2.16}$$

the metric transforms as

$$g_{\mu\nu} \to g'_{\mu\nu}(x') = \Lambda_\mu^\rho \Lambda_\nu^\sigma g_{\rho\sigma} = \eta_{\mu\nu} + \Lambda_\mu^\rho \Lambda_\nu^\sigma h_{\rho\sigma}(x) = \eta_{\mu\nu} + h'_{\mu\nu}(x'), \tag{2.17}$$

thus, $h_{\mu\nu}$ is a tensor under Lorentz transformations. It is straightforward to prove that $h_{\mu\nu}$ is also invariant under translations. In conclusions, linearized theory is invariant under the Poincaré group and under the transformation $x^\mu \to x^\mu + \xi^\mu$ with $|\partial_\nu \xi^\mu| \ll 1$.

## 2.2. Wave equation

Let us now linearize Einstein equations posing $g_{\mu\nu} = \eta_{\mu\nu} + h_{\mu\nu}$. At linear order in $h_{\mu\nu}$ the affine connections and curvature tensor read

$$\Gamma^\nu_{\mu\rho} = \frac{1}{2}\eta^{\nu\lambda}(\partial_\rho h_{\lambda\mu} + \partial_\mu h_{\lambda\rho} - \partial_\lambda h_{\mu\rho}), \tag{2.18}$$

$$R^\nu_{\mu\rho\sigma} = \partial_\rho \Gamma^\nu_{\mu\sigma} - \partial_\sigma \Gamma^\nu_{\mu\rho} + \mathcal{O}(h^2), \tag{2.19}$$

more explicitly

$$R_{\mu\nu\rho\sigma} = \frac{1}{2}(\partial_\rho\nu h_{\mu\sigma} + \partial_\sigma\mu h_{\nu\rho} - \partial_\rho\mu h_{\nu\sigma} - \partial_\sigma\nu h_{\mu\rho}). \tag{2.20}$$

Using the above equations, it is straightforward to show that the linearized Riemann tensor is invariant under the transformation $h_{\mu\nu} \to h_{\mu\nu} - \partial_\mu \xi_\nu - \partial_\nu \xi_\mu$. Equation (2.20) can be greatly simplified if we introduce the so-called *trace-reverse* tensor

$$\overline{h}^{\mu\nu} = h^{\mu\nu} - \frac{1}{2}\eta^{\mu\nu} h, \tag{2.21}$$

where $h = \eta_{\alpha\beta} h^{\alpha\beta}$ and $\overline{h} = -h$, which explains the name. Some algebra leads to

$$\Box\overline{h}_{\nu\sigma} + \eta_{\nu\sigma} \partial^\rho \partial^\lambda \overline{h}_{\rho\lambda} - \partial^\rho \partial_\nu \overline{h}_{\rho\sigma} - \partial^\rho \partial_\sigma \overline{h}_{\rho\nu} + \mathcal{O}(h^2) = -\frac{16\pi G}{c^4} T_{\nu\sigma}, \tag{2.22}$$

where the wave operator $\Box = \eta_{\rho\sigma} \partial^\rho \partial^\sigma$. To further simplify Eq. (2.22) we can impose the Lorenz gauge (also denoted in the literature as harmonic or De Donder gauge)

$$\partial_\nu \overline{h}^{\mu\nu} = 0, \tag{2.23}$$

and obtain

$$\Box \overline{h}_{\nu\sigma} = -\frac{16\pi G}{c^4} T_{\nu\sigma}. \tag{2.24}$$

If $\overline{h}^{\mu\nu}$ does not satisfy the Lorenz gauge, i.e. $\partial_\mu \overline{h}^{\mu\nu} = q_\nu$, we can introduce a coordinate transformation such that $\overline{h}'_{\mu\nu} = \overline{h}_{\mu\nu} - \xi_{\mu,\nu} - \xi_{\nu,\mu} + \eta_{\mu\nu}(\partial_\rho \xi^\rho)$ and impose $\Box \xi_\nu = q_\nu$, obtaining $\partial_\mu \overline{h}'^{\mu\nu} = 0$.

Summarizing, the Lorenz gauge imposes 4 conditions that allow to reduce the 10 independent components of the $4 \times 4$ symmetric tensor $h_{\mu\nu}$ to 6 independent components. Note that we also have the condition $\partial_\mu T^{\mu\nu} = 0$, which is the conservation of the energy-momentum tensor of the matter in linearized theory. By contrast in the full theory $T^{\mu\nu}_{;\nu} = 0$.

## 2.3. Transverse-traceless gauge

We want to study the propagation of GWs once they have been generated. We set $T_{\mu\nu} = 0$ in Eq. (2.24) and obtain the wave equation in vacuum

$$\Box \overline{h}_{\mu\nu} = 0. \tag{2.25}$$

GWs propagate at the speed of light. Within the Lorenz gauge we can always consider coordinate transformations such that $\Box \xi_\mu = 0$. The trace-reverse tensor transforms as $\overline{h}'_{\mu\nu} = \overline{h}_{\mu\nu} + \xi_{\mu\nu}$ with $\xi_{\mu\nu} = \eta_{\mu\nu} \partial_\rho \xi^\rho - \xi_{\mu,\nu} - \xi_{\nu,\mu}$. Using $\Box \xi_{\mu\nu} = 0$, we can subtract 4 of the 6 components of $\overline{h}_{\mu\nu}$. More specifically, we can choose $\xi^0$ such that $\overline{h} = 0$ and $\xi^i$ such that $h^{i0} = 0$, thus $\partial_0 h^{00} = 0$. The GW being a time-dependent field, we can set $h^{00} = 0$. We denote the field $h_{ij}$ which satisfies the following transverse and traceless gauge conditions,

$$h^{00} = 0, \quad h^{0i} = 0, \quad \partial_i h^{ij} = 0, \quad h^{ii} = 0, \tag{2.26}$$

the *transverse-traceless* tensor $h_{ij}^{\mathrm{TT}}$. Note that for a single plane wave with wave vector $\mathbf{k}$ and propagation direction $\mathbf{n} = \mathbf{k}/k$, the transversality condition reduces to $n^i h_{ij}^{\mathrm{TT}} = 0$. Without loosing in generality, we can assume that the plane wave propagates along the $z$-axis, thus

$$h_{ij}^{\mathrm{TT}}(t, z) = \begin{pmatrix} h_+ & h_\times & 0 \\ h_\times & -h_+ & 0 \\ 0 & 0 & 0 \end{pmatrix} \cos\left[\omega\left(t - \frac{z}{c}\right)\right], \tag{2.27}$$

where we indicate with $h_+$ and $h_\times$ the two independent polarization states. Following [31, 35], we can introduce the projector operator $P_{ij}(\mathbf{n}) = \delta_{ij} - n_i n_j$, which satisfies the conditions

$$P_{ij} = P_{ji}, \quad n^i P_{ij} = 0, \quad P_{ij} P^{jk} = P_i^k. \quad P_{ii} = 2, \tag{2.28}$$

and the $\Lambda$-operator

$$\Lambda_{ijkl}(\mathbf{n}) = P_{ik} P_{jl} - \frac{1}{2} P_{ij} P_{kl}, \tag{2.29}$$

and obtain the TT field for a generic propagation direction

$$h_{ij}^{\mathrm{TT}} = \Lambda_{ij,kl} h_{kl}, \tag{2.30}$$

where $h_{kl}$ is given in the Lorenz gauge but not necessarily in the TT gauge.

The GW is described in the TT gauge by a $2 \times 2$ matrix in the plane orthogonal to the direction of propagation $\mathbf{n}$. If we perform a rotation $\psi$ about the axis $\mathbf{n}$, we obtain

$$h_{\times} \pm i h_{+} \rightarrow e^{\mp 2i \psi} (h_{\times} \pm i h_{+}). \tag{2.31}$$

In particle physics we call helicity the projection of the (total) angular momentum along the propagation direction: $\mathcal{H} = \mathbf{J} \cdot \mathbf{n} = \mathbf{S} \cdot \mathbf{n}$, being $\mathbf{S}$ the particle's spin. Under a rotation $\psi$ about the propagation direction the helicity states transform as $h \rightarrow e^{i \mathcal{H} \psi} h$. Thus, Eq. (2.31) states that $h_{\times} - i h_{+}$ are the helicity states and that the graviton is a spin-2 particle.

## 3. Interaction of gravitational waves with point particles

### 3.1. Newtonian and relativistic description of tidal gravity

Let us consider two point particles, labeled A and B, falling freely through 3-D Euclidean space under the action of an external Newtonian potential $\Phi$. We assume that at time $t = 0$ the particles are separated by a small distance $\xi$ and have initially the same speed $\mathbf{v}_A(t = 0) = \mathbf{v}_B(t = 0)$. Since the two particles are at slightly different positions, they experience a slightly different gravitational potential $\Phi$ and a different acceleration $\mathbf{g} = -\nabla \Phi$. At time $t > 0$, $\mathbf{v}_A(t) \neq \mathbf{v}_B(t)$. Let us introduce the separation vector $\boldsymbol{\xi} = \mathbf{x}_A - \mathbf{x}_B$ in 3-D Euclidean space. We have

$$\frac{d^2 \xi^i}{dt^2} = -\left(\frac{\partial \Phi}{dx^i}\right)_B + \left(\frac{\partial \Phi}{dx^i}\right)_A \simeq -\left(\frac{\partial^2 \Phi}{\partial x^i \partial x^j}\right) \xi^j \equiv \mathcal{E}_j^i \xi^j, \tag{3.1}$$

where the second equality is obtained by Taylor expanding around the position of particle A. The tensor $\mathcal{E}_j^i$ is called the *Newtonian tidal-gravity* tensor [42], it measures the inhomogeneities of Newtonian gravity. It is the tensor responsible of the Moon's tides on the Earth's ocean.

We now generalize the above Newtonian discussion to Einstein theory. In general relativity, nonspinning test particles move along geodesics

$$\frac{d^2x^\mu}{d\tau^2} + \Gamma^\mu_{\rho\sigma}(x) \frac{dx^\rho}{d\tau} \frac{dx^\sigma}{d\tau} = 0, \tag{3.2}$$

where $\Gamma^\mu_{\rho\sigma}(x)$ is given by Eq. (2.19). Let us consider two nearby geodesics, labeled A and B, parametrized by $x^\mu(\tau)$ and $x^\mu(\tau) + \xi^\mu(\tau)$, with $|\xi^\mu|$ smaller than the typical scale on which the gravitational field varies. By expanding the geodesic equation of particle B around the position of particle A, and subtracting it from the geodesic equation of particle A, we get

$$\nabla_u \nabla_u \xi^\mu = -R^\mu_{\ \nu\rho\sigma} \xi^\rho \frac{dx^\nu}{d\tau} \frac{dx^\sigma}{d\tau}, \tag{3.3}$$

$u^\beta = dx^\beta/d\tau$ being the four-velocity and where we introduce the covariant derivative along the curve $x^\mu(\tau)$

$$\nabla_u \xi^\mu = \frac{d\xi^\mu}{d\tau} + \Gamma^\mu_{\rho\sigma} \xi^\rho \frac{dx^\sigma}{d\tau}. \tag{3.4}$$

Thus, two nearby time-like geodesics experience a tidal gravitational *force* proportional to the Riemann tensor.

## 3.2. Description in the transverse-traceless gauge

In this section we describe the interaction of a GW with a point particle in the TT gauge. Let us consider a test particle A at rest at time $\tau = 0$. Using the geodesic equation, we have

$$\frac{d^2x^i}{d\tau^2}\Big|_{\tau=0} = -\left(\Gamma^i_{\rho\sigma} \frac{dx^\rho}{d\tau} \frac{dx^\sigma}{d\tau}\right)\Big|_{\tau=0} = -\left(\Gamma^i_{00} \frac{dx^0}{d\tau} \frac{dx^0}{d\tau}\right)\Big|_{\tau=0}. \tag{3.5}$$

Because the particle is initially at rest $(dx^\mu/d\tau)_{\tau=0} = (c, 0)$ and

$$\Gamma^i_{00} = \frac{1}{2}\eta^{ij}(\partial_0 h_{0j} + \partial_0 h_{j0} - \partial_j h_{00}). \tag{3.6}$$

In the TT gauge $h_{00} = 0$ and $h_{0j} = 0$, so $(\Gamma^i_{00})_{\tau=0} = 0$. Thus, we conclude that in the TT gauge, if at time $\tau = 0$, $dx^i/d\tau = 0$, also $d^2x^i/d\tau^2 = 0$ and a particle at rest before the GW arrives, remains at rest. The coordinates in the TT gauge stretch themselves when the GW arrives so that the coordinate position of the point particles, initially at rest, does not vary. What varies is the proper distance between the two particles and physical effects are monitored by proper distances.

For a wave propagating along the $z$-axis, the metric is [see Eq. (2.27)]

$$ds^2 = -c^2\,dt^2 + dz^2 + dy^2\left[1 - h_+\,\cos\omega\left(t - \frac{z}{c}\right)\right]$$

$$+ dx^2\left[1 + h_+\,\cos\omega\left(t - \frac{z}{c}\right)\right] + 2dx\,dy\,h_\times\,\cos\omega\left(t - \frac{z}{c}\right). \quad (3.7)$$

If particles A and B set down initially along the $x$-axis, we have

$$s \simeq L\left(1 + \frac{h_+}{2}\,\cos\omega t\right), \tag{3.8}$$

where $L$ is the initial, unperturbed distance between particles A and B.

## 3.3. Description in the free-falling frame

It is always possible to perform a change of coordinates such that at a given space-time point $\mathcal{Q}$, we can set $\Gamma^\mu_{\rho\sigma}(\mathcal{Q}) = 0$ and $(d^2x^\mu/d\tau^2)_\mathcal{Q} = 0$. In this frame, at one moment in space and one moment in time, the point particle is *free falling* (FF). This frame can be explicitly constructed using *Riemann normal co-ordinates* [30]. Actually, it is possible to build a frame such that the point particle is free-falling all along the geodesics using *Fermi normal coordinates* [30].

Let us introduce a FF frame attached to particle A with spatial origin at $x^j = 0$ and coordinate time equal to proper time $x^0 = \tau$. By definition of a FF frame, the metric reduces to Minkowski metric at the origin and all its derivatives vanish at the origin, that is

$$ds^2 = -c^2\,dt^2 + d\mathbf{x}^2 + \mathcal{O}\left(\frac{|\mathbf{x}|^2}{\mathcal{R}^2}\right), \tag{3.9}$$

where $\mathcal{R}$ is the curvature radius $\mathcal{R}^{-2} = |R_{\mu\nu\rho\sigma}|$. Doing explicitly the calculation at second order in $x$, one finds [44]

$$ds^2 = -c^2\,dt^2\left[1 + R_{i0j0}\,x^i\,x^j\right] - 2c\,dt\,dx^i\left(\frac{2}{3}R_{0jik}\,x^j\,x^k\right)$$

$$+ dx^i\,dx^j\left[\delta_{ij} - \frac{1}{3}R_{ijkl}\,x^k\,x^l\right]. \tag{3.10}$$

For GW experiments located on the Earth, the interferometer is not in free fall with respect to the Earth gravity. The detector is subjected to an acceleration $\mathbf{a} = -\mathbf{g}$ with respect to a local inertial frame and it rotates with respect to local gyroscopes. Thus, in general the effect of GWs on point particles compete with other effects. We shall restrict our discussion to the frequency band ($10$–$10^3$ Hz) in which the other effects are subdominant and/or static.

Let us compute the equation of geodesic deviation in the FF frame attached to particle A. We have

$$\nabla_u \nabla_u \xi^\alpha = u^\beta \nabla_\beta (u^\lambda \nabla_\lambda \xi^\alpha) = u^\beta u^\lambda (\partial_{\beta\lambda} \xi^\alpha + \Gamma^\alpha_{\lambda\sigma,\beta} \xi^\sigma), \tag{3.11}$$

where in the last equality we use $\Gamma^\alpha_{\lambda\sigma} = 0$. Since we assume that the particles are initially at rest, $u^\beta = \delta^\beta_0$. Using $\xi^0 = 0$ and the fact that $\Gamma^j_{0k,0}$ can be neglected when computed at position A, we have

$$\nabla_u \nabla_u \xi^j = \frac{d^2 \xi^j}{d\tau^2} \quad \Rightarrow \quad \frac{d^2 \xi^j}{d\tau^2} = -R^j_{0k0} \xi^k. \tag{3.12}$$

To complete the calculation we need to evaluate the Riemann tensor $R^j_{0k0}$. As discussed in Sec. 2.2, in linearized theory the Riemann tensor is invariant under change of coordinates, so we can compute it in the TT gauge. Using Eq. (2.20), we obtain

$$R^{TT}_{j0k0} = -\frac{1}{2c^2} \ddot{h}^{TT}_{jk}. \tag{3.13}$$

Thus,

$$\frac{d^2 \xi^j}{dt^2} = \frac{1}{2} \ddot{h}^{TT}_{jk} \xi^k. \tag{3.14}$$

In conclusion, in the FF frame the effect of a GW on a point particle of mass $m$ can be described in terms of a *Newtonian force* $F_i = (m/2) \ddot{h}^{TT}_{ij} \xi^j$. Note that in the FF frame, coordinate distances and proper distances coincide, and we recover immediately Eq. (3.8).

The description in the FF frame is useful and simple as long as we can write the metric as $g_{\mu\nu} = \eta_{\mu\nu} + \mathcal{O}(x^2/\mathcal{R}^2)$, i.e. as long as we can disregard the corrections $x^2/\mathcal{R}^2$. Since $\mathcal{R}^{-2} = |R_{i0j0}| \sim \ddot{h} \sim h/\lambda^2_{GW}$, we have $x^2/\mathcal{R}^2 \simeq L^2 h/\lambda^2_{GW}$, and comparing it with $\delta L/L \sim h$, we find $L^2/\lambda^2_{GW} \ll 1$. This condition is satisfied by ground-based detectors because $L \sim 4$ km and $\lambda_{GW} \sim 3000$ km, but not by space-based detectors which have $L \sim 5 \times 10^6$ km and will observe GWs with wavelength shorter than $L$. [For a recent thorough analysis and a proof of the equivalence between the TT and FF description, see, e.g., Ref. [45].]

### 3.4. *Key ideas underlying gravitational-wave detectors*

To illustrate the effect of GWs on FF particles, we consider a ring of point particles initially at rest with respect to a FF frame attached to the center of the ring,

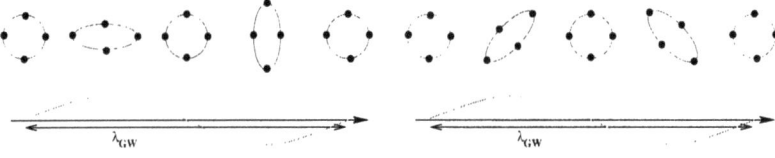

Fig. 1. We show how point particles along a ring move as a result of the interaction with a GW propagating in the direction perpendicular to the plane of the ring. The left panel refers to a wave with + polarization, the right panel with × polarization.

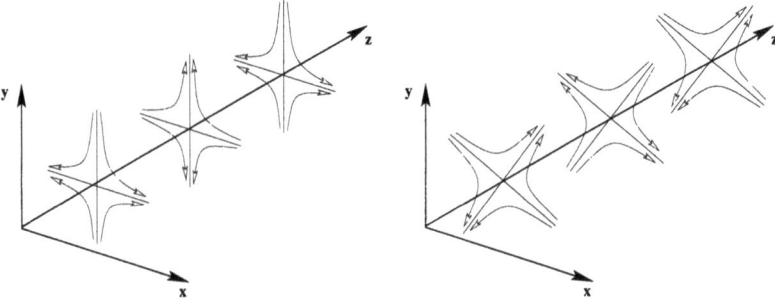

Fig. 2. Lines of force associated to the + (left panel) and × (right panel) polarizations.

as shown in Fig. 1. We determine the motion of the particles considering the + and × polarizations separately. If only the + polarization is present, we have

$$h_{ij}^{\mathrm{TT}} = h_+ \begin{pmatrix} 1 & 0 \\ 0 & -1 \end{pmatrix} \sin \omega t, \quad \xi_i = [x_0 + \delta x(t), y_0 + \delta y(t)], \tag{3.15}$$

where $x_0$ and $y_0$ are the unperturbed position at time $t = 0$. Thus

$$\delta x(t) = \frac{h_+}{2} x_0 \sin \omega t \qquad \delta y(t) = -\frac{h_+}{2} y_0 \sin \omega t. \tag{3.16}$$

If only the $x$ polarization is present, a straightforward calculation gives

$$\delta x(t) = \frac{h_\times}{2} y_0 \sin \omega t \qquad \delta y(t) = \frac{h_\times}{2} x_0 \sin \omega t. \tag{3.17}$$

The + and × polarizations differ by a rotation of $45^o$. In Fig. 2 we show the lines of force associated to the + and × polarizations.

The simplest GW detector we can imagine is a body of mass $m$ at a distance $L$ from a fiducial laboratory point, connected to it by a spring of resonant frequency $\Omega$ and quality factor $Q$. Einstein equation of geodesic deviation predicts that the infinitesimal displacement $\Delta L$ of the mass along the line of separation from its equilibrium position satisfies the equation [36] (valid for wavelengths $\gg L$ and in the FF frame of the observer at the fiducial laboratory point)

$$\ddot{\Delta L}(t) + 2\frac{\Omega}{Q}\dot{\Delta L}(t) + \Omega^2 \Delta L(t) = \frac{L}{2}\left[F_+ \ddot{h}_+(t) + F_\times \ddot{h}_\times(t)\right], \qquad (3.18)$$

where $F_{+,\times}$ are coefficients of order unity which depend on the direction of the source [see Eqs. (6.31), (6.31) below] and the GW polarization angle.

Laser-interferometer GW detectors are composed of two perpendicular km-scale arm cavities with two test-mass mirrors hung by wires at the end of each cavity. The tiny displacements $\Delta L$ of the mirrors induced by a passing GW are monitored with very high accuracy by measuring the relative optical phase between the light paths in each interferometer arm. The mirrors are pendula with quality factor $Q$ quite high and resonant frequency $\Omega$ much lower ($\sim 1$ Hz) than the typical GW frequency ($\sim 100$ Hz). In this case Eq. (3.18), written in Fourier domain, reduces to $\Delta L/L \sim h$. The typical amplitude, at 100 Hz, of GWs emitted by binary systems in the VIRGO cluster of galaxies ($\sim 20$ Mpc distant), which is the largest distance the first-generation ground-based interferometers can probe, is $\sim 10^{-21}$. This means $\Delta L \sim 10^{-18}$ m, a very tiny number. It may appear rather discouraging, especially if we think to monitor the test-mass motion with light of wavelength nearly $10^{12}$ times larger. However, this precision is currently be demonstrated experimentally.

The electromagnetic signal leaking out the interferometer's dark-port contains the GW signal but also noise — for example the thermal noise from the suspension system and the mirror itself, can shake the mirror mimicking the effect of a GW. The root-mean-square of the noise is generally expressed in terms of the noise power per unit frequency $S_h$ through the relation $h \sim \sqrt{S_h(f)\,\Delta f} \sim \Delta L/L$, $\Delta L$ being the mirror displacement induced by noise and $\Delta f$ the frequency bandwidth. In Fig. 3 we plot the noise curves of LIGOs (June 2006). The interferometers are currently operating at design sensitivity for almost the entire frequency band.

## 4. Effective stress-energy tensor of gravitational waves

Until now we have defined the GWs as fluctuations of a flat spacetime. Here, we want to be more general and consider GWs as perturbations of a generic background $\bar{g}_{\mu\nu}$, that is

$$g_{\mu\nu} = \bar{g}_{\mu\nu} + h_{\mu\nu}, \qquad |h_{\mu\nu}| \ll 1. \qquad (4.1)$$

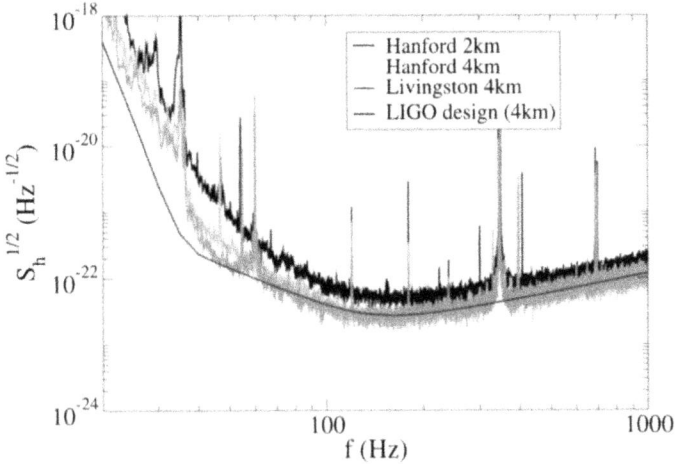

Fig. 3. We plot the square root of the noise spectral density versus frequency for the three LIGO detectors together with the LIGO noise curve at designed sensitivity. The noise curves refer to June 2006, during the fifth scientific run.

We need a criterion to define what is the background and what is the perturbation. In general there are two cases [30, 35]:

1. $\bar{g}_{\mu\nu}$ has typical scale $L_B$ and $h_{\mu\nu}$ has typical wavelength $\lambda$ with $\lambda \ll L_B$, i.e. $h_{\mu\nu}$ is a small *ripple* on a smooth background;

2. $\bar{g}_{\mu\nu}$ has frequencies only up to $f_B$ and $h_{\mu\nu}$ is different from zero around $f$ with $f \gg f_B$, i.e. the background is static.

Let us expand $R_{\mu\nu}$ through $\mathcal{O}(h^2)$. Note that we have now two small parameters $h$ and $\lambda/L_B$ (or $f_B/f$), we have

$$R_{\mu\nu} = \underbrace{\bar{R}_{\mu\nu}}_{\text{low freq}} + \underbrace{R^{(1)}_{\mu\nu}}_{\text{high freq}} + \underbrace{R^{(2)}_{\mu\nu}}_{\text{low and high freq.}} + \cdots . \tag{4.2}$$

Quantities having a bar are computed using the background metric $\bar{g}_{\mu\nu}$; they contain only low-frequency modes. The superscript (1) [(2)] in Eq. (4.2) refers to quantities computed at linear (quadratic) order in $h$. Using the Einstein equations we get

$$\bar{R}_{\mu\nu} = -\left[R^{(2)}_{\mu\nu}\right]^{\text{low freq}} + \frac{8\pi G}{c^4}\left[T_{\mu\nu} - \frac{1}{2}g_{\mu\nu}T\right]^{\text{low freq}} . \tag{4.3}$$

We introduce a scale $\ell$ such that $\lambda \ll \ell \ll L_B$, and average over a spatial volume $\ell^3$ [35, 46]. We denote the average as $\langle\rangle$. Short-wave modes average to zero, whereas modes with wavelength $L_B$ are constant. We can rewrite Eq. (4.3) as

$$\overline{R}_{\mu\nu} = -\langle R^{(2)}_{\mu\nu}\rangle + \frac{8\pi G}{c^4}\left\langle T_{\mu\nu} - \frac{1}{2}g_{\mu\nu}T\right\rangle$$

$$\equiv -\langle R^{(2)}_{\mu\nu}\rangle + \frac{8\pi G}{c^4}\left(\overline{T}_{\mu\nu} - \frac{1}{2}\overline{g}_{\mu\nu}\overline{T}\right). \tag{4.4}$$

Defining the *effective* stress-energy tensor of GWs

$$t_{\mu\nu} = -\frac{c^4}{8\pi G}\left\langle R^{(2)}_{\mu\nu} - \frac{1}{2}\overline{g}_{\mu\nu}R^{(2)}\right\rangle, \tag{4.5}$$

we have

$$\overline{R}_{\mu\nu} - \frac{1}{2}\overline{g}_{\mu\nu}\overline{R} = \frac{8\pi G}{c^4}(\overline{T}_{\mu\nu} + t_{\mu\nu}). \tag{4.6}$$

An explicit calculation carried on far from the source gives

$$t_{\mu\nu} = \frac{c^4}{32\pi G}\langle\partial_\mu h_{\alpha\beta}\,\partial_\nu h^{\alpha\beta}\rangle. \tag{4.7}$$

For a plane wave, using the TT gauge

$$t_{00} = \frac{c^2}{32\pi G}\langle\dot{h}^{TT}_{ij}\,\dot{h}^{TT}_{ij}\rangle = \frac{c^2}{16\pi G}\langle\dot{h}^2_+ + \dot{h}^2_\times\rangle, \tag{4.8}$$

and the GW energy flux per unit area is

$$\frac{dE}{dt\,dA} = \frac{c^3}{16\pi G}\langle\dot{h}^2_+ + \dot{h}^2_\times\rangle. \tag{4.9}$$

For a supernovae $dE/(dt\,dA) \sim c^3 f^2 h^2/(16\pi G) \sim 400\text{erg}/(\text{cm}^2\text{sec})$, where we set $h = 10^{-21}$ and $f = 1$ kHz. The GW burst has a duration of a few msec. It is telling to compare it with the neutrino energy flux $\sim 10^5\text{erg}/(\text{cm}^2\text{sec})$ and the photon energy flux (optical radiation) $\sim 10^{-4}\text{erg}/(\text{cm}^2\text{sec})$ from a supernovae. Neutrinos and optical radiation are emitted during a few seconds and one week, respectively.

## 5. Generation of gravitational waves

### 5.1. Sources in slow motion, weak-field and negligible self-gravity

In this section we evaluate the leading order contribution to the metric perturbations under the assumption that the internal motions of the source are slow compared to the speed of light. We also assume that the source's self-gravity is negligible. Henceforth, we shall discuss how to extend those results to sources with non-negligible self-gravity. We start from

$$\Box \bar{h}_{\mu\nu} = -\frac{16\pi G}{c^4} T_{\mu\nu}, \quad \partial_\mu \bar{h}^{\mu\nu} = 0, \quad \partial_\mu T^{\mu\nu} = 0, \tag{5.1}$$

and introduce retarded Green functions

$$G(x - x') = -\frac{1}{4\pi} \frac{1}{|\mathbf{x} - \mathbf{x}'|} \delta\left(t - \frac{|\mathbf{x} - \mathbf{x}'|}{c} - t'\right), \tag{5.2}$$

which satisfy $\Box_x G(x - x') = \delta^{(4)}(x - x')$. The solution of Eq. (5.1) can be written as

$$\bar{h}_{\mu\nu}(x) = -\frac{16\pi G}{c^4} \int d^4x' \, G(x - x') \, T_{\mu\nu}(x'). \tag{5.3}$$

Outside the source, using the TT gauge, we have

$$\bar{h}_{ij}^{\text{TT}}(t, \mathbf{x}) = \Lambda_{ij,kl}(\mathbf{n}) \frac{4G}{c^4} \int d^3x' \frac{1}{|\mathbf{x} - \mathbf{x}'|} T_{kl}\left(t - \frac{|\mathbf{x} - \mathbf{x}'|}{c}; \mathbf{x}'\right). \tag{5.4}$$

Denoting by $d$ the typical size of the source, assuming to be far from the source, i.e. $r \gg d$, we can write $|\mathbf{x} - \mathbf{x}'| = r - \mathbf{x}' \cdot \mathbf{n} + \mathcal{O}(d^2/r)$, and Eq. (5.3) becomes

$$\bar{h}_{ij}^{\text{TT}}(t, \mathbf{x}) \simeq \frac{1}{r} \frac{4G}{c^4} \Lambda_{ij,kl}(\mathbf{n}) \int_{|\mathbf{x}'| < d} d^3x' \, T_{kl}\left(t - \frac{r}{c} + \frac{\mathbf{x}' \cdot \mathbf{n}}{c}; \mathbf{x}'\right). \tag{5.5}$$

We can simplify the above equations if we assume that typical velocities inside the sources are much smaller than the speed of light $c$. If $\omega$ is the typical frequency associated to the source motion, typical source velocities are $v \sim \omega d$. As we shall see in the following, the GW signal is determined by the leading multipole moments, thus $\omega_{\text{GW}} \sim \omega \sim v/d$ and $\lambda_{\text{GW}} \sim (c/v)\, d$. If $v/c \ll 1$, we have $\lambda_{\text{GW}} \gg d$.

Applying a Fourier decomposition, we can write

$$T_{kl}\left(t - \frac{r}{c} + \frac{\mathbf{x}' \cdot n}{c}; \mathbf{x}'\right) = \int \frac{d^4k}{(2\pi)^4} \tilde{T}_{kl}(\omega, \mathbf{k}) \times e^{-i\omega\left(t - \frac{r}{c} + \frac{\mathbf{x}' \cdot \mathbf{n}}{c}\right) + i\mathbf{k} \cdot \mathbf{x}'}, \tag{5.6}$$

using $\omega \mathbf{x}' \cdot \mathbf{n} \sim \omega d/c \ll 1$, expanding the exponential and Taylor-expanding $T_{kl}$ we get

$$
\begin{aligned}
h_{ij}^{\mathrm{TT}}(t, \mathbf{x}) \simeq \frac{1}{r} \frac{4G}{c^4} \Lambda_{ij,kl}(\mathbf{n}) \Bigg[ & \int d^3x\, T^{kl}(t, \mathbf{x}) \\
& + \frac{1}{c} n_m \frac{d}{dt} \int d^3x\, T^{kl}(t, \mathbf{x})\, x^m \\
& + \frac{1}{2c^2} n_m n_p \frac{d^2}{dt^2} \int d^3x\, T^{kl}(t, \mathbf{x})\, x^m\, x^p + \cdots \Bigg]_{t-r/c} .
\end{aligned}
\tag{5.7}
$$

The above expression is valid in linearized gravity and for negligible self-gravity sources, i.e. for sources whose dynamics is not determined by gravitational forces. We notice that in Eq. (5.7) the higher multipoles are suppressed by a factor $v/c$. To make Eq. (5.7) more transparent, we can express the momenta $T^{ij}$ in terms of the momenta of $T^{00}$ and $T^{0i}$. Let us first introduce the momenta of the mass density

$$
M = \frac{1}{c^2} \int d^3x\, T^{00}(t, \mathbf{x}),
\tag{5.8}
$$

$$
M^i = \frac{1}{c^2} \int d^3x\, T^{00}(t, \mathbf{x})\, x^i,
\tag{5.9}
$$

$$
M^{ij} = \frac{1}{c^2} \int d^3x\, T^{00}(t, \mathbf{x})\, x^i\, x^j,
\tag{5.10}
$$

$$
\tag{5.11}
$$

and impose the conservation law $\partial_\mu T^{\mu\nu} = 0$ valid in linearized gravity. Setting $\nu = 0$, we have $\partial_0 T^{00} + \partial_i T^{i0} = 0$, integrating this equation in a volume containing the source, we obtain the conservation of the mass $\dot{M} = 0$. Similarly, one can prove the conservation of the momentum $\dot{M}^i = 0$ Moreover, we have

$$
\begin{aligned}
c\, \dot{M}^{ij} &= \int_V d^3x\, x^i\, x^j\, \partial_0 T^{00} = -\int_V d^3x\, x^i\, x^j\, \partial_k T^{0k} \\
&= \int_V d^3x\, (x^j\, T^{0i} + x^i\, T^{0j}),
\end{aligned}
\tag{5.12}
$$

where the second line is obtained after integrating by parts. Finally,

$$
\ddot{M}^{ij} = 2 \int_V d^3x\, T^{ij}.
\tag{5.13}
$$

Thus, the leading term in Eq. (5.7), can be rewritten as

$$
h_{ij}^{\mathrm{TT}}(t, \mathbf{x}) = \frac{1}{r} \frac{2G}{c^4} \Lambda_{ij,kl}(\mathbf{n})\, \ddot{M}^{kl}\left(t - \frac{r}{c}\right),
\tag{5.14}
$$

where $M^{kl}$ is given by Eq. (5.11). The quantity $T^{00}/c^2$ in Eq. (5.11) is a mass density. Besides the rest-mass contribution, it can contain terms due to the kinetic energy and the potential energy. For sources having strong gravitational field, as NSs and BHs, the mass density can depend also on the binding energy. Only for weak fields and small velocities, which is the assumption so far made, $T^{00}/c^2$ reduces to the rest-mass density $\rho$.

Henceforth, we shall discuss some applications to binary systems and pulsars, so it is convenient to compute explicitly $h_+$ and $h_\times$. Assuming that the GW propagates along the direction $\mathbf{n} = (\cos\phi \sin\theta, \sin\phi \sin\theta, \cos\theta)$, a straight calculations gives:

$$h_+ = \frac{G}{r\,c^4} \left\{ \ddot{M}_{11}(\sin^2\phi - \cos^2\theta \cos^2\phi) + \ddot{M}_{22}(\cos^2\phi - \cos^2\theta \sin^2\phi) \right.$$
$$- \ddot{M}_{33} \sin^2\theta - \ddot{M}_{12} \sin 2\phi (1 + \cos^2\theta) + \ddot{M}_{13} \cos\phi \sin 2\theta$$
$$\left. + \ddot{M}_{23} \sin 2\theta \sin\phi \right\}, \tag{5.15}$$

$$h_\times = \frac{2G}{r\,c^4} \left\{ \frac{1}{2}(\ddot{M}_{11} - \ddot{M}_{22}) \cos\theta \sin 2\phi - \ddot{M}_{12} \cos\theta \cos 2\phi \right.$$
$$\left. - \ddot{M}_{13} \sin\theta \sin\phi + \ddot{M}_{23} \cos\phi \sin\theta \right\}. \tag{5.16}$$

### 5.2. Sources in slow motion and weak-field, but non-negligible self-gravity

As already stated above, the derivation in linearized gravity of the quadrupole formula (5.14) cannot be applied to systems like binary stars whose dynamics is dominated by gravitational forces. In fact, because of the conservation-law valid in linearized gravity $\partial_\mu T_{\mu\nu} = 0$, the two bodies move along geodesics in Minkowski spacetime. The extension to the case in which self-gravity is non-negligible can be done as follows [47].

In full general relativity one can define the field $H_{\mu\nu}$ such that

$$\sqrt{-g}\, g^{\mu\nu} = \eta^{\mu\nu} - H^{\mu\nu}, \tag{5.17}$$

where in the weak-field limit $H^{\mu\nu}$ coincides with the reverse-trace tensor that we introduced above. Imposing the harmonic gauge $\partial_\mu H^{\mu\nu} = 0$, one derives

$$\Box H_{\mu\nu} = -\frac{16\pi G}{c^4} \left[(-g)\, T_{\mu\nu} + \tau_{\mu\nu}\right], \tag{5.18}$$

where $\tau_{\mu\nu}$ is the pseudotensor depending on $H_{\mu\nu}$ that can be read explicitly from Refs. [14, 29]. The conservation law reads in this case

$$\partial_\mu \left[(-g)\, T^{\mu\nu} + \tau^{\mu\nu}\right] = 0. \tag{5.19}$$

We can redo the derivation in linearized gravity (see Sec. 2.2), but replace $T_{\mu\nu} \rightarrow (-g) T_{\mu\nu} + \tau_{\mu\nu}$, obtaining for the leading term in Eq. (5.14) $\int T_{00} x^i x^j d^3x \rightarrow \int (T_{00} + \tau_{00}) x^i x^j d^3x$. For sources characterized by weak gravity $\tau_{00}$ is negligible with respect to $T_{00}$. Even though at the end one obtains the same formula, it is crucial to take into account the second order corrections in $h_{\mu\nu}$, i.e. the field $\tau_{\mu\nu}$, in the conservation law. Otherwise the sources would be obliged to move along geodesics in Minkowski spacetime, instead of moving in a bounded orbit.

### 5.3. *Radiated energy, angular momentum and linear momentum*

Using the results of Sec. 4, we can compute the power radiated at leading order

$$\frac{dP}{d\Omega} = \frac{r^2 c^3}{32\pi G} \langle \dot{h}_{ij}^{\text{TT}} \dot{h}_{ij}^{\text{TT}} \rangle = \frac{G}{8\pi c^5} \Lambda_{kl,mp}(\mathbf{n}) \langle \dddot{Q}_{kl} \dddot{Q}_{mp} \rangle, \tag{5.20}$$

where we introduce the traceless quadrupole tensor

$$Q_{ij} = M_{ij} - \frac{1}{3} \delta_{ij} M_{kk}. \tag{5.21}$$

Using the following relations

$$\int \frac{d\Omega}{4\pi} n_i n_j = \frac{1}{3} \delta_{ij}, \tag{5.22}$$

$$\int \frac{d\Omega}{4\pi} n_i n_j n_k n_l = \frac{1}{15} (\delta_{ij} \delta_{kl} + \delta_{ik} \delta_{jl} + \delta_{il} \delta_{jk}), \tag{5.23}$$

we derive for the total power radiated

$$P = \frac{G}{5c^5} \langle \dddot{Q}_{ij} \dddot{Q}_{ij} \rangle. \tag{5.24}$$

In the literature Eq. (5.24) is generally denoted as the *quadrupole formula*.

GWs not only carry away from the source the energy, but also angular momentum and linear momentum. At leading-order the angular-momentum radiated is [35]

$$\frac{dL^i}{dt} = \frac{2G}{5 c^5} \epsilon^{ijk} \langle \ddot{Q}_{jl} \dddot{Q}_{lk} \rangle \tag{5.25}$$

while the linear momentum radiated is given by

$$\frac{dP^i}{dt} = -\frac{G}{8\pi c^5} \int d\omega \, \dddot{Q}_{jk}^{\text{TT}} \partial^i \ddot{Q}_{jk}^{\text{TT}}. \tag{5.26}$$

Under parity, $\mathbf{x} \rightarrow -\mathbf{x}$, the mass quadrupole does not vary, and the integral in (5.26) is overall odd and vanishes. The first nonzero contribution comes at order $\mathcal{O}(1/c^7)$ from the interference between the mass quadrupole and the sum of the octupole and current quadrupole. As a consequence of the loss of linear momentum through GW emission, the BH formed by the coalescence of a BH binary can acquire a kick or recoil velocity. The recoil velocity is astrophysically significant. If it were too large, the BH can be ejected from the host galaxy with important consequences on BH's mass growth through hierarchical mergers. Recently, there have been a plethora of numerical [48] and analytic [49] predictions.

The above discussion on energy, angular momentum and linear-momentum can be made more rigorous and applicable to sources with non-negligible self gravity thanks to the work of Bondi in the 50s [50]. Let us consider a sphere $S$ of volume $\mathcal{V}$ and radius $r$ containing the source. Be $r$ much larger than the source dimension and the gravitational wavelength (far zone). It can be proven [29, 50] that $P^\mu$, defined by[2]

$$P^\mu = \int \tau^{\mu 0}\, d^3 x, \tag{5.27}$$

is a four vector with respect to Lorentz transformations. Using the relation $\partial_\mu \tau^{\mu\nu} = 0$, we can write

$$\frac{dP^\mu}{dt} = \int_\mathcal{V} \partial_0 \tau^{\mu 0}\, d^3 x = -\oint_S \tau^{\mu i}\, n^i\, dS. \tag{5.28}$$

For $\mu = 0$ the above equation gives the conservation of the energy

$$\frac{dP^0}{dt} = -r^2 \oint d\Omega\, \tau^{0i}\, n_i, \tag{5.29}$$

for $\mu = j$ Eq. (5.28) gives the conservation of the linear momentum

$$\frac{dP^j}{dt} = -r^2 \oint d\Omega\, \tau^{ji}\, n_i. \tag{5.30}$$

## 6. Application to binary systems

### 6.1. Inspiral waveforms at leading Newtonian order

Let us consider a binary system with masses $m_1$ and $m_2$, total mass $M = m_1 + m_2$, reduced mass $\mu = m_1 m_2 / (m_1 + m_2)$ and symmetric mass-ratio $\nu = \mu / M$. We first assume that the two bodies are rather separated and move along a circular orbit. In the center-of-mass frame we can write for the relative coordinates

$$X(t) = R \cos\omega t, \quad Y(t) = R \sin\omega t, \quad Z(t) = 0, \tag{6.1}$$

---

[2]Note that the integration is done over a constant-time hypersurface.

$R$ being the relative distance between the two bodies. The only nonzero components of the tensor $M^{ij} = \mu X^i X^j$ are

$$M_{11} = \frac{1}{2} \mu R^2 (1 + \cos 2\omega t), \tag{6.2}$$

$$M_{22} = \frac{1}{2} \mu R^2 (1 - \cos 2\omega t), \tag{6.3}$$

$$M_{12} = \frac{1}{2} \mu R^2 \sin 2\omega t. \tag{6.4}$$

Taking time-derivatives and plugging the above expressions in Eqs. (5.14), (5.15) we obtain

$$h_+(t) = \frac{1}{r} \frac{4G}{c^4} \mu R^2 \omega^2 \frac{(1 + \cos^2 \theta)}{2} \cos(2\omega t), \tag{6.5}$$

$$h_\times(t) = \frac{1}{r} \frac{4G}{c^4} \mu R^2 \omega^2 \cos\theta \sin(2\omega t), \tag{6.6}$$

where we shift the origin of time to get rid of the dependence on $\phi$ and trade the retarded time with $t$.

For $\theta = 0$, i.e. along the direction perpendicular to the orbital plane, $h_+$ and $h_\times$ are both different from zero, and $h_\times \pm i h_+ \propto \pm i e^{-2i\omega t}$, thus the wave is circularly polarized. For $\theta = \pi/2$, i.e. along the orbital plane, only $h_+$ is different from zero and the wave is linearly polarized.

The angular distribution of the radiated power is given by Eq. (5.24). It reads

$$\left(\frac{dP}{d\Omega}\right) = \frac{2G \mu^2 R^4 \omega^6}{\pi c^5} \mathcal{P}(\theta), \tag{6.7}$$

$$\mathcal{P}(\theta) = \frac{1}{4}(1 + 6 \cos^2 \theta + \cos^4 \theta). \tag{6.8}$$

The maximum power is emitted along the direction perpendicular to the orbital plane, $\theta = 0$, where $\mathcal{P}(\pi/2) = 2$. Since in the case of a binary system, there is always a component of the source's motion perpendicular to the observation direction, the power radiated does not vanish in any direction. Integrating over the total solid angle, we obtain the total power radiated,

$$P = \frac{32}{5} \frac{G \mu^2 R^4 \omega^6}{c^5}. \tag{6.9}$$

If we consider the binary system composed of Jupiter and the Sun, using $m_J = 1.9 \times 10^{30}$ g, $R = 7.8 \times 10^{13}$ cm and $\omega = 1.68 \times 10^{-7}$ Hz, we get $P = 5 \times 10^3$ Joules/sec. This value is tiny, especially when compared to the luminosity of the

Sun in electromagnetic radiation $P_\odot = 3.9 \times 10^{26}$ Joules/sec. If the binary moves along an eccentric orbit the power radiated is [51]

$$P = \frac{32}{5} \frac{G^4 \mu^2 M^2}{a^5 c^5} \frac{1}{(1-e^2)^{7/2}} \left(1 + \frac{73}{24}e^2 + \frac{37}{96}e^4\right), \qquad (6.10)$$

where $a$ is the semi-major axis and $e$ the eccentricity. Plugging in Eq. (6.10) the values for the Hulse-Taylor binary pulsar [52], $a = 1.95 \times 10^{11}$ cm, $m_1 = 1.441 M_\odot$, $m_2 = 1.383 M_\odot$, $e = 0.617$, we get that the power radiated is $7.35 \times 10^{24}$ Joules/sec, which is about 2% of the luminosity of the Sun in electromagnetic radiation.

The emission of GWs costs energy and to compensate for the loss of energy, the radial separation $R$ between the two bodies must decrease. We shall now derive how the orbital frequency and GW frequency change in time, using Newtonian dynamics and the balance equation

$$\frac{dE_{\text{orbit}}}{dt} = -P. \qquad (6.11)$$

At Newtonian order, $E_{\text{orbit}} = -G m_1 m_2/(2R)$ and $\omega^2 = G M/R^3$. Thus, $\dot{R} = -2/3(R\,\omega)(\dot\omega/\omega^2)$. As long as $\dot\omega/\omega^2 \ll 1$, the radial velocity is smaller than the tangential velocity and the binary's motion is well approximated by an adiabatic sequence of quasi-circular orbits. Equation (6.11) implies that the orbital frequency varies as

$$\frac{\dot\omega}{\omega^2} = \frac{96}{5} \nu \left(\frac{GM\omega}{c^3}\right)^{5/3}, \qquad (6.12)$$

and the GW frequency $f_{\text{GW}} = 2\omega$,

$$\dot{f}_{\text{GW}} = \frac{96}{5} \pi^{8/3} \left(\frac{G \mathcal{M}}{c^3}\right)^{5/3} f_{\text{GW}}^{11/3}, \qquad (6.13)$$

where $\mathcal{M} = \mu^{3/5} M$ is the so-called *chirp* mass. Introducing the time to coalescence $\tau = t_{\text{coal}} - t$, and integrating Eq. (6.13), we get

$$f_{\text{GW}} \simeq 130 \left(\frac{1.21 M_\odot}{\mathcal{M}}\right)^{5/8} \left(\frac{1\text{sec}}{\tau}\right)^{3/8} \text{Hz}, \qquad (6.14)$$

where $1.21 M_\odot$ is the chirp mass of a NS-NS binary. Equation (6.14) predicts coalescence times of $\sim 17$ min, 2 sec, 1 msec, for $f_{\text{GW}} \sim 10$, 100, $10^3$ Hz. Using the above equations, it is straightforward to compute the relation between the radial separation and the GW frequency, we find

$$R \simeq 300 \left( \frac{M}{2.8 M_\odot} \right)^{1/3} \left( \frac{100\,\text{Hz}}{f_{\text{GW}}} \right)^{2/3} \text{km}. \tag{6.15}$$

Finally, a useful quantity is the number of GW cycles, defined by

$$\mathcal{N}_{\text{GW}} = \frac{1}{\pi} \int_{t_{\text{in}}}^{t_{\text{fin}}} \omega(t)\, dt = \frac{1}{\pi} \int_{\omega_{\text{in}}}^{\omega_{\text{fin}}} \frac{\omega}{\dot{\omega}}\, d\omega. \tag{6.16}$$

Assuming $\omega_{\text{fin}} \gg \omega_{\text{in}}$, we get

$$\mathcal{N}_{\text{GW}} \simeq 10^4 \left( \frac{\mathcal{M}}{1.21 M_\odot} \right)^{-5/3} \left( \frac{f_{\text{in}}}{10\text{Hz}} \right)^{-5/3}. \tag{6.17}$$

### 6.2. Inspiral waveform including post-Newtonian corrections

Already in the early developments of Einstein theory, an approximation method called post-Newtonian (PN) method was developed by Einstein, Droste and De Sitter. This method allowed theorists to draw quickly several observational consequences, and within one year of the discovery of general relativity, led to the predictions of the relativistic advance of the perihelion of planets, the gravitational redshift and the deflection of light.

The PN method involves an expansion around the Newtonian limit keeping terms of higher order in the small parameter [12, 14]

$$\epsilon \sim \frac{v^2}{c^2} \sim |h_{\mu\nu}| \sim \left| \frac{\partial_0 h}{\partial_i h} \right|^2 \sim \left| \frac{T^{0i}}{T^{00}} \right| \sim \left| \frac{T^{ij}}{T^{00}} \right|. \tag{6.18}$$

In order to be able to determine the dynamics of binary systems with a precision acceptable for detection, it has been necessary to compute the force determining the motion of the two bodies and the amplitude of the gravitational radiation with a precision going beyond the quadrupole formula (5.14). For nonspinning BHs, the two-body equations of motion and the GW flux are currently known through 3.5PN order [13]. If we restrict the discussion to circular orbits, as Eq. (6.12) shows, there exists a natural *adiabatic* parameter $\dot{\omega}/\omega^2 = \mathcal{O}[(v/c)^5]$. Higher-order PN corrections to Eq. (6.12) have been computed [13, 15], yielding ($G = 1 = c$)

$$\frac{\dot{\omega}}{\omega^2} = \frac{96}{5} \nu v_\omega^{5/3} \sum_{k=0}^{7} \omega_{(k/2)\text{PN}}\, v_\omega^{k/3} \tag{6.19}$$

where we define $v_\omega \equiv (M\,\omega)^{1/3}$ and

$$\omega_{0PN} = 1, \tag{6.20}$$

$$\omega_{0.5PN} = 0, \tag{6.21}$$

$$\omega_{1PN} = -\frac{743}{336} - \frac{11}{4}\,v, \tag{6.22}$$

$$\omega_{1.5PN} = 4\pi + \left[ -\frac{47}{3}\frac{S_\ell}{M^2} - \frac{25}{4}\frac{\delta m}{M}\frac{\Sigma_\ell}{M^2} \right], \tag{6.23}$$

$$\omega_{2PN} = \frac{34\,103}{18\,144} + \frac{13\,661}{2\,016}\,v + \frac{59}{18}\,v^2 - \frac{1}{48}\,v\,\chi_1\chi_2\big[247(\widehat{\mathbf{S}}_1 \cdot \widehat{\mathbf{S}}_2) - 721(\hat{\boldsymbol{\ell}} \cdot \widehat{\mathbf{S}}_1)(\hat{\boldsymbol{\ell}} \cdot \widehat{\mathbf{S}}_2)\big], \tag{6.24}$$

$$\omega_{2.5PN} = -\frac{1}{672}(4\,159 + 15\,876\,v)\,\pi + \left[ \left( -\frac{31811}{1008} + \frac{5039}{84}\,v \right) \frac{S_\ell}{M^2} \right. $$
$$\left. + \left( -\frac{473}{84} + \frac{1231}{56}\,v \right) \frac{\delta m}{M}\frac{\Sigma_\ell}{M^2} \right], \tag{6.25}$$

$$\omega_{3PN} = \left( \frac{16\,447\,322\,263}{139\,708\,800} - \frac{1\,712}{105}\,\gamma_E + \frac{16}{3}\pi^2 \right) + \left( -\frac{56\,198\,689}{217\,728} \right. $$
$$\left. + \frac{451}{48}\pi^2 \right)v + \frac{541}{896}\,v^2 - \frac{5\,605}{2\,592}\,v^3 - \frac{856}{105}\,\log[16v^2], \tag{6.26}$$

$$\omega_{3.5PN} = \left( -\frac{4\,415}{4\,032} + \frac{358\,675}{6\,048}\,v + \frac{91\,495}{1\,512}\,v^2 \right)\pi. \tag{6.27}$$

We denote $\boldsymbol{\ell} = \mu\,\mathbf{X} \times \mathbf{V}$ the Newtonian angular momentum (with $\mathbf{X}$ and $X$ the two-body center-of-mass radial separation and relative velocity), and $\hat{\boldsymbol{\ell}} = \boldsymbol{\ell}/|\boldsymbol{\ell}|$; $\mathbf{S}_1 = \chi_1\,m_1^2\,\widehat{\mathbf{S}}_1$ and $\mathbf{S}_2 = \chi_2\,m_2^2\,\widehat{\mathbf{S}}_2$ are the spins of the two bodies (with $\widehat{\mathbf{S}}_{1,2}$ unit vectors, and $0 < \chi_{1,2} < 1$ for BHs) and

$$\mathbf{S} \equiv \mathbf{S}_1 + \mathbf{S}_2, \quad \Sigma \equiv M\left[ \frac{\mathbf{S}_2}{m_2} - \frac{\mathbf{S}_1}{m_1} \right]. \tag{6.28}$$

Finally, $\delta m = m_1 - m_2$ and $\gamma_E = 0.577\ldots$ is Euler's constant.

It is instructive to compute the relative contribution of the PN terms to the total number of GW cycles accumulating in the frequency band of LIGO/VIRGO. In Table 1, we list the figures obtained by plugging Eq. (6.19) into Eq. (6.16). As final frequency we use the innermost stable circular orbit (ISCO) of a point particle in Schwarzschild [$f_{GW}^{ISCO} \simeq 4400/(M/M_\odot)$ Hz].

Table 1

Post-Newtonian contributions to the number of GW cycles accumulated from $\omega_{in} = \pi \times 10\,\text{Hz}$ to $\omega_{fin} = \omega^{ISCO} = 1/(6^{3/2}\,M)$ for binaries detectable by LIGO and VIRGO. We denote $\kappa_i = \widehat{\mathbf{S}}_i \cdot \hat{\boldsymbol{\ell}}$ and $\xi = \widehat{\mathbf{S}}_1 \cdot \widehat{\mathbf{S}}_2$.

|  | $(10 + 10)M_\odot$ | $(1.4 + 1.4)M_\odot$ |
|---|---|---|
| Newtonian | 601 | 16034 |
| 1PN | $+59.3$ | $+441$ |
| 1.5PN | $-51.4 + 16.0\,\kappa_1\,\chi_1 + 16.0\,\kappa_2\,\chi_2$ | $-211 + 65.7\,\kappa_1\,\chi_1 + 65.7\,\kappa_2\,\chi_2$ |
| 2PN | $+4.1 - 3.3\,\kappa_1\,\kappa_2\,\chi_1\,\chi_2 + 1.1\,\xi\,\chi_1\,\chi_2$ | $+9.9 - 8.0\,\kappa_1\,\kappa_2\,\chi_1\,\chi_2 + 2.8\,\xi\,\chi_1\,\chi_2$ |
| 2.5PN | $-7.1 + 5.5\,\kappa_1\,\chi_1 + 5.5\,\kappa_2\,\chi_2$ | $-11.7 + 9.0\,\kappa_1\,\chi_1 + 9.0\,\kappa_2\,\chi_2$ |
| 3PN | $+2.2$ | $+2.6$ |
| 3.5PN | $-0.8$ | $-0.9$ |

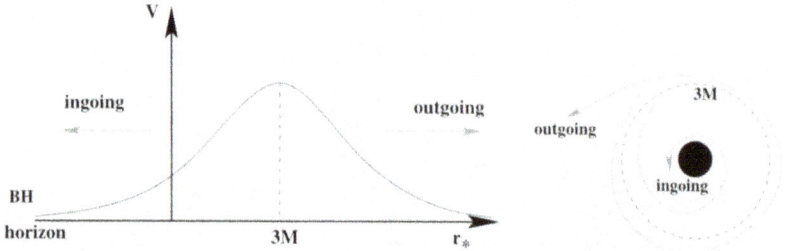

Fig. 4. We sketch the effective (radial) potential as function of the tortoise coordinate $r^*$ associated to metric perturbations of a Schwarzschild BH. The potential peaks at the last unstable orbit for a massless particle (the light ring). Ingoing modes propagate toward the BH horizon, whereas outgoing modes propagate away from the source.

## 6.3. Merger and ring-down waveforms

After the two BHs merge, the system settles down to a Kerr BH and emits quasi-normal modes (QNMs), as originally predicted by Ref. [53, 54]. This phase is commonly known as the ring-down (RD) phase. Since the QNMs have complex frequencies totally determined by the BH's mass and spin, the RD waveform is a superposition of damped sinusoidals. The inspiral and RD waveforms can be computed analytically. What about the merger? Since the nonlinearities dominate, the merger would be described at *best* and *utterly* through numerical simulations of Einstein equations. However, before NR results became available, some analytic approaches were proposed. In the test-mass limit, $\nu \ll 1$, Refs. [54, 55] realized a long time ago that the basic physical reason underlying the presence of a universal merger signal was that when a test particle falls below $3M$ (the unstable light storage ring of Schwarzschild), the GW it generates is strongly filtered by the effective potential barrier centered around it (see Fig. 4).

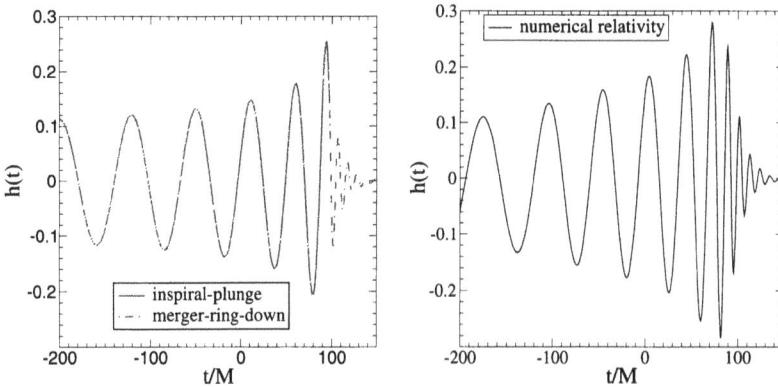

Fig. 5. On the left panel we show the GW signal from an equal-mass nonspinning BH binary as predicted at 2.5PN order by Buonanno and Damour (2000) in Ref. [16]. The merger is assumed almost instantaneous and one QNM is included. On the right panel we show the GW signal from an equal-mass BH binary with a small spin $\chi_1 = \chi_2 = 0.06$ obtained in full general relativity by Pretorius [58].

For the equal-mass case $\nu = 1/4$, Price and Pullin [56] proposed the so-called close-limit approximation, which consists in switching from the two-body description to the one-body description (perturbed-BH) close to the light-ring location. Based on this observations, the effective-one-body resummation scheme [16] provided a first *example* of full waveform by modeling the merger as a very short (instantaneous) phase and by matching the end of the plunge (around the light-ring) with the RD phase (see Ref. [57] where similar ideas were developed also in NR). The matching was initially done using *only* the least damped QNM whose mass and spin were determined by the binary BH energy and angular momentum at the end of the plunge. An example of full waveform is given in Fig. 5. Today, with the spectacular results in NR, we are in the position of assessing the closeness of analytic to numerical waveforms for inspiral, merger and RD. In Fig. 6, we show some first-order comparisons between analytic and numerical waveforms [58] (see also Ref. [59]). Similar results for the inspiral phase but using PN theory [13, 15] (without resummation) at 3.5PN order are given in Refs. [58, 59]. So far, the agreement is good, but more accurate simulations, starting with the BHs farther apart, are needed to draw robust conclusions.

Those comparisons are suggesting that it should be possible to design hybrid numerical/analytic templates, or even purely analytic templates with the full numerics used to guide the patching together of the inspiral and RD waveforms. This is an important avenue to template construction as eventually hundreds of

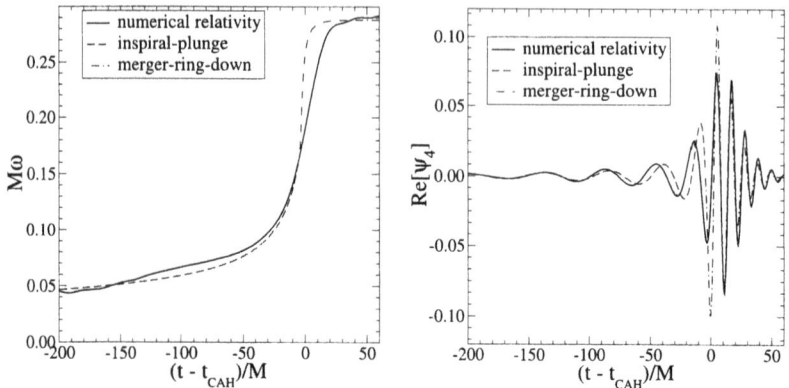

Fig. 6. Comparison between inspiral-merger-ring-down frequency (left panel) and waveform (right panel) for an equal-mass BH binary with spin $\chi_1 = \chi_2 = 0.06$, as predicted at 3.5PN order in Ref. [16] and as obtained in a numerical simulation by Pretorius [58]. In the analytic model the merger is assumed almost instantaneous and three QNMs are included [58]. $\Psi_4$ is the Weyl tensor proportional to the second derivative of $h$; we denote with $t_{CAH}$ the time when the apparent common horizon forms.

thousands of waveform templates may be needed to extract the signal from the noise, an impossible demand for NR itself.

### 6.4. *Templates for data analysis*

The search for GWs from coalescing binaries with laser interferometer GW detectors is based on the matched-filtering technique, which requires accurate knowledge of the waveform (or template) of the incoming signal. In this section we want to derive the GW template in Fourier domain using the stationary phase approximation (SPA). Those templates are currently used to search for inspiraling binary with LIGO/VIRGO/GEO/TAMA detectors. Henceforth, we use $G = 1 = c$.

The detector response to a GW signal is given by [24]

$$h(t) = h_+(t)\, F_+ + h_\times(t)\, F_\times, \tag{6.29}$$

$$F_+ = \frac{1}{2}(1 + \cos^2\Theta)\cos 2\Phi \cos 2\Psi - \cos\Theta \sin 2\Phi \sin 2\Psi, \tag{6.30}$$

$$F_\times = \frac{1}{2}(1 + \cos^2\Theta)\cos 2\Phi \sin 2\Psi + \cos\Theta \sin 2\Phi \cos 2\Psi, \tag{6.31}$$

being $\Theta$, $\Phi$, $\psi$ the angles defining the relative orientation of the binary with respect to the detector [24]. It is convenient to introduce the variables

$$\tilde{F}_+ \equiv \frac{(1 + \cos^2 \Theta) F_+}{[(1 + \cos^2 \Theta)^2 F_+^2 + 4 \cos^2 \Theta F_\times^2]^{1/2}}, \tag{6.32}$$

$$\tilde{F}_\times \equiv \frac{4 \cos^2 \Theta F_\times}{[(1 + \cos^2 \Theta)^2 F_+^2 + 4 \cos^2 \Theta F_\times^2]^{1/2}}. \tag{6.33}$$

Noticing that $\tilde{F}_+^2 + \tilde{F}_\times^2 = 1$, we can define $\cos \xi \equiv \tilde{F}_+$ and $\sin \xi \equiv \tilde{F}_\times$, and

$$\mathcal{A}(\Theta, \Phi, \Psi; \theta) \equiv \left[ (1 + \cos^2 \theta)^2 F_+^2 + 4 \cos^2 \theta F_\times^2 \right]^{1/2}, \tag{6.34}$$

$$\tan \xi (\Theta, \Phi, \Psi; \theta) \equiv \frac{4 \cos^2 \theta F_\times}{(1 + \cos^2 \theta) F_+}. \tag{6.35}$$

The GW signal for an inspiraling binary (6.29) can be rewritten in the simpler form

$$h(t) = \frac{2\mathcal{M}}{r} \mathcal{A}(\Theta, \Phi, \Psi; \theta)[\mathcal{M} \omega(t)]^{2/3} \cos[2\phi(t) + 2\phi_0 - \xi]. \tag{6.36}$$

If we are not interested in recovering the binary's orientation with respect to the detector (the so-called *inverse* problem), we can absorb $\xi$ into $\phi_0$, and average over the angles $(\Theta, \Phi, \Psi, \theta)$ obtaining [24] $\overline{\mathcal{A}^2} = 16/25$.

Let us now compute the Fourier transform of the GW signal

$$\tilde{h}(f) = \int_{-\infty}^{+\infty} e^{2\pi i f t} h(t) \, dt,$$

$$= \frac{1}{2} \int_{-\infty}^{+\infty} dt \, A(t) \left[ e^{2\pi i f t + i\phi_{GW}(t)} + e^{2\pi i f t - i\phi_{GW}(t)} \right], \tag{6.37}$$

where $A(t) = (2\mathcal{M}/R) \sqrt{\mathcal{A}^2} [\mathcal{M} \omega(t)]^{2/3}$ and $\phi_{GW}(t) = 2\phi(t) + 2\Phi_0$. We compute the integral as follows. In Eq. (6.37) The dominant contribution comes from the vicinity of the *stationary* points in the phase. Assuming $f > 0$, we pose $\psi(t) \equiv 2\pi f t - \phi_{GW}$ and impose $(d\psi/dt)_{t_f} = 0$, that is $(d\phi_{GW}/dt)_{t_f} = 2\pi f = 2\pi F(t_f)$. Expanding the phase up to quadratic order

$$\psi(t_f) = 2\pi f t_f - \phi_{GW}(t_f) - \pi \dot{F}(t_f)(t - t_f)^2, \tag{6.38}$$

we get

$$\tilde{h}_{SPA}(f) = \frac{1}{2} \frac{A(t_f)}{\sqrt{\dot{F}(t_f)}} e^{i[2\pi f t_f - \phi_{GW}(t_f)] - i\pi/4}. \tag{6.39}$$

To compute $\phi_{GW}(t_f)$ and $\dot{F}(t_f)$, we need to solve the following equations

$$v^3 = \phi_{GW} \frac{M}{2}, \quad \frac{dE}{dt}(v) = -\mathcal{F}(v), \tag{6.40}$$

where $E$ is the center-of-mass energy and $\mathcal{F}$ the GW energy flux. A direct calculation yields

$$t(v) = t_c + M \int_v^{v_c} dv \, \frac{E'(v)}{\mathcal{F}(v)}, \tag{6.41}$$

$$\phi_{GW}(v) = \phi_c + 2 \int_v^{v_c} dv \, v^3 \frac{E'(v)}{\mathcal{F}(v)}, \tag{6.42}$$

thus,

$$\psi(f) = 2\pi f \, t_c - \phi_c - \frac{\pi}{4} + 2 \int_v^{v_c} (v_c^3 - v^3) \frac{E'(v)}{\mathcal{F}(v)} \, dv, \tag{6.43}$$

Using Eq. (6.12), we have $\dot{F}(t_f) \equiv \dot{\omega}/\pi = (96/5)(1/\pi) \, v \, M^{5/3} \, \omega^{11/3}$, and we obtain [60]

$$\tilde{h}_{SPA}(f) = \mathcal{A}_{SPA}(f) \, e^{i \psi_{SPA}(f)}, \tag{6.44}$$

$$\mathcal{A}_{SPA}(f) = \frac{\sqrt{\mathcal{A}^2}}{r} \frac{1}{\pi^{2/3}} \left( \frac{5}{96} \right)^{1/2} \mathcal{M}^{5/6} \, f^{-7/6}, \tag{6.45}$$

$$\psi_{SPA}(f) = 2\pi f \, t_c - \phi_c - \frac{\pi}{4} + \frac{3}{128 \, v \, v_f^5} \sum_{k=0}^{7} \psi_{(k/2)PN} \, v_f^k, \tag{6.46}$$

where we denote $v_f = (\pi M f)^{1/3}$. The coefficients $\psi_{(k/2)PN}$'s, $k = 0, \ldots, 7$, (with $N = 7$ at 3.5PN order) in the Fourier phase are given by

$$\psi_{0PN} = 1, \tag{6.47a}$$

$$\psi_{0.5PN} = 0, \tag{6.47b}$$

$$\psi_{1PN} = \left( \frac{3715}{756} + \frac{55}{9} v \right), \tag{6.47c}$$

$$\psi_{1.5PN} = -16\pi + 4\beta, \tag{6.47d}$$

$$\psi_{2PN} = \left( \frac{15293365}{508032} + \frac{27145}{504} v + \frac{3085}{72} v^2 \right) - 10\sigma, \tag{6.47e}$$

$$\psi_{2.5PN} = \pi \left( \frac{38645}{756} - \frac{65}{9} v \right) (1 + 3 \log v_f), \tag{6.47f}$$

$$\psi_{3PN} = \left( \frac{11583231236531}{4694215680} - \frac{640\,\pi^2}{3} - \frac{6848\,\gamma_E}{21} \right)$$
$$+ v \left( -\frac{15737765635}{3048192} + \frac{2255\,\pi^2}{12} \right)$$
$$+ \frac{76055}{1728} v^2 - \frac{127825}{1296} v^3 - \frac{6848}{21} \log(4\,v), \tag{6.47g}$$

$$\psi_{3.5PN} = \pi \left( \frac{77096675}{254016} + \frac{378515}{1512} v - \frac{74055}{756} v^2 \right), \tag{6.47h}$$

where

$$\beta = \sum_{i=1}^{2} \left( \frac{113}{12} \frac{m_i^2}{M^2} + \frac{25}{4} v \right) \chi_i \kappa_i, \tag{6.48}$$

$$\sigma = v \left( \frac{721}{48} \chi_1 \kappa_1 \chi_2 \kappa_2 - \frac{247}{48} \xi \right). \tag{6.49}$$

In alternative theories of gravity [61], such as Brans-Dicke theory or massive graviton theories, the SPA phase (6.46) contains the terms

$$\psi_{SPA}^{alt\,th}(f) = \frac{3}{128\,v\,v_f^5} \left[ -\frac{5\mathcal{S}^2}{84\omega_{BD}} v_f^{-2} - \frac{128\,\pi^2 D\,v\,M}{3\,\lambda_g^2(1+z)} v_f^2 \right]. \tag{6.50}$$

The first term in the square brace is the contribution of dipole gravitational radiation in Brans-Dicke theory. The scalar charge of the $i$-th body is $\alpha_i = \bar{\alpha}\hat{\alpha}_i = \bar{\alpha}(1 - 2s_i)$, where $\bar{\alpha}^2 = 1/(2\omega_{BD}+3) \sim (2\omega_{BD})^{-1}$ in the limit $\omega_{BD} \gg 1$, and $s_i$ is called the *sensitivity* of the $i$-th body (a measure of the self-gravitational binding energy per unit mass). The coefficient in the dipole term is $\mathcal{S} = (\hat{\alpha}_1 - \hat{\alpha}_2)/2$. The fact that it is dipole radiation means that it is proportional to $v^{-2}$ compared to the quadrupole term, but the small size of $\mathcal{S}$ and the large current solar-system bound on $\omega_{BD} > 40,000$ make this a small correction. The second term in the square brace of Eq. (6.50) is the effect of a massive graviton, which alters the arrival time of waves of a given frequency, depending on the size of the graviton Compton wavelength $\lambda_g$ and on a distance quantity $D$, defined in Ref. [62]. It is possible to put interesting bounds on $\omega_{BD}$ and $\lambda_g$ using LISA [61,62].

Finally, the sky-average signal-to-noise ratio (SNR) for SPA waveforms is

$$\sqrt{\overline{\left(\frac{S}{N}\right)^2}} = \frac{1}{r} \frac{1}{\pi^{2/3}} \sqrt{\frac{2}{15}} \mathcal{M}^{5/6} \left[ \int_0^{f_{fin}} \frac{f^{-7/3}}{S_n(f)} \right]^{1/2}. \tag{6.51}$$

In the left panel of Fig. 7, we show the sky average SNRs versus total mass, for an equal-mass nonspinning binary at 100 Mpc using the noise-spectral density of

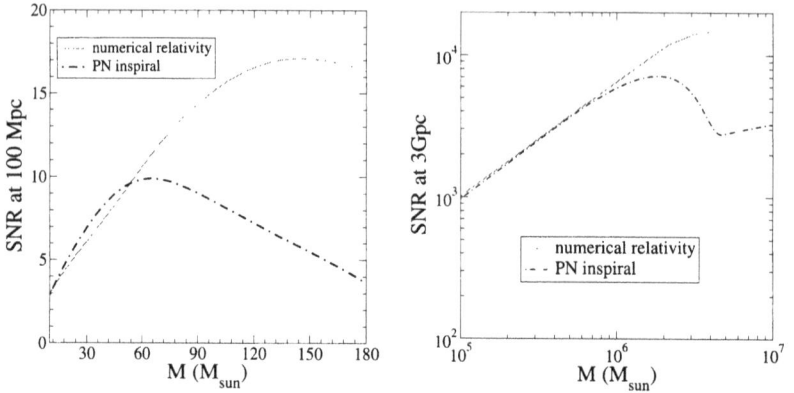

Fig. 7. We show the sky-averaged SNR for an equal-mass nonspinning binary when either the PN inspiral waveform or the full NR waveform are included. The left panel uses the noise spectral density of LIGO, whereas the right panel the noise spectral density of LISA.

LIGO. Astrophysical observations and theoretical predictions suggest that stellar mass BHs have a total mass ranging between $6$–$30 M_\odot$. If binary BHs of larger total mass exist, they could be detected by LIGO with high SNR. In the right panel of Fig. 7 we show the average SNR for one Michelson LISA configuration versus the total (redshifted) mass for an equal-mass nonspinning binary at 3 Gpc. The dip in the plot is due to the white-dwarf binary confusion noise. Due to the inclusion of merger and ring-down phases, the SNR increases considerably for total masses larger than $2 \times 10^6 M_\odot$.

## 7. Other astrophysical sources

### 7.1. Pulsars

The production of GWs from a rotating body is of considerable importance, in particular for application to isolated pulsars. For simplicity we shall examine the case the rigid body rotates around one of its principal axis.

Let us denote by $(X'_1, X'_2, X'_3)$ the coordinates in the reference frame attached to the body, the so-called *body frame*, and introduce a *fixed frame* with coordinates $(X_1, X_2, X_3)$, oriented such that $X'_3 = X_3$. In both frames the origin of the axes coincides with the center-of-mass of the body. The two frames are related by a time-dependent rotation matrix

$$\mathcal{R}_{ij} = \begin{pmatrix} \cos \omega t & \sin \omega t & 0 \\ -\sin \omega t & \cos \omega t & 0 \\ 0 & 0 & 1 \end{pmatrix}, \tag{7.1}$$

where $\omega$ is the rotation frequency. The inertia tensor of a rigid body is defined by

$$I^{ij} = \int d^3X \, \rho(\mathbf{X})(R^2 \, \delta^{ij} - X^i \, X^j), \tag{7.2}$$

$\rho$ being the mass density. Since any hermitian matrix can be diagonalized, there exists an orthogonal frame where $I_{ij}$ is diagonal. The eigenvalues are called principal moments of inertia. The frame where the inertia tensor is diagonal is the body frame. We denote by $I'_{ij} = \text{diag}(I_1, I_2, I_3)$ the inertia tensor in the $(X'_1, X'_2, X'_3)$ coordinate system, and by $I_{ij}$ its components in the $(X_1, X_2, X_3)$ frame. Using the relation $I = \mathcal{R}^t \, I' \, \mathcal{R}$, it is straightforward to derive

$$I_{11} = 1 + \frac{1}{2}(I_1 - I_2) \cos(2\omega t), \tag{7.3}$$

$$I_{12} = \frac{1}{2}(I_1 - I_2) \sin(2\omega t), \tag{7.4}$$

$$I_{22} = 1 - \frac{1}{2}(I_1 - I_2) \cos(2\omega t), \tag{7.5}$$

$$I_{33} = I_3, \tag{7.6}$$

$$I_{13} = I_{23} = 0. \tag{7.7}$$

To get the leading-order GW signal, we need to compute the second-time derivative of the quadrupole tensor $M_{ij} = \int d^3X \, \rho(\mathbf{X}) \, X^i \, X^j$, i.e. $\ddot{M}_{ij} = -\ddot{I}_{ij}$. We obtain

$$M_{11} = -\frac{1}{2}(I_1 - I_2) \cos(2\omega t), \tag{7.8}$$

$$M_{12} = -\frac{1}{2}(I_1 - I_2) \sin(2\omega t), \tag{7.9}$$

$$M_{22} = \frac{1}{2}(I_1 - I_2) \cos(2\omega t). \tag{7.10}$$

Thus, if the body rotates around the principal axis $X'_3$, there is a time-varying second mass moment only if $I_1 \neq I_2$. Plugging in Eqs. (5.14), (5.15) the above expressions, yields

$$h_+ = \frac{1}{r} \frac{4G\,\omega^2}{c^4}(I_1 - I_2) \frac{(1 + \cos\theta)}{2} \cos(2\omega t), \tag{7.11}$$

$$h_\times = \frac{1}{r} \frac{4G\,\omega^2}{c^4}(I_1 - I_2) \cos\theta \, \sin(2\omega t). \tag{7.12}$$

The GW signal is emitted at twice the pulsar rotation frequency. It is common to define the ellipticity $\epsilon = (I_1 - I_2)/I_3$ and $h_0 = (4\pi^2 G/c^4)(I_3 f_{GW}^2/r)\epsilon$. The value of the ellipticity depends on the NS properties, in particular the maximum strain that can be supported by its crust. Pulsars are thought to form in supernova explosions. The outer layers of the star crystallize as the newborn pulsar cools by neutrino emission. Anisotropic stresses during this phase could lead to values $\epsilon \lesssim 10^{-6}$ [39], although with exotic equation of state $\epsilon \simeq 10^{-5}$–$10^{-4}$ [63]. Plugging numbers we find

$$h_0 \simeq 10^{-25} \left( \frac{\epsilon}{10^{-6}} \right) \left( \frac{I_3}{10^{45} \text{ g cm}^2} \right) \left( \frac{10\text{kpc}}{r} \right) \left( \frac{f_{\text{GW}}}{1\text{kHz}} \right)^2. \tag{7.13}$$

Using Eq. (5.24), we can compute the total power radiated

$$P = \frac{32}{5} \frac{G}{c^5} \epsilon^3 I_3^2 \omega^6. \tag{7.14}$$

Due to GW emission, the rotational energy of the star decreases as

$$\frac{dE}{dt} = -\frac{32}{5} \frac{G}{c^5} \epsilon^3 I_3^2 \omega^6. \tag{7.15}$$

Since the rotational energy of a star rotating around its principal axis is

$$E = \frac{1}{2} I_3 \omega^2, \tag{7.16}$$

if the GW emission were the dominant mechanism for the loss of rotational energy, the pulsar frequency should decrease as

$$\dot{\omega} = -\frac{32}{5} \frac{G}{c^5} \epsilon^2 I_3 \omega^5. \tag{7.17}$$

From electromagnetic observations one finds instead $\dot{\omega} \propto -\omega^n$, where $n \simeq 2$–3. Thus, the GW emission is not the major mechanism of energy loss for a rotating pulsar.

If the rotation axis does not coincide with a principal axis, the pulsar motion is a combination of rotation around the principal axis and precession. New features appear in the GW signal, as discussed in detail in Ref. [64].

The detection of continuous, monochromatic frequency waves is achieved by constructing power spectrum estimators and searching for statistically significant peaks at fixed frequencies for very long time. If $T$ is the observation time, the signal-to-noise ratio grows like $\sqrt{T}$. The detection is complicated by the fact that the signal received at the detector is not perfectly monochromatic due to the Earth's motion. Because of Doppler shifts in frequency, the spectral lines of fixed frequency sources spread power into many Fourier bins about the observed frequency. Given the possibility that the strongest sources of continuous GWs may be electromagnetically invisible or previously undiscovered, an all sky, all frequency search for such unknown sources is very important, though computationally very expensive [25].

## 7.2. Supernovae

Supernovae are triggered by the violent collapse of a stellar core which forms a NS or BH. The core collapse proceeds extremely fast, lasting less than a second and the dense fluid undergoes motions with relativistic speeds. Small deviations from spherical symmetry during this phase can generate GWs.

From electromagnetic observations we know that stars with mass $M > 8M_\odot$ end their evolution in core collapse and that 90% of them have mass between $8$–$20M_\odot$. During the collapse most of the material is ejected and if the progenitor star has a mass $M \leq 20M_\odot$, it leaves behind a NS. If $M \geq 20M_\odot$, more than 10% of the material falls back and pushes the proto-NS above the maximum NS mass, leading to a BH. If the progenitor star has a mass $M \geq 40M_\odot$, no supernovae takes place, the star collapses directly to a BH. Numerical simulations [65] have predicted strains on the order of

$$h = 6 \times 10^{-17} \sqrt{\eta_{\text{eff}}} \left(\frac{M}{M_\odot}\right)^{1/2} \left(\frac{10\text{kpc}}{r}\right)^{1/2} \left(\frac{1\text{kHz}}{f}\right) \left(\frac{10\text{msec}}{\tau_{\text{collapse}}}\right)^{1/2}, \quad (7.18)$$

where $\Delta E_{\text{GW}} = \eta_{\text{eff}} M c^2$, $\eta_{\text{eff}} \sim 10^{-9}$. Reference [65] pointed out that GWs could also be produced by neutrino emission during the supernovae explosion. The GW signal would extend toward lower frequencies $\sim 10$ Hz. Moreover, the superposition of independent GW signals from cosmological supernovae may give rise to a stochastic GW background. While the estimates remain uncertain within several orders of magnitude, this background may become detectable by second-generation space-based interferometers operating around $\sim 0.1$ Hz [66].

Note that after a supernovae explosion or a collapsar a significant amount of the ejected material can fall back, subsequently spinning and heating the NS or BH. Quasi-normal modes can be excited. There is also the possibility that the collapsed material might fragment into clumps which orbit for some cycles like a a a binary system or form bar-like structures, which also produced GW signals [67].

## 8. Cosmological sources

In this section we want to review stochastic GW backgrounds. Depending on its origin, the stochastic background can be broadly divided into two classes: the astrophysically generated background due to the incoherent superposition of gravitational radiation emitted by large populations of astrophysical sources that cannot be resolved individually [68], and the primordial GW background generated by processes taking place in the early stages of the Universe. A primordial component of such background is especially interesting, since it would carry unique information about the state of the primordial Universe.

The energy and spectral content of such radiation is encoded in the spectrum, defined as follows

$$\Omega_{GW} = \frac{1}{\rho_c} \, f \tilde{\rho}_{GW}(f),$$ (8.1)

where $f$ is the frequency, $\rho_c$ is the critical energy density of the Universe ($\rho_c = 3H_0^2/8\pi G$) and $\tilde{\rho}_{GW}$ is the GWs energy density per unit frequency, i.e.

$$\rho_{GW} = \int_0^\infty df \, \tilde{\rho}_{GW}(f).$$ (8.2)

Before discussing the mechanisms that might have generated a primordial GW background, we review the main phenomenological bounds.

### 8.1. Phenomenological bounds

The theory of big-bang nucleosynthesis (BBN) predicts rather successfully the primordial abundances of light elements. If at $t_{BBN}$, the contribution of primordial GWs (or any other extra energy component) to the total energy density were to large, then the expansion rate of the Universe $H$ will be too large and the freeze-out temperature which determines the relative abundance of neutrons and protons will be too high. As a consequence, neutrons will be more available and light elements will be overproduced spoiling BBN predictions. Detailed calculations provide the following bound on the energy density [69]

$$\int d \ln f \, h_0^2 \, \Omega_{GW}(f) \le 5.6 \times 10^{-6}(N_\nu - 3),$$ (8.3)

where $N_\nu$ is the *effective number of neutrino species* at $t_{BBN}$ [38]. The above bound extends to frequency (today) greater than $\sim 10^{-10}$ Hz. More recently, Ref. [70] derived a similar bound by constraining the primordial GW energy density at the time of decoupling $t_{dec}$. The latter bound extends to lower frequency (today) $\sim 10^{-15}$ Hz, being determined by the comoving horizon size at the time of decoupling.

Another important bound is the so-called COBE bound, which comes from the measurement of temperature fluctuations in the CMB produced by the Sachs-Wolfe effect. If $\delta T$ is the temperature fluctuation

$$\Omega_{GW}(f) \le \left(\frac{H_0}{f}\right)^2 \left(\frac{\delta T}{T}\right)^2 \qquad 3 \times 10^{-18} \text{Hz} < f < 10^{-16} \text{Hz}.$$ (8.4)

The lower frequency is fixed by demanding that the fluctuations should be inside the Hubble radius today, whereas the higher frequency by imposing that fluctuations were outside the Hubble radius at the time of the last scattering surface. Detailed analysis give [71]

$$h_0^2 \, \Omega_{GW}(f) \leq 7 \times 10^{-11} \left( \frac{H_0}{f} \right)^2. \tag{8.5}$$

GWs could saturate this bound if the contribution from scalar perturbations is subdominant and this depends on the specific inflationary model.

The very accurate timing of millisec pulsars can constrain $\Omega_{GW}$. If a GW passes between us and the pulsar, the time of arrival of the pulse is Doppler shifted. Many years of observation yield to the bound [72] (see also Ref. [10])

$$h_0^2 \, \Omega_{GW} \lesssim 4.8 \times 10^{-9} \left( \frac{f}{f_*} \right)^2 \quad f > f_* = 4.4 \times 10^{-9} \, \text{Hz}. \tag{8.6}$$

## 8.2. Gravitational waves produced by causal mechanisms

Two features determine the typical frequency of GWs of cosmological origin produced by causal mechanism: (i) the dynamics, which is model dependent and (ii) the kinematics, i.e. the redshift from the production time. Let us assume that a graviton is produced with frequency $f_*$ at time $t_*$ during matter or radiation era. What is the frequency today? We have

$$f = f_* \frac{a_*}{a_0}, \tag{8.7}$$

assuming that the Universe evolved adiabatically, so $g_S(T_*) \, T_*^3 \, a_*^3 = g_S(T_0) \times T_0^3 \, a_0^3$, using $T_0 = 2.73$ K and $g_S(T_0) = 3.91$, where $g_S(T_*)$ is the number of degrees of freedom at temperature $T_*$, we get

$$f \simeq 10^{-13} \, f_* \left( \frac{100}{g_{S*}} \right)^{1/3} \left( \frac{1 \text{GeV}}{T_*} \right). \tag{8.8}$$

What is $f_*$? Since the size of the Hubble radius is the scale beyond which causal microphysics cannot operate, we can say that from causality considerations, the characteristic wavelength of gravitons produced at time $t_*$ is $H_*^{-1}$ or smaller. Thus, we set

$$\lambda_* = \frac{\epsilon}{H_*} \quad \epsilon \lesssim 1. \tag{8.9}$$

If the GW signal is produced during the radiation era,

$$H_*^2 = \frac{8\pi G}{3}\rho = 8\pi^3 g_* T_*^4 \frac{1}{90}\frac{1}{M_{\text{Pl}}^2},$$ (8.10)

and we find

$$f \simeq 10^{-7}\frac{1}{\epsilon}\left(\frac{T_*}{1\text{GeV}}\right)\left(\frac{g_*}{100}\right)^{1/6}\text{Hz}.$$ (8.11)

From the above equation, we obtain that millisec pulsars can probe physics at the $\sim$ MeV scale, LISA at the $\sim$ TeV scale, ground-based detectors at $10^3$–$10^6$ TeV and detectors in the GHz bandwidth [73] at GUT or Planck scale.

As an application of GWs produced by causal mechanisms, let us consider GW signals from first-order phase transitions. In the history of the Universe several phase transitions could have happened. The quantum-chromodynamics (QCD) phase transition takes place at $T \sim 150$ MeV. Around $T \sim 100$ GeV the electroweak (EW) phase transition happens (and $SU(2) \times U(1)$ breaks to $U_{\text{em}}(1)$ through the Higgs mechanism). Let us assume that $V(\phi, T)$ is the potential associated to the phase transition, where $\phi$ is the order (field) parameter. As the Universe cools down, the *true* and *false* vacuum are separated by a potential barrier. Through quantum tunneling, bubbles of true vacuum can nucleate. The difference of energy between the true and false vacua is converted in kinetic energy (speed) of the bubble's wall. In order to start expanding, the bubbles must have the right size, so that the volume energy overcomes the shrinking effects of the surface tension. The relevant parameter is the nucleation rate $\Gamma = \Gamma_0 e^{\beta t}$. If $\Gamma$ is large enough, the bubbles can collide within the Hubble radius, and being the collision nonsymmetric, they produce GWs [74]. The parameter $\beta$ fixes the frequency at the time of production.

So, we can write for the parameter $\epsilon$ defined in Eq. (8.9), $\epsilon = H_*/f_* = H_*/\beta$ and the peak of the GW spectrum occurs at [75,77,79]

$$f_{\text{peak}} \simeq 10^{-8}\left(\frac{\beta}{H_*}\right)\left(\frac{T_*}{1\text{GeV}}\right)\left(\frac{g_*}{100}\right)^{1/6}\text{Hz},$$ (8.12)

where $T_*$ is the temperature at which the probability that a bubble is nucleated within the horizon size is $\mathcal{O}(1)$.

In the case of EW phase transitions, we have $\beta/H_* \simeq 10^2$–$10^3$ and $T_* \sim 10^2$ GeV, thus $f_{\text{peak}} \simeq 10^{-4}$–$10^{-3}$ Hz, which is in the frequency band of LISA. The GW spectrum can be computed semianalytically, it reads [75,77,79]

$$h_0^2\Omega_{\text{GW}} \simeq 10^{-6}k^2\frac{\alpha^2}{(1+\alpha)^2}\frac{v_b^3}{(0.24+v_b^3)}\left(\frac{H_*}{\beta}\right)^2\left(\frac{100}{g_*}\right)^{1/3},$$ (8.13)

where $\alpha$ is the ratio between the false vacuum energy density and the energy density of the radiation at the transition temperature $T_*$; $\kappa$ quantifies the fraction of latent heat that is transformed into bubble-wall kinetic energy and $v_b$ is the bubble expansion velocity.

Nonperturbative calculations obtained using lattice field theory have shown that there is no first-order phase transition in the standard model if the Higgs mass is larger than $M_W$. In minimal supersymmetric standard model, if Higgs mass is $\sim$110–115 GeV, the transition is first-order but $h_0 \Omega_{GW} \sim 10^{-19}$ at $f \sim 10$ mHz. In next-to-minimal supersymmetric standard models, there exist regions of the parameter space in which $h_0 \Omega_{GW} \sim 10^{-15}$–$10^{-10}$ at $f \sim 10$ mHz. Note that for frequencies $10^{-4}$–$3 \times 10^{-3}$ Hz the stochastic GW background from white-dwarf binaries could *hide* part of the GW spectrum from first-order phase transitions. More recently, Ref. [78] pointed out that the new models of EW symmetry breaking that have been proposed have typically a Higgs potential different from the one in the Standard Model. Those potentials could lead to a stronger first-order phase transition, thus to a more promising GW signature in the milliHz frequency range.

A stochastic GW background could be also produced during a first-order phase transition from turbulent (anisotropic) eddies generated in the background fluid during the fast expansion and collision of the true-vacuum bubbles [76, 77, 79]. In the next-to-leading supersymmetric standard model there exist regions of the parameter space where [77] $h_0^2 \Omega_{GW} \sim 10^{-10}$ with peak frequency in the mHz. Reference [80] evaluated the stochastic GW background generated by cosmic turbulence before neutrino decoupling, i.e. much later than EW phase transition, and at the end of a first-order phase transition if magnetic fields also affect the turbulent energy spectrum. The observational perspectives of those scenarios are promising for LISA.

Long time ago Turner and Wilczek [81] pointed out that if inflation ends with bubble collisions, as in extended inflation, the GW spectrum produced has a peak in the frequency range of ground-based detectors. Subsequent analyses have shown that in two-field inflationary models where a field performs the first-order phase transition and a second field provides the inflationary slow rolling (so-called first-order or false vacuum inflation [82]), if the phase transition occurs well before the end of inflation, a GW spectrum peaked around $10$–$10^3$ Hz, can be produced [83], with an amplitude large enough, depending on the number of e-foldings left after the phase transition, to be detectable by ground-based interferometers. A successful detection of such a spectrum will allow to distinguish between inflation and other cosmological phase transitions, like QCD or EW, which have a different peak frequency.

Another mechanism that could have produced GWs is parametric amplification after preheating [84]. During this phase classical fluctuations produced by the oscillations of the inflaton field $\phi$ can interact back, via parametric resonance, on the oscillating background producing GWs. In the model where the inflaton potential contains also the interaction term $\sim\phi^2 \chi^2$, $\chi$ being a scalar field, the authors of Ref. [84] estimated $\Omega_{GW} \sim 10^{-12}$ at $f_{min} \sim 10^5$ Hz, while in pure chaotic inflation $\Omega_{GW} \leq 10^{-11}$ at $f_{min} \sim 10^4$ Hz. [See Fig. 3 in Ref. [84] for the GW spectrum in the range $10^6$–$10^8$ Hz and Refs. [85] for a thorough recent reanalysis.] Unfortunately, the predictions lay in the frequency range where no GW detectors exist, but some proposals have been made [73]

### 8.3. Gravitational waves produced by cosmic and fundamental strings

Topological defects could have been produced during symmetry-breaking phase transitions in the early Universe. Since the 80s they received significant attention as possible candidates for seeding structure formation. Recently, more accurate observations of CMB inhomogeneities on smaller angular scales and compatibility with the density fluctuation spectrum on scales of $100\ h_0^{-1}$ Mpc, restrict the contribution of topological defects to less than 10%.

Cosmic strings are characterized by a single dimensional scale, the mass-per-unit length $\mu$. The string length is defined as the energy of the loop divided by $\mu$. Cosmic string are stable against all types of decay, except from the emission of GWs. Let us assume that a network of cosmic strings did form during the evolution of the Universe. In this network the only relevant scale is the Hubble length. Small loops (smaller than Hubble radius length) oscillate, emit GWs and disappear, but they are all the time replaced by small loops broken off very long loops (longer than Hubble radius). The wavelength of the GW is determined by the length of the loop, and since in the network there are loops of all sizes, the GW spectrum is (almost) flat in a large frequency band, extending from $f \sim 10^{-8}$ Hz to $f \sim 10^{10}$ Hz.

In 2001, Damour and Vilenkin [86] (see also Ref. [87]), worked out that strong bursts of GWs could be produced at cusps (where the string reaches a speed very close to the light speed) and at kinks along the string loop. As a consequence of these bursts the GW background emitted by a string network is strongly non-Gaussian. The most interesting feature of these GW bursts is that they could be detectable for a large range of values of $G\mu$, larger than the usually considered search for the Gaussian spectrum. GW bursts can be also produced by fundamental strings, as pointed out in Ref. [88]. For a detailed analysis of the prospects of detecting the stochastic GW background and the GW bursts with ground and space-based detectors, and millisec pulsars see Refs. [10, 89].

A GW burst emitted at the cusp of cosmic or fundamental strings can be detected using matched filtering. In Fourier domain the signal is [89]

$$h(f) = A|f|^{-4/3} \Theta(f_h - f) \Theta(f - f_e), \tag{8.14}$$

where $A \sim (G\mu L^{2/3})/r$, $f_e$ is determined by the size $L$ of the feature that produces the cusp, but also by the low-frequency cutoff frequency of the detector and $f_h \sim 2/(\theta^3 L)$, $\theta$ being the angle between the line of sight and the direction of the cusp.

### 8.4. Gravitational waves produced during inflation

The amplification of quantum vacuum fluctuations is a common mechanism in quantum field theory in curved space time [90]. In the 70s Grishchuk and Starobinsky [91] applied it to cosmology, predicting a stochastic GW background which today would span a very large frequency band $10^{-17}$–$10^{10}$ Hz. Henceforth, we shall compute the GW spectrum using semiclassical arguments, and refer the reader to Refs. [92, 93] for more detailed computations.

The background field dynamics is described by the action

$$S = \frac{1}{16\pi G} \int d^4x \sqrt{-g}\, R + S_{\mathrm{m}}. \tag{8.15}$$

We assume an isotropic and spatially homogeneous Friedmann-Lamaitre-Robertson-Walker metric with scale factor $a$,

$$ds^2 = -dt^2 + a^2\, d\mathbf{x}^2 = g_{\mu\nu}\, dx^\mu\, dx^\nu. \tag{8.16}$$

For what we have learned in previous lectures, we can derive the free-linearized wave equations for the TT metric perturbations $\delta g_{\mu\nu} = h_{\mu\nu}$, with $h_{\mu 0} = 0$, $h_\mu^\mu = 0$, $h^\mu_{\nu;\mu} = 0$ and $\delta T^\nu_\mu = 0$, obtaining

$$\Box h_i^j = \frac{1}{\sqrt{-g}}\, \partial_\mu \left(\sqrt{-g}\, g^{\mu\nu}\, \partial_\nu\right) h_i^j = 0, \tag{8.17}$$

where we disregard any anisotropic stresses. Introducing the conformal time $\eta$, with $d\eta = dt/a(t)$, we can write

$$h_i^j(\eta, \mathbf{x}) = \sqrt{8\pi G} \sum_{A=+,\times} \sum_{\mathbf{k}} h_{\mathbf{k}}^A(\eta)\, e^{i\mathbf{k}\cdot\mathbf{x}}\, e_i^{A\,j}(\mathbf{n}), \tag{8.18}$$

$e_i^{A\,j}(\mathbf{n})$ being the polarization tensors. Since we assume isotropic and spatially homogeneous metric perturbations $h_{\mathbf{k}} = h_k$, and each polarization mode satisfies the equation

$$h_k''(\eta) + 2\frac{a'}{a}\, h_k'(\eta) + k^2\, h_k(\eta) = 0, \tag{8.19}$$

where we denote with a prime the derivative with respect to the conformal time. By introducing the canonical variable $\psi_k(\eta) = a\, h_k(\eta)$ we can recast Eq. (8.19) in the much simpler form

$$\psi_k''(\eta) + \left[k^2 - U(\eta)\right]\psi_k = 0, \qquad U(\eta) = \frac{a''}{a}, \tag{8.20}$$

which is the equation of an harmonic oscillator in a time-depedent potential $U(\eta)$. We want to solve the above equation and study the properties of the solutions.

For simplicity, we consider a De Sitter inflationary era, $a = -1/(H_{DS}\,\eta)$ and $a''/a = 2/\eta^2$, and make the crude assumption that the De Sitter era is followed instantaneously by the radiation era, $a(\eta) = (\eta - 2\eta_*)/(H_{DS}\,\eta_*^2)$ and $a''(\eta) = 0$.

If $k^2 \gg U(\eta)$, i.e. $k\eta \gg 1$ or $a/k \ll |a\eta| = H_{DS}^{-1}$, the mode is inside the Hubble radius or (in jargon) is over the potential barrier $U(\eta)$ and the (positive frequency) solution reads

$$\psi_k \sim \frac{1}{\sqrt{2k}}\, e^{-ik\eta} \quad \Rightarrow \quad h_k \sim \frac{1}{2k}\frac{1}{a}\, e^{-ik\eta}, \tag{8.21}$$

thus $h_k$ decreases while inside the Hubble radius. If $k^2 \ll U(\eta)$, i.e. $k\eta \ll 1$, $a/k \gg |a\eta| = H_{DS}^{-1}$, the mode is outside the Hubble radius or (in jargon) under the potential barrier. In this case the solution reads

$$\psi_k \sim a\left[A_k + B_k \int \frac{d\eta}{a^2(\eta)}\right] \quad \Rightarrow \quad h_k \sim A_k + B_k \int \frac{d\eta}{a^2(\eta)}. \tag{8.22}$$

Since during the De Sitter era the scale factor gets larger and larger, the term proportional to B in the above equation becomes more and more negligible. Thus, the perturbation $h_k$ remains (almost) constant while outside the Hubble radius. So, the longer the tensorial-perturbation mode remains outside the Hubble radius, the more it gets amplified (with respect to the case it stayed always inside the Hubble radius). During the RD era, the mode is again under the barrier, and the solution is

$$\psi_k = \alpha_k\, e^{-ik\eta} + \beta_k\, e^{+ik\eta}, \tag{8.23}$$

and contains both positive and negative modes. Even normalizing the initial state to positive frequency mode, the final state is a mixture of positive and frequency modes. In a quantum field theory language, such a mixing represents a process of pair production from vacuum. The coefficients $\alpha_k$ and $\beta_k$ are called Bogoliubov coefficients and can be obtained imposing the continuity of the tensorial perturbation and its first time derivative at the transition between cosmological phases.

The intensity of the stochastic GW background can be expressed in terms of the number of gravitons per cell of the phase space $n_f$ with $f = |\mathbf{k}|/(2\pi)$. For an isotropic stochastic GW background $\rho_{GW} = 2 \int d^3k/(2\pi)^3 (k\, n_k)$, thus

$$\Omega_{GW}(f) = \frac{1}{\rho_c} 16\pi^2 n_f f^4. \tag{8.24}$$

where $n_f = |\beta_f|^2$.

The stochastic GW spectrum produced during *slow-roll* inflation, decreases as $f^{-2}$ in the frequency window $10^{-18}$–$10^{-16}$ Hz, and then slowly decreases up to a frequency corresponding to modes whose physical frequency becomes less than the maximum causal distance during the reheating phase (which is of order of a few GHz). Its magnitude depends on both the value of the Hubble parameter during inflation and a number of features characterizing the Universe evolution after the inflationary era — for example tensor anisotropic stress due to free-streaming relativistic particles, equations of state, etc. [93, 94]. An upper bound on the spectrum can be obtained from the measurement of the quadrupole anisotropy of the CMB, as seen in Sec. 8.1. Since for a generic slow-roll inflationary model the spectrum is (weakly) decreasing with frequency this implies an upper bound $h_0^2 \Omega_{GW} \sim 5 \times 10^{-16}$ at frequencies around $f \sim 100$ Hz, where ground-based detectors reach the best sensitivity. The spectrum predicted by the class of single-field inflationary models is then too low to be observed by ground-based and also space-based detectors. It is therefore evident that a background satisfying the bound imposed by the observed amount of CMB anisotropies at large scales could be detected by GW interferometers provided that its spectrum grows significantly with frequency. This could happen in bouncing Universe cosmologies, such as pre–big-bang scenario [95, 96], the ekpyrotic models [97] (although the amplitude of the GW spectrum is too low to be observable) and in quintessential inflation [98].

Finally, a stochastic GW background can be detected by correlating two GW interferometers [99]. The upper limit on a flat-spectrum set by the LIGO Scientific Collaboration is $h_0^2\Omega_{GW} \simeq 6.5 \times 10^{-5}$ in the frequency band 70–156 Hz [28]. For frequency-independent spectra, the expected upper limit for the current LIGO configuration is $h_0^2 \Omega_{GW} < 5 \times 10^{-6}$, while for advanced LIGO project is $h_0^2 \Omega_{GW} \sim 8 \times 10^{-9}$.

## Acknowledgments

I wish to thank the organizers of the Les Houches School for having invited me to such a pleasant and stimulating school and all the students for their interesting questions. I acknowledge support from NSF grant PHY-0603762 and from the Alfred Sloan Foundation. I wish to thank also Francis Bernardeau and Christophe Grojean for their patience.

## References

[1] A. Abramovici et al., Science **256**, 325 (1992); http://www.ligo.org.

[2] B. Caron et al., Class. Quant. Grav. **14**, 1461 (1997); http://www.virgo.infn.it.

[3] H. Lück et al., Class. Quant. Grav. **14**, 1471 (1997); http://www.geo600.uni-hannover.de.

[4] M. Ando et al., Phys. Rev. Lett. **86**, 3950 (2001); http://tamago.mtk.nao.ac.jp.

[5] B.J. Meers and K. Strain, Phys. Rev. A **44**, 4693 (1991); J. Mizuno et al., Phys. Lett. A **175**, 273 (1993); A. Buonanno and Y. Chen, Phys. Rev. D **64**, 042006 (2001).

[6] V.B. Braginsky and F.Ya. Khalili, Phys. Lett. A **147**, 251 (1990); V.B. Braginsky, M.L. Gorodet-sky, F.Ya. Khalili, and K.S. Thorne, Phys. Rev. A **61**, 044002 (2000); P. Purdue, Phys. Rev. D **66**, 022001 (2001); P. Purdue and Y. Chen, Phys. Rev. D **66**, 122004 (2002); J. Harms et al. Phys. Rev. D **68**, 042001 (2003); A. Buonanno and Y. Chen, Phys. Rev. D **69** 102004 (2004).

[7] http://www.lisa-science.org.

[8] P. Astone et al., Europhys. Lett. **12**, 5 (1990); G. Pallottino, in "Gravitational Waves, Sources and Detectors," p 159 (Singapore, World Scientific, 1997); E. Mauceli et al., Phys. Rev. **D54**, 1264 (1996); D. Blair et al., Phys. Rev. Lett. **74**, 1908 (1995); M. Cerdonio et al., Class. Quantum Grav. **14**, 1491 (1997).

[9] M. Kamionkowski, A. Kosowsky, and A. Stebbins, Phys. Rev. **D55**, 7368 (1997); U. Seljak and M. Zaldarriaga, Phys. Rev. **D78**, 2054 (1997); M. Kamionkowski and A. Kosowsky, Phys. Rev. **D57**, 685 (1998).

[10] See. e.g., F.A. Jenet et al., Astrophys. J. **653**, 1571 (2006), and references therein.

[11] http://www.skatelescope.org.

[12] T. Damour, *The problem of motion in Newtonian and Einsteinian gravity*, in *300 Years of Gravitation* ed. by S.W. Hawking and W. Israel (Cambridge University Press, Cambridge, 1987).

[13] P. Jaranowski, and G. Schäfer, Phys. Rev. D **57**, 7274 (1998); Erratum-ibid D **63** 029902; T. Damour, P. Jaranowski, and G. Schäfer, Phys. Lett. B **513**, 147 (2001); L. Blanchet, G. Faye, B.R. Iyer, and B. Joguet, Phys. Rev. D **65**, 061501(R) (2002); Erratum-ibid D **71**, 129902 (2005); L. Blanchet, T. Damour, G. Esposito-Farese, and B.R. Iyer, Phys. Rev. Lett. **93**, 091101 (2004).

[14] See, e.g., L. Blanchet, Living Rev. Rel. **5**, 3 (2002) and references therein. http://relativity.livingreviews.org.

[15] L. Kidder, C. Will, and A. Wiseman, Phys. Rev. D **47**, 4183 (1993). L. Kidder, Phys. Rev. D **52**, 821 (1995). B. Owen, H. Tagoshi, and Ohashi, Phys. Rev. D **57**, 6168 (1998). H. Tagoshi, Ohashi, and B. Owen, Phys. Rev. D **63**, 044006 (2001). G. Faye, L. Blanchet, and A. Buonanno, Phys. Rev. D **74**, 104033 (2006); L. Blanchet, A. Buonanno, and G. Faye, Phys. Rev. D **74**, 104034 (2006).

[16] A. Buonanno and T. Damour, Phys. Rev. D **59**, 084006 (1999); ibid. D **62**, 064015 (2000); A. Buonanno and T. Damour, Proceedings of IX[th] Marcel Grossmann Meeting (Rome, July 2000) [gr-qc/0011052]; T. Damour, P. Jaranowski, and G. Schäfer, Phys. Rev. D **62**, 084011 (2000); T. Damour, Phys. Rev. D **64**, 124013 (2001); A. Buonanno, Y. Chen, and T. Damour, Phys. Rev. D **74**, 104005 (2006).

[17] T. Damour, B.R. Iyer, and B.S. Sathyaprakash, Phys. Rev. D **57**, 885 (1998).

[18] W.D. Goldberger and I.Z. Rothstein, Phys. Rev. D **73**, 104029 (2006). R.A. Porto and I.Z. Rothstein, Phys. Rev. Lett. **97**, 021101 (2006).

[19] F. Pretorius, Phys. Rev. Lett. **95**, 121101 (2005); M. Campanelli, C.O. Lousto, P. Marronetti, and Y. Zlochower, Phys. Rev. Lett. **96**, 111101 (2006); J. Baker, J. Centrella, D. Choi, M. Koppitz, and J. van Meter, Phys. Rev. Lett. **96**, 111102 (2006).

[20] See, e.g., E. Poisson, Living Rev. Rel. **6**, 3 (2004), and references therein. http://relativity.livingreviews.org.

[21] Y. Mino, Prog. Theor. Phys. **113**, 733 (2005); ibid. Prog. Theor. Phys. **115**, 43 (2006); A. Pound, E. Poisson, and B.G. Nickel, Phys. Rev. D **72**, 124001 (2005); L. Barack and C. Lousto, Phys. Rev. D **72**, 104026 (2005); L. Barack and N. Sago, Phys. Rev. D **75**, 064021 (2007).

[22] L. S. Finn and K. S. Thorne, Phys. Rev. D **62**, 124021 (2000); S. Babak et al. Phys. Rev. D **75**, 024005 (2007); L. Barack and C. Cutler, Phys. Rev. D **70**, 122002 (2004); ibid. Phys. Rev. D **70**, 122002 (2004); G. Sigl, J. Schnittman, and A. Buonanno, Phys. Rev. D **75**, 024034 (2007).

[23] C. Cutler et al., Phys. Rev. Lett. **70**, 2984 (1993); T. Damour, B. Iyer, and B. Sathyaprakash, Phys. Rev. D **63**, 044023 (2001); ibid. D **66**, 027502 (2002); T. Damour, B.R. Iyer, and B.S. Sathyaprakash, Phys. Rev. D **67**, 064028 (2003); A. Buonanno, Y. Chen, and M. Vallisneri, Phys. Rev. D **67**, 104025 (2003); Erratum-ibid. D **74**, 029904 (2006); A. Buonanno, Y. Chen, and M. Vallisneri, Phys. Rev. D **67**, 024016 (2003); Erratum-ibid. D **74**, 029903 (2006); T. Damour, B. Iyer, P. Jaranowski, and B. Sathyaprakash, Phys. Rev. D **67**, 064028 (2003); A. Buonanno, Y. Chen, and M. Vallisneri, Phys. Rev. D **67**, 104025 (2003); Erratum-ibid. D **74**, 029904 (2006); Y. Pan, A. Buonanno, Y. Chen and M. Vallisneri, Phys. Rev. D **69** 104017 (2004).

[24] L.S. Finn, Phys. Rev. D **46**, 5236 (1992); L. S. Finn and D.F. Chernoff, Phys. Rev. D **47**, 2198 (1993); É.E. Flanagan and S.A. Hughes, Phys. Rev. D **57**, 4535 (1998).

[25] P.R. Brady, T. Creighton, C. Cutler, and B.F. Schutz, Phys. Rev. D **57**, 2101 (1998); P.R. Brady and T. Creighton, Phys. Rev. D **61**, 082001 (2000); M. Tinto, F.B. Estabrook, and J.W. Armstrong, Phys. Rev. D **65**, 082003 (2002); B. Krishnan et al. Phys. Rev. D **70**, 082001 (2004); C. Cutler, I. Gholami, and B. Krishnan, Phys. Rev. D **72**, 042004 (2005); S. Chatterji et al., Phys. Rev. D **74**, 082005 (2006).

[26] B. Abbott et al. (LIGO Scientific Collaboration), Phys. Rev. D **72** 082002 (2005); ibid. D **72** 122004 (2005); ibid. D **72** 102004 (2005); ibid. D **73** 062001 (2006); ibid. D **73** 102002 (2006).

[27] B. Abbott et al. (LIGO Scientific Collaboration), *Coherent searches for periodic gravitational waves from unknown isolated sources and Scorpius X-1*, [gr-qc/0605028].

[28] B. Abbott et al. (LIGO Scientific Collaboration), Phys. Rev. Lett. **95** 221101 (2005); *Searching for a stochastic background of gravitational waves with LIGO*, [astro-ph/0608606].

[29] L.D. Landau and E.M. Lifshits, *The Classical Theory of Field*, (Pergamon Press, Oxford, 1975).

[30] C. Misner, K.S. Thorne, and J.A. Wheeler, *Gravitation* (W.H. Freeman and Company, New York, 1973).

[31] S. Weinberg, *Gravitation and cosmology* (John Wiley and Sons, New York, 1972).

[32] R.M. Wald, *General relativity* (The University of Chicago Press, Chicago, 1983).

[33] B. Schutz, *A first course in general relativity* (Cambridge University Press, Cambridge, 1985).

[34] S. Carroll *Spacetime and geometry: an introduction to general relativity* (Addison-Wesly, 2003).

[35] M. Maggiore, *Gravitational waves: Theory and Experiments* (Oxford University Press, Oxford, 2007).

[36] K.S. Thorne, *Gravitational radiation*, in *300 Years of Gravitation* ed. by S.W. Hawking and W. Israel (Cambridge University Press, Cambridge, 1987).

[37]  B. Allen, *Lectures at Les Houches School* (1996) [gr-qc/9604033].

[38]  M. Maggiore, Phys. Rep. **331**, 283 (2000).

[39]  C. Cutler and K.S. Thorne, *An overview of gravitational-wave sources* (2002) [gr-qc/0204090] and references therein.

[40]  A. Buonanno, TASI *lectures on gravitational-wave from the early Universe* (2003) [gr-qc/0303085].

[41]  É.E. Flanagan and S.A. Hughes, *The basics of gravitational-wave theory*, New J.Phys. **7**, 204 (2005).

[42]  K.S. Thorne, http://elmer.caltech.edu/ph237/.

[43]  A. Einstein, Sitzber. Preuss. Akad. Wiss., 688 (1916); see also Preuss. Akad. Wiss., 154 (1918).

[44]  W.T. Ni and M. Zimmermann, Phys. Rev. D **17**, 1473 (1978).

[45]  M. Rakhamanov, Phys. Rev. D **71**, 084003 (2005).

[46]  D.R. Brill and J.B. Hartle, Phys. Rev. **135**, B271 (1963). R.A. Isaacson, Phys. Rev. **166**, 1263 (1968); **166**, 1272 (1968).

[47]  M. Walker and C.M. Will, Astrophys. J. **242**, L129 (1980); M. Walker and C.M. Will, Phys. Rev. Lett. **45**, 1741 (1980); T. Damour, Phys. Rev. Lett. **51**, 1019 (1983).

[48]  F. Herrmann, D. Shoemaker, and P. Laguna, [gr-qc/0601026]; J.G. Baker, J. Centrella, D. Choi, M. Koppitz, J. R. van Meter, and M.C Miller, Astrophys. J **653**, L93 (2006); M. Campanelli, C.O. Lousto, and Y. Zlochower, Phys. Rev. D **74**, 041501 (2006); ibid. D **74**, 084023 (2006); U. Sperhake, [gr-qc/0606079]; J.A. Gonzalez, U. Sperhake, B. Bruegmann, M. Hannam, and S. Husa, Phys.Rev.Lett.**98**,091101 (2007); M. Koppitz, D. Pollney, C. Reisswig, L. Rezzolla, J. Thornburg, P. Diener, and E. Schnetter, [gr-qc/0701163]; M. Campanelli, C.O. Lousto, Y. Zlochower, and D. Merritt, [gr-qc/0701164]; F. Herrmann, I. Hinder, D. Shoemaker, P. Laguna, and R.A. Matzner [gr-qc/0701143]; J.A. Gonzalez, M.D. Hannam, U. Sperhake, B. Brugmann, and S. Husa, [gr-qc/0702052]; J.G. Baker, W.D. Boggs, J.Centrella, B.J. Kelly, S.T. McWilliams, M.C. Miller, and J.R. van Meter, [astro-ph/0702390]; W. Tichy and P. Marronetti, [gr-qc/0703075].

[49]  L. Blanchet, M.S.S. Qusailah, and C.M. Will, Astrophys. J. **635**, 508 (2006); T. Damour and A. Gopakumar, Phys. Rev. D **73**, 124006 (2006); C.F. Sopuerta, N. Yunes, and P. Laguna (2006) [astro-ph/0611110]; J. Schnittman and A. Buonanno (2007), [astro-ph/0702641].

[50]  H. Bondi, Nature **179**, 1072 (1957); ibid. **186**, 535 (1960).

[51]  P.C. Peters and J. Mathews, Phys. Rev. **131**, 435 (1963); P.C. Peters, Phys. Rev. **136**, B1224 (1964).

[52]  R. Hulse and J. Taylor, Astrophys. J., **324** (1975).

[53]  C.V. Vishveshwara, Nature **227**, 936 (1970); B. Schutz and C.M. Will, Astrophys. J. **291**, 33 (1985); S. Chandrasekhar and S. Detweiler, Proc. R. Soc. Lond. A **344**, (1975) 441; B. Mashhoon, Phys. Rev. D **31**, 290 (2005).

[54]  W. Press, Astrophys J. Letters **170**, L105 (1971).

[55]  M. Davis, R. Ruffini, W.H. Press, and R.H. Price, Phys. Rev. Lett. **27**, 1466 (1971); M. Davis, R. Ruffini, and J. Tiomno, Phys. Rev. D **5**, 2932 (1972).

[56]  R.H. Price and J. Pullin, Phys. Rev. Lett. **72**, 3297 (1994); A. M. Abraham and G.B. Cook, Phys. Rev. D **50**, R2364 (1994); R.J. Gleiser, C.O. Nicasio, R. Price, and J. Pullin, Class. Quant. Grav. **13**, L117 (1996); Phys. Rev. Lett. **77**, 4483 (1996); J. Pullin, *The close limit of colliding black holes: an update*, Talk given at the Yukawa International Symposium at Kyoto, Japan, 1999 [gr-qc/9909021] and references therein; P. Anninos, D. Hobill, E. Seidel, L. Smarr, and W.M. Suen, Phys. Rev. Lett. **71**, 2851 (1993); Z. Andrade and R. H. Price, Phys.Rev.D **56**, 6336 (1997).

[57]  J. Baker, B. Brügmann, M. Campanelli, C.O. Lousto, and R. Takahashi, Phys. Rev. Lett. **87**, 121103 (2001).

[58] A. Buonanno, G. Cook, and F. Pretorius (2006), [gr-qc/0610122].

[59] J. Baker, J. van Meter, S. McWilliams, J. Centrella, and B. Kelly (2006), [gr-qc/0612024].

[60] C. Cutler and É.E. Flanagan, Phys. Rev. D **49**, 2658 (1994); A. Krolak, K. D. Kokkotas, and G. Schafer, Phys. Rev. D **52**, 2089 (1995); E. Poisson and C.M. Will, Phys. Rev. D **52**, 848 (1995); Arun et al., Phys. Rev. D **71**, 084008 (2005); Erratum-ibid.D **72**, 069903 (2005).

[61] C.M. Will, Phys. Rev. D **50**, 6058 (1994); C.M. Will, Phys. Rev. D **57**, 2061 (1998); T. Damour and G. Esposito-Farése, Phys. Rev. D **58**, 042001 (1998); P.D. Scharre and C.M. Will, Phys. Rev. D **65**, 042002 (2002); C.M. Will and N. Yunes, Class. Quantum Gravit. **21**, 4367 (2004).

[62] E. Berti, A. Buonanno, and C.S. Will, Phys. Rev. D **71** 084025 (2005).

[63] B.J. Owen, Phys. Rev. Lett. **95**, 211101 (2005).

[64] M. Zimmermann and E. Szedenits, Phys. Rev. D **20**, 351 (1979); M. Zimmermann and E. Szedenits, Phys. Rev. D **21**, 891 (1980); C. Cutler and D.I. Jones, Phys. Rev.D **63**, 024002 (2001).

[65] H. Dimmelmeier, J. A. Font, and E. Mueller, Astron. Astrophys. **393**, 523 (2002); E. Müller, M. Rampp, R. Buras, H.-T. Janka, and D. H. Shoemaker, Astrophys. J. **603**, 221 (2004); C. L. Fryer, D. E. Holz, and S. A. Hughes, Astrophys. J. **609**, 288 (2004); A. Burrows et al., Astrophys. J. **640**, 878 (2006); H. Dimmelmeier et al., [astro-ph/0702305].

[66] A. Buonanno, G. Sigl, G. Raffelt, T. Janka, and E. Mueller, Phys. Rev. D **72** 084001 (2005).

[67] S. Kobayashi and P. Meszaros, Astrophys. J. **589**, 861 (2003).

[68] A.J. Farmer and E.S. Phinney, Mon. Not. Roy. Astron. Soc. **346**, 1197 (2003).

[69] C.J. Copi, D.N. Schramm, and M.S. Turner, Phys. Rev. **D55**, 3389 (1997).

[70] T. L. Smith, E. Pierpaoli, and M. Kamionkowski, Phys. Rev. Lett. **97**, 021301 (2006).

[71] B. Allen and S. Koranda, Phys. Rev. **D50**, 3713 (1994).

[72] S. Thorsett and R. Dewey, Phys. Rev. **D53**, 3468 (1996).

[73] V.B. Braginski, L.P. Grishchuck, A.G. Doroshkevich, Ya.B. Zeldovich, I.D. Novikov and M.V. Sazhin, Sov. Phys. JETP **38**, 865 (1974); E. Iacopini, E. Picasso, F. Pegoraro, and L.A. Radicati, Phys. Lett **73A**, 140 (1979); C.M. Caves, Phys. Lett. **80B**, 323 (1979).

[74] C.J. Hogan, Phys. Lett. **B133**, 172 (1983); E. Witten, Phys. Rev. **D30**, 272 (1984); C.J. Hogan, Mon. Not. R. Astr. Soc. **218**, 629 (1986).

[75] A. Kosowsky, M.S. Turner, and R. Watkins, Phys. Rev. **D45**, 4514 (1992); Phys. Rev. Lett. **69**, 2026 (1992); A. Kosowsky and M.S. Turner, Phys. Rev. **D47**, 4372 (1993); M. Kamionkowski, A. Kosowsky, and M.S. Turner, Phys. Rev. **D49**, 2837 (1994).

[76] See M. Kamionkowski, A. Kosowsky, and M.S. Turner in Ref. [75].

[77] R. Apreda, M. Maggiore, A. Nicolis, and A. Riotto, Class. Quant. Grav. **18**, L155-L162 (2001); Nucl. Phys. **B631**, 342 (2002); A. Nicolis, PhD Thesis 2002, Scuola Normale Superiore, Pisa, Italy.

[78] C. Grojean and G. Servant, Phys. Rev. D **75**, 043507 (2007).

[79] A. Kosowsky, A. Mack, and T. Kahniashvili, Phys. Rev. **D66**, 024030 (2002).

[80] A.D. Dolgov and D. Grasso, Phys. Rev. Lett. **88**, 011301 (2002); A.D. Dolgov, D. Grasso, and A. Nicolis, Phys. Rev. **D66**, 103505 (2002).

[81] M.S. Turner and F. Wilczeck, Phys. Rev. Lett **65**, 3080 (1990).

[82] See, e.g., M.S. Turner, E.J. Weinberg, and L.M. Widrow, Phys. Rev. **D46**, 2384 (1992); E.J. Copeland, A.R. Liddle, D.H. Lyth, E.D. Stewart, and D. Wands, Phys. Rev. **D49**, 6410 (1994).

[83] C. Baccigalupi, L. Amendola, P. Fortini, and F. Occhionero, Phys. Rev. **D56**, 4610 (1997).

[84] S. Khlebnikov and I. Tkachev, Phys. Rev. **D56**, 653 (1997).

[85] R. Easther, J.T. Giblinn, and E.A. Lim, [astro-ph/0612294]; J. Garcia-Bellido and D.G. Figueroa, Phys. Rev. Lett. **98**, 061302 (2007).

[86] T. Damour and A. Vilenkin, Phys. Rev. Lett. **85**, 3761 (2000); Phys. Rev. **D64**, 064008 (2001); ibid. Phys. Rev. D **71**, 063510 (2005).

[87] V. Berezinsky, B. Hnatyk, and A. Vilenkin, [astro-ph/0001213].

[88] E.J. Copeland, R.C. Myers, and J. Polchinski, JHEP **0406**, 013 (2004); M. G. Jackson, N.T. Jones, and J. Polchinski, JHEP **0510**, 013 (2005); J. Polchinski, [hep-th/0410082].

[89] X. Siemens et al., Phys. Rev. D **73**, 105001 (2006); X. Siemens, V. Mandic, and J. Creighton, Phys. Rev. Lett. **98**, 111101 (2007).

[90] N.D. Birrell and P.C.W. Davis, *Quantum fields in curved space*, (Cambridge, Cambridge University Press, 1982).

[91] L. Grishchuk, Sov. Phys. JETP **40**, 409 (1974); A. Starobinsky, JETP Lett. **30**, 682 (1979); Class. Quantum Grav. **10**, 2449 (1993).

[92] L.F. Abbott and D.D. Harari, Nucl. Phys. **B264**, 487 (1986); B. Allen, Phys. Rev. **D37**, 2078 (1988); V. Sahni, Phys. Rev. **D42**, 453 (1990); M.S. Turner, Phys. Rev. **D55**, R435 (1996); L. Krauss and M. White, Phys. Rev. Lett. **69**, 869 (1992); L. Hui and W.H. Kinney, Phys. Rev. **D65**, 103507 (2002).

[93] L. Boyle, P. Steinhardt, and N. Turok, Phys. Rev. Lett. **96**, 111301 (2006); L. Boyle and P. Steinhardt, [astro-ph/0512014]; T. Smith, A. Cooray, and M. Kamionkoswki, Phys. Rev. D **73**, 023504 (2006).

[94] S. Weinberg, Phys. Rev. D **69**, 023503 (2004).

[95] M. Gasperini and G. Veneziano, Phys. Rep. **373**, 1 (2003).

[96] R. Brustein, G. Gasperini, G. Giovannini, and G. Veneziano, Phys. Lett. **B361**, 45 (1995); M. Gasperini, [hep-th/9604084]; A. Buonanno, M. Maggiore, and C. Ungarelli, Phys. Rev. **D55**, 3330 (1997); M. Gasperini, Phys. Rev. **D56**, 4815 (1997); A. Buonanno, K. Meissner, C. Ungarelli, and G. Veneziano, JHEP **001**, 004 (1998).

[97] J. Khoury, B.A. Ovrut, P.J. Steinhardt, and N. Turok, Phys. Rev. **D64**, 123522 (2001); J. Khoury, B.A. Ovrut, N. Seiberg, P.J. Steinhardt, and N. Turok, Phys. Rev. **D65**, 086007 (2002); L. Boyle, P. Steinhardt, and N. Turok, Phys. Rev. D **69**, 127302 (2004).

[98] P.J.E. Peebles and A. Vilenkin, Phys. Rev. **D59**, 063505 (1999); M. Giovannini, Phys. Rev. **D60**, 123511 (1999).

[99] N. Christensen, Phys. Rev. **D46**, 5250 (1992); É.E. Flanagan, Phys. Rev. **D48**, 2389 (1993); B. Allen and J.D. Romano, Phys. Rev. **D59**, 102001 (1999); S. Drasco and E.E. Flanagan, Phys. Rev. D **67**, 082003 (2003).

Course 2

# BARYOGENESIS

## James M. Cline

*McGill University, Department of Physics, Montreal, Qc H3A 2T8, Canada*

*F. Bernardeau, C. Grojean and J. Dalibard, eds.*
*Les Houches, Session LXXXVI, 2006*
*Particle Physcis and Cosmology: The Fabric of Spacetime*
© *2007 Published by Elsevier B.V.*

# Contents

## 1. Observational evidence for the BAU

In everyday life it is obvious that there is more matter than antimatter. In nature we see antimatter mainly in cosmic rays, *e.g.*, $\bar{p}$, whose flux expressed as a ratio to that of protons, is

$$\frac{\bar{p}}{p} \sim 10^{-4} \tag{1.1}$$

consistent with no ambient $\bar{p}$'s, only $\bar{p}$'s produced through high energy collisions with ordinary matter. On earth, we need to work very hard to produce and keep $e^+$ (as in the former LEP experiment) or $\bar{p}$ (as at the Tevatron).

We can characterize the asymmetry between matter and antimatter in terms of the baryon-to-photon ratio

$$\eta \equiv \frac{n_B - n_{\bar{B}}}{n_\gamma} \tag{1.2}$$

where

$$n_B = \text{number density of baryons}$$
$$n_{\bar{B}} = \text{number density of antibaryons}$$
$$n_\gamma = \text{number density of photons}$$
$$= \frac{\zeta(3)}{\pi^2} g_* T^3, \qquad g_* = 2 \text{ spin polarizations} \tag{1.3}$$

and $\zeta(3) = 1.20206\ldots$ This is a useful measure because it remains constant with the expansion of the universe, at least at late times. But at early times, and high temperatures, many heavy particles were in thermal equilibrium, which later annihilated to produce more photons but not baryons. In this case, the entropy density $s$ is a better quantity to compare the baryon density to, and it is convenient to consider

$$\frac{n_B - n_{\bar{B}}}{s} = \frac{1}{7.04} \eta \tag{1.4}$$

where the conversion factor $1/7$ is valid in the present universe, since the epoch when neutrinos went out of equilibrium and positrons annihilated.

Historically, $\eta$ was determined using big bang nucleosynthesis. Abundances of $^3$He, $^4$He, D, $^6$Li and $^7$Li are sensitive to the value of $\eta$; see the Particle Data Group [1] for a review. The theoretical predictions and experimental measurements are summarized in figure 1, similar to a figure in [2]. The boxes represent the regions which are consistent with experimental determinations, showing that different measurements obtain somewhat different values. The smallest error bars are for the deuterium (D/H) abundance, giving

$$\eta = 10^{-10} \times \begin{cases} 6.28 \pm 0.35 \\ 5.92 \pm 0.56 \end{cases} \tag{1.5}$$

for the two experimental determinations which are used. These values are consistent with the $^4$He abundance, though marginally inconsistent with $^7$Li (the "Lithium problem").

In the last few years, the Cosmic Microwave Background has given us an independent way of measuring the baryon asymmetry [3]. The relative sizes of the Doppler peaks of the temperature anisotropy are sensitive to $\eta$, as illustrated in fig. 2. The Wilkinson Microwave Anisotropy Probe first-year data fixed the combination $\Omega_b h^2 = 0.0224 \pm 0.0009$, corresponding to [2]

$$\eta = (6.14 \pm 0.25) \times 10^{-10} \tag{1.6}$$

which is more accurate than the BBN determination. Apart from the Lithium problem, this value is seen to be in good agreement with the BBN value. The CMB-allowed range is shown as the vertical band in figure 1.

We can thus be confident in our knowledge of the baryon asymmetry. The question is, why is it not zero? This would be a quite natural value; a priori, one might expect the big bang—or in our more modern understanding, reheating following inflation—to produce equal numbers of particles and antiparticles.

One might wonder whether the universe could be baryon-symmetric on very large scales, and separated into regions which are either dominated by baryons or antibaryons. However we know that even in the least dense regions of the universe there are hydrogen gas clouds, so one would expect to see an excess of gamma rays in the regions between baryon and antibaryon dominated regions, due to annihilations. These are not seen, indicating that such patches should be as large as the presently observable universe. There seems to be no plausible way of separating baryons and antibaryons from each other on such large scales.

It is interesting to note that in a homogeneous, baryon-symmetric universe, there would still be a few baryons and antibaryons left since annihilations aren't

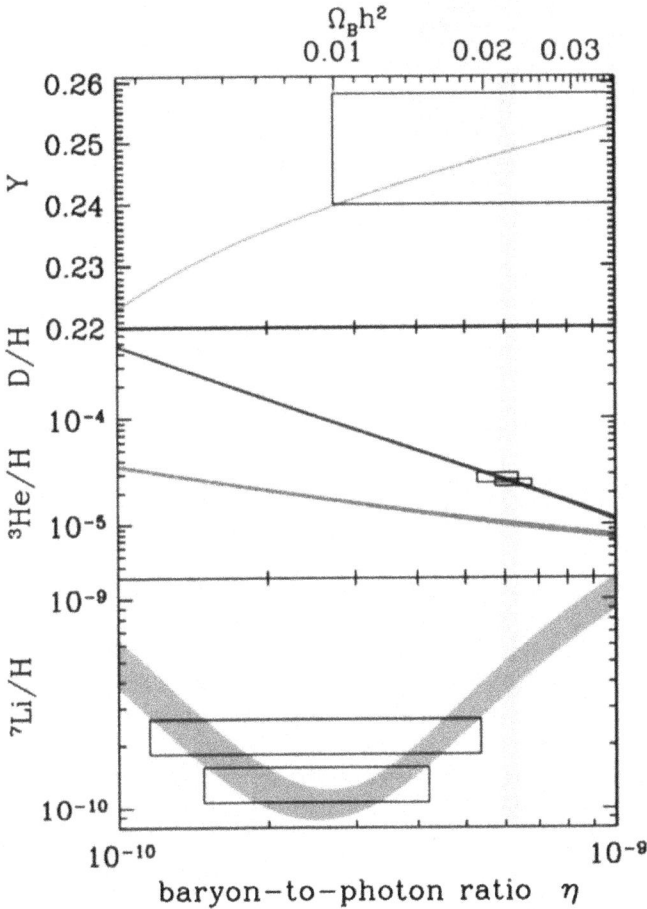

Fig. 1. Primordial abundances versus $\eta$, courtesy of R. Cyburt.

perfectly efficient. But the freeze-out abundance is

$$\frac{n_B}{n_\gamma} = \frac{n_{\bar{B}}}{n_\gamma} \approx 10^{-20} \tag{1.7}$$

(see ref. [4], p. 159), which is far too small for the BBN or CMB.

In the early days of big bang cosmology, the baryon asymmetry was considered to be an initial condition, but in the context of inflation this idea is no longer tenable. Any baryon asymmetry existing before inflation would be diluted to a negligible value during inflation, due to the production of entropy during reheating.

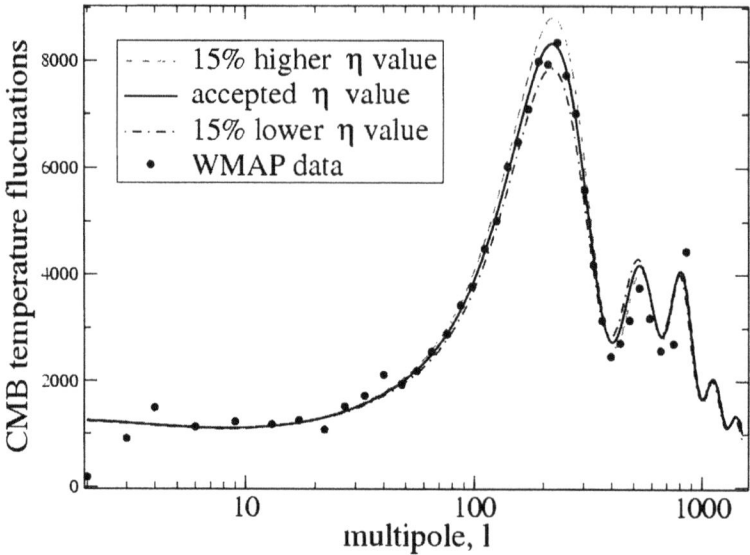

Fig. 2. Dependence of the CMB Doppler peaks on $\eta$.

It is impressive that A. Sakharov realized the need for dynamically creating the baryon asymmetry in 1967 [5], more than a decade before inflation was invented. The idea was not initially taken seriously; in fact it was not referenced again, with respect to the idea of baryogenesis, until 1979 [6]. Now it has 1040 citations (encouragement to those of us who are still waiting for our most interesting papers to be noticed!). It was only with the advent of grand unified theories, which contained the necessary ingredients for baryogenesis, that interest in the subject started to grow dramatically. I have documented the rate of activity in the field from 1967 until now in figure 3.

## 2. Sakharov's conditions for baryogenesis

It is traditional to start any discussion of baryogenesis with the list of three necessary ingredients needed to create a baryon asymmetry where none previously existed:

1. B violation

2. Loss of thermal equilibrium

3. C, CP violation

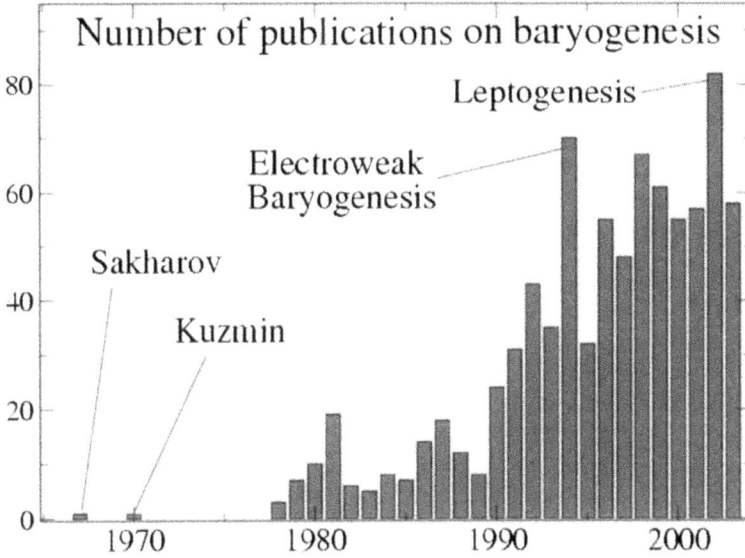

Fig. 3. Number of publications on baryogenesis as a function of time.

Although these principles have come to be attributed to Sakharov, he did not enunciate them as clearly in his three-page paper as one might have been led to think, especially the second point. (Sakharov describes a scenario where a universe which was initially contracting and with equal and opposite baryon asymmetry to that existing today goes through a bounce at the singularity and reverses the magnitude of its baryon asymmetry.) It is easy to see why these conditions are necessary. The need for B (baryon) violation is obvious. Let's consider some examples of B violation.

### 2.1. B violation

In the standard model, B is violated by the triangle anomaly, which spoils conservation of the left-handed baryon + lepton current,

$$\partial_\mu J^\mu_{B_L + L_L} = \frac{3g^2}{32\pi^2} \epsilon_{\alpha\beta\gamma\delta} W_a^{\alpha\beta} W_a^{\gamma\delta} \tag{2.1}$$

where $W_a^{\alpha\beta}$ is the SU(2) field strength. As we will discuss in more detail in section 4, this leads to the nonperturbative sphaleron process pictured in fig. 4. It involves 9 left-handed (SU(2) doublet) quarks, 3 from each generation, and 3

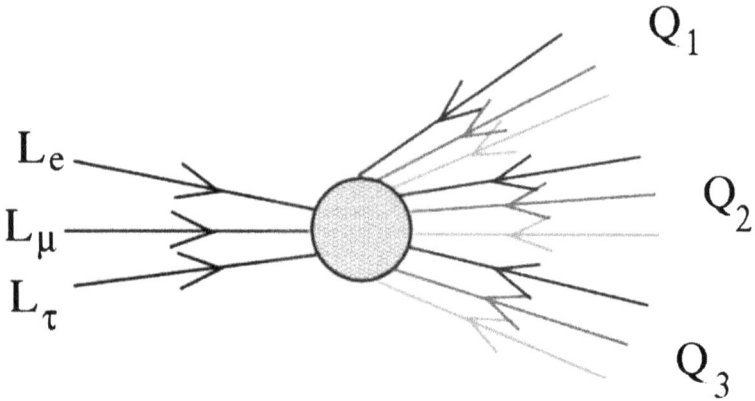

Fig. 4. The sphaleron.

left-handed leptons, one from each generation. It violates B and L by 3 units each,

$$\Delta B = \Delta L = \pm 3 \tag{2.2}$$

In grand unified theories, like SU(5), there are heavy gauge bosons $X^\mu$ and heavy Higgs bosons $Y$ with couplings to quarks and leptons of the form

$$Xqq, \quad X\bar{q}\bar{l} \tag{2.3}$$

and similarly for $Y$. The simultaneous existence of these two interactions imply that there is no consistent assignment of baryon number to $X^\mu$. Hence B is violated.

In supersymmetric models, one might choose not to impose R-parity. Then one can introduce dimension-4 B-violating operators, like

$$\tilde{t}_R^a b_R^b s_R^c \epsilon_{abc} \tag{2.4}$$

which couple the right-handed top squark to right-handed quarks. One usually imposes R-parity to forbid proton decay, but the interaction (2.4) could be tolerated if for some reason lepton number was still conserved, since then the decay channels $p \to \pi^+ \nu$ and $p \to \pi^0 e^+$ would be blocked.

## 2.2. *Loss of thermal equilibrium*

The second condition of Sakharov (as I am numbering them), loss of thermal equilibrium, is also easy to understand. Consider a hypothetical process

$$X \to Y + B \tag{2.5}$$

where $X$ represents some initial state with vanishing baryon number, $Y$ is a final state, also with $B = 0$, and $B$ represents the excess baryons produced. If this process is in thermal equilibrium, then by definition the rate for the inverse process, $Y + B \rightarrow X$, is equal to the rate for (2.5):

$$\Gamma(Y + B \rightarrow X) = \Gamma(X \rightarrow Y + B) \tag{2.6}$$

No net baryon asymmetry can be produced since the inverse process destroys B as fast as (2.5) creates it.

The classic example is out-of-equilibrium decays, where $X$ is a heavy particle, such that $M_X > T$ at the time of decay, $\tau = 1/\Gamma$. In this case, the energy of the final state $Y + B$ is of order $T$, and there is no phase space for the inverse decay: $Y + B$ does not have enough energy to create a heavy $X$ boson. The rate for $Y + B \rightarrow X$ is Boltzmann-suppressed, $\Gamma(Y + B \rightarrow X) \sim e^{-M_X/T}$.

### 2.3. C, CP violation

The most subtle of Sakharov's requirements is C and CP violation. Consider again some process $X \rightarrow Y + B$, and suppose that C (charge conjugation) is a symmetry. Then the rate for the C-conjugate process, $\bar{X} \rightarrow \bar{Y} + \bar{B}$, is the same:

$$\Gamma(\bar{X} \rightarrow \bar{Y} + \bar{B}) = \Gamma(X \rightarrow Y + B) \tag{2.7}$$

The net rate of baryon production goes like the difference of these rates,

$$\frac{dB}{dt} \propto \Gamma(\bar{X} \rightarrow \bar{Y} + \bar{B}) - \Gamma(X \rightarrow Y + B) \tag{2.8}$$

and so vanishes in the case where C is a symmetry.

However, even if C is violated, this is not enough. We also need CP violation. To see this, consider an example where X decays into two left-handed or two right-handed quarks,

$$X \rightarrow q_L q_L, \qquad X \rightarrow q_R q_R \tag{2.9}$$

Under CP,

$$CP: \quad q_L \rightarrow \bar{q}_R \tag{2.10}$$

where $\bar{q}_R$ is the antiparticle of $q_R$, which is *left*-handed (in my notation, L and R are really keeping track of whether the fermion is an SU(2) doublet or singlet), whereas under C,

$$C: \quad q_L \rightarrow \bar{q}_L \tag{2.11}$$

Therefore, even though $C$ violation implies that

$$\Gamma(X \to q_L q_L) \neq \Gamma(\bar{X} \to \bar{q}_L \bar{q}_L) \tag{2.12}$$

CP conservation would imply

$$\Gamma(X \to q_L q_L) = \Gamma(\bar{X} \to \bar{q}_R \bar{q}_R) \tag{2.13}$$

and also

$$\Gamma(X \to q_R q_R) = \Gamma(\bar{X} \to \bar{q}_L \bar{q}_L) \tag{2.14}$$

Then we would have

$$\Gamma(X \to q_L q_L) + \Gamma(X \to q_R q_R) = \Gamma(\bar{X} \to \bar{q}_R \bar{q}_R) + \Gamma(\bar{X} \to \bar{q}_L \bar{q}_L) \tag{2.15}$$

As long as the initial state has equal numbers of $X$ and $\bar{X}$, we end up with no net asymmetry in quarks. The best we can get is an asymmetry between left- and right-handed quarks, but this is not a baryon asymmetry.

Let us recall exactly how C, P and CP operate on scalar fields, spinors, and vector fields. For complex scalars,

$$C: \quad \phi \to \phi^* \tag{2.16}$$
$$P: \quad \phi(t, \vec{x}) \to \pm\phi(t, -\vec{x})$$
$$CP: \quad \phi(t, \vec{x}) \to \pm\phi^*(t, -\vec{x})$$

For fermions,

$$C: \quad \psi_L \to i\sigma_2 \psi_R^*, \quad \psi_R \to -i\sigma_2 \psi_L^*, \quad \psi \to i\gamma^2 \psi^* \tag{2.17}$$
$$P: \quad \psi_L \to \psi_R(t, -\vec{x}), \quad \psi_R \to \psi_L(t, -\vec{x}), \quad \psi \to \gamma^0 \psi(t, -\vec{x})$$
$$CP: \quad \psi_L \to i\sigma_2 \psi_R^*(t, -\vec{x}), \quad \psi_R \to -i\sigma_2 \psi_L^*(t, -\vec{x}),$$
$$\psi \to i\gamma^2 \psi^*(t, -\vec{x})$$

For vectors,

$$C: \quad A^\mu \to -A^\mu \tag{2.18}$$
$$P: \quad A^\mu(t, \vec{x}) \to (A^0, -\vec{A})(t, -\vec{x})$$
$$CP: \quad A^\mu(t, \vec{x}) \to (-A^0, \vec{A})(t, -\vec{x})$$

To test your alertness, it is amusing to consider the following puzzle. Going back to the example of $X \to qq$, let us suppose that C and CP are both violated and that

$$\Gamma(X \to qq) \neq \Gamma(\bar{X} \to \bar{q}\bar{q}) \tag{2.19}$$

where now we ignore the distinction between $q_R$ and $q_L$. An astute reader might object that if all $X$'s decay into $qq$ and all $\bar{X}$'s decay into $\bar{q}\bar{q}$, with $n_X = n_{\bar{X}}$ initially, then eventually we will still have equal numbers of $q$ and $\bar{q}$, even though there was temporarily an excess. To avoid this outcome, there must exist at least one competing channel, $X \to Y$, $\bar{X} \to \bar{Y}$, such that

$$\Gamma(X \to Y) \neq \Gamma(\bar{X} \to \bar{Y}) \tag{2.20}$$

and with the property that $Y$ has a different baryon number than $qq$. One can then see that a baryon asymmetry will develop by the time all $X$'s have decayed. Why is it then, that we do not have an additional fourth requirement, that a competing decay channel with the right properties exists? The reason is that this is guaranteed by the CPT theorem, combined with the requirement of B violation. CPT assures us that the total rates of decay for $X$ and $\bar{X}$ are equal, which in the present example means

$$\Gamma(X \to qq) + \Gamma(X \to Y) = \Gamma(\bar{X} \to \bar{q}\bar{q}) + \Gamma(\bar{X} \to \bar{Y}) \tag{2.21}$$

The stipulation that B is violated tells us that $Y$ has different baryon number than $qq$. Otherwise we could consistently assign the baryon number 2/3 to $X$ and there would be no B violation.

## 2.4. *Examples of CP violation*

We now consider the conditions under which a theory has CP violation. Generally CP violation exists if there are complex phases in couplings in the Lagrangian which can't be removed by field redefinitions. Let's start with scalar theories. For example

$$\mathcal{L} = |\partial\phi|^2 - m^2|\phi|^2 - \lambda|\phi|^4 - (\mu^2\phi^2 + g\phi^4 + \text{h.c.}) \tag{2.22}$$

where the top line is CP conserving, and the bottom line is (potentially) CP-violating. This can be seen by applying the CP transformation $\phi \to \phi^*$ and observing that the bottom line is not invariant:

$$\mathcal{L}_{CPV} \to -(\mu^2\phi^{*2} + g\phi^{*4} + \text{h.c.}) \tag{2.23}$$

Of course, this lack of invariance only occurs if $\mu^2$ and $g$ are not both real. Parametrize

$$\mu^2 = |\mu|^2 e^{i\phi_\mu}, \qquad g = |g|e^{i\phi_g} \tag{2.24}$$

We can perform a field refefinition to get rid of one of these phases, *e.g.* $\phi_\mu$, by letting

$$\phi \to e^{-i\phi_\mu/2}\phi \tag{2.25}$$

so that

$$\mathcal{L}_{CPV} = -\left(|\mu^2|\phi^2 + |g|e^{i(\phi_g - 2\phi_\mu)}\phi^4 + \text{h.c.}\right) \tag{2.26}$$

Now under CP, only the $\phi^4$ term changes. We can see that it is only the combination

$$\phi_{\text{inv}} = \phi_g - 2\phi_\mu = \text{"invariant phase"} \tag{2.27}$$

which violates CP, and which is independent of field redefinitions. If $\phi_{\text{inv}} = 0$, the theory is CP-conserving. We can also write

$$\phi_{\text{inv}} = \arg\left(\frac{g}{\mu^4}\right) \tag{2.28}$$

No physical result can depend on field redefinitions, since these are arbitrary. Therefore any physical effect of CP violation must manifest itself through the combination $\phi_{\text{inv}}$. This can sometimes provide a valuable check on calculations.

An equivalent way of thinking about the freedom to do field redefinitions is the following. One can also define the CP transformation with a phase $\phi \to e^{i\alpha}\phi^*$. As long as there exists any value of $\alpha$ for which this transformation is a symmetry, we can say that this is the real CP transformation, and CP is conserved.

Next let's consider an example with fermions. We can put a complex phase into the mass of a fermion (here we consider Dirac),

$$\begin{aligned}\mathcal{L} &= \bar{\psi}\left(i\slashed{\partial} - m(\cos\theta + i\sin\theta\gamma_5)\right)\psi \\ &= \bar{\psi}\left(i\slashed{\partial} - me^{i\theta\gamma_5}\right)\psi \end{aligned} \tag{2.29}$$

[In Weyl components, the mass term has the form $m(\bar{\psi}_L e^{i\theta}\psi_R + \bar{\psi}_R e^{-i\theta}\psi_L)$.] Under CP,

$$\begin{aligned}\psi &\to i\gamma^2\gamma^0\psi^*, \quad \psi^* \to -i(-\gamma^2)\gamma^0\psi, \\ \psi^\dagger &\to \psi^T\gamma^0(-\gamma^2)(-i), \\ \psi^T &\to \psi^\dagger\gamma^0\gamma^2 i = \bar{\psi}\gamma^2 i \end{aligned} \tag{2.30}$$

Also, by transposing the fields, $\bar{\psi}e^{i\theta\gamma_5}\psi = -\psi^T e^{i\theta\gamma_5}\gamma^0\psi^*$. Putting these results together, we find that the complex mass term transforms under CP as

$$\begin{aligned}\bar{\psi}e^{i\theta\gamma_5}\psi &\to -\bar{\psi}\gamma^2 i e^{i\theta\gamma_5}\gamma^0 i\gamma^2\gamma^0\psi \\ &= -\bar{\psi}\gamma^2 e^{i\theta\gamma_5}\gamma^2\psi \\ &= \bar{\psi}e^{-i\theta\gamma_5}\psi \end{aligned} \tag{2.31}$$

So ostensibly $\theta$ is CP-violating. However, we can remove this phase using the chiral field redefinition

$$\psi \to e^{-i(\theta/2)\gamma_5}\psi \tag{2.32}$$

To get an unremovable phase, we need to consider a more complicated theory. For example, with two Dirac fermions we can have

$$\mathcal{L} = \sum_{i=1}^{2} \bar{\psi}_i(i\slashed{\partial} - m_i e^{i\theta_i\gamma_5})\psi_i + \mu(\bar{\psi}_1 e^{i\alpha\gamma_5}\psi_2 + \text{h.c.}) \tag{2.33}$$

After field redefinitions we find that

$$\mathcal{L} \to \sum_{i=1}^{2} \bar{\psi}_i(i\slashed{\partial} - m_i)\psi_i + \mu(\bar{\psi}_1 e^{i(\alpha-\theta_1/2-\theta_2/2)\gamma_5}\psi_2 + \text{h.c.}) \tag{2.34}$$

so the invariant phase is $\alpha - \frac{1}{2}(\theta_1 + \theta_2)$.

It is also useful to think about this example in terms of 2-component Weyl spinors, where the mass term takes the form

$$\sum_{i=1}^{2} \psi_{L,i}^\dagger e^{i\theta_i}\psi_{R,i} + \mu\psi_{L,1}^\dagger e^{i\alpha}\psi_{R,2} + \text{h.c.} \tag{2.35}$$

One might have thought that there are 4 field redefinitions, corresponding to independent rephasings of $\psi_{L,i}$ and $\psi_{R,i}$, so there would be enough freedom to remove all the phases. However, it is only the axial vector transformations, where $L$ and $R$ components are rotated by equal and opposite phases, that can be used to remove $\theta_i$. Half of the available field redefinitions are useless as far as removing phases from the mass terms. We might more accurately state the general rule as "the number of invariant phases equals the total number of phases in the Lagrangian parameters, minus the number of *relevant* field redefinitions which can be done."

A third example of CP violation is the famous $\theta$ term in QCD,

$$\theta_{QCD} F_{\mu\nu}^a \tilde{F}^{a,\mu\nu} = \frac{1}{2}\theta_{QCD}\epsilon^{\mu\nu\alpha\beta} F_{\mu\nu}^a F_{\alpha\beta}^a = -2\theta_{QCD}\vec{E}^a \cdot \vec{B}^a \tag{2.36}$$

This combination is P-odd and C-even, thus odd under CP. Its smallness is the strong CP problem of QCD: if $\theta_{QCD} \gtrsim 10^{-10}$, the electric dipole moment of the neutron would exceed experimental limits. The EDM operator is

$$\begin{aligned}
\text{EDM} = -\frac{i}{2}d_n\bar{n}\sigma_{\mu\nu}\gamma_5 F^{\mu\nu}n &= d_n\big(n_L^\dagger\vec{\sigma}\cdot\vec{E}n_R + n_R^\dagger\vec{\sigma}\cdot\vec{E}n_L \\
&\quad + i\big(n_L^\dagger\vec{\sigma}\cdot\vec{B}n_R - n_R^\dagger\vec{\sigma}\cdot\vec{B}n_L\big)\big)
\end{aligned} \tag{2.37}$$

which is mostly $\vec{\sigma} \cdot \vec{E}$, since there is a cancellation between left and right components in the bottom line. Contrast this with the magnetic dipole moment operator,

$$\text{MDM} = \frac{1}{2}\mu_n \bar{n}\sigma_{\mu\nu}F^{\mu\nu}n = \mu_n\left(n_L^\dagger\vec{\sigma}\cdot\vec{B}n_R + n_R^\dagger\vec{\sigma}\cdot\vec{B}n_L \right.$$
$$\left. + i\left(n_L^\dagger\vec{\sigma}\cdot\vec{E}n_R - n_R^\dagger\vec{\sigma}\cdot\vec{E}n_L\right)\right) \tag{2.38}$$

which is mostly $\vec{\sigma} \cdot \vec{B}$. To verify the CP properties of these operators, notice that under CP, $\vec{\sigma} \to \vec{\sigma}$, $L \leftrightarrow R$, $\vec{E} \to -\vec{E}$, $\vec{B} \to \vec{B}$. EDM's will play an important role in constraining models of baryogenesis, since the phases needed for one can also appear in the other. The current limit on the neutron EDM is [7]

$$|d_n| < 3 \times 10^{-26} e \,\text{cm} \tag{2.39}$$

In 1979 it was estimated that [8]

$$d_n = 5 \times 10^{-16}\theta_{QCD} e \,\text{cm} \tag{2.40}$$

leading to the stringent bound on $\theta_{QCD}$.

Other particle EDM's provide competitive constraints on CP violation. The electron EDM is constrained [9],

$$|d_e| < 1.6 \times 10^{-27} e \,\text{cm} \tag{2.41}$$

as well those of Thallium and Mercury atoms [10],

$$|d_{Tl}| < 9 \times 10^{-25} e \,\text{cm}$$
$$|d_{Hg}| < 2 \times 10^{-28} e \,\text{cm} \tag{2.42}$$

### 2.5. More history

I conclude this section on "Sakharov's Laws" with a bit more about the early history of baryogenesis. The earliest papers on baryogenesis from the era when GUT's were invented were not at first aware of Sakharov's contribution. Yoshimura was the first to attempt to get baryogenesis from the SU(5) theory, in 1978 [11]. However, this initial attempt was flawed because it used only scatterings, not decays, and therefore did not incorporate the loss of thermal equilibrium. This defect was pointed out by Barr [6], following work of Toussaint *et al.* [12], and corrected by Dimopoulos and Susskind [13], who were the first to use the out-of-equilibrium decay mechanism. None of these authors knew about the Sakharov paper. Weinberg [14] and Yoshimura [15] quickly followed Dimopoulos and Susskind with more quantitative calculations of the baryon asymmetry in GUT models.

The first paper to attribute to Sakharov the need for going out of equilibrium, by Ignatiev *et al.,* [16] was not in the context of GUTS. They constructed a lower-energy model of baryogenesis, having all the necessary ingredients, by a modification of the electroweak theory. Their idea was not as popular as GUT baryogenesis, but it did make people aware of Sakharov's early paper.

## 3. Example: GUT baryogenesis

I will now illustrate the use of Sakharov's required ingredients in the SU(5) GUT theory [17]. This is no longer an attractive theory for baryogenesis because it requires a higher reheating temperature from inflation than is desired, from the point of view of avoiding unwanted relics, like monopoles and heavy gravitinos. Nevertheless, it beautifully illustrates the principles, and it is also useful for understanding leptogenesis, since the two approaches are mathematically quite similar.

In SU(5) there exist gauge bosons $X^\mu$ whose SU(3), SU(2) and U(1)$_y$ quantum numbers are $(3, 2, -\frac{5}{6})$, as well as Higgs bosons $Y$ with quantum numbers $(3, 1, \frac{1}{3})$, and whose couplings are similar to those of the vectors. The couplings to quarks and leptons are shown in figure 5. We see that the requirement of B violation is satisfied in this theory, and the CP-violating decays of any of these scalars can lead to a baryon asymmetry. To get CP violation, we need the matrix Yukawa couplings $h_{ij}, y_{ij}$ to be complex, where $i, j$ are generation indices. It will become apparent what the invariant phases are.

Let's consider the requirement for $X^\mu$ or $Y$ to decay out of equilibrium. The decay rate is

$$\Gamma_D \cong \alpha m N / \gamma \tag{3.1}$$

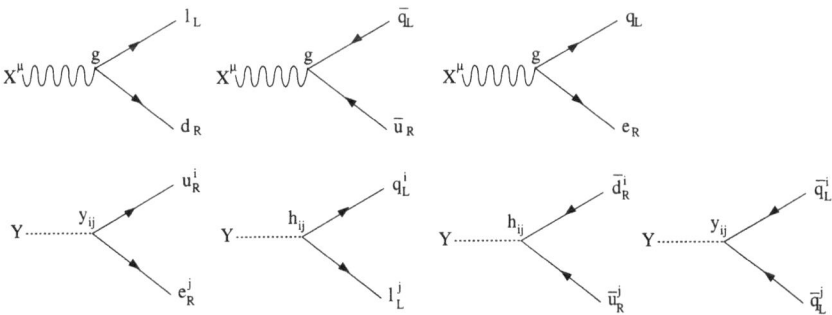

Fig. 5. GUT couplings.

where $\alpha = g^2/(4\pi)$ or $\sim (y+h)^2/(4\pi)$, for $X^\mu$ or $Y$, resepctively, $m$ is the mass, $N$ is the number of decay channels, and the Lorentz gamma factor can be roughly estimated at high temperature as

$$\gamma = \frac{\langle E \rangle}{m} \cong (1 = 9T^2/m^2)^{1/2} \tag{3.2}$$

considering that $\langle E \rangle \cong 3T$ for highly relativistic particles. The $\gamma$ factor will actually not be necessary for us, because we are interested in cases where the particle decays at low temperatures compared to its mass, so that the decays will occur out of equilibrium. Thus we can set $\gamma = 1$. The age of the universe is $\tau \equiv 1/H = 1/\Gamma_D$ at the time of the decay, so we set

$$\Gamma_D = H \cong \sqrt{g_*}\frac{T^2}{M_p} = \alpha m N$$
$$\longrightarrow (\alpha N m M_p/\sqrt{g_*})^{1/2} \ll m \tag{3.3}$$

where $g_*$ is the number of relativistic degrees of freedom at the given temperature. Therefore we need $\alpha \ll (m/M_p)(\sqrt{g_*}/N)$. Let us first consider the decays of the $X$ gauge boson. Since it couples to everything, the number of decay channels $N$ is of the same order as $g_*$, and we get

$$\frac{g^2}{4\pi} \ll \frac{m}{M_p\sqrt{g_*}} \tag{3.4}$$

But unification occurs at the scale $m \cong 10^{16}$ GeV, for values of $g^2$ such that $\alpha \cong 1/25$, so this condition cannot be fulfilled. On the other hand, for the Higgs bosons, $Y$, we get the analogous bound

$$\frac{h^2 \text{ or } y^2}{4\pi} \ll \frac{m}{M_p\sqrt{g_*}} \tag{3.5}$$

which can be satisfied by taking small Yukawa couplings. Therefore the Higgs bosons are the promising candidates for decaying out of equilibrium.

For clarity, I will now focus on a particular decay channel, $Y \to u_R e_R$, whose interaction Lagrangian is

$$y_{ij} Y^a \bar{u}_{R\,a,i} e^c_{R,j} + \text{h.c.} \tag{3.6}$$

At tree level, we cannot generate any difference between the squared matrix elements, and we find that

$$\left| \mathcal{M}_{Y \to e_R u_R} \right|^2 = \left| \mathcal{M}_{\bar{Y} \to \bar{e}_R \bar{u}_R} \right|^2 \tag{3.7}$$

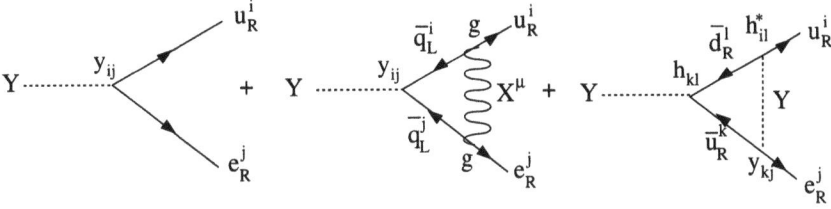

Fig. 6. Tree plus 1-loop contributions to $Y \rightarrow e_R u_R$.

The complex phases are irrelevant at tree level. To get a CP-violating effect, we need interference between the tree amplitude and loop corrections, as shown in figure 6.

The one-loop diagrams develop imaginary parts which interfere with the phase of the tree diagram in a CP-violating manner. We can write the amplitude as

$$-i\mathcal{M} = -i\left(y_{ij} + y_{ij} F_X\left(M_X^2/p^2\right) + (h^* H^T y)_{ij} F_Y\left(M_X^2/p^2\right)\right) \tag{3.8}$$

where $p$ is the 4-momentum of the decaying $Y$ boson,

$$
\begin{aligned}
F_Y &= i^5 \int \frac{d^4 q}{(2\pi)^4} \frac{1}{q^2 - M_Y^2} \frac{1}{\not{q} + \frac{1}{2}\not{p} + i\epsilon} \frac{1}{\not{q} - \frac{1}{2}\not{p} + i\epsilon} \\
&\equiv R_Y + i I_Y
\end{aligned} \tag{3.9}
$$

and $F_X$ is similar, but with $M_Y \rightarrow M_X$ and $\gamma_\mu$ factors in between the propagators. (It will turn out that the $X$ exchange diagram does not contribute to the CP-violating interference.) In the second line we have indicated that the loop integral has real and imaginary parts. The latter arise as a consequence of unitarity,

$$-i\left[\mathcal{M}(p_i \rightarrow p_f) - \mathcal{M}^*(p_f \rightarrow p_i)\right] = \sum_a d\Pi_a \mathcal{M}(p_i \rightarrow p_a)\mathcal{M}^*(p_f \rightarrow p_a)$$

where the sum is over all possible on-shell intermediate states and $d\Pi_a$ is the Lorentz-invariant phase space measure. If $\mathcal{M}^*(p_f \rightarrow p_i) = \mathcal{M}^*(p_i \rightarrow p_f)$, then the left-hand-side is the imaginary part of $\mathcal{M}$. The right-hand-side implies that an imaginary part develops if intermediate states can go on shell:

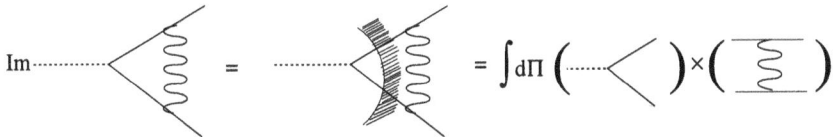

This will be the case as long as the decaying particle is heavier than . the intermediate states.

To see how a rate asymmetry comes about, schematically we can write

$$\mathcal{M}(Y \to eu) = y + yF_X + \tilde{y}F_Y$$
$$\mathcal{M}(\bar{Y} \to \bar{e}\bar{u}) = y^* + y^*F_X + \tilde{y}^*F_Y \qquad (3.10)$$

where $\tilde{y} \equiv h^* h^T y$. Then the difference between the probability of decay of $Y$ and $\bar{Y}$ goes like

$$
\begin{aligned}
|\mathcal{M}_{Y \to eu}|^2 - |\mathcal{M}_{\bar{Y} \to \bar{e}\bar{u}}|^2 &= \left[ y^*(1 + F_X^*) + \tilde{y}^* F_Y^* \right] \left[ y(1 + F_X) + \tilde{y} F_Y \right] \\
&\quad - \left[ y(1 + F_X^*) + \tilde{y} F_Y^* \right] \left[ y^*(1 + F_X) + \tilde{y}^* F_Y \right] \\
&= y^* \tilde{y} F_Y + \tilde{y}^* y F_Y^* - y \tilde{y}^* F_Y - \tilde{y} y^* F_Y^* \\
&= [\tilde{y} y^* - \tilde{y}^* y] 2 i \, I_Y \\
&= -4 \mathrm{Im}[\tilde{y} y^*] I_Y \\
&= -4 \mathrm{Im} \, \mathrm{tr}(h^* h^T y y^\dagger) I_y \qquad (3.11)
\end{aligned}
$$

(You can check that the $F_X$ term does not contribute to this difference, since $g$ is real.) Now, unfortunately, it is easy to see using the properties of the trace that $\mathrm{tr}(h^* H^T y y^\dagger)$ is purely real, so this vanishes. However, there is a simple generalization which allows us to get a nonvanishing result. Suppose there are two or more $Y$ bosons with different Yukawa matrices; then we get

$$|\mathcal{M}_{Y_A \to eu}|^2 - |\mathcal{M}_{\bar{Y}_A \to \bar{e}\bar{u}}|^2 = -4 \sum_B \mathrm{Im} \, \mathrm{tr}\left(h_B^* h_A^T y_B y_A^\dagger\right) I_y \left( \frac{M_B^2}{M_A^2} \right) \qquad (3.12)$$

which no longer vanishes. The function $I_Y$ can be found in [17], who estimate it to be of order $10^{-2} - 10^{-3}$ for typical values of the parameters.

Let us now proceed to estimate the baryon asymmetry. Define the rate asymmetry

$$r_A = \frac{|\mathcal{M}_{Y_A \to eu}|^2 - |\mathcal{M}_{\bar{Y}_A \to \bar{e}\bar{u}}|^2}{\sum_f |\mathcal{M}_{Y_A \to f}|^2} \qquad (3.13)$$

which is the fraction of decays that produce a baryon excess. The sum in the denominator is over all possible final states. The resulting baryon density is

$$n_B - n_{\bar{B}} = \sum_A n_{Y_A} r_A B_q \qquad (3.14)$$

where $B_q$ is the amount of baryon number produced in each decay ($B_q = 1/3$ for the channel we have been considering). Of course, we should also add the corresponding contributions from all the competing $B$-violating decay channels, as was emphasized in the previous section. For simplicity we have shown how it works for one particular decay channel. Up to factors of order unity, this channel by itself gives a good estimate of the baryon asymmetry.

To compute the baryon asymmetry, it is convenient to take the ratio of baryons to entropy first, since as we have noted in section 1, entropy density changes in the same way as baryon density as the universe expands.

$$\frac{n_B - n_{\bar{B}}}{s} = \frac{\frac{1}{3}n_Y \sum_A r_A}{\frac{4\pi^2}{45}g_* T^2} = \frac{90\zeta(3)}{4\pi^4 g_*} \sum_A r_A \tag{3.15}$$

where we used $n_Y = (2\zeta(3)/\pi^2)T^3 \times 3$, the final factor of 3 counting the color states of $Y$. The number of degrees of freedom is

$$g_* = \left(\sum_{\text{bosons}} + \frac{7}{8} \sum_{\text{fermions}}\right) \text{(spin states)}$$

$$= \begin{cases} 106.75 & \text{standard model} \\ 160.75 & \text{SU(5)} \end{cases} \times (\sim 2 \text{ if SUSY}) \tag{3.16}$$

(these numbers can be found in the PDG Big Bang Cosmology review [1]). We can now use (1.4) to convert this to $\eta$, by multiplying by a factor of 7. Suppose we have SUSY SU(5); then

$$\eta \cong \frac{1.2}{320} \sum r_A \cong 10^{-2} \sum r_A \tag{3.17}$$

Thus we only need $\sum r_A \cong 6 \times 10^{-8}$, a quite small value, easy to arrange since

$$r_A \sim \frac{y^2 h^2}{y^2 + h^2} \epsilon I_y \tag{3.18}$$

where $\epsilon$ is the CP violating phase, and we noted that $I_y \sim 10^{-2} - 10^{-3}$. According to our estimate of the size of $y$ or $h$ needed to decay out of equilibrium, $y^2 \sim 4\pi \times (10^{-4} - 10^{-6})$. We can easily make $\eta$ large enough, or even too large.

## 3.1. Washout processes

The foregoing was the first quantitative estimate of the baryon asymmetry, but it is missing an important ingredient which reduces $\eta$: B-violating rescattering and inverse decay processes. We have ignored diagrams like

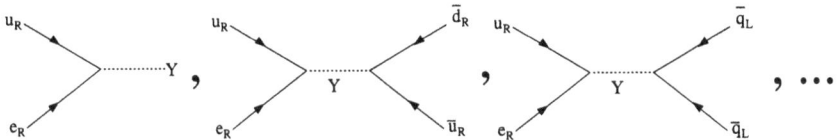

The first of these, inverse decay, was assumed to be negligible by our assumption of out-of-equilibrium decay, but more generally we could include this process and quantify the degree to which it is out of equilibrium. Any of the washout processes shown would cause the baryon asymmetry to relax to zero if they came into equilibrium.

To quantify these effects we need the Boltzmann equation for the evolution of baryon number. This was first carried out by Kolb and Wolfram [18], who were at that time postdoc and graduate student, respectively, at Caltech. The form of the Boltzmann equations is

$$\frac{dn_{u_R}}{dt} + 3Hn_{u_R} \cong \int d\Pi_i \left( f_Y |\mathcal{M}(Y \to ue)|^2 - f_u f_e |\mathcal{M}(ue \to Y)|^2 \right)$$

$$+ \int d\Pi_i \left[ f_{q_L}^2 |\mathcal{M}(qq \to ue)|^2 - f_u f_e |\mathcal{M}(ue \to qq)|^2 \right] + \ldots \quad (3.19)$$

where $f_x$ is the distribution function for particle $x$, $\int d\Pi_i$ is the phase space integral over all final and initial particles,

$$\prod_i d\Pi_i \equiv \left[ \prod_i \left( \frac{d^4 p_i}{(2\pi)^3} \delta(p_i^2 - m_i^2) \right) \right] \times (2\pi)^4 \delta^{(4)} \left( \sum_{\text{final}} p_a - \sum_{\text{initial}} p_a \right)$$

$$(3.20)$$

and I have ignored final-state Pauli-blocking or Bose-enhancement terms, $(1 \pm f)$. We should include similar equations for any other particles which are relevant for the baryon asymmetry ($X^\mu$, $Y$, quarks, leptons), and solve the coupled equations numerically to determine the final density of baryons. Depending on the strength of the scattering, the washout effect can be important. We can estimate its significance without solving the Boltzmann equations numerically.

Define the abundance $Y_i \equiv n_i/n_\gamma$ and consider a simplified model where $X \to bb$, $\bar{b}\bar{b}$ (baryons or antibaryons), with

$$\Gamma_{X \to bb} \equiv \frac{1}{2}(1 + \epsilon)\Gamma_X$$

$$\Gamma_{X \to \bar{b}\bar{b}} \equiv \frac{1}{2}(1 - \epsilon)\Gamma_X$$

$$\Gamma_{\bar{X} \to bb} \equiv \frac{1}{2}(1 - \bar{\epsilon})\Gamma_X$$

$$\Gamma_{\bar{X} \to \bar{b}\bar{b}} \equiv \frac{1}{2}(1 + \bar{\epsilon})\Gamma_X \qquad (3.21)$$

Kolb and Wolfram found that the time rate of change of the baryon abundance (which we could also have called $\dot{\eta}$) is

$$\dot{Y}_B = \Gamma_X(\epsilon - \bar{\epsilon})[Y_X - Y_{eq}] - 2Y_B \left[\Gamma_X Y_{X,eq} + \Gamma_{bb \leftrightarrow \bar{b}\bar{b}}\right] \tag{3.22}$$

On the right-hand-side, the first term in brackets represents baryon production due to CP-violating decays, while the second term is due to the washout induced by inverse decays and scatterings. The scattering term is defined by the thermal average

$$\Gamma_{bb \leftrightarrow \bar{b}\bar{b}} = n_\gamma \left(\langle v\sigma_{bb \to \bar{b}\bar{b}}\rangle + \langle v\sigma_{\bar{b}\bar{b} \to bb}\rangle\right) \tag{3.23}$$

In the treatment of Nanopoulos and Weinberg, the effect of $\Gamma_{bb \leftrightarrow \bar{b}\bar{b}}$ was neglected, but we can estimate it using dimensional analysis as

$$\Gamma_{bb \leftrightarrow \bar{b}\bar{b}} \sim \frac{\alpha^2 T^5}{(M_X^2 + 9T^2)^2} \tag{3.24}$$

(considering $s$-channel exchange of the $X$ boson), again using $\langle E \rangle \sim 3T$. Then if we imagine the initial production of the baryon asymmetry happens quickly (by the time $t \equiv 0$), followed by gradual relaxation, then

$$Y_B(t) \sim Y_B(0) \, e^{-\int \Gamma_{bb \leftrightarrow \bar{b}\bar{b}} dt} \tag{3.25}$$

Using $H \sim 1/t \sim g_* T^2/M_P$, $dt \sim (M_P/g_*)(dT/T^3)$,

$$\int \Gamma dt \sim \frac{M_P}{g_*} \int \Gamma \frac{dT}{T^3} \cong \frac{\pi \alpha^2 M_P}{100 M_X} \tag{3.26}$$

so the maximum BAU gets diluted by a factor $\sim e^{-(\pi\alpha^2 M_P/(100 M_X))}$. We thus need $\alpha^2 \lesssim 30 M_X/M_P$, hence $\alpha \lesssim 0.05$, using $M_X/M_P \sim 10^{-4}$. This agrees with the quantitative findings of Kolb and Turner, schematically shown in figure 7. Notice that this value of $\alpha$ is consistent with the unification value, so GUT baryogenesis works.

This concludes our discussion of GUT baryogenesis. As noted earlier, it is disfavored because it needs a very high reheat temperature after inflation, which could introduce monopoles or gravitinos as relics that would overclose the universe. Another weakness, shared by many theories of baryogenesis, is a lack of testability. There is no way to distinguish whether baryogenesis happened via GUT's or some other mechanism. We now turn to the possibility of baryogenesis at lower temperatures, with greater potential for testability in the laboratory.

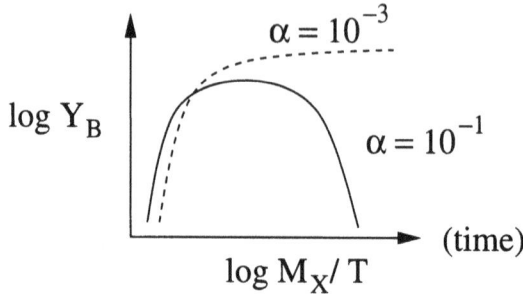

Fig. 7. Evolution of baryon asymmetry with with B-violating rescattering processes included.

## 4.  B and CP violation in the standard model

During the development of GUT baryogenesis, it was not known that the standard model had a usable source of B violation. In 1976, 't Hooft showed that the triangle anomaly violates baryon number through a nonperturbative effect [19]. In (2.1) we saw that the baryon current is not conserved in the presence of external $W$ boson field strengths. However this violation of B is never manifested in any perturbative process. It is associated with the vacuum structure of SU(N) gauge theories with spontaneously broken symmetry. To explain this, we need to introduce the concept of Chern-Simons number,

$$N_{CS} = \int d^3x \, K^0 \tag{4.1}$$

where the current $K^\mu$ is given by

$$K^\mu = \frac{g^2}{32\pi^2} \epsilon^{\mu\nu\alpha\beta} \left( F^a_{\nu\alpha} A^a_\beta - \frac{g}{3} \epsilon_{abc} A^a_\nu A^b_\alpha A^c_\beta \right) \tag{4.2}$$

(note: there are a number of references [20, 21] in which the factor $g/3$ is incorrectly given as $2g/3$; this seems to be an error that propagated without being checked). This current has the property that

$$\partial_\mu K^\mu = \frac{g^2}{32\pi^2} F^a_{\mu\nu} \widetilde{F}^{a,\mu\nu} \tag{4.3}$$

Chern-Simons number has a topological nature which can be seen when considering configurations which are pure gauge at some initial and final times, $t_0$ and $t_1$. It can be shown that

$$N_{CS}(t_1) - N_{CS}(t_0) = \int_{t_0}^{t_1} dt \int d^3x \, \partial_\mu K^\mu = \nu \tag{4.4}$$

E

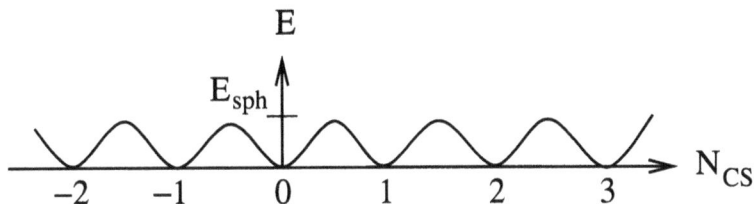

Fig. 8. Energy of gauge field configurations as a function of Chern-Simons number.

an integer, which is a winding number. The gauge field is a map from the physical space to the manifold of the gauge group. If we consider an SU(2) subgroup of SU(N), and the boundary of 4D space compactified on ball, both manifolds are 3-spheres, and the map can have nontrivial homotopy.

We are interested in the vacuum structure of SU(2) gauge theory. Consider a family of static gauge field configurations with continuously varying $N_{CS}$. Those configuration with integer values turn out to be pure gauge everywhere, hence with vanishing field strength and zero energy. But to interpolate between two such configurations, one must pass through other configurations whose field strength and energy are nonvanishing. The energy versus Chern-Simons number has the form shown in figure 8.

Each minimum is a valid perturbative vacuum state of the theory. They are called $n$-vacua. The height of the energy barrier is

$$E_{sph} = f\left(\frac{\lambda}{g^2}\right)\frac{4\pi v}{g} \cong \frac{8\pi v}{g} = \frac{2M_W}{\alpha_W}f\left(\frac{\lambda}{g^2}\right) \tag{4.5}$$

where $v = 174$ GeV is the Higgs field VEV, $\lambda$ is the Higgs quartic coupling, $\alpha_W = g^2/4\pi \cong 1/30$, and the function $f$ ranges between $f(0) = 1.56$ and $f(\infty) = 2.72$.

't Hooft discovered that tunneling occurs between $n$-vacua through field configurations now called instantons (Physical Review tried to suppress that name at first, so the term "pseudoparticle" appears in the early literature, but it soon succumbed to the much more popular name.) The relevance of this for B violation is through the relation between the divergence of the left-handed baryon plus lepton current to the divergence of $K^\mu$:

$$\partial_\mu J^\mu_{B+L} = N_f \partial_\mu K^\mu \tag{4.6}$$

where $N_f = 3$ is the number of families. Integrating this relation over space and time, the spatial divergence integrates to zero and we are left with

$$3\frac{d}{dt}N_{CS} = \frac{d}{dt}B = \frac{d}{dt}L \tag{4.7}$$

Hence each instanton transition violates B and L by 3 units each—there is spontaneous production of 9 quarks and 3 leptons, with each generation represented equally.

However, the tunneling amplitude is proportional to

$$\mathcal{A} \sim e^{-8\pi^2/g^2} \sim 10^{-173} \tag{4.8}$$

which is so small as to never have happened during the lifetime of the universe. For this reason, anomalous B violation in the SM was not at first considered relevant for baryogenesis. But in 1985, Kuzmin, Rubakov and Shaposhnikov realized that at high temperatures, these transitions would become unsuppressed, due to the availability of thermal energy to hop over the barrier, instead of tunneling through it [23]. This occurs when $T \gtrsim 100$ GeV. The finite-$T$ transitions are known as sphaleron processes, a term coined by Klinkhamer and Manton from the Greek, meaning "ready to fall:" it is the field configuration of Higgs and $W^\mu$ which sits at the top of the energy barrier between the $n$-vacua [20]. A thermal transition between $n$-vacua must pass through a configuration which is close to the sphaleron, unless $T \gg E_{\text{sph}}$. The sphaleron is a static, saddle point solution to the field equations.

Evaluating the path integral for a sphaleron transition semiclassically, one finds that the amplitude goes like

$$\mathcal{A} \sim e^{-E\text{sph}/T} \tag{4.9}$$

To find the actual rate of transitions, one must do a more detailed calculation including the fluctuation determinant around the saddle point [21]. Khlebnikov and Shaposhnikov obtained the rate per unit volume of sphaleron transitions

$$\frac{\Gamma}{V} = \text{const} \left(\frac{E_{\text{sph}}}{T}\right)^3 \left(\frac{m_W(T)}{T}\right)^4 T^4 e^{-E\text{sph}/T} \tag{4.10}$$

where the constant is a number of order unity and $m_W(T)$ is the temperature dependent mass of the W boson. But this formula is only valid at temperatures $T < E_{\text{sph}}$. In fact, at $T \gtrsim m_W$, electroweak symmetry breaking has not yet occurred, and $m_W(T) = 0$. The Higgs VEV vanishes in this symmetry-restored phase, and $E_{\text{sph}} = 0$. There is no longer any barrier between the $n$-vacua at these high temperatures.

Above the electroweak phase transition (EWPT), the rate of sphaleron interactions cannot be computed analytically; lattice computations are required. For many years it was believed that the parametric dependence was of the form

$$\frac{\Gamma}{V} = c\alpha_w^4 T^4 \tag{4.11}$$

with $c \sim 1$ and $\alpha_w = g_2^4/(4\pi) \cong 1/30$. This is based on the idea that the transverse gauge bosons acquire a magnetic thermal mass of order $g^2 T$, which is the only relevant scale in the problem, and therefore determines (4.11) by dimensional analysis. The origin of this scale can be readily understood [25] by the following argument. We are looking for field configurations with $N_{CS} \sim 1$, so using the definition of Chern-Simons number,

$$\int d^3 x (g^2 F A \text{ or } g^3 A^3) \sim 1 \tag{4.12}$$

but the energy of the configuration is determined by the temperature,

$$\int d^3 x \, F^2 \sim T \tag{4.13}$$

We can therefore estimate for a configuration of size $R$ that

$$R^3 \left( g^2 \frac{A^2}{R^2} \text{ or } g^3 A^3 \right) \sim 1$$

$$R^3 \frac{A^2}{R^2} \sim T \tag{4.14}$$

which fixes the size of $A$ and $R$, giving $R \sim 1/(g^2 T)$. Hence the estimate that $\Gamma/V \sim (g^2 T)^4$. However it was later shown [26] that this is not quite right— the time scale for sphaleron transitions is actually $g^4 T$, not $g^2 T$, giving $\Gamma/V \sim \alpha_w^5 T^4$. Careful measurements in lattice gauge theory fixed the dimensionless coefficient to be [27] $\Gamma/V = (29 \pm 6)\alpha_w^5 T^4$, and more recently [28]

$$\frac{\Gamma}{V} = (25.4 \pm 2.0)\alpha_w^5 T^4 = (1.06 \pm 0.08) \times 10^{-6} T^4 \tag{4.15}$$

Ironically the prefactor is such that the old $\alpha_w^4 T^4$ estimate was good using $c = 1$.

We can now determine when sphalerons were in thermal equilibrium in the early universe. To compute a rate we must choose a relevant volume. We can take the thermal volume, $1/T^3$, which is the average space occupied by a particle in the thermal bath. Then

$$\Gamma = 10^{-6} T \tag{4.16}$$

which must be compared to the Hubble rate, $H \sim \sqrt{g_*} T^2/M_p$. At very high temperatures, sphalerons are *out* of equilibrium, as illustrated in figure 9. They

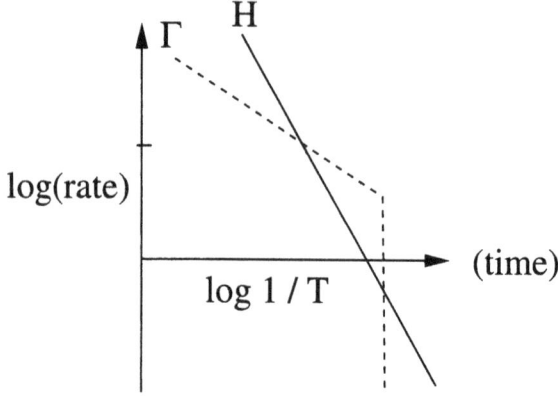

Fig. 9. Sphaleron rate and Hubble rate versus time. The sharp drop in the sphaleron rate occurs at the EWPT.

come into equilibrium when

$$\Gamma = H \quad \Rightarrow \quad 10^{-6}T = \sqrt{g_*}\frac{T^2}{M_p}$$

$$T = 10^{-6}g_*^{-1/2}M_p \sim 10^{-5}M_p \sim 10^{13}\text{GeV} \tag{4.17}$$

Hence GUT baryogenesis is affected by sphalerons, after the fact. Of course when $T$ falls below the EWPT temperature $\sim 100$ GeV, the sphaleron rate again falls below $H$.

### 4.1. CP violation in the SM

Since the SM provides such a strong source of B violation, it is natural to wonder whether baryogenesis is possible within the SM. We must investigate whether the other two criteria of Sakharov can be fulfilled.

It is well known that CP violation exists in the CKM matrix of the SM,

$$V_{CKM} = \begin{pmatrix} V_{ud} & V_{us} & V_{ub} \\ V_{cd} & V_{cs} & V_{cb} \\ V_{td} & V_{ts} & V_{tb} \end{pmatrix} = \begin{pmatrix} c_1 & -s_1c_3 & -s_1s_3 \\ s_1c_2 & \begin{matrix}c_1c_2c_3\\-s_2s_3e^{i\delta}\end{matrix} & \begin{matrix}c_1c_2s_3\\+s_2c_3e^{i\delta}\end{matrix} \\ s_1s_2 & \begin{matrix}c_1s_2c_3\\+c_2s_3e^{i\delta}\end{matrix} & \begin{matrix}c_2s_2s_3\\-c_2c_3e^{i\delta}\end{matrix} \end{pmatrix} \tag{4.18}$$

This is the original parametrization, whereas in the Wolfenstein parametrization, $V_{ub}$ and $V_{td}$ contain the CP-violating phase. Where the phase resides in the CKM matrix can be changed by field redefinitions. How do we express the invariant phase? There is no unique answer: it depends upon which physical process is manifesting the CP violation. However C. Jarlskog tried to address the question in a rather general way [29]. She showed that one possible invariant is the combination

$$J = (m_t^2 - m_c^2)(m_t^2 - m_u^2)(m_c^2 - m_u^2)$$
$$(m_b^2 - m_s^2)(m_b^2 - m_d^2)(m_s^2 - m_d^2) \, K$$

where $K = s_1^2 s_2 s_3 c_1 c_2 c_3 \sin \delta$

$$= \mathrm{Im} \, V_{ii} V_{jj} V_{ij}^* V_{ji}^* \text{ for } i \neq j \qquad (4.19)$$

This is derived by computing the determinant of the commutator of the up- and down-type quark mass matrices (squared):

$$J = \det \left[ m_u^2, \, m_d^2 \right] \qquad (4.20)$$

and therefore is invariant under rotations of the fields.

The form of $J$ has been used to argue that CP violation within the standard model cannot be large enough for baryogenesis. One should find a dimensionless measure of the strength of CP violation, using the relevant temperature of the universe, which must be at least of order 100 GeV for sphalerons to be effective. One then finds that

$$\frac{J}{(100 \text{ GeV})^{12}} \sim 10^{-20} \qquad (4.21)$$

which is much too small to account for $\eta \sim 10^{-10}$.

One might doubt whether this argument is really robust. For example, in the original PRL of Jarlskog [29], $J$ was defined in terms of the linear quark mass matrices, which yielded a formula like (4.19), but with linear rather than squared mass differences. In the subsequent paper it was argued that actually the sign of a fermion mass has no absolute physical significance, so any physical quantity should depend on squares of masses. However, this seems to have no bearing on the mathematical fact that the original linear-mass definition of $J$ is a valid invariant characterization of the CP phase. If this were the physically correct definition, then (4.21) would be revised to read $J \sim 10^{-10}$, which is in the right ballpark.

A more specific criticism of the argument was given in [30], who noted that the argument cannot be applied to $K \bar{K}$ mixing in the neutral kaon system, coming from the box diagram

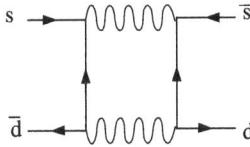

Farrar and Shaposhnikov point out that this CP-violating effect is not proportional to $J$, and that the relevant scale is much smaller than 100 GeV—it is the mass of $K^0$. The idea that $J/T^{12}$ is the correct measure of CP violation only makes sense if all the ratios of mass to temperature can be treated as perturbatively small. This is clearly not the case for the ratio of the top quark mass to the $K^0$ mass. Could there not also be some scale lower than 100 GeV playing a role in the baryogenesis mechanism? Ref. [30] attempted to construct such a mechanism within the SM, which was refuted by [31]. There is no theorem proving that it is impossible to find some other mechanism that does work, but so far there are no convincing demonstrations, and most practitioners of baryogenesis agree that CP violation in the SM is too weak; one needs new sources of CP violation and hence new physics beyond the SM. We will see that it is rather easy to find such new sources; for example the MSSM has many new phases, such as in the term $W = \mu H_1 H_2$ in the superpotential, and in the gaugino masses.

## 5. Electroweak phase transition and electroweak baryogenesis

In the previous section we showed that B violation is present in the standard model, and new sources of CP violation are present in low-energy extensions of the SM. But the remaining requirement, going out of thermal equilibrium, is not so easy to achieve at low energies. Recall the condition for out-of-equilibrium decay, (3.4),

$$\alpha \ll \frac{m}{M_p} \tag{5.1}$$

If $m \sim 100$ GeV, we require extraordinarily weak couplings which are hard to justify theoretically, and which would also be hard to verify experimentally.

But the EWPT can provide a departure from thermal equilibrium if it is sufficiently strongly first order. The difference between first and second order is determined by the behavior of the Higgs potential at finite temperature, as shown in figure 10. In a first order transition, the potential develops a bump which separates the symmetric and broken phases, while in a second order transition or a smooth cross-over there is no bump, merely a change in sign of the curvature of the potential at $H = 0$. The critical temperature $T_c$ is defined to be the temperature at which the two minima are degenerate in the first order case, or the temperature at which $V''(0) = 0$ in the second order case.

A first order transition proceeds by bubble nucleation (fig. 11), where inside the bubbles the Higgs VEV and particle masses are nonzero, while they are still vanishing in the exterior symmetric phase. The bubbles expand to eventually

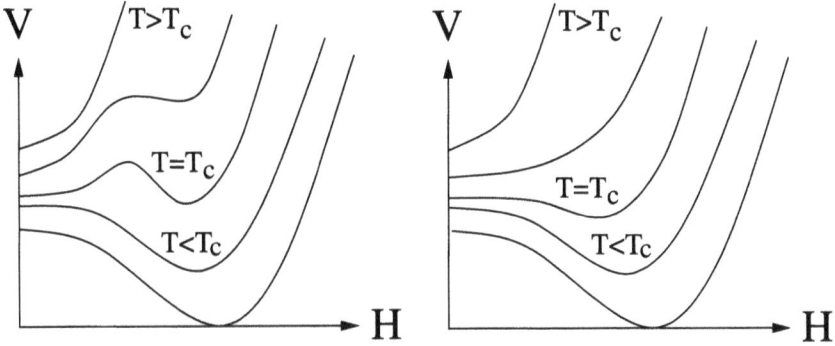

Fig. 10. Schematic illustration of Higgs potential evolution with temperature for first (left) and second (right) order phase transition.

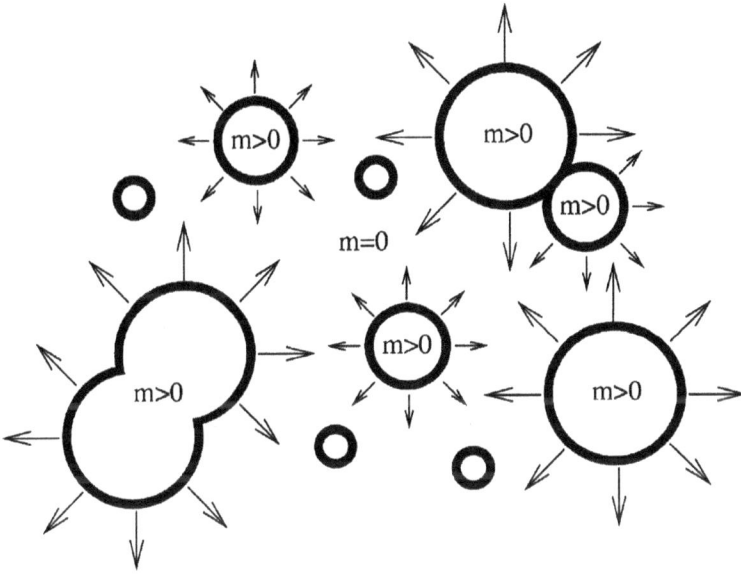

Fig. 11. Bubble nucleation during a first-order EWPT.

collide and fill all of space. If the Higgs VEV $v$ is large enough inside the bubbles, sphalerons can be out of equilibrium in the interior regions, while still in equilibrium outside of the bubbles. A rough analogy to GUT baryogenesis is that sphalerons outside the bubbles correspond to B-violating $Y$ boson decays, which are fast, while sphalerons inside the bubbles are like the B-violating inverse Y de-

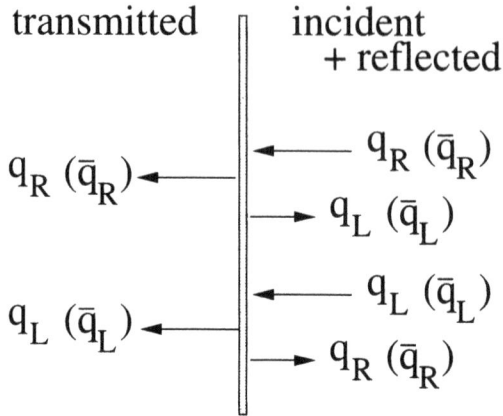

Fig. 12. CP-violating reflection and transmission of quarks at the moving bubble wall.

cays. The latter should be slow; otherwise they will relax the baryon asymmetry back to zero.

In a second order EWPT, even though the sphalerons go from being in equilibrium to out of equilibrium, they do so in a continuous way, and uniformly throughout space. To see why the difference between these two situations is important, we can sketch the basic mechanism of electroweak baryogenesis, due to Cohen, Kaplan and Nelson [32]. The situation is illustrated in figure 12, which portrays a section of a bubble wall moving to the right. Because of CP-violating interactions in the bubble wall, we get different amounts of quantum mechanical reflection of right- and left-handed quarks (or of quarks and antiquarks). This leads to a chiral asymmetry in the vicinity of the wall. There is an excess of $q_L + \bar{q}_R$ relative to $q_R + \bar{q}_L$ in front of the wall, and a compensating deficit of this quantity on the other side of the wall. This CP asymmetry is schematically shown in figure 13.

Sphalerons interact only with $q_L$, not $q_R$, and they try to relax the CP-asymmetry to zero. Diagrammatically,

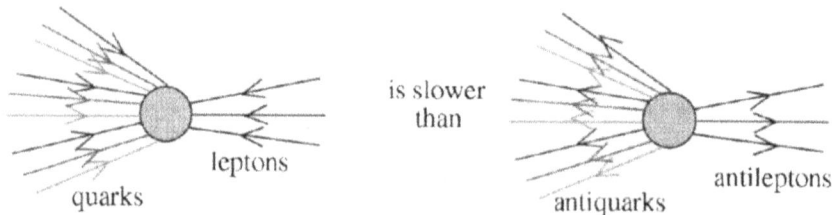

## CP asymmetry

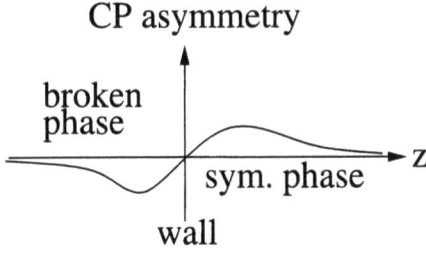

Fig. 13. The CP asymmetry which develops near the bubble wall.

## B asymmetry

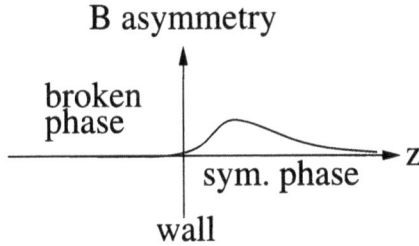

Fig. 14. The B asymmetry which initially develops in front of the bubble wall.

simply because there are more $\bar{q}_L$ than $q_L$ in front of the wall. But the first interaction violates baryon number by $-3$ units while the second has $\Delta B = 3$. Therefore the CP asymmetry gets converted into a baryon asymmetry in front of the wall (but not behind, since we presume that sphaleron interactions are essentially shut off because of the large Higgs VEV). Schematically the initial baryon asymmetry takes the form of figure 14.

If the baryon asymmetry remained in front of the wall, eventually sphalerons would cause it to relax to zero, because there are other processes besides sphalerons in the plasma which can relax the CP asymmetry, for example *strong* SU(3) sphalerons which change chirality $Q_5$ by 12 units, 2 for each flavor, as shown in figure 15. (See [33] for a lattice computation of the strong sphaleron rate.) The combination of weak and strong sphalerons would relax $Q_5$ and $B + L$ to zero if the wall was not moving. But due to the wall motion, there is a tendency for baryons to diffuse into the broken phase, inside the bubble. If $E_{sph}/T$ is large enough, $\Gamma_{sph}$ is out of equilibrium and $B$ violation is too slow to relax $B$ to zero. This is the essence of electroweak baryogenesis.

### 5.1. Strength of the phase transition

Now that we appreciate the importance of the first order phase transition, we can try to compute whether it occurs or not. The basic tool for doing so is the finite-

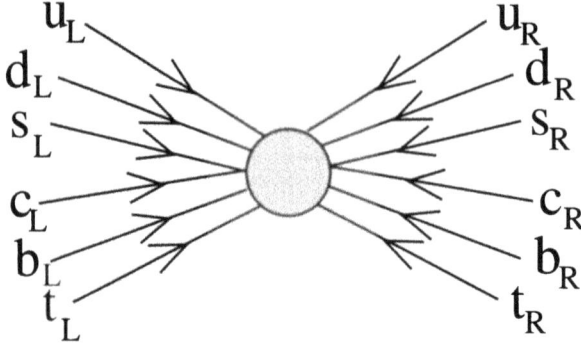

Fig. 15. The strong sphaleron.

temperature effective potential of the Higgs field, defined by the path integral

$$e^{-\beta \int d^3x\, V_{\text{eff}}(H)} = \int \prod_i \mathcal{D}\phi_i\, e^{-\int_0^\beta d\tau \int d^3x\, \mathcal{L}[H,\phi_i]} \tag{5.2}$$

where $\beta = 1/T$, $H$ is the background Higgs field, and $\phi_i$ are fluctuations of all fields which couple to the Higgs, including $H$ itself. Here $\tau$ is imaginary time and the fields are given periodic (antiperiodic) boundary conditions between $\tau = 0$ and $\beta$ if they are bosons (fermions). For a compact introduction to field theory at finite temperature, see ref. [34].

To evaluate $V$eff, we can use perturbation theory. Since there are no external legs, $V_{\text{eff}}$ is given by a series of vacuum bubbles,

$$V_{\text{eff}} = \bigcirc + \{ \infty + \oslash \} + \dots$$

The one-loop contribution has a familiar form,

$$V_{1-\text{loop}} = T \sum_i \pm \int \frac{d^3p}{(2\pi)^3} \ln\left(1 \mp e^{-\beta\sqrt{p^2 + m_i^2(H)}}\right) \begin{cases} \text{bosons} \\ \text{fermions} \end{cases} \tag{5.3}$$

This is the free energy of a relativistic gas of bosons or fermions. Recall that the partition function for a free nonrelativistic gas of $N$ particles in a volume $V$ is

$$F = -T \ln Z;$$

$$Z = \left(\frac{V}{h} \int d^3p\, e^{-\beta p^2/2m}\right)^N \tag{5.4}$$

Eq. (5.3) is the relativistic generalization, taking into account that a fermion loop comes with a factor of $-1$. The nontrivial aspect is that we must evaluate the particle masses as a function of the Higgs VEV $H$.

In the standard model, the important particles contributing to the thermal partition function, and their field-dependent masses, are

$$\text{top quark: } m_t = yH \tag{5.5}$$

$$\text{gauge bosons, } W^{\pm}, W^3, B : m^2 = \frac{1}{2}H^2 \begin{pmatrix} g^2 & & & \\ & g^2 & & \\ & & g^2 & gg' \\ & & gg' & g^2 \end{pmatrix} \tag{5.6}$$

(One must of course find the eigenvalues of the gauge boson mass matrix.) For the Higgs bosons, we need to refer to the tree-level Higgs potential

$$V = \lambda \left( |H|^2 - \frac{1}{2}v^2 \right)^2 = \frac{1}{4}\lambda \left( \phi^2 + \sum_i \chi_i^2 - v^2 \right)^2 \tag{5.7}$$

where the real components of the Higgs doublet are $H = \frac{1}{\sqrt{2}}(\phi + i\chi_1, \chi_2 + i\chi_3)$. $\phi$ is the massive component, while $\chi_i$ are the Goldstone bosons which get "eaten" by the $W$ gauge bosons. In the effective potential, we set $\chi_i = 0$ so that the VEV is given by $H = \frac{1}{\sqrt{2}}\phi$, but to find the masses of the Goldstone bosons, we must temporarily include their field-dependence in $H$, take derivatives with respect to $\chi_i$, and then set $\chi_i = 0$. Then

$$\text{Higgs boson, } H : m_H^2 = \frac{\partial^2 V}{\partial \phi^2} = \lambda(3\phi^2 - v^2) = \lambda(6H^2 - v^2) \tag{5.8}$$

$$\text{Goldstone bosons, } \chi_i : m_i^2 = \frac{\partial^2 V}{\partial \chi_i^2} = \lambda(\phi^2 - v^2) = \lambda(2H^2 - v^2) \tag{5.9}$$

We must use these field-dependent masses in the 1-loop thermal effective potential. Unfortunately the momentum integrals cannot be done in any enlightening closed form, except in the high- or low-temperature limits. We will be interested in the high-$T$ case, where $V_{1-\text{loop}}$ can be expanded as

$$V_{1-\text{loop}} = \sum_{i \in B,F} \frac{m_i^2 T^2}{48} \times \begin{Bmatrix} 2, & \text{each real B} \\ 4, & \text{each Dirac F} \end{Bmatrix} - \frac{m_i^3 T}{12\pi} \begin{Bmatrix} 1, & \text{B} \\ 0, & \text{F} \end{Bmatrix}$$

$$+ \frac{m_i^4}{64\pi^2} \left( \ln \frac{m_i^2}{T^2} - c_i \right) \times \begin{Bmatrix} -1, & \text{B} \\ +4, & \text{Dirac F} \end{Bmatrix} + O\left( \frac{m_i^5}{T} \right) \tag{5.10}$$

where

$$c_i = \begin{cases} \frac{3}{2} + 2\ln 4\pi - 2\gamma_E \cong 5.408, & B \\ c_B - 2\ln 4 \cong 2.635, & F \end{cases} \qquad (5.11)$$

Let's examine the effects of these terms, starting with the term of order $m_i^2 T^2$. This is the source of a $T$-dependent squared mass for the Higgs boson, and explains why $V''(0) > 0$ at high $T$. Focusing on just the $H^2$ terms, the contributions to the potential are

$$\begin{aligned} V_{1-\text{loop}} &= H^2 T^2 \left[ \frac{1}{2}\lambda + \frac{3}{16}(3g^2 + g'^2) + \frac{1}{4}y^2 \right] + O(H^3) \\ &\equiv aT^2 H^2 + O(H^3) \qquad\qquad\qquad\qquad\qquad (5.12) \\ V_{\text{tree}} &= -\lambda v^2 H^2 + O(H^4) \qquad\qquad\qquad\qquad\quad (5.13) \end{aligned}$$

Putting these together, we infer that the $T$-dependent Higgs mass is given by

$$m_H^2(T) = -\lambda v^2 + aT^2 \qquad (5.14)$$

and we can find the critical temperature of the phase transition (if it was second order) by solving for $m_H^2 = 0$:

$$T_c = \sqrt{\frac{\lambda}{a}}\, v \cong 2\frac{\sqrt{\lambda}}{y}v \qquad (5.15)$$

If the transition is first order, it is due to the "bump" in the potential, which can only come about because of the $H^3$ term in $V_{1-\text{loop}}$. The cubic term is

$$V_{\text{cubic}} \cong -\frac{TH^3}{12\sqrt{2\pi}} \left( 3g^3 + \frac{3}{2}(g^2 + g'^2)^{3/2} + O(\lambda) \right) \equiv -ETH^3 \qquad (5.16)$$

where the $O(\lambda)$ contributions have been neglected, since we will see that $\lambda \ll g^2$ is necessary for having a strongly first order phase transition.

With the cubic term, the potential takes the form

$$V_{\text{tot}} \cong m_H^2(T)H^2 - ETH^3 + \lambda H^4 \qquad (5.17)$$

and at the critical temperature it becomes

$$V_{\text{tot}} = \lambda H^2 \left( H - \frac{v_c}{\sqrt{2}} \right)^2 \qquad (5.18)$$

since this is the form which has degenerate minima at $H = 0$ and $H = v_c$. Notice that the VEV at the critical temperature, $v_c$, is not the same as the zero-temperature VEV, $v$. By comparing (5.17) and (5.18) we find that

$$m^2(T) = \frac{1}{2}\lambda v_c^2, \quad ET_c = \sqrt{2}\lambda v_c \tag{5.19}$$

which allows us to solve for $v_c$ and $T_c$. The result is

$$T_c = \frac{\lambda v}{\sqrt{a\lambda - \frac{1}{4}E^2}} \tag{5.20}$$

and

$$\frac{v_c}{T_c} = \frac{E}{2\sqrt{\lambda}} \cong \frac{3g^3}{16\pi\lambda} \tag{5.21}$$

The ratio $v_c/T_c$ is a measure of the strength of the phase transition. It determines how strongly the sphalerons are suppressed inside the bubbles. Recall that

$$\frac{E_{\text{sph}}}{T} \sim \frac{8\pi}{g}\left(\frac{v}{T}\right) \tag{5.22}$$

appears in $\Gamma_{\text{sph}} \sim e^{-E_{\text{sph}}/T}$, so the bigger $v/T$ is at the critical temperature, the less washout of the baryon asymmetry will be caused by sphalerons. The total dilution of the baryon asymmetry inside the bubbles is the factor

$$e^{-\int_{t_c}^{\infty} \Gamma_{\text{sph}} dt} \tag{5.23}$$

where $t_c$ is the time of the phase transition. Requiring this to be not too small gives a bound [35]

$$\frac{v_c}{T_c} \gtrsim 1 \tag{5.24}$$

which we can use to infer a bound on the Higgs mass [36], by combining (5.21) and (5.24). Given that

$$\alpha_w = \frac{g^2}{4\pi} \cong \frac{8M_w^2 G_F}{4\sqrt{2}\pi} \cong \frac{1}{29.5} \tag{5.25}$$

the weak coupling constant is $g = 0.65$, so (5.24) implies $\lambda < 3g^3/16\pi = 0.128$, and therefore

$$m_H = \sqrt{\frac{\lambda}{2}}v < 32 \text{ GeV} \tag{5.26}$$

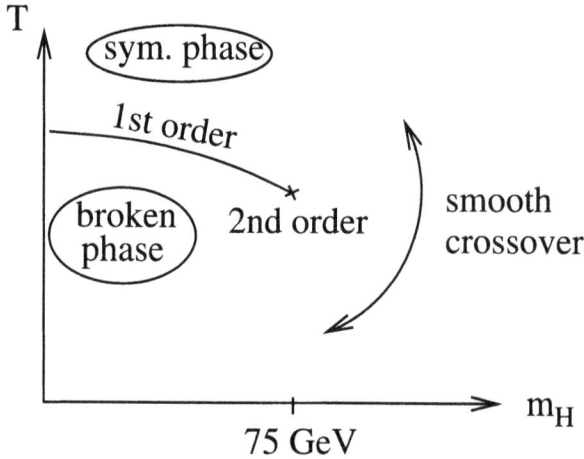

Fig. 16. Phase diagram for the EWPT, from ref. [38].

This is far below the current LEP limit $m_H > 115$ GeV, so it is impossible to have a strongly first order electroweak phase transition in the standard model.

The expressions we have derived give the impression that, regardless of the value of the Higgs mass, the phase transition is first order, even if very weakly. This is not the case, as explained by P. Arnold [37]: at high temperatures, the perturbative expansion parameter is not just made from dimensionless couplings, but rather

$$\frac{g^2 T}{m_W} \sim \frac{\lambda}{g^2} \sim \left. \frac{m_H^2}{m_W^2} \right|_{T=0} \tag{5.27}$$

due to the fact that the high-$T$ theory is effectively 3-dimensional, and is more infrared-sensitive to the gauge boson masses than is the 4D ($T = 0$) theory. In the regime of large $\lambda$, nonperturbative lattice methods must be used to study the phase transition. It was found in [38] that in the $T$-$m_H$ phase diagram, there is a line of 1st order phase transitions ending at a critical point with a 2nd order transition at $m_H = 75$ GeV. For larger $m_H$, there is no phase transition at all, but only a smooth crossover. This is illustrated in figure 16.

Most attempts to improve this situation have relied on new scalar particles coupling to the Higgs field. Ref. [39] considered a singlet field $S$ with potential

$$V = 2\zeta^2 |H|^2 |S|^2 + \mu^2 |S|^2 \tag{5.28}$$

including a large coupling to the Higgs field, so

$$m_s^2 = \mu^2 + 2\zeta^2 |H|^2 \tag{5.29}$$

If $\mu^2$ is not too large ($\mu \lesssim 90$ GeV) then the cubic term $-\frac{1}{12\pi} T(\mu^2 + 2\zeta^2 H^2)^{3/2}$ is sufficiently enhanced so that $v/T \sim 1$ for heavy Higgs. If $\mu^2$ is too large, this does not function as a cubic term, since it can be Taylor-expanded in $H^2$.

## 5.2. EWPT in the MSSM

It is possible to implement this mechanism in the MSSM because top squarks (stops) couple strongly to the Higgs field [40]. The MSSM is a model with two Higgs doublets, $H_1$ and $H_2$, of which one linear combination is relatively light, and plays the role of the SM Higgs, as illustrated in figure 17.

The stops come in two kinds, $\tilde{t}_L$ and $\tilde{t}_R$, and have mass matrix

$$m_{\tilde{t}}^2 \simeq \begin{pmatrix} m_Q^2 + y^2 H_2^2 & y(A_t^* H_2^* + \mu H_1) \\ y(A_t H_2 + \mu^* H_1^*) & m_U^2 + y^2 H_2^2 \end{pmatrix} \tag{5.30}$$

where $m_Q^2$, $m_U^2$ and $A_t$ are soft SUSY-breaking parameters. If we ignore the off-diagonal terms (possible if we choose $A_t \tan \beta = \mu$) it is easier to see what is happening—there are two unmixed bosons coupling to $H$. However, we cannot make both of them light. This is because $m_H^2$ gets major contributions from stops at 1 loop, from the diagrams shown in fig. 18.

These give

$$m_H^2 \simeq m_Z^2 + c \frac{m_t^4}{v^2} \ln\left(\frac{m_{\tilde{t}_L} m_{\tilde{t}_R}}{m_t^2}\right) \tag{5.31}$$

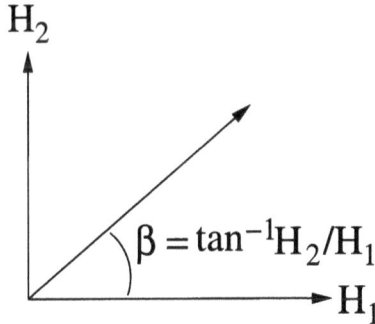

Fig. 17. Light effective Higgs direction in the MSSM (in the limit where the $A^0$ is very heavy).

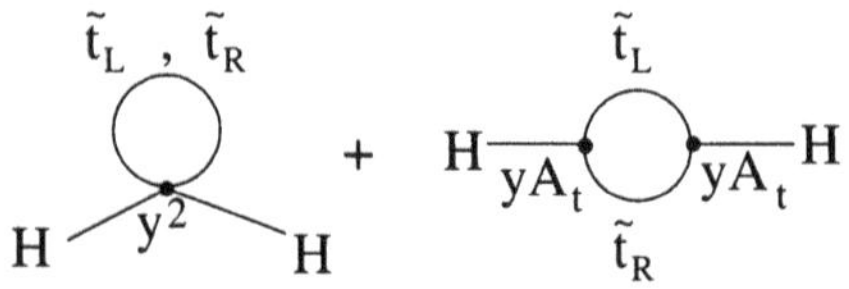

Fig. 18. Stop loop contributions to the Higgs mass.

If both $\tilde{t}_L$ and $\tilde{t}_R$ are light, it is impossible to make the Higgs heavy enough to satisfy the LEP constraint. Precision electroweak constraints dicatate that $\tilde{t}_L$ should be the heavy one. Otherwise corrections to the $\rho$ parameter,

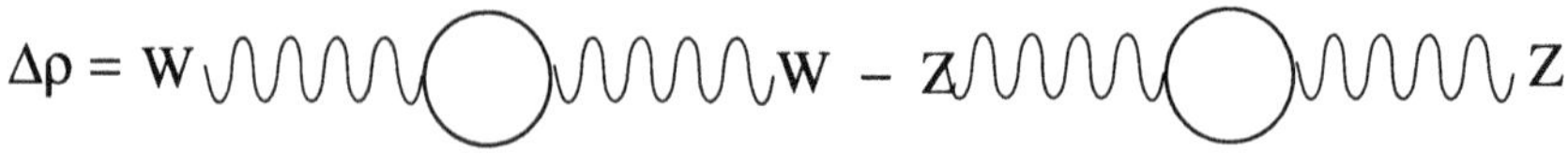

are too large, since $\tilde{t}_L$ couples more strongly to $W$ and $Z$ than does $\tilde{t}_R$.

Although we have only discussed thermal field theory at one loop, two-loop effects are also important in the MSSM. The $\tilde{t}_R$ diagrams

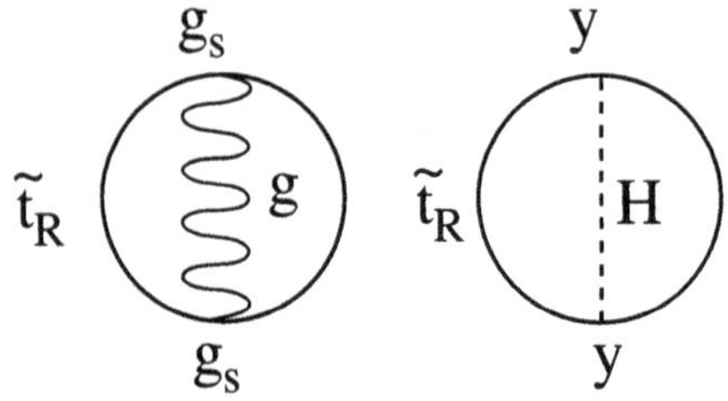

with gluon and Higgs exchange give a contribution of the form

$$\Delta V_{\text{eff}} = -cT^2 H^2 \ln \frac{H}{T} \tag{5.32}$$

which is different from any arising at one loop, and which shift the value of $v_c/T_c$ by

$$\frac{v_c}{T_c} = \frac{E}{2\sqrt{2}\lambda} + \sqrt{\frac{E^2}{8\lambda^2} + \frac{2c}{\lambda}} \tag{5.33}$$

Detailed studies of the strength of the phase transition have been done using dimensional reduction [41], two-loop perturbation theory (see for example [42]) and lattice gauge theory [43] which show that *negative* values of $m_U^2$ (the soft

SUSY-breaking mass parameter for $\tilde{t}_R$) are required. This may seem strange; why would we want a cubic term of the form

$$\mathcal{L}_{\text{cubic}} \sim -\frac{T}{6\pi}\left(-|m_U^2| + y^2|H_2|^2\right)^{3/2}? \tag{5.34}$$

To understand this, we must consider thermal corrections to the *squark* mass as well,

$$m_{\tilde{t}_R}^2 = -|m_U^2| + y^2|H_2|^2 + c_R T^2 \tag{5.35}$$

where $c_R = \frac{4}{9}g_s^2 + \frac{1}{6}y^2(1 + \sin^2\beta)$, which can be computed similarly to the thermal Higgs mass: one must calculate $V_{\text{eff}}(H, \tilde{t}_R)$. The insertion of 1-loop thermal masses into the cubic term corresponds to resumming a class of diagrams called daisies:

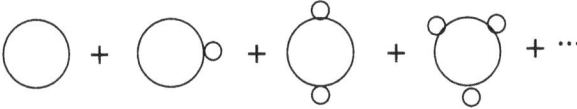

The most prominent effect for strengthening the phase transition comes when $-|m_U^2| + c_R T^2 \cong 0$. The negative values of $m_U^2$ correspond to a light stop, $m_{\tilde{t}_R} < 172$ GeV.

It is often said that a relatively light Higgs is also needed for a strong phase transition, but this is just an indirect effect of eq. (5.31). If we were willing to make $m_Q$ arbitrarily heavy, we could push $m_H$ to higher values. The relation is [42]

$$m_Q \cong 100 \text{ GeV } \exp\left[\frac{1}{9.2}\left(\frac{m_H}{\text{GeV}} - 85.9\right)\right] \tag{5.36}$$

For example, for $m_H = 120$ GeV, $m_Q = 4$ TeV. This is unnaturally high since SUSY should be broken near the 100 GeV scale to avoid fine-tuning. And it looks strange to have $m_Q \gg |m_U|$.

The conclusion is that it is possible, but not natural, to get a strong enough phase transition in the MSSM. However it is easy in the NMSSM, which includes an additional singlet Higgs, with superpotential

$$W = \mu H_1 H_2 + \lambda S H_1 H_2 + \frac{k}{3}S^3 + rS \tag{5.37}$$

giving a potential

$$V = \sum_i \left|\frac{\partial W}{\partial H_i}\right|^2 + \left|\frac{\partial W}{\partial S}\right|^2 + \text{ soft SUSY-breaking terms} \tag{5.38}$$

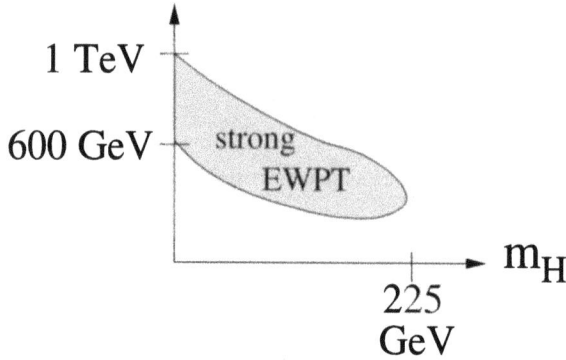

Fig. 19. Region of strong EWPT with a $|H|^6/\Lambda^2$ operator.

Ref. [44] finds large regions of parameter space where the phase transition is strong enough.

Similar studies have been done for the general 2-Higgs doublet model (2HDM) [45, 46], with

$$
\begin{aligned}
V &= -\mu_1^2|H_1|^2 - \mu_2^2|H_2|^2 - \left(\mu_3^2 H_1^\dagger H_2 + \text{h.c.}\right) + \frac{1}{2}\lambda_1|H_1|^4 + \frac{1}{2}\lambda_2|H_2|^4 \\
&\quad + \lambda_3|H_1|^2|H_2|^2 + \lambda_4|H_1^\dagger H_2|^2 + \frac{1}{2}\lambda_5\big[(H_1^\dagger H_2)^2 + \text{h.c.}\big]
\end{aligned} \tag{5.39}
$$

There are several extra scalar degrees of freedom coupling to the light Higgs in this model, giving a large region of parameter space where the EWPT is strong.

A model-independent approach to increasing the strength of the EWPT has been presented in [47], where the effects of heavy particles interacting with the Higgs are parametrized by a low-energy effective potential,

$$
V = \lambda\left(|H|^2 - \frac{v^2}{2}\right)^2 + \frac{1}{\Lambda^2}|H|^6 \tag{5.40}
$$

It is found that $v/T > 1$ is satisfied in a sizeable region of parameter space, as roughly depicted in figure 19. This could provide another way of understanding the enhancement of the strength of the phase transition in models like the NMSSM or the 2HDM, in cases where the new particles coupling to $H$ are not as light as one might have thought, based on the idea of increasing the cubic coupling.

## 6. A model of electroweak baryogenesis: 2HDM

Let us now go through a detailed example of how electroweak baryogenesis could work. The simplest example is the two Higgs doublet model, just mentioned in the previous section, where it is assumed that the top quark couples only to one of the Higgs fields. This assumption is useful for avoiding unwanted new contributions to flavor-changing neutral currents (FCNC's).

The basic idea is as follows. There are phases in the potential (5.39), in the complex parameters $\mu_3^2$ and $\lambda_5$. One of these can be removed by a field redefinition, leaving an invariant phase

$$\phi \equiv \arg\left(\frac{\mu_3^2}{\lambda_5^{1/2}}\right) \tag{6.1}$$

The bubble wall will be described by a Higgs field profile sketched in figure 20,

$$H_i(z) = \frac{1}{\sqrt{2}} h_i(z) e^{i\theta_i(z)} \tag{6.2}$$

where

$$h_i(z) \cong \begin{pmatrix} \cos\beta \\ \sin\beta \end{pmatrix} h(z),$$

$$h(z) \cong \frac{v_c}{2}\left(1 - \tanh\left(\frac{z}{L}\right)\right) \tag{6.3}$$

This shape can be understood by considering the SM, where

$$\mathcal{L} = |\partial H|^2 - \lambda\left(|H|^2 - \frac{v^2}{2}\right)^2$$

$$E = \frac{1}{2}(\partial_z h)^2 + \begin{cases} \frac{\lambda}{4}(h^2 - v^2)^2 \text{ at } T = 0 \\ \frac{\lambda}{4}h^2(h - v_c)^2 \text{ at } T = T_c \end{cases} \tag{6.4}$$

At the critical temperature, the energy is minimized when

$$\partial_z^2 h = 2\lambda h(h - v_c)(2h - v_c) \tag{6.5}$$

Substituting the ansatz (6.3) for $h$ into (6.5), one finds that it is a solution provided that

$$L = \frac{1}{\sqrt{\lambda} v_c} \sim \frac{1}{m_h} \tag{6.6}$$

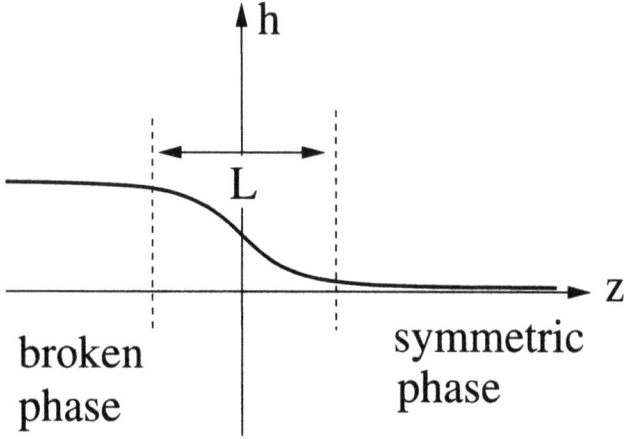

Fig. 20. Shape of Higgs field in the bubble wall.

We can also solve for the phases $\theta_i$. In fact, only the difference

$$\theta = -\theta_1 + \theta_2 \tag{6.7}$$

appears in the potential (5.39), due to $U(1)_y$ gauge invariance; it can be rewritten as

$$
\begin{aligned}
V(h_1, h_2, \theta) &= -\frac{1}{2} \sum_i \mu_i^2 h_i^2 - \mu_3^2 \cos(\theta + \phi) h_1 h_2 \\
&= \frac{1}{8} \sum_i \lambda_i h_i^4 + \frac{1}{4} (\lambda_3 + \lambda_4 + \lambda_5 \cos 2\theta) h_1^2 h_2^2
\end{aligned} \tag{6.8}
$$

where now all parameters are real, and the invariant CP phase $\phi$ has been shown explicitly. (Of course, we could move it into the $\lambda_5$ term if we wanted to, by shifting $\theta \to \theta - \phi$.) If $h_i = \left(\begin{smallmatrix} c_\beta \\ s_\beta \end{smallmatrix}\right) h$, then

$$
\begin{aligned}
c_\beta^2 \partial_z \left( h^2 \partial_z \theta_1 \right) &= \frac{\partial V}{\partial \theta_1} \\
s_\beta^2 \partial_z \left( h^2 \partial_z \theta_2 \right) &= \frac{\partial V}{\partial \theta_2} = -\frac{\partial V}{\partial \theta_1}
\end{aligned} \tag{6.9}
$$

and $c_\beta^2 \theta_1 + s_\beta^2 \theta_2 = 0$, while $\partial_z (h^2 \partial_z \theta) = (c_\beta^{-2} + s_\beta^{-2}) \frac{\partial V}{\partial \theta}$. These equations allow us to solve for $\theta_1$ and $\theta_2$ in the vicinity of the bubble wall. Notice that the trivial solution $\theta = $ constant is prevented if $\phi \neq 0$ since

$$\frac{\partial V}{\partial \theta} = \mu_3^2 h_1 h_2 \sin(\theta + \phi) - \frac{1}{2}\lambda_5 \sin 2\theta h_1^2 h_2^2 \tag{6.10}$$

vanishes at different values of $\theta$ depending on the value of $z$. In fact, as $z \to \infty$, $\theta \to -\phi$, while for $z \to -\infty$, $\theta$ approaches a different value due to the nonvanishing of the quartic term in the broken phase. This guarantees that there will be a nontrivial profile for $\theta$ as well as $h$ in the bubble wall.

The result is a complex phase for the masses of the quarks inside the wall. Focusing on the top quark since it couples most strongly to the Higgs, this mass term is given by

$$\frac{y}{\sqrt{2}}h_2(z)\bar{t}e^{i\theta_2\gamma_5}t = \frac{y}{\sqrt{2}}h_2(z)\bar{t}\,(\cos\theta + i\gamma_5 \sin\theta_2)\,t \tag{6.11}$$

This means that the propagation of the quark inside the wall will have CP violating effects. We can no longer remove $\theta_2$ by a field redefinition, as we did in section 2, because $\theta_2$ is a function of $z$, not just a constant. If we try to remove it using $t \to e^{-i\theta_2\gamma_5/2}t$, although this removes $\theta_2$ from the mass term, it also generates a new interaction from the kinetic term,

$$\bar{t}i\partial t \to \bar{t}i\partial t + \frac{1}{2}\bar{t}\partial\theta_2\gamma_5 t \tag{6.12}$$

Any physics we derive must the be same using either description, but we will keep the phase in the mass term as this has a clearer interpretation.

Now let's consider the Dirac equation, which can be written with the help of the chiral projection operators $P_{L,R} = \frac{1}{2}(1 \mp \gamma_5)$ as

$$(i\partial - mP_L - m^* P_R)\psi = 0 \tag{6.13}$$

using the complex mass

$$m = \frac{y}{\sqrt{2}}h_2(z)e^{i\theta_2(z)} \tag{6.14}$$

Decompose the Dirac spinor into its chiral components as

$$\psi = e^{-iEt}\begin{pmatrix} R_s \\ L_s \end{pmatrix} \otimes \chi_s \tag{6.15}$$

where $\chi_s$ is a 2-component spinor and $s = \pm 1$ labels spin up and down along the direction of motion, which we can take to be $\hat{z}$. The Dirac equation becomes two coupled equations,

$$\begin{aligned}(E - is\partial_z)L_s &= mR_s \\ (E + is\partial_z)R_s &= m^*L_s\end{aligned} \tag{6.16}$$

These can be converted to uncoupled second-order equations,

$$\left[(E + is\partial_z)\frac{1}{m}(E - is\partial_z)\right] L_s = 0$$

$$\left[(E - is\partial_z)\frac{1}{m^*}(E + is\partial_z)\right] R_s = 0 \qquad (6.17)$$

We see that the LH and RH components can propagate differently in the bubble wall—a signal of CP violation. We can think of it as quantum mechanical scattering,

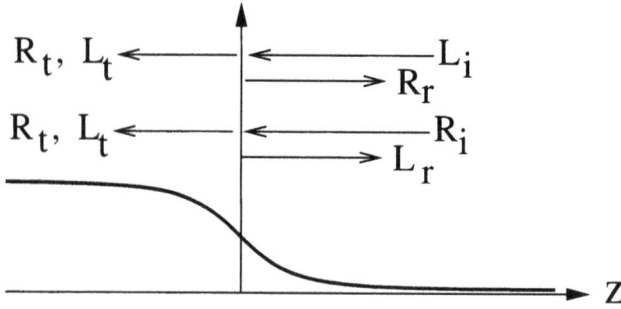

where left-handed components incident from the symmetric phase reflect into right-handed components and vice-versa. One can see that as $z \to \infty$, $m \to 0$, so it is consistent to have RH and LH components moving in opposite directions. On the broken-phase side of the wall, both components must be present because the mass mixes them.

However, quantum mechanical (QM) refelection is only a strong effect if the potential is sharp compared to the de Broglie wavelength. In our case,

$$\lambda \sim \frac{1}{3T}, \quad L \sim \frac{1}{m_h} \qquad (6.18)$$

while for strong scattering we need

$$\frac{L}{\lambda} = \frac{3T}{m_h} < 1 \qquad (6.19)$$

which is not true, except in low-energy part of the particle distribution functions. Thus only some low-momentum fraction of the quarks in the thermal plasma will experience significant QM scattering. However there is a classical effect which applies to the dominant momentum components with $p \sim 3T$: a CP-violating force. It can be deduced by solving the Dirac equation in the WKB approximation. Making the ansatz

$$L_s = A_s(z)e^{i\int^z p(z')dz' - iEt} \tag{6.20}$$

and substituting it into the equation of motion, we can solve for $p(z)$ in an expansion in derivatives of the background fields $|m(z)|$ and $\theta(z)$. This approach was pioneered in ref. [48] and refined in [49]. We find that the canonical momentum is given by

$$p(z) \cong p_0 + s_c \frac{sE \pm p_0}{2p_0}\theta' + O\left(\frac{d^2}{dz^2}\right) \tag{6.21}$$

where

$$p_0 = \text{sign } p\sqrt{E^2 - |m^2(z)|} \tag{6.22}$$

is the kinetic momentum, $s_c = +1 \, (-1)$ for particles (antiparticles), and the $\pm$ in (6.21) is $+$ for LH and $-$ for RH states—but it will be shown that this term has no physical significance.

The dispersion relation corresponding to (6.21) is

$$E = \sqrt{(p \pm s_c\theta'/2)^2 + |m|^2} - \frac{1}{2}ss_c\theta' \tag{6.23}$$

and the group velocity is

$$v_g = \left(\frac{\partial E}{\partial p}\right)_z = \frac{p_0}{E}\left(1 + ss_c\frac{|m|^2\theta'}{2E^3}\right) \tag{6.24}$$

This is a physically meaningful result: particles with different spin or $s_c$ are speeded up or slowed down due to the CP-violating effect. Notice that the effect vanishes if $\theta' = 0$ or $m = 0$. Furthermore the effect of the $\pm\theta'/2$ term in (6.21) has dropped out, showing that indeed it is unphysical. It can be thought of as a being like a gauge transformation. One can understand this from the fact that it has no dependence on $m$. In the limit that $m \to 0$, the phase $\theta$ has no meaning, so any physical effects of $\theta$ must be accompanied by powers of $m$. The same can be said concerning derivatives of $\theta$. We know that if $\theta$ is constant, it can be removed by a field redefinition, hence any physical effects can only depend on $\theta'$ or higher derivatives.

We can furthermore identify a CP-violating force which is responsible for the above effect. The Hamilton equation which gives the force is

$$\dot{p} = -\left(\frac{\partial E}{\partial z}\right)_p = -\frac{(m^2)'}{2E} + ss_c\frac{(m^2\theta')'}{2E^2} \tag{6.25}$$

The first term on the r.h.s. is the CP-conserving force due to the spatial variation of the mass, while the second term is the CP-violating force. This is the main result which allows us to compute the baryon asymmetry, since $\dot{p}$ appears in the Boltzmann equation,

$$\frac{\partial f}{\partial t} + \frac{p}{m}\frac{\partial f}{\partial z} + \dot{p}\frac{\partial f}{\partial p} = \text{collision terms} \qquad (6.26)$$

To set up the Boltzmann equations we make an ansatz for the distribution function of particle species $i$, in the rest frame of the bubble wall,

$$f_i = \frac{1}{e^{\beta(\gamma(E_i + v_w p_z) - \mu_i)} \pm 1} + \delta f_i \qquad (6.27)$$

which can be understood as follows. $\gamma(E_i + v_w p_z)$ is the Lorentz transformation of the energy when we go to the rest frame of the wall, which is moving at speed $v_w$, and $\mu_i$ is the chemical potential. The first term in (6.27) tells us about the departure from chemical equilibrium induced by the CP violating force, while the term $\delta f_i$ is included to parametrize the departure from kinetic equilibrium. We are free to impose the constraint

$$\int d^3x\, \delta f_i = 0 \qquad (6.28)$$

since $\mu_i$ already accounts for the change in the overall normalization of $f_i$.

The idea is to make a perturbative expansion where

$$\mu_i \sim \delta f_i \sim m^2\theta' \qquad (6.29)$$

We will thus linearize the Boltzmann equations in $\mu_i$ and $\delta f_i$, and take the differences between particles and antiparticles. The next step is to make a truncation of the full Boltzmann equations (BE) by taking the first two moments,

$$\int d^3p\, (BE) \quad \text{and} \quad \int d^3p\, p_z\, (BE) \qquad (6.30)$$

The reason for taking two moments is that we have two functions to determine, $\mu(t,z)$ and $\delta f(t,z)$; a more accurate treatment would involve more parameters in the ansatz for the distribution functions and taking higher moments of the BE. In any case, the truncation is necessary because the full BE is an integro-differential equation, which we do not know how to solve, whereas the moments are conventional first-order PDE's. Moreover in the rest frame of the wall, the equations no longer depend on time and therefore become ODE's. We can solve for $\delta f$ in terms of $\mu$ and further convert the coupled first-order ODE's into uncoupled second-order ODE's for the chemical potentials. These are known as diffusion equations, and they have the form

$$-\left(D_i \xi_i'' + v_w \xi_i'\right) + \sum_k \Gamma_k(\pm \xi_i \pm \xi_j \pm \ldots) + \ldots = S_i \qquad (6.31)$$

where $D_i$ is the diffusion constant, defined in terms of the total scattering rate $\Gamma_i$ for particle $i$ (in the thermal plasma) as

$$D_i = \frac{\langle v_z^2 \rangle}{\Gamma_i} \qquad (6.32)$$

with the thermal average $\langle \cdot \rangle$ applied to the $z$-component of the particle's squared velocity; $v_w$ is the wall velocity, usually in the range $0.1 - 0.01$ depending on details of the model; $\xi_i$ is the rescaled chemical potential

$$\xi_i = \frac{\mu_i(z)}{T} \qquad (6.33)$$

related to the particle densities by

$$n_i - n_{\bar{i}} = \frac{T^3}{6} \xi \begin{cases} 1, & \text{fermions} \\ 2, & \text{bosons} \end{cases} \qquad (6.34)$$

and $\Gamma_k$ is the $k$th reaction rate for particle in which the particle $i$ is either produced or consumed—for example, if $i + j \rightarrow l + m$, then we would write $\Gamma_k(\xi_l + \xi_m - \xi_i - \xi_j)$; and $S_i$ is the source term due to the CP violating force, which turns out to be

$$S_i \cong \frac{v_w D_i}{\langle v_z^2 \rangle T} \left\langle v_z \frac{(m^2 \theta')'}{2E^2} \right\rangle' \qquad (6.35)$$

The source term has an asymmetric shape similar to figure 21, when $m(z)$ and $\theta(z)$ have kink-like profiles.

We can solve the diffusion equation using Green's functions,

$$\left(D \frac{d^2}{dz^2} + v_w \frac{d}{dz} + \Gamma\right) G(z - z_0) = \delta(z - z_0) \qquad (6.36)$$

where

$$G(x) = \frac{D^{-1}}{k_+ - k_-} \begin{cases} e^{-k_+ x}, & x > 0, \\ e^{-k_- x}, & x < 0 \end{cases}$$

$$k_\pm = \frac{v_w}{2D}\left(1 \pm \sqrt{1 + \frac{2\Gamma D}{3 v_w^2}}\right) \qquad (6.37)$$

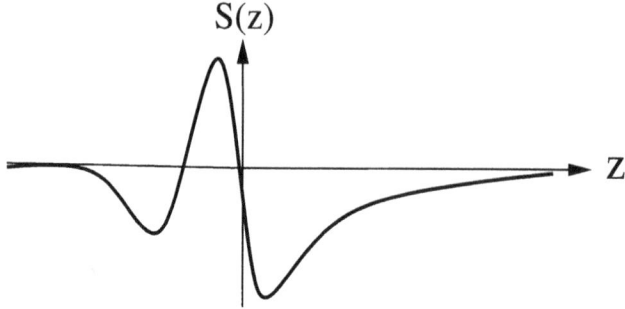

Fig. 21. Typical shape of source term in diffusion equation.

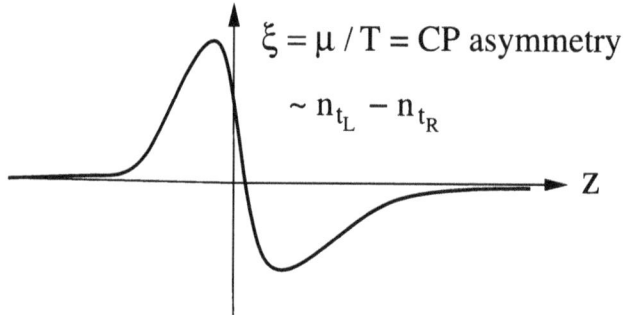

Fig. 22. Typical shape of the CP asymmetry from solving the diffusion equations.

so

$$\xi(z) = \int_{\infty}^{\infty} dz_0 \, G(z - z_0) S(z_0) \tag{6.38}$$

This tells us the asymmetry in one helicity of $t$ quarks, say $(+)$, which must be equal and opposite to that for the other helicity $(-)$. So far we have calculated the CP asymmetry near the wall. It has a shape similar to that shown in figure 22.

The final step is to compute the baryon asymmetry. Since the $q_L$ asymmetry biases sphalerons, we get

$$\frac{dn_B}{dt} \sim 3\Gamma_{sph}\xi - c\Gamma_{sph}\frac{n_B}{T^2} \tag{6.39}$$

where $3\Gamma_{sph}\xi$ is the term which causes the initial conversion of the CP asymmetry into a baryon asymmetry, and $-c\Gamma_{sph}n_B/T^2$ is the washout term, which would eventually relax $B$ to zero if sphalerons did not go out of equilibrium inside the bubbles. In this equation, we should think of $\Gamma_{sph}$ as being a function of $t$ and $z$, which at any given position $z$ abruptly decreases at the time $t$ when the wall passes $z$, so the position of interest goes from being in the symmetric phase to being in the broken phase. (The constant $c$ is approximately 10.) In the approximation that the sphaleron rate is zero in the broken phase, and ignoring the washout term, we should integrate (6.39) at a given point in space until the moment $t_w$ when the wall passes. Trading the time integral for an integral over $z$, we get

$$n_B = \int_{-\infty}^{t_w} dt\, \frac{dn_B}{dt} = \int_0^\infty dz\, \frac{1}{\dot{z}}\frac{dn_B}{dt} = \frac{3\Gamma_{sph}}{v_w} \int_0^\infty \xi(z) \tag{6.40}$$

This can be generalized to the case where the washout term is not ignored, giving an additional factor $e^{-z(c/v_w)(\Gamma_{sph}/T^2)}$ in the integrand. The entire analysis for the 2HDM has been carried out in [46].

The above derivation of the transport equations gives a clear intuitive picture of the physics, but more rigorous treatments have been given in [50].

### 6.1. EDM constraints

An interesting experimental prediction is that CP violation in EWBG should also lead to an observable EDM for the neutron and perhaps other particles. One needs the invariant CP phase $\phi$ to be in the range $10^{-2} - 1$ for sufficient baryogenesis, which leads to mixing between CP-even and odd Higgs bosons (scalars and pseudoscalars). Weinberg [51] notes that if any of the propagators $\langle H_2^+ H_1^{+*}\rangle$, $\langle H_2^+ H_1^{+*}\rangle$, $\langle H_2^+ H_1^{+*}\rangle$, $\langle H_2^+ H_1^{+*}\rangle$, $\langle H_2^+ H_1^{+*}\rangle$ have imaginary parts then CP is violated, giving rise to quark EDM's at one loop, through the diagram

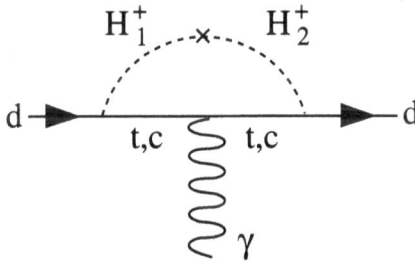

The CP-violating propagators occur because the invariant phase $\phi$ appears in the mass matrix of the Higgs bosons. However this diagram with charged scalars requires 3 or more Higgs doublets since one of the charged fields is a Goldstone boson, and can be set to zero by going to unitary gauge. A similar diagram with neutral Higgs exchange could work, but it is suppressed for the relevant quarks

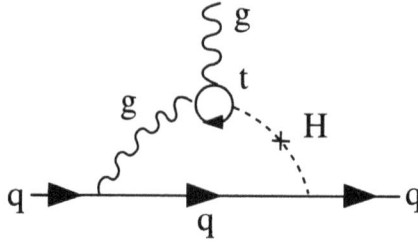

Fig. 23. The Barr-Zee [53] contribution to the neutron EDM.

($u$ and $d$, which are valence quarks of the neutron) because the Yukawa couplings are small. Moreover we should avoid coupling $H_1$ and $H_2$ to the same flavor of quarks because this tends to give large FCNC contributions.

It was realized that two-loop diagrams give the largest contribution (see [52], following work of [53]), which look like figure 23, where the cross on the Higgs line is the CP-violating mass insertion. These contribute to the *chromo*electric dipole moment of the quark,

$$\text{CEDM} = -\frac{i}{2} g_q^g \bar{q} \sigma_{\mu\nu} \gamma_5 G^{\mu\nu} q \tag{6.41}$$

where $G^{\mu\nu}$ is the gluon field strength. The quark CEDM's in turn contribute to the neutron EDM

$$d_n = (1 \pm 0.5)\left(0.55e\, d_u^g + 1.1e\, d_d^g\right) + O\left(d_u^\gamma,\ d_d^\gamma\right) \tag{6.42}$$

where the factor $(1 \pm 0.5)$ is due to hadronic uncertainties. There are various ways of estimating the quark contribution to the neutron EDM; (6.42) uses QCD sum rules. See [54] for a review.

Ref. [46] finds that the experimental limit on $d_n$ is close to being saturated for values of $\phi$ which give enough baryogenesis: $d_n = (1.2 - 2.2) \times 10^{-26} e$ cm, which should be compared to the experimental limit $|d_n| < 3 \times 10^{-26} e$ cm.

There are extra Higgs particles with masses $\sim 300$ GeV whose strong couplings to the light Higgs give rise to a first order phase transition, as we have discussed in the previous section. These are not CP eigenstates, so they have interactions which could distinguish them from the Higgs sector of the MSSM.

## 7. EWBG in the MSSM

We have already discussed how a light $\tilde{t}_R$ can give a strongly first order EWPT in the MSSM. But what is the mechanism for baryogenesis? We cannot use the same one as in the 2HDM because in the MSSM, the Higgs Lagrangian has no $(H_1^\dagger H_2)^2$ coupling at tree level. Although such a coupling is generated at one loop, its coefficient is too small to get enough baryon production.

The principal mechanism which has been considered is chargino-wino reflection at the bubble wall. The first computation of this effect using the WKB approach, described in the previous section, was done by ref. [55]. (For earlier approaches, based on quantum mechanical scattering or the quantum Boltzmann equation, see [56] or [57], respectively.) The WKB method has been carefully scrutinized and verified by [44] and a number of other papers [58]. Charginos/winos have a $2 \times 2$ Dirac mass term

$$
\left( \overline{\widetilde{W}_R^+}, \overline{\widetilde{h}_{1,R}^+} \right) \begin{pmatrix} \tilde{m}_2 & g H_2(z) \\ g H_1(z) & \mu \end{pmatrix} \begin{pmatrix} \widetilde{W}_L^+ \\ \tilde{h}_{2,L}^+ \end{pmatrix} + \text{h.c.} \tag{7.1}
$$

where $\mu$ and $\tilde{m}_2$ are complex. Denoting the $2 \times 2$ Dirac mass matrix by $\mathcal{M}$, we diagonalize $\mathcal{M}$ locally in the bubble wall using a $z$-dependent similarity transformation

$$
U^\dagger \mathcal{M} V = \begin{pmatrix} m_+ e^{i\theta_+(z)} & \\ & m_- e^{i\theta_-(z)} \end{pmatrix} \tag{7.2}
$$

so similarly to the top quark in the 2HDM, charginos and winos experience a CP-violating force in the wall, where

$$
m_\pm^2 \partial_z \theta_\pm = \pm g^2 \frac{\text{Im}(\tilde{m}_2 \mu)}{m_+^2 - m_-^2} \frac{\partial}{\partial z} \left( H_1(z) H_2(z) \right) \tag{7.3}
$$

However the resulting chiral asymmetry in $\tilde{h}$ or $\widetilde{W}$ does not have a direct effect on sphalerons. Instead, we must take into account decays and scatterings such as

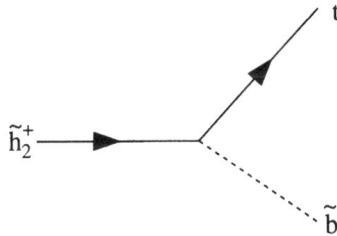

which transfer part of the $\tilde{h}$ CP asymmetry into quarks. One must solve the coupled Boltzmann equations for all the species to find the quark CP asymmetry which biases sphalerons.

Because of the indirectness of the effect, baryon generation is less efficient in the MSSM than in the 2HDM. The effect is largest when $m_+^2 \cong m_-^2$, which

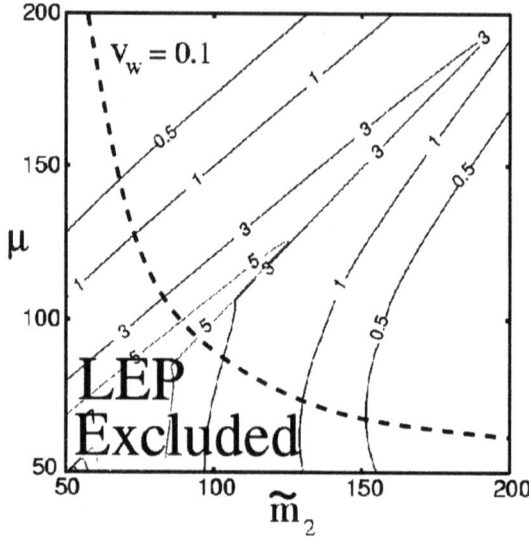

Fig. 24. Contours of constant $\eta \times 10^{10}$ from baryogenesis in the MSSM.

happens when $m_2 \cong \mu$, as shown in figure 24, taken from [59]. The figure shows the contours of $\eta$ in the plane of $\mu$ and $\tilde{m}_2$, with the LEP limit on the chargino mass superimposed, $m_{\chi^\pm} > 104$ GeV. This figure was prepared before the WMAP determination of $\eta$, at which time the preferred value was $3 \times 10^{-10}$. Hence the contours only go up to $5 \times 10^{-10}$ in the LEP-allowed region. It is possible to choose other values of the MSSM parameters to slightly increase the baryon asymmetry to the measured value; for example taking $\tan\beta \lesssim 3$ helps, and assuming a wall thickness (which we have not tried to compute carefully, but is only parametrized) $L \sim 6/T$ increases $\eta$. The conclusion is that all relevant parameters, $\arg(\tilde{m}_2\mu)$, $v_w$, $L$, $\tan\beta$, must be at their optimal values to get a large enough baryon asymmetry. The MSSM is nearly ruled out for baryogenesis.

In fact some additional tuning is required since the large CP-violating phases which are needed would lead to large EDM's through the diagram

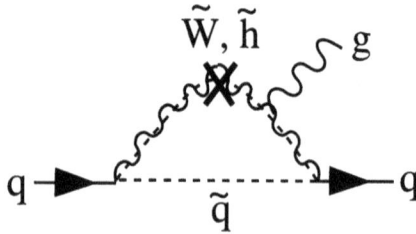

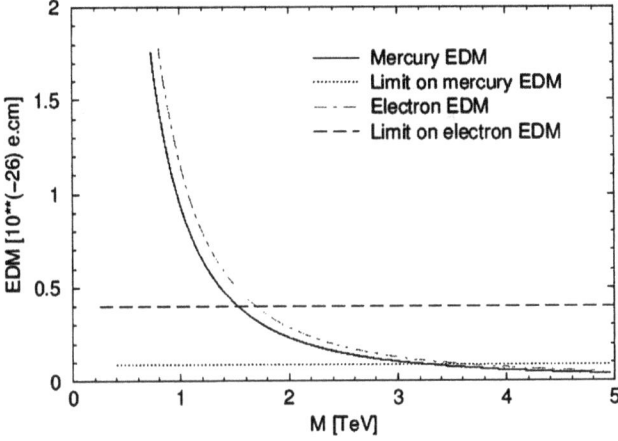

Fig. 25. Bound on squark masses from EDM's assuming maximal phase in the $\mu$ parameter, from [60].

If the squark $\tilde{q}$ is light, the invariant phase (which we can refer to as the phase of $\mu$) must be small, $\theta_\mu \lesssim 10^{-2}$, whereas we need $\theta_\mu \sim 1$. To suppress the EDM's it is necessary to take the squarks masses in the range $m_{\tilde{q}} \gtrsim$ TeV. Roughly, the bound goes like [54]

$$\theta_\mu \left( \frac{1 \text{ TeV}}{m_{\tilde{q}}} \right)^2 < 1 \tag{7.4}$$

In more detail, [60] finds a bound of $m_{\tilde{q}} > 3$ TeV from the EDM of Hg, if $\theta_\mu$ is maximal, as shown in figure 25. This does not conflict with the need to keep $\tilde{b}_R$ light because we only need to make $d_u^g$ and $d_d^g$ (the CEDM's of the valence quarks of the neutron) small, while the CEDM's of higher generation quarks might be large.

Assuming heavy first and second generation squarks, the largest contribution to EDM's can come from two-loop diagrams of the Barr-Zee type (figure 23), where the top loop is replaced by a chargino loop [61]. This contribution does not decouple when the squarks are heavy, and it gives stringent constraints on EWBG in the MSSM through the electron EDM. A heavy charged Higgs is needed to suppress this diagram enough to allow for maximal CP violation in the $\mu$ parameter.

However we can loosen the constraints on the MSSM by adding a singlet Higgs field $S$ with the superpotential (5.37) and potential (5.38), where the soft SUSY-breaking terms include CP violating terms of the form $\lambda A S H_1 H_2$ and

$kAS^3$. This model has much more flexibility to get a strong EWPT and large CP violation [44]. We no longer need 1-loop effects to generate cubic terms in the Higgs potential; already at tree level there are cubic terms $H_1 H_2 S$ and $S^3$. If $S$ participates in the EWPT by getting a VEV, these cubic terms help to strengthen the phase transition, and it is possible to have $v_c/T_c > 1$ even with heavy Higgses.

The NMSSM also provides many new CP-violating phases in the Higgs potential due to complex couplings $k$, $\mu$ and $A_k$, and the model works similarly to the 2HDM, generating a CP asymmetry directly in the top quark.

## 8. Other mechanisms; Leptogenesis

Unfortunately there was not enough time to cover other interesting ideas for baryogenesis, including the very elegant idea of Affleck and Dine which makes use of flat directions in supersymmetric models to generate baryon number very efficiently [62].

A very popular mechanism for baryogenesis today is leptogenesis, invented by Fukugita and Yanagida in 1986 [63]; see [64] for a recent review. I am not able to do justice to the subject here, but it deserves mention. Leptogenesis is a very natural mechanism, which ties in with currently observed properties of neutrinos; hence its popularity. Unfortunately, the simplest versions of leptogenesis occur at untestably high energies, similar to GUT baryogenesis. Although it is possible to bring leptogenesis down to the TeV scale [65], it requires a corresponding decrease in Yukawa couplings which renders the theory still untestable by direct laboratory probes.

In leptogenesis, the first step is to create a lepton asymmetry, which can be done without violating electric charge neutrality of the universe since the asymmetry can initially reside in neutrinos. We know however that sphalerons violate $B + L$, while conserving $B - L$. Therefore, an excess in $L$ will bias sphalerons to produce baryon number, just as an excess in left-handed $B$ did within electroweak baryogenesis. Roughly, we can expect that any initial asymmetry $L_i$ in $L$ will be converted by sphalerons into a final asymmetry in $B$ and $L$ given by

$$B_f \sim -\frac{1}{2}L_i, \quad L_f \sim \frac{1}{2}L_i \tag{8.1}$$

This estimate is not bad. A more detailed analysis gives [24, 66]

$$B_f = -L_i \left\{ \begin{array}{l} \frac{28}{79}, \text{ SM} \\ \frac{8}{23}, \text{ MSSM} \end{array} \right\} \sim -\frac{1}{3}L_i \tag{8.2}$$

since sphalerons couple to $B_L + L_L$ and we must consider the conditions of thermal equilibrium between all species.

Leptogenesis is conceptually very similar to GUT baryogenesis, but it relies on the decays of heavy (Majorana) right-handed neutrinos, through the diagrams

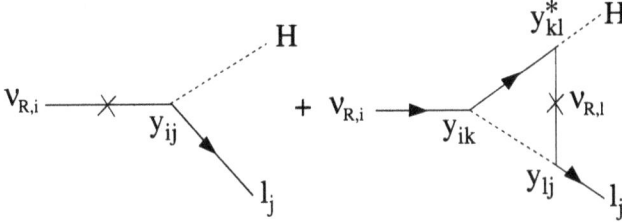

The cross represents an insertion of the $L$-violating heavy neutrino Majorana mass, which makes it impossible to assign a lepton number to $\nu_R$, and hence makes the decay diagram $L$-violating. The principles are the same as in GUT baryogenesis. Physically this is a highly motivated theory because (1) heavy $\nu_R$ are needed to explain the observed $\nu_L$ masses, and (2) $\nu_R$'s are predicted by SO(10) GUT's.

The first point is well-known; it is the seesaw mechanism, based on the neutrino mass terms

$$y_{ij} \bar{\nu}_{R,i} H L_j + \text{h.c.} - \frac{1}{2} \left( M_{ij} \bar{\nu}^c_{R,i} \nu_{R,j} + \text{h.c.} \right) \tag{8.3}$$

When the Higgs gets its VEV, the mass matrix in the basis $\nu_L$ and $\nu_R$ is

$$\begin{pmatrix} 0 & m_D \\ m_D^T & M \end{pmatrix} \tag{8.4}$$

where the Dirac mass matrix is $(m_D)_{ij} = v y_{ij}$ (and we must take the complex conjugate of (8.3) to get the mass matrix for the conjugate fields $\nu_R^c$, $\nu_L^c$). This can be partially diagonalized to the form

$$\begin{pmatrix} -m_D M^{-1} m_D^T & 0 \\ 0 & M \end{pmatrix} \tag{8.5}$$

so the light neutrino masses are $O(m_D^2/M) \sim y^2 v^2/M$. More exactly, they are the eigenvalues of the matrix $v^2 (y M^{-1} y^T)_{ij}$, up to phases. We can estimate the size of the $\nu_R$ mass scale:

$$M \sim \frac{y^2 v^2}{m_\nu} \sim y^2 \times 10^{14} \text{ GeV} \tag{8.6}$$

where we used the tau neutrino mass, $m_\nu = 0.05$ eV, as measured through atmospheric neutrino mixing. Since $y$ can be small, it is possible to make $M$ smaller than the gravitino bound on the reheat temperature after inflation.

It is usually assumed that there is a hierarchy such that $M_1 \ll M_2, M_3$, since this simplifies the calculations, and it seems like a natural assumption. Then $\nu_{R,1}$ is the last heavy neutrino to decay out of equilibrium, and it requires the lowest reheat temperature to be originally brought into equilibrium.

To find the lepton asymmetry, we need the dimensionless measure of CP violation

$$\epsilon_1 = \frac{|\mathcal{M}_{\nu_{R,1} \to lH}|^2 - |\mathcal{M}_{\nu_{R,1} \to \bar{l}\bar{H}}|^2}{|\mathcal{M}_{\nu_{R,1} \to \text{anything}}|^2} = -\frac{3}{16\pi} \sum_i \frac{\text{Im}(y^\dagger y)_{i1}^2}{(y^\dagger y)_{11}} \frac{M_1}{M_i} \tag{8.7}$$

It can be shown that $\epsilon_1$ has a maximum value of $\frac{3}{16\pi} \frac{M_1 m_3}{v^2}$.

The baryon asymmetry can be expressed as

$$\eta = 10^{-2} \epsilon_1 \kappa \tag{8.8}$$

where $\kappa$ is the efficiency factor, which takes into account the washout processes

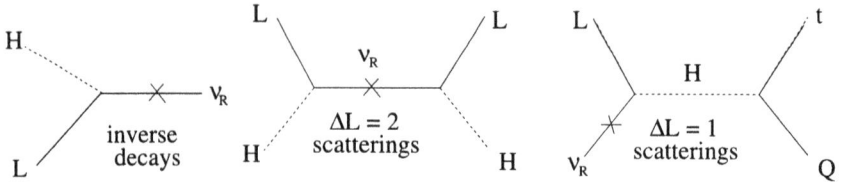

It is interesting that leptogenesis can be related to the measured neutrino masses. The rate of heavy neutrino decays is given by

$$\Gamma_D = \frac{\tilde{m}_1 M_1^2}{8\pi v^2} \tag{8.9}$$

where

$$\tilde{m}_1 = \frac{(m_D^\dagger m_D)_{11}}{M_1} \tag{8.10}$$

is an "effective" neutrino mass; it is not directly related to the neutrino mass eigenvalues (which requires the transpose of $m_D$ instead of the Hermitian conjugate, and also (8.10) is a matrix element rather than an eigenvalue), though it should be of the same order as the light neutrino masses. It can be shown that

$m_1 < \tilde{m}_1 < m_3$, and that the condition for $M_1$ to decay out of equilibrium is satisfied if

$$\tilde{m}_1 < m_* \equiv \frac{16\pi^{5/2}}{2\sqrt{5}} \frac{v^2}{M_p} = 1.1 \times 10^{-3} \text{eV} \tag{8.11}$$

The ratio $\tilde{m}_1/m_*$ enters into the efficiency factor $\kappa$, and for $\tilde{m}_1 > m_*$,

$$\kappa \cong 2 \times 10^{-2} \left( \frac{0.01 \text{ eV}}{\tilde{m}_1} \right) \tag{8.12}$$

Another interesting connection to the neutrino masses comes through the $\Delta L = 2$ scattering diagram, whose rate does relate to the actual light masses,

$$\Gamma_{\Delta L=2} = \frac{T^3}{\pi^2 v^4} \sum_{i=e,\mu,\tau} m_{\nu_i}^2 \equiv \frac{T^3}{\pi^2 v^4} \bar{m}^2 \tag{8.13}$$

There is a bound on the r.m.s. neutrino mass from leptogenesis, due to this washout process,

$$\bar{m} \lesssim 0.3 \text{ eV} \tag{8.14}$$

which intriguingly is consistent with and close to the bound on the sum of the neutrino masses from the CMB (see for example [67]).

It is also possible to derive a lower bound on the lightest right-handed neutrino mass, $M_1 > 2 \times 10^9$ GeV, assuming maximum value of the efficiency factor [68]. This follows from expressing the maximum value of $\epsilon_1$ in the form

$$\epsilon_1^{\max} < \frac{3}{16\pi} \frac{M_1}{v^2} \left( \frac{\Delta m_{\text{atm}}^2 + \Delta m_{\text{sol}}^2}{m_3} \right) \tag{8.15}$$

and using $\Delta m_{\text{atm}}^2 = 2.5 \times 10^{-3} \text{ eV}^2$, $m_3 \geq \Delta m_{\text{atm}}^2$.

These results which are consistent with the known neutrino masses (and marginally consistent with the gravitino bound) are suggestive that leptogenesis could indeed be the right theory. Unfortunately, there is no way to find out for sure through independent laboratory probes of the particle physics, as we hope to do in electroweak baryogenesis. On the other hand, it is much easier to quantitatively predict the baryon asymmetry in leptogenesis than in electroweak baryogenesis; the Boltzmann equations are much simpler. One need only solve two coupled equations, for the heavy decaying neutrino density, and the produced lepton number,

$$\frac{dN_{N_1}}{dz} = -(D + S)\big(N_{N_1} - N_{N_1}^{eq}\big)$$

$$\frac{dL}{dz} = -\epsilon_1 D\big(N_{N_1} - N_{N_1}^{eq}\big) - WL \tag{8.16}$$

where $z = M_1/T$, $D$ stands for the rate of decay and inverse decay, $S$ is the rate of $\Delta = 1$ scatterings (which can also produce part of the lepton asymmetry), and $W$ is the washout terms which includes $\Delta L = 2$ scatterings; see [69]. Care must be taken when defining the $s$-channel $\Delta L = 2$ processes which contribute to the washout because when the exchanged heavy neutrino goes on shell, these are the same as inverse decays followed by decays, which have been separately counted; see [70].

## Acknowledgments

I thank the organizers and students of Les Houches for a very enjoyable and stimulating week. Thanks to S. Dimopoulos, S. Huber, and S. Weinberg for useful information while preparing these lectures, and to R. Cyburt for providing figure 1. I am grateful to B. Grzadkowski, A. Pilaftsis, P. Mathews, S. Menary, M. Schmidt, A. Strumia and H. Yamashita for pointing out corrections and improvements to the manuscript. I also thank K. Kainulainen for many years of productive collaboration on this subject.

## References

[1] Particle Data Group, http://pdg.lbl.gov/2006/reviews/contents_sports.html#astroetc.

[2] R. H. Cyburt, "Primordial Nucleosynthesis for the New Cosmology: Determining Uncertainties and Examining Concordance," Phys. Rev. D **70**, 023505 (2004) [arXiv:astro-ph/0401091].

[3] W. Hu and S. Dodelson, "Cosmic Microwave Background Anisotropies," Ann. Rev. Astron. Astrophys. **40**, 171 (2002) [arXiv:astro-ph/0110414].

[4] E. W. Kolb and M. S. Turner, "The Early universe," Front. Phys. **69**, 1 (1990).

[5] A. D. Sakharov, "Violation of CP Invariance, C Asymmetry, and Baryon Asymmetry of the Universe," Pisma Zh. Eksp. Teor. Fiz. **5**, 32 (1967) [JETP Lett. **5**, 24 (1967 SOPUA,34,392-393.1991 UFNAA,161,61-64.1991)].

[6] S. M. Barr, "Comments On Unitarity And The Possible Origins Of The Baryon Asymmetry Of The Universe," Phys. Rev. D **19**, 3803 (1979).

[7] C. A. Baker *et al.*, "An improved experimental limit on the electric dipole moment of the neutron," arXiv:hep-ex/0602020.

[8] R. J. Crewther, P. Di Vecchia, G. Veneziano and E. Witten, "Chiral Estimate Of The Electric Dipole Moment Of The Neutron In Quantum Chromodynamics," Phys. Lett. B **88**, 123 (1979) [Erratum-ibid. B **91**, 487 (1980)].

[9] B. C. Regan, E. D. Commins, C. J. Schmidt and D. DeMille, "New limit on the electron electric dipole moment," Phys. Rev. Lett. **88**, 071805 (2002).

[10] M. V. Romalis, W. C. Griffith, J. P. Jacobs and E. N. Fortson, "New limits on CP violation from a search for a permanent electric dipole moment of Hg-199," *Prepared for International Nuclear Physics Conference (INPC 2001): Nuclear Physics and the 21st Century, Berkeley, California, 30 Jul – 3 Aug 2001.*

[11] M. Yoshimura, "Unified Gauge Theories And The Baryon Number Of The Universe," Phys. Rev. Lett. **41**, 281 (1978) [Erratum-ibid. **42**, 746 (1979)].

[12] D. Toussaint, S. B. Treiman, F. Wilczek and A. Zee, "Matter – Antimatter Accounting, Thermodynamics, And Black Hole Radiation," Phys. Rev. D **19**, 1036 (1979).

[13] S. Dimopoulos and L. Susskind, "Baryon Asymmetry In The Very Early Universe," Phys. Lett. B **81**, 416 (1979); "On The Baryon Number Of The Universe," Phys. Rev. D **18**, 4500 (1978).

[14] S. Weinberg, "Cosmological Production Of Baryons," Phys. Rev. Lett. **42**, 850 (1979).

[15] M. Yoshimura, "Origin Of Cosmological Baryon Asymmetry," Phys. Lett. B **88**, 294 (1979).

[16] A. Y. Ignatiev, N. V. Krasnikov, V. A. Kuzmin and A. N. Tavkhelidze, "Universal CP Noninvariant Superweak Interaction And Baryon Asymmetry Of The Universe," Phys. Lett. B **76**, 436 (1978).

[17] D. V. Nanopoulos and S. Weinberg, "Mechanisms For Cosmological Baryon Production," Phys. Rev. D **20**, 2484 (1979).

[18] E. W. Kolb and S. Wolfram, "The Development Of Baryon Asymmetry In The Early Universe," Phys. Lett. B **91**, 217 (1980). "Baryon Number Generation In The Early Universe," Nucl. Phys. B **172**, 224 (1980) [Erratum-ibid. B **195**, 542 (1982)].

[19] G. 't Hooft, "Symmetry breaking through Bell-Jackiw anomalies," Phys. Rev. Lett. **37**, 8 (1976).

[20] F. R. Klinkhamer and N. S. Manton, "A Saddle Point Solution In The Weinberg-Salam Theory," Phys. Rev. D **30**, 2212 (1984).

[21] P. Arnold and L. D. McLerran, "Sphalerons, Small Fluctuations And Baryon Number Violation In Electroweak Theory," Phys. Rev. D **36**, 581 (1987); "The Sphaleron Strikes Back."

[22] M. Laine and K. Rummukainen, Phys. Rev. Lett. **80**, 5259 (1998) [arXiv:hep-ph/9804255]. Phys. Rev. D **37**, 1020 (1988).

[23] V. A. Kuzmin, V. A. Rubakov and M. E. Shaposhnikov, "On The Anomalous Electroweak Baryon Number Nonconservation In The Early Universe," Phys. Lett. B **155**, 36 (1985).

[24] S. Y. Khlebnikov and M. E. Shaposhnikov, "The Statistical Theory of Anomalous Fermion Number Nonconservation," Nucl. Phys. B **308**, 885 (1988).

[25] G.D. Moore, private communication.

[26] P. Arnold, D. Son and L. G. Yaffe, "The hot baryon violation rate is $O(\alpha_w^5 T^4)$," Phys. Rev. D **55**, 6264 (1997) [arXiv:hep-ph/9609481].

[27] G. D. Moore, C. r. Hu and B. Muller, "Chern-Simons number diffusion with hard thermal loops," Phys. Rev. D **58**, 045001 (1998) [arXiv:hep-ph/9710436].

[28] D. Bodeker, G. D. Moore and K. Rummukainen, "Chern-Simons number diffusion and hard thermal loops on the lattice," Phys. Rev. D **61**, 056003 (2000) [arXiv:hep-ph/9907545].

[29] C. Jarlskog, "Commutator Of The Quark Mass Matrices In The Standard Electroweak Model And A Measure Of Maximal CP Violation," Phys. Rev. Lett. **55**, 1039 (1985); "A Basis Independent Formulation Of The Connection Between Quark Mass Matrices, CP Violation And Experiment," Z. Phys. C **29**, 491 (1985).

[30] G. R. Farrar and M. E. Shaposhnikov, "Baryon Asymmetry Of The Universe In The Standard Electroweak Theory," Phys. Rev. D **50**, 774 (1994) [arXiv:hep-ph/9305275].

[31] M. B. Gavela, P. Hernandez, J. Orloff and O. Pene, "Standard model CP violation and baryon asymmetry," Mod. Phys. Lett. A **9**, 795 (1994) [arXiv:hep-ph/9312215].

[32] A. G. Cohen, D. B. Kaplan and A. E. Nelson, "Weak Scale Baryogenesis," Phys. Lett. B **245**, 561 (1990); "Baryogenesis at the weak phase transition," Nucl. Phys. B **349**, 727 (1991).

[33] G. D. Moore, "Computing the strong sphaleron rate," Phys. Lett. B **412**, 359 (1997) [arXiv:hep-ph/9705248].

[34] J. I. Kapusta and P. V. Landshoff, "Finite Temperature Field Theory" J. Phys. G **15**, 267 (1989).

[35] M. E. Shaposhnikov, "Possible Appearance of the Baryon Asymmetry of the Universe in an Electroweak Theory," JETP Lett. **44**, 465 (1986) [Pisma Zh. Eksp. Teor. Fiz. **44**, 364 (1986)]; "Baryon Asymmetry of the Universe in Standard Electroweak Theory," Nucl. Phys. B **287**, 757 (1987).

[36] A. I. Bochkarev and M. E. Shaposhnikov, "Electroweak Production Of Baryon Asymmetry And Upper Bounds On The Higgs And Top Masses," Mod. Phys. Lett. A **2**, 417 (1987).

[37] P. Arnold, "The Electroweak phase transition: Part 1. Review of perturbative methods," arXiv:hep-ph/9410294.

[38] K. Kajantie, M. Laine, K. Rummukainen and M. E. Shaposhnikov, "Is there a hot electroweak phase transition at m(H) > approx. m(W)?," Phys. Rev. Lett. **77**, 2887 (1996) [arXiv:hep-ph/9605288]; "The universality class of the electroweak theory," Nucl. Phys. B **532**, 283 (1998) [arXiv:hep-lat/9805013].

[39] G. W. Anderson and L. J. Hall, "The Electroweak Phase Transition And Baryogenesis," Phys. Rev. D **45**, 2685 (1992).

[40] M. Carena, M. Quiros and C. E. M. Wagner, "Opening the Window for Electroweak Baryogenesis," Phys. Lett. B **380**, 81 (1996) [arXiv:hep-ph/9603420].

[41] J. M. Cline and K. Kainulainen, "Supersymmetric Electroweak Phase Transition: Beyond Perturbation Theory," Nucl. Phys. B **482**, 73 (1996) [arXiv:hep-ph/9605235].
     M. Laine, "Effective theories of MSSM at high temperature," Nucl. Phys. B **481**, 43 (1996) [Erratum-ibid. B **548**, 637 (1999)] [arXiv:hep-ph/9605283].
     M. Losada, "High temperature dimensional reduction of the MSSM and other multi-scalar models," Phys. Rev. D **56**, 2893 (1997) [arXiv:hep-ph/9605266].

[42] J. M. Cline and G. D. Moore, "Supersymmetric electroweak phase transition: Baryogenesis versus experimental constraints," Phys. Rev. Lett. **81**, 3315 (1998) [arXiv:hep-ph/9806354].

[43] M. Laine and K. Rummukainen, "A strong electroweak phase transition up to m(H) approx. 105-GeV," Phys. Rev. Lett. **80**, 5259 (1998) [arXiv:hep-ph/9804255]; "The MSSM electroweak phase transition on the lattice," Nucl. Phys. B **535**, 423 (1998) [arXiv:hep-lat/9804019].

[44] S. J. Huber and M. G. Schmidt, "Electroweak baryogenesis: Concrete in a SUSY model with a gauge singlet," Nucl. Phys. B **606**, 183 (2001) [arXiv:hep-ph/0003122].

[45] J. M. Cline and P. A. Lemieux, "Electroweak phase transition in two Higgs doublet models," Phys. Rev. D **55**, 3873 (1997) [arXiv:hep-ph/9609240].

[46] L. Fromme, S. J. Huber and M. Seniuch, "Baryogenesis in the two-Higgs doublet model," arXiv:hep-ph/0605242.

[47] C. Grojean, G. Servant and J. D. Wells, "First-order electroweak phase transition in the standard model with a low cutoff," Phys. Rev. D **71**, 036001 (2005) [arXiv:hep-ph/0407019].

[48] M. Joyce, T. Prokopec and N. Turok, "Electroweak baryogenesis from a classical force," Phys. Rev. Lett. **75**, 1695 (1995) [Erratum-ibid. **75**, 3375 (1995)] [arXiv:hep-ph/9408339]; "Nonlocal electroweak baryogenesis. Part 2: The Classical regime," Phys. Rev. D **53**, 2958 (1996) [arXiv:hep-ph/9410282].

[49]  J. M. Cline, M. Joyce and K. Kainulainen, "Supersymmetric electroweak baryogenesis," JHEP **0007**, 018 (2000) [arXiv:hep-ph/0006119].

[50]  T. Prokopec, M. G. Schmidt and S. Weinstock, "Transport equations for chiral fermions to order h-bar and electroweak baryogenesis," Annals Phys. **314**, 208 (2004) [arXiv:hep-ph/0312110]; "Transport equations for chiral fermions to order h-bar and electroweak baryogenesis. II," Annals Phys. **314**, 267 (2004) [arXiv:hep-ph/0406140].

T. Konstandin, T. Prokopec, M. G. Schmidt and M. Seco, Nucl. Phys. B **738**, 1 (2006) [arXiv:hep-ph/0505103].

T. Konstandin, T. Prokopec and M. G. Schmidt, "Kinetic description of fermion flavor mixing and CP-violating sources for baryogenesis," Nucl. Phys. B **716**, 373 (2005) [arXiv:hep-ph/0410135].

[51]  S. Weinberg, "Larger Higgs Exchange Terms in the Neutron Electric Dipole Moment," Phys. Rev. Lett. **63**, 2333 (1989).

[52]  J. F. Gunion and D. Wyler, "Inducing a large neutron electric dipole moment via a quark chromoelectric dipole moment," Phys. Lett. B **248**, 170 (1990).

[53]  S. M. Barr and A. Zee, "Electric Dipole Moment of the Electron and of the Neutron," Phys. Rev. Lett. **65**, 21 (1990) [Erratum-ibid. **65**, 2920 (1990)].

[54]  M. Pospelov and A. Ritz, "Electric dipole moments as probes of new physics," Annals Phys. **318**, 119 (2005) [arXiv:hep-ph/0504231].

[55]  J. M. Cline, M. Joyce and K. Kainulainen, "Supersymmetric electroweak baryogenesis in the WKB approximation," Phys. Lett. B **417**, 79 (1998) [Erratum-ibid. B **448**, 321 (1999)] [arXiv:hep-ph/9708393].

[56]  P. Huet and A. E. Nelson, "Electroweak baryogenesis in supersymmetric models," Phys. Rev. D **53**, 4578 (1996) [arXiv:hep-ph/9506477].

[57]  M. Carena, M. Quiros, A. Riotto, I. Vilja and C. E. M. Wagner, "Electroweak baryogenesis and low energy supersymmetry," Nucl. Phys. B **503**, 387 (1997) [arXiv:hep-ph/9702409].

[58]  M. Joyce, K. Kainulainen and T. Prokopec, "The semiclassical propagator in field theory," Phys. Lett. B **468**, 128 (1999) [arXiv:hep-ph/9906411]; "Fermion propagator in a nontrivial background field," JHEP **0010**, 029 (2000) [arXiv:hep-ph/0002239]

K. Kainulainen, T. Prokopec, M. G. Schmidt and S. Weinstock, "First principle derivation of semiclassical force for electroweak baryogenesis," JHEP **0106**, 031 (2001) "Semiclassical force for electroweak baryogenesis: Three-dimensional derivation," Phys. Rev. D **66**, 043502 (2002) [arXiv:hep-ph/0202177].

[59]  J. M. Cline and K. Kainulainen, "A new source for electroweak baryogenesis in the MSSM," Phys. Rev. Lett. **85**, 5519 (2000) [arXiv:hep-ph/0002272].

[60]  T. Falk, K. A. Olive, M. Pospelov and R. Roiban, "MSSM predictions for the electric dipole moment of the Hg-199 atom," Nucl. Phys. B **560**, 3 (1999) [arXiv:hep-ph/9904393].

[61]  "Higgs-mediated electric dipole moments in the MSSM: An application to baryogenesis and Higgs searches," Nucl. Phys. B **644** (2002) 263 [arXiv:hep-ph/0207277].

[62]  I. Affleck and M. Dine, "A New Mechanism For Baryogenesis," Nucl. Phys. B **249**, 361 (1985).

[63]  M. Fukugita and T. Yanagida, "Baryogenesis Without Grand Unification," Phys. Lett. B **174**, 45 (1986).

[64]  W. Buchmuller, P. Di Bari and M. Plumacher, "Leptogenesis for pedestrians," Annals Phys. **315**, 305 (2005) [arXiv:hep-ph/0401240].

[65]  A. Pilaftsis and T. E. J. Underwood, "Electroweak-scale resonant leptogenesis," Phys. Rev. D **72**, 113001 (2005) [arXiv:hep-ph/0506107].

[66]  J. A. Harvey and M. S. Turner, "Cosmological baryon and lepton number in the presence of electroweak fermion number violation," Phys. Rev. D **42**, 3344 (1990).

[67] S. Hannestad and G. G. Raffelt, "Neutrino masses and cosmic radiation density: Combined analysis," arXiv:astro-ph/0607101.

[68] S. Davidson and A. Ibarra, "A lower bound on the right-handed neutrino mass from leptogenesis," Phys. Lett. B **535**, 25 (2002) [arXiv:hep-ph/0202239].

[69] M. Plumacher, "Leptogenesis," J. Phys. G **29**, 1561 (2003).

[70] G. F. Giudice, A. Notari, M. Raidal, A. Riotto and A. Strumia, "Towards a complete theory of thermal leptogenesis in the SM and MSSM," Nucl. Phys. B **685**, 89 (2004) [arXiv:hep-ph/0310123].

Course 3

# STRING COSMOLOGY

## James M. Cline

*McGill University, Department of Physics, Montreal, Qc H3A 2T8, Canada*

*F. Bernardeau, C. Grojean and J. Dalibard, eds.*
*Les Houches, Session LXXXVI, 2006*
*Particle Physcis and Cosmology: The Fabric of Spacetime*
© *2007 Published by Elsevier B.V.*

# WKB approx.

ansatz

$$L_5 = A_5^{(z)} \left( \ldots \right) dz - Et$$

subst. into Dirac

find

# Contents

# 1. Dark energy

## 1.1. The problem of scales

In preparing a set of lectures on string cosmology, I originally set out to explain why I am going to focus exclusively on inflation from string theory, ignoring the important question of the dark energy of the universe, which is another very exciting area on the observational side. The basic reason is that the string scale is so much greater than the scale of the dark energy that it seems implausible that a stringy description should be needed for any physics that is occuring at the milli-eV scale.

Naively, one can generate small energy scales from string-motivated potentials; for example exponential potentials often arise, so that one might claim a quintessence-like Lagrangian of the form

$$\mathcal{L} = (\partial \phi)^2 - \Lambda e^{-\alpha \phi} \tag{1.1}$$

could be derived from string theory, and the small scale of the current vacuum energy could be a result of $\phi$ having rolled to large values. The problem with this kind of argument is that it is ruled out by 5th force constraints, unless $\phi$ happens to be extremely weakly coupled to matter. But in string theory, there is no reason for any field to couple with a strength that is suppressed compared to the gravitational coupling, so it seems difficult to get around this problem. The Eöt-Wash experiment [1] bounds the strength of this particle's coupling to matter as a function of its inverse mass (the range of the 5th force which it mediates) as sketched in figure 1. The current limit restricts a fifth force of gravitational strength to have a range less than 0.044 mm, corresponding to a mass greater than 4.4 milli-eV. On the other hand a field that is still rolling today must have a mass less than the current Hubble parameter, $\sim 10^{-33}$ eV.

## 1.2. The string theory landscape

On the other hand, if we accept the simplest hypothesis, which is also in good agreement with the current data, that the dark energy is just a cosmological constant $\Lambda$, then string theory does have something to tell us. The vast landscape of string vacua [2], combined with the selection effect that we must be able to exist

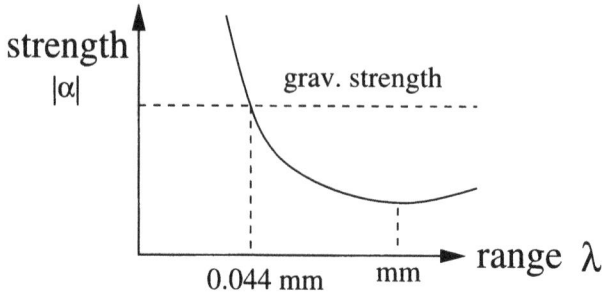

Fig. 1. Schematic limit on coupling strength versus range of a fifth force.

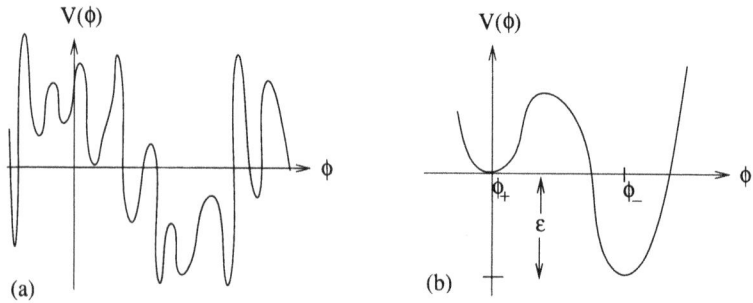

Fig. 2. (a) Scalar field potential which is a toy model for the landscape. (b) A part of the potential illustrating the failure of tunneling, *à la* Coleman-DeLuccia.

in a given vacuum in order to observe it (the anthropic principle) has given the only plausible explanation of the smallness of $\Lambda$ to date.

We can illustrate the idea starting with a toy model, of a scalar field with Lagrangian

$$\mathcal{L} = (\partial\phi)^2 - V(\phi) \tag{1.2}$$

whose potential has many minima, as shown in figure 2(a). Naively one might guess that eventually the universe must tunnel to the lowest minimum, which would generically be a negative energy anti-de Sitter space ending in a catastrophic big crunch. This is wrong, for two reasons. The first reason is that the lifetime of a metastable state may be longer than the age of our universe, in which case it is as good as stable. More than that, a positive energy vacuum which would be unstable to tunneling to a negative energy vacuum in the absence of gravity can actually be stable when gravity is taken into account. As shown by Coleman and De Luccia [3], if the difference $\epsilon$ in vacuum energies between a zero-energy minimum and one with negative energy is too great, tunneling is inhibited. Referring to figure 2(b), define

$$S_1 = \int_{\phi_-}^{\phi_+} d\phi \sqrt{2(V_0(\phi) - V_0(\phi_+)}$$  (1.3)

where, roughly speaking, $V_0$ is what $V$ would look like if $\epsilon$ were set to zero. This is well-defined when $\epsilon$ is small, and the bubbles of true vacuum which nucleate during the tunneling consequently have a thin wall. The initial radius of the bubbles in this case turns out to be $\bar{\rho}_0 = 3S_1/\epsilon$. On the other hand, there is a distance scale $\Lambda = \sqrt{3/(\kappa\epsilon)}$ (where $1/\kappa = 8\pi G$) which is the size of a bubble whose radius equals its Schwarzschild radius. Coleman and De Luccia show that if $\bar{\rho}_0 \geq 2\Lambda$, there is no vacuum decay.

The second reason to consider vacua with large energies is eternal inflation [4]. In regions where the potential is large, so are the de Sitter quantum fluctuations of the field, $\delta\phi \sim H/2\pi$. If they exceed the distance by which the field classically rolls down its potential during a Hubble time $1/H$, then the quantum effects can keep the field away from its minimum indefinitely, until some chance fluctuations send it in the direction of the classical motion in a sufficiently homogeneous spatial region. The classical motion during a Hubble time, $\Delta\phi$, can be estimated using the slow-roll approximation to the field equation, $3H\dot{\phi} = -V'$, so $\delta\phi = -V'/3H^2$. Comparing this to the quantum excursion, we see that the condition for eternal inflation is

$$H^3 \gtrsim |V'|$$  (1.4)

In this picture, the global universe consists of many regions undergoing inflation for indefinitely long periods, occasionally giving rise to regions in which inflation ends and a subuniverse like ours can emerge. Thus any minimum of the full $\phi$ potential which is close to values of $\phi$ where (1.4) can be fulfilled will eventually be populated parts of the landscape.

Now we must make contact with the cosmological constant, whose value locally depends on which of the many minima a given subuniverse finds itself in. Suppose there are $N$ which are accessible through classical evolution or tunneling. It is as good a guess as any to assume that the values of the minima of the potential, $V_{min}$ are uniformly distributed over the range $\sim[-M_p^4, M_p^4]$ with spacing $\Delta V \sim M_p^4/N$. If $N \gtrsim 10^{120}$ then there will exist some vacua with $V_{min}$ close to the observed value. The possible values of $\Lambda$ will be distributed according to some intrinsic probability distribution function $P_i(\Lambda)$ which depends on the details of the potential $V$, but in the absence of reasons to the contrary should be roughly uniform, as in figure 3.

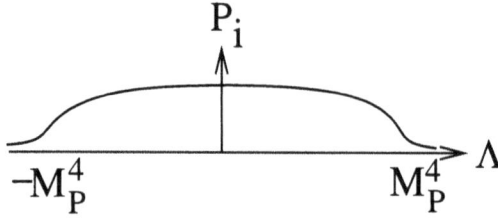

Fig. 3. Intrinsic probability distribution for values of $\Lambda$ coming from minima of the landscape potential.

This theory explains why it is possible to find ourselves in a vacuum with the observed value of $\Lambda$. To understand why we were so lucky as to find it so unnaturally close to zero, we need to consider the conditional probability

$$P_{\text{tot}}(\Lambda) = P_i(\Lambda) P_{\text{obs}}(\Lambda) \tag{1.5}$$

where the second factor is the probability that an observer could exist in a universe with the given value of $\Lambda$. This part of the problem was considered by S. Weinberg originally in [5] in also in the review article [6]. In the abstract of the [5], he says that the anthropic bound on $\Lambda$ is too weak to explain the observed value of $\Lambda$, but in the later work he took the more positive view; living in a universe whose probability is 1% is far less puzzling than one where $P \sim 10^{-120}$. There are actually two anthropic bounds. For $\Lambda > 0$, we must have $\Lambda \lesssim 10^2 \Lambda_{\text{obs}}$; otherwise the universe expands too quickly to have structure formation by redshifts of $z \sim 4$. For $\Lambda < 0$, we need $\Lambda \gtrsim -\Lambda_{\text{obs}}$ to avoid the recollapse of the universe before structure formation.

It is worth pointing out that this was a *successful prediction* of $\Lambda_{\text{obs}}$ before it was determined to be nonzero through observations of distant supernovae [7]. From the anthropic point of view, there is no reason for $\Lambda$ to be zero, so its most natural value is of the same order of magnitude as the bound—assuming of course that $P_i$ is roughly uniform over the range where $P_{\text{obs}}$ is nonnegligible.

### 1.3. The Bousso-Polchinski (Brown-Teitelboim) mechanism

We can now consider how the setting for this idea can be achieved within string theory. The most concrete example is due to Bousso and Polchinski [8], who found a stringy realization of the Brown-Teitelboim (BT) mechanism [9]. This is based on the existence of 4-form gauge field strengths $F^{(4)}_{\alpha\beta\gamma\delta}$ whose action is

$$S = \int d^4x \left[ -\sqrt{-g} F^{(4)}_{\alpha\beta\gamma\delta} F^{\alpha\beta\gamma\delta}_{(4)} + 8\partial_\mu \left( \sqrt{-g} F^{\mu\alpha\beta\gamma}_{(4)} A_{\alpha\beta\gamma} \right) \right] \tag{1.6}$$

leading to the equation of motion $\partial_\mu \left( \sqrt{-g} F_{(4)}^{\mu\alpha\beta\gamma} \right) = 0$, with solution $F_{(4)}^{\mu\alpha\beta\gamma} = c\epsilon^{\mu\alpha\beta\gamma}$ where $c$ is a constant. The contribution to the action is

$$S = + \int d^4x \sqrt{-g} \, (-4!) \, c^2 \tag{1.7}$$

The sign $+$ in front of the integral in (1.7) would have been $-$ had it not been for the total derivative term in (1.6) which must be there for consistency; otherwise the value of the vacuum energy appearing in the action has the opposite sign to that appearing in the equations of motion. This negative contribution to the action is a positive contribution to the vacuum energy density.

Although in field theory the constant $c$ is arbitrary, in string theory it is quantized. The particular setting used by Bousso and Polchinski is M-theory, which is 11 dimensional and has no elementary string excitations, but it does have M2 and M5 branes. The low-energy effective action for gravity and the 4-form is

$$S = 2\pi M_{11}^9 \int d^{11}x \sqrt{-g_{11}} \left( R - F_{(4)}^2 \right) \tag{1.8}$$

The 4-form is electrically sourced by the M2-branes and magnetically sourced by the M5-branes, in a way which we will discuss shortly. To understand the quantization of $c$, it is useful to note that $F_{(4)}$ is Hodge-dual to a 7-form in 11 dimensions,

$$F_{(4)} = *F_{(7)} \tag{1.9}$$

(in terms of indices, this means contracting $F_{(7)}$ with the 11D Levi-Civita tensor). We can say that $F_{(7)}$ is electrically sourced by M5 branes. In the same way that electric charge in 4D is quantized if magnetic monopoles exist, there is a generalized Dirac quantization condition in higher dimensions which applies here because the wave function of an M5 brane picks up a phase when transported around an M2 brane, or *vice versa*. For the wave function to be single valued, it is necessary that the $F_{(7)}$ flux which is sourced by the M2 brane, when integrated over the 7D compactification manifold $K$, is quantized,

$$2\pi M_{11}^6 \int_K F_{(7)} = 2\pi n \tag{1.10}$$

This quantization also implies the quantization of $F_{(4)}$ through the duality relation (1.9). With this identification, we can evaluate the contribution to $\Lambda$ from (1.7) as

$$\Delta\Lambda = 2n \frac{(2\pi M_{11}^3)^2}{2\pi M_{11}^9 V_7} = \frac{4\pi n}{M_{11}^3 V_7} \sim \frac{n M_{11}^6}{M_p^2} \tag{1.11}$$

where $V_7$ is the volume of $K$. This can be taken as the typical spacing between nearby values of the cosmological constant. Unfortunately, even if $M_{11}$ is as small as the TeV scale using large extra dimensions, one finds that $\Delta\Lambda/M_p^4 \sim (\text{TeV}/10^{15}\ \text{TeV})^6 \sim 10^{-90}$, and this spacing is too large to solve the cosmological constant problem, which needs $10^{-120}$. In fact the problem is even worse, because if the bare value of $\Lambda$ which needs to be canceled is negative and of order $-M_{11}^4$, a large value of $n$ is required,

$$
\Lambda_{\text{obs}} = \Lambda_{\text{bare}} + O\left(\frac{n^2 M_{11}^6}{M_p^2}\right)
$$

$$
\implies n \sim \frac{M_p |\Lambda_{\text{bare}}|^{1/2}}{M_{11}^3} \tag{1.12}
$$

in which case $\Delta\Lambda/M_p^4 \sim M_{11}^3 |\Lambda_{\text{bare}}|^{1/2}/M_p^5 \sim 10^{-75}$ assuming $|\Lambda_{\text{bare}}| \sim M_{11}^4$.

However, string theory has a nice solution to this problem. There is not just one 4-form, but one for each nontrivial 3-cycle in $K$. Then

$$
F_{(7)} = *F_{(4,N+1)} \wedge \epsilon_1(y) + \sum_{i=1}^{N} F_{(4,i)} \wedge \omega_{3,i}(y) \tag{1.13}
$$

where $F_{(4,N+1)}$ can be thought of as the original 4-form we started with, while $F_{(4,i)}$ is the $i$th additional one which arises for each nontrivial 3-cycle. Here $\epsilon_1(y)$ is the volume form on $K$ (whose coordinates are taken to be $y$) and $\omega_{3,i}(y)$ is the harmonic 3-form on the $i$th 3-cycle. There can easily be a large number of 3-cycles. For example if $K$ is a 7-torus, the number of inequivalent triples of 1-cycles is $\binom{7}{3} = 35$. If $K$ has more interesting topology the number can be larger. We no longer have a single flux integer $n$ but a high-dimensional lattice, $n_i$, such that

$$
\Lambda_{\text{obs}} = \Lambda_{\text{bare}} + \frac{1}{2}\sum_i n_i^2 q_i^2 \tag{1.14}
$$

where $q_i = M_{11}^{3/2} V_{3,i}/\sqrt{V_7}$ and $V_{3,i}$ is the volume of the $i$th 3-cycle. We can visualize a spherical shell in the space of the flux integers $n_i$, which is bounded by surfaces where $\Lambda = \Lambda_{\text{obs}}$ and $\Lambda = \Lambda_{\text{obs}} + \Delta\Lambda$, as illustrated in figure 4. As long as the lattice is fine enough so that at least one point is contained within the shell, then there exists a set of fluxes which can give a finely-enough tuned cancellation of $\Lambda_{\text{bare}}$. The number of such lattice points is just the volume of the shell with radius $\sqrt{\frac{1}{2}\sum_i n_i^2 q_i^2} = |\Lambda_{\text{bare}}|^{\frac{1}{2}}$ and thickness $\Delta\Lambda^{\frac{1}{2}} \sim \Lambda_{\text{obs}}^{\frac{1}{2}}$, in

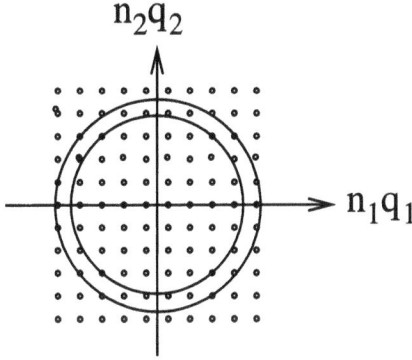

Fig. 4. Lattice of allowed 4-form flux integers, and a hyperspherical shell of values consistent with canceling a large negative $\Lambda_{\text{bare}}$ to get a small $\Lambda_{\text{obs}}$.

the $N$ dimensional space whose $i$th axis is the magnitude of $n_i q_i$. This fixes the magnitude of $q_i$ via

$$\frac{\Delta \Lambda}{M_p^4} \sim q_i^{N/2} \Gamma(N/2) \pi^{-N/2} \tag{1.15}$$

For example, if $N \sim 35$ we need $q_i \sim 10^{-3}$ while if $N \sim 100$, we need $q_i \sim 10^{-2/3}$. Let us suppose that the compactification and 3-cycle volumes are given in terms of a distance scale $R$, as $V_7 \sim R^7$, $V_{3,i} \sim R^3$. Then

$$q_i \sim \frac{M_{11}^{3/2} R^{-1/2}}{M_p^2} \sim \frac{M_{11}^{3/2} R^{-1/2}}{M_{11}^9 R^7} = \frac{1}{(M_{11}R)^{13/2}} \tag{1.16}$$

so the required small values of $q_i$ can be achieved by taking extra dimensions which are moderately larger than the fundamental distance scale.

To complete the picture, we must describe how tunneling between the different vacuum states (labeled by the flux quanta) takes place. In the toy-model field-theory example, bubbles of the new vacuum phase with $\Lambda_-$ nucleate within the old one with $\Lambda_+$. In M-theory there is no scalar field, but we do have M2 and M5 branes. The M2 branes can serve as the walls of the bubbles since they are spatially two-dimensional, as illustrated in figure 5. Furthermore there is a natural way to couple the M2 branes to the 4-forms, which is a generalization of the coupling of a charged relativistic point particle to the Maxwell gauge field,

$$S = -\int d\tau \, J^\mu A_\mu \tag{1.17}$$

defined as an integral along the worldline of the particle, where $\tau$ is the proper time and $J^\mu = q\, dx^\mu/d\tau$. The generalization to strings is

$$S = -q \int d\tau d\sigma \, \frac{\partial x^\mu}{\partial \tau} \frac{\partial x^\nu}{\partial \sigma} A_{\mu\nu} \tag{1.18}$$

and for 2-branes it is

$$S = -q \int d\tau d\sigma_1 d\sigma_2 \, \frac{\partial x^\mu}{\partial \tau} \frac{\partial x^\nu}{\partial \sigma_1} \frac{\partial x^\rho}{\partial \sigma_2} A_{\mu\nu\rho} \tag{1.19}$$

where $A_{\mu\nu}$ and $A_{\mu\nu\rho}$ are totally antisymmetric gauge potentials. This shows why M2-branes are sources of 4-form field strengths, since the latter are related to the 3-index gauge potential by

$$F_{(4)} = dA_{(3)}, \quad F_{\mu\nu\alpha\beta} = A_{[\mu\nu\alpha,\beta]} \tag{1.20}$$

(the brackets indicate total antisymmetrization on the indices). Adding the Chern-Simons action (1.19) to the kinetic term for $F_{(4)}$ results in the equation of motion

$$\partial_\mu\left(\sqrt{-g}\, F^{\mu\nu\alpha\beta}\right) = q \sum_\mu \delta^{(1)}\left(x^\mu - X^\mu(\tau,\sigma_i)\right)\epsilon^{\mu\nu\alpha\beta} \tag{1.21}$$

For example if the M2 brane is in the $x$-$y$ plane and the spacetime is Minkowskian, then

$$\partial_\mu F^{\mu\nu\alpha\beta} = q\delta(z)\epsilon^{z\nu\alpha\beta} \tag{1.22}$$

whose solution is

$$F^{\mu\nu\alpha\beta} = \epsilon^{\mu\nu\alpha\beta} \begin{cases} c + q, & z > 0 \\ c, & z < 0 \end{cases} \tag{1.23}$$

Thus the value of the 4-form changes by the charge of the M2 brane when going from one side of the brane to the other.

In the BT mechanism, the value of $\Lambda$ decreases dynamically by the nucleation of bubbles whose surface has charge $q$ (figure 5). We now see why these bubble walls can be identified with the M2 branes of M-theory, which carry the quantized charge $q$. Having introduced a large number $N$ of 4-forms, we need to find separate sources with different charges $q_i$, one for each different $F_{(4,i)}$, yet M-theory only has one kind of M2 brane, with a unique charge. The source of the new $F_{(4,i)}$'s is the M5 brane, whose extra three dimensions can be wrapped on the different 3-cycles. Each one of these provides effectively a new kind of M2 brane. The fact that these branes must come in integer multiples also explains why the $F_{(4,i)}$ fluxes are quantized: since they are sourced by M2 branes, their strength must be proportional to the number of source branes. We can now better

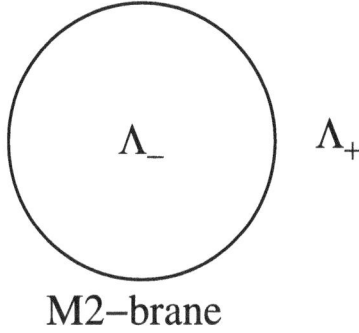

Fig. 5. Nucleation of bubble of new $\Lambda$ whose wall is an M2-brane.

understand eq. (1.15): if we start from a situation with no fluxes and nucleate a stack of M2 branes, where $n_i$ copies of the $i$th kind of M2 brane are coincident, the new value of $F_{(4,i)}$ (after normalizing it in a convenient way) is given by $n_i q_i$, and the corresponding contributions to $\Lambda$ add in quadrature.

This construction establishes that string theory contains the necessary ingredients for realizing the original BT idea to explain the smallness of $\Lambda$. It is a concrete, detailed, and quantitative treatment, which gives a clear explanantion of how many closely-spaced values of $\Lambda$ can be generated. It is a nontrivial feat that it is possible to make the numbers work out favorably. There are also further hurdles to pass: the desired endpoint must have a lifetime greater than the age of the present universe since it is only metastable—this can be arranged since tunneling is typically exponentially suppressed—and the final universe (ours) must be born in an excited state so that inflation and reheating can take place subsequent to the nucleation. The latter requirement is less generic to satisfy, but possible, due to eternal inflation. If the inflaton is at an eternally inflating value during the tunneling, the daughter universe can also be eternally inflating. This puts a lower bound on $M_{11}$ around the GUT scale.

It is likely that other corners of the string theory parameter space can also provide settings for this basic idea. The theory we will focus on for inflation, type IIB string theory with flux compactification, gives exponentially large ($10^{\sim 500}$) numbers of vacua [10]. (See [11] for a discussion of the BTBP mechanism in this theory.)

It is worth emphasizing that the use of the anthropic principle is not merely optional once we accept the existence of many vacuum states. Unless there is a dynamical mechanism singling out one or a few of the possible states, we are forced to consider all of them, and then to restrict attention to those which satisfy the prior that observers can exist.

*1.4. Caveats to the landscape approach*

It has been noted that the anthropic bound on $\Lambda$ is relaxed if the amplitude of density perturbations, $Q \sim 10^{-5}$ in our universe, is also allowed to vary [12]:

$$\text{bound on } \Lambda \sim Q^3 \qquad\qquad (1.24)$$

Rees and Tegmark argue that if $Q > 10^{-4}$, galaxies would be too dense for life to evolve, which loosens Weinberg's bound by a factor of $10^3$. Nevertheless, a one part in $10^5$ fine-tuning is much less daunting than one in $10^{120}$. Furthermore, there could be reasons for $Q$ not being "scanned" by the possible vacua of string theory, when we understand them better.

Another problem is the lack of additional predictions which would allow us to test whether the anthropic explanation is really the right one. However, we are still in the early days of understanding the landscape, so it may be premature to give up on making such predictions. Work along these lines is continuing at a steady pace [13, 14]. One hope is that by studying large anthropically allowed regions of the landscape, there might exist generic predictions for other quantities correlated with a small value of $\Lambda$, for example the scale of SUSY breaking. However the predictions are not yet clear, with some authors arguing for a small scale of SUSY breaking [15] while others suggest a high scale, for example the interesting possibility of "split SUSY," in which all the bosonic superpartners are inaccessibly heavy and only the fermionic ones can be produced at low energies [16, 17]. In such scenarios the low Higgs mass would arise from the anthropic need for fine tuning instead of any mechanism like SUSY for suppressing the large loop contributions [13]. Unpalatable as some might find such a possibility, it is nonetheless a distinctive prediction, which if shown to be true would lend more weight to the landscape resolution of the cosmological constant problem.

## 2. Inflation

String theory brings some qualitatively new candidates for the inflaton, as well as some possible observable signatures that are distinct from typical field theoretic inflation. The inflaton could be the separation between two branes within the compact dimensions, or it could be moduli like axions associated with the Calabi-Yau (C-Y) compactification manifold. The new effects include large non-gaussianity and inflation without the slow-roll conditions being satisfied, as in the DBI inflation scenario [18, 19]. There are also new issues connected with reheating in models with warped throats.

## 2.1. Brane-antibrane inflation

D$p$-branes in type II string theory are dynamical $(p + 1)$-dimensional objects (with $p$ spatial dimensions) on which the ends of open strings can be confined, hence giving Dirichlet boundary conditions to the string coordinates transverse to the brane. There has been considerable progress in achieving stable compactifications within type IIB string theory in the last few years, using warped Klebanov-Strassler throats with fluxes [20, 21].

To put these developments into perspective, let us recall some differences between types of string theories. Type I theory describes both open and closed, unoriented strings with SO(32) gauge group. Its low energy effective theory is that of $N = 1$ supergravity. Type II theories are of closed strings only (not counting the open string excitations which live on the D-branes themselves), with $N = 2$ SUGRA effective theories. Type IIA theories contain D$p$-branes with $p$ even, while IIB contains odd-dimensional branes. Both of the type II theories have the same action from the NS-NS sector—the sector where the superpartners to the string coordinates $\psi^\mu$ have antiperiodic boundary conditions [22]:

$$S_{NS\text{-}NS} = \frac{1}{\kappa_{10}^2} \int d^{10}x \sqrt{-g}\, e^{-2\Phi} \left( R + 4(\partial\Phi)^2 - \frac{1}{2}|H_3|^2 \right) \tag{2.1}$$

where $H_3$ is the 3-index Kalb-Ramond field strength, and $\Phi$ is the dilaton, which determines the string coupling via $g_s = e^\Phi$. The Ramond-Ramond (R-R) sector, where $\psi^\mu$ has periodic boundary conditions, exhibits the differences between type IIA and IIB theories:

$$S_{R\text{-}R} = -\frac{1}{4\kappa_{10}^2} \int d^{10}x \begin{cases} |F_2|^2 + |F_4|^2, & IIA \\ |F_1|^2 + |F_3|^2 + \frac{1}{2}|F_5|^2, & IIB \end{cases} \tag{2.2}$$

where $F_n$ is an $n$-index antisymmetric field strength. These expressions could have been written in a more symmetric way by including higher values of $n$, and taking coefficients of $\frac{1}{2}$ for each term, because of the duality relations

$$F_6 = *F_4, \quad F_8 = *F_2, \quad F_5 = *F_5, \quad etc. \tag{2.3}$$

$F_n$ couples to the D$p$-brane with $p = n - 2$, as in eqs. (1.18, 1.19). Branes which have the wrong dimension for the theory in which they appear are unstable and quickly decay into closed strings.

The Chern-Simons couplings of the branes to the gauge fields show that branes are charged objects, and not simply delta-function sources of stress-energy as is often assumed in phenomenological brane-world models. This puts restrictions on the placement of branes within compact spaces. Since the gauge fields $F_n$ obey Gauss' law, the net charge of the branes in the compact volume must vanish.

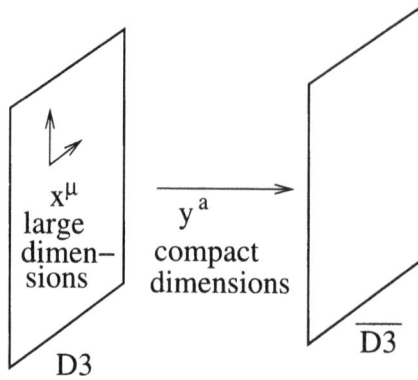

Fig. 6. Parallel D3 brane and $\overline{D3}$ antibrane.

Consider for example a D3 brane parallel to a $\overline{D3}$ antibrane, which share parallel dimensions $x^\mu$ and are separated by a vector $y^a$ in the transverse directions, lying in the compact space. This is illustrated in figure 6. They are sources of $F_5$ through their Chern-Simons coupling, which can be more compactly written as

$$\sum_i \mu_{D_3}^{(i)} \int_{M_4^{(i)}} A_4 \tag{2.4}$$

where $i$ labels the brane or antibrane having charge $\mu_{D_3}^{(i)}$ and world volume $M_4^{(i)}$. The equation of motion for $F_5$ is

$$\partial_a \left( \sqrt{-g} F^{a\alpha\beta\gamma\delta} \right) + \sum_i \mu_{D_3}^{(i)} \delta^{(6)} (y^a - y_i^a) \epsilon^{\alpha\beta\gamma\delta} \tag{2.5}$$

where $a = 5, \ldots, 9$ and $\alpha\beta\gamma\delta = 0, 1, 2, 3$ or some permutation. Integrating this over the compact dimensions gives

$$\int dF + \sum_i \mu_{D_3}^{(i)} = 0 \tag{2.6}$$

Since the integral must vanish, so must the sum of the charges.

We can also use Gauss's law to find the force associated with the $F_5$ field, by considering noncompact extra dimensions, and integrating them over a 6D spherical region $R$ surrounding a single D3 brane. Then

$$\int_R dF_5 = \int_{\partial R} F_5 = |F_5| \times (\text{area of 5-sphere}) \sum \mu_{D_3} \tag{2.7}$$

This tells us that $|F_5| \sim \mu_{D_3}/r^5$ (the force) and $|A_4| \sim \mu_{D_3}/r^4$ (the potential), analogous to the $1/r$ Coulomb potential in 3D.

Similarly, the gravitational potential for a D3 brane goes like $1/r^4$, so the total potential for a D3-D3 or D3-$\overline{\text{D3}}$ system with separation $r$ in the compact dimensions is

$$V_{\text{tot}} = \frac{G_{10}}{\frac{1}{4}\pi^2 r^4}\left(-\tau_3^2 \pm \frac{\mu_3^2}{g_s^2}\right) \begin{cases} \text{upper sign for D3-D3} \\ \text{lower sign for D3-}\overline{\text{D3}} \end{cases} \tag{2.8}$$

where $\tau_3$ is the tension, generally given by

$$\tau_p = \frac{M_s^{p+1}}{(2\pi)^p g_s} \tag{2.9}$$

for a D$p$-brane, $M_s$ is the string mass scale, the 10D Newton constant is given by $G_{10} = (2\pi)^6 g_s^2/(8M_s^8)$, and the charge is related to the tension by $\mu_p = g_s\tau_p$. Because of this, the potential vanishes for D3-D3, but not for D3-$\overline{\text{D3}}$. We now consider whether this potential can be used to drive inflation.

We would like to use the distance $r$ between D3 and $\overline{\text{D3}}$ as the inflaton. This indeed is a field, since the branes are not perfectly rigid objects, but have transverse fluctuations. Hence at any longitudinal position $x^\mu$, the distance between the brane and antibrane is

$$r(x^\mu) = \left(\sum_a \left(y^a(x^\mu) - \bar{y}^a(x^\mu)\right)^2\right)^{1/2} \tag{2.10}$$

We know the potential for $r$ from (2.8); all that remains is to determine its kinetic term. This is the Dirac-Born-Infeld (DBI) action which for a D$p$-brane in Minkowski space is

$$S = -\tau_p \int d^{p+1}x \left|\det\left(\eta_{\mu\nu} + \partial_\mu y^a \partial_\nu y^a\right)\right|^{1/2} \tag{2.11}$$

in coordinates where $x^0, \ldots, x^p$ are in the world-volume of the brane and $a = p+1, \ldots, 9$ label the transverse coordinates.

For inflation we are interested in homogeneous backgrounds where $y^a = y^a(t)$ and so

$$\eta_{\mu\nu} + \partial_\mu y^a \partial_\nu y^a = \text{diag}\left(-1 + \sum_a (\dot{y}^a)^2, +1, \ldots, +1\right) \tag{2.12}$$

Hence

$$\left|\det\left(\eta_{\mu\nu} + \partial_\mu y^a \partial_\nu y^a\right)\right|^{1/2} = \sqrt{1 - \sum_a (\dot{y}^a)^2} \tag{2.13}$$

In the limit of small velocities we can Taylor-expand the square root to get the kinetic part of the action into the form

$$S = \frac{1}{2}\tau_p \sum_a (\dot{y}^a)^2 + O(\dot{y}^4) \tag{2.14}$$

It follows that the canonically normalized inflaton field is

$$\phi = \sqrt{\tau_3}|\vec{y} - \bar{\vec{y}}| \tag{2.15}$$

and the Lagrangian for D3 brane-antibrane inflation becomes

$$\mathcal{L} = \frac{1}{2}\dot{\phi}^2 + 2\left(\tau_3 - \frac{c}{\phi^4}\right) \tag{2.16}$$

where $c = \frac{4}{\pi^2}G_{10}\tau_3^4$. This should not be taken literally for $\phi^4 < c/\tau_3$ however; the Coulomb-like potential is only a large-$r$ approximation. The energy density never becomes negative at small $r$. A more careful calculation shows that $V$ remains finite as $r \to 0$, but there a tachyonic instability occurs in another field when the brane-antibrane system reaches a critical separation [23]

$$y_c^2 = \frac{2\pi^2}{M_s^2} \tag{2.17}$$

This is the separation at which the brane-antibrane system becomes unstable to annihilation into closed strings. The tachyon can be seen as the ground state of the string which stretches between the D3 and the $\overline{D3}$, whose mass is

$$m_T^2 = M_s^2\left(\frac{y^2}{y_c^2} - 1\right) \tag{2.18}$$

The full potential for $\phi$ and the tachyonic field $T$ resembles that of hybrid inflation, fig. 7.

Unfortunately, $V(\phi)$ is not flat enough to be suitable for inflation, unless the separation $r$ is greater than the size of the extra dimensions [24], an impossible requirement [25]. To see how this problem comes about, we compute the slow-roll parameters

$$\epsilon = \frac{1}{2}M_p^2\left(\frac{V'}{V}\right)^2 \cong \frac{1}{2}M_p^2\left(\frac{4c}{\tau_3\phi^5}\right)^2 \tag{2.19}$$

$$\eta = M_p^2\frac{V''}{V} \cong -M_p^2\frac{20c}{\tau_3\phi^6} \tag{2.20}$$

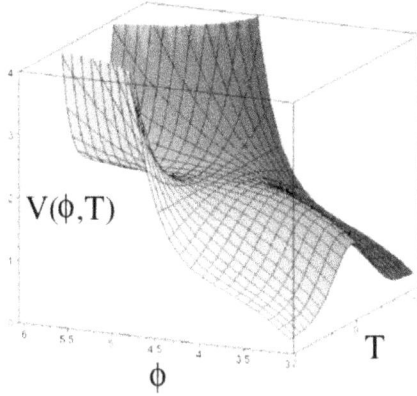

Fig. 7. The hybrid-like potential for the brane separation $\phi$ and the tachyon $T$.

where $M_p = 2.4 \times 10^{18}$ GeV $= (8\pi G_N)^{-1/2}$. What is the relation between $G_N$ and $G_{10}$? This can be found by integrating out the extra dimensions from the 10D SUGRA action:

$$S_{NS\text{-}NS} = \frac{1}{2\kappa_{10}^2 g_s^2} \int d^{10}x \sqrt{-g_{10}} R_{10} \implies \frac{1}{16\pi G_N} \int d^4x \sqrt{-g} R \quad (2.21)$$

The coefficient of the 10D action can also be written as $(16\pi G_{10})^{-1}$, and after integrating over the extra dimensions, assumed to have volume $L^6$, the 10D integral becomes $L^6 \int d^4x \sqrt{-g} R$. Hence

$$\frac{L^6}{G_{10}} = \frac{1}{G_N} \quad (2.22)$$

and

$$c = \frac{4}{\pi^2} L^6 G_N \tau_3^4 = \frac{L^6 \tau_3^4}{2\pi^2 M_p^2} \quad (2.23)$$

This allows us to evaluate the slow-roll parameters as

$$\epsilon \sim \frac{M_s^{24} L^{12}}{M_p^2 \phi^{10}}, \quad \eta \sim \frac{M_s L^6}{\phi^6} \quad (2.24)$$

To get enough inflation, and to get the observed spectral index

$$n_s = 2\eta - 6\epsilon = 0.95 \pm 0.02 \quad (2.25)$$

as observed by WMAP [26], it is necessary to have small values of $\epsilon$ and $\eta$. The condition $\eta < 1$ is the more restrictive, and implies that the brane separation be

$$r = y \sim \frac{\phi}{M_s^2} \gg L \tag{2.26}$$

which is the impossible situation of the branes being farther from each other than the size of the extra dimensions [25]. One can show that asymmetric compactifications do not help the situation. There were a number of attempts to solve this problem, but these could not be considered complete because of the additional unsolved problem of how to stabilize the moduli of the compactification. The early papers had to assume that the extra dimensions were stabilized somehow, in a way that would not interfere with efforts to keep the inflaton potential flat. However, this is a big assumption, as it later proved.

## 2.2. Warped compactification

The problem of compactification was advanced significantly by Giddings, Kachru and Polchinski (GKP) [21], building on work of Klebanov and Strassler (KS) [20]. The KS construction involved putting branes at a conifold singularity, in order to reduce the number of supersymmetries of the effective action on the D3-brane world volume from $N = 4$ down to the more realistic case of $N = 1$. We need to understand this construction before studying its applications to inflationary model building.

The conifold is a 6D Calabi-Yau space, which can be described in terms of 4 complex coordinates $w_i$, constrained by the complex condition

$$\sum_{i=1}^{4} w_i^2 = 0 \tag{2.27}$$

This looks similar to a cone, fig. 8 and the singularity is at the tip, where $w_i = 0$. The base of the cone has the topology of $S_2 \times S_3$, a 5D manifold. As one approaches the tip, the $S_3$ shrinks to a point, and the topology of the tip is the

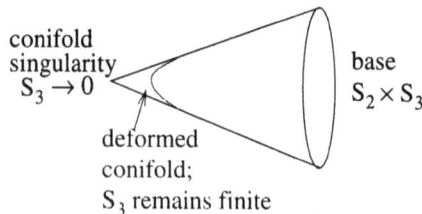

Fig. 8. The conifold and deformed conifold.

remaining $S_2$. The $S_3$ subspace is a 3-cycle, which is referred to as the A-cycle. There is also another (dual) 3-cycle, namely the $S_2$ times a circle which is extended along the radial direction, and which we call the B-cycle. The whole 6D manifold times 4D Minkowski space is a solution to Einstein's equations in 10D, and so it is a suitable background for string theory.

It is possible to deform the conifold so that it is no longer singular, by taking a more general condition than (2.27),

$$\sum_{i=1}^{4} w_i^2 = z \tag{2.28}$$

Here $z$ is known as the complex structure modulus. For $z \neq 0$, the tip of the deformed conifold becomes a smooth point, at which the $S_3$ is no longer singular but has a size determined by $z$. The deformed conifold becomes the solution to Einstein's equations when certain gauge fields of string theory are given nonzero background values. These are the RR field $F_{(3)}$ of type IIB theory, and the NS-NS Kalb-Ramond field $H_{(3)}$. Since they are 3-forms, they can have nonvanishing values when their indices are aligned with some of the 3-cycles mentioned above, similar to turning on an electric field along the 1-cycle (a circle) in the Schwinger model (electrodynamics in 1+1 dimensions). The lines of flux circulate and so obey Gauss's law in this way. In a similar way to the 4-form flux of heterotic M-theory discussed in the previous section, these fluxes are also quantized, due to the generalized Dirac quantization argument. There are integers $M$ and $K$ such that

$$\frac{1}{2\pi\alpha'} \int_A F_3 = 2\pi M, \qquad \frac{1}{2\pi\alpha'} \int_B H_3 = -2\pi K \tag{2.29}$$

where the slope parameter $\alpha'$ is related to the string mass scale by $\alpha' = 1/M_s^2$, and A,B label the 3-cycles mentioned above. It can be shown that these 3-cycles are specified by the conditions $\sum_{i=1}^{4} x_i^2 = z$ for the A-cycle and $x_4^2 - \sum_{i=1}^{3} y_i^2 = z$ if $w_n = x_n + iy_n$. The size of the $S_3$ at the tip of the conifold gets stabilized by the presence of the fluxes; the complex structure modulus takes the value

$$z = e^{-2\pi K/(Mg_s)} \equiv a^3(r_0) \tag{2.30}$$

We will discuss the meaning of $a(r_0)$ shortly.

When the fluxes are turned on, they also warp the geometry of the conifold. The line element for the full 10D geometry is

$$ds^2 = \frac{dx^\mu dx_\mu}{\sqrt{h(r)}} + \sqrt{h(r)}\left(dr^2 + r^2 ds_{T_{1,1}}^2\right) \tag{2.31}$$

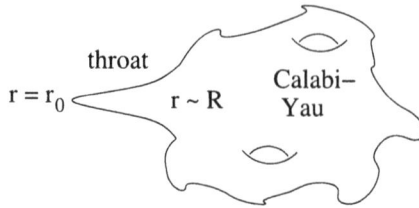

Fig. 9. The KS throat glued to the larger Calabi-Yau manifold where the radial coordinate $r \sim R$.

which is approximately $AdS_5 \times T_{1,1}$. The factor $T_{1,1}$ is the 5D base of the cone, known as an Einstein-Sasaki space, which for our purposes is some compact angular space whose details will not be important. The warp factor $h(r)$ is approximately of the form

$$h(r) = \frac{R^4}{r^4}\left(1 + g_s \frac{M}{K} \times (\ln r \text{ correction})\right)$$

$$\cong \frac{R^4}{r^4}, \qquad R^4 = \frac{27}{4}\pi g_s N \alpha'^2, \qquad N = MK \qquad (2.32)$$

This is a stringy realization of the Randall-Sundrum (RS) model [27], known as the Klebanov-Strassler throat, where the bottom of the throat (the tip of the conifold) is at $r = r_0$, such that

$$\frac{r_0}{R} = z^{1/3} = a(r_0) = \text{warp factor} \qquad (2.33)$$

This explains the introduction of $a^3(r_0)$ in (2.30). The approximation $h = (R/r)^4$ gives the simple $AdS_5$ geometry for 4D Minkowski space times radial direction, and $R$ is the curvature scale of the $AdS_5$. The AdS part of the metric (2.31) can be converted to the RS form $ds^2 = e^{\pm 2ky}dx^2 + dy^2$ through the change of variables $dy = \mp(R/r)dr$, $y = \mp R \ln r$, $r = e^{\mp ky}$, where $k = 1/R$. As shown in fig. 9, the top of the throat is understood to be smoothly glued onto the bulk of the greater Calabi-Yau manifold, whose geometry could be quite different from that of (2.31). The gluing is done at some radius $r \sim R$ where $h(r) \sim 1$.

Let us take stock of what the fluxes have done for us before we continue with the search for inflation. First, they stabilized the complex structure modulus $z$, which without the fluxes was a massless field, with an undetermined value. In a similar way, they stabilize the dilaton, which is essential for any realistic string model since it determines the string coupling, and would give a ruled-out 5th force if massless. Second, they have given us warping, which introduces interesting possibilities for generating hierarchies of scales and generally having another parameter to tune.

Now we consider how the KS throat can be relevant for brane-antibrane infla- tion. We will see that a D3 brane by itself feels no force in the throat, whereas the $\overline{D3}$ antibrane sinks to the bottom. This is because of the combination of gravita- tional and gauge forces—they cancel for D3 but add for $\overline{D3}$. To understand how this comes about, recall that the D3 brane is the source of the $F_{(5)}$ field strength, which comes from the $C_{(4)}$ gauge potential. The fluxes create a background of $C_{(4)}$ or $F_{(5)}$ because there is contribution to the $F_{(5)}$ equation of motion (which we did not previously mention) of the form

$$dF_5 \sim H_3 \wedge F_3 \tag{2.34}$$

Thus the complete action for a brane or an antibrane located at $r = r_1(x^\mu)$ is

$$S = -\tau_3 \int d^4x \frac{1}{h(r_1)} \sqrt{1 - h(r_1)(\partial r_1)^2} \pm \tau_3 \int d^4x (C_4)_{0123} \tag{2.35}$$

where the sign is $+$ for D3 and $-$ for $\overline{D3}$. The form of the DBI part of the action can be understood from the more general definition,

$$S_{DBI} = -\tau_3 \int d^4x \sqrt{-G} \tag{2.36}$$

where the induced metric on the D3-brane is given by

$$G_{\mu\nu} = G_{AB} \frac{\partial X^A}{\partial x^\mu} \frac{\partial X^B}{\partial x^\nu} = \frac{1}{\sqrt{h}} \eta_{\mu\nu} - \sqrt{h}\, \partial_\mu r\, \partial_\nu r \tag{2.37}$$

in the case where $r$ is the only one of the 6 extra dimensions which depends on $x^\mu$. Furthermore the equation of motion for the RR field has the solution

$$(C_4)_{\alpha\beta\gamma\delta} = \frac{1}{h(r_1)} \epsilon_{\alpha\beta\gamma\delta} \tag{2.38}$$

Consider the case where the transverse fluctuations of the brane vanish, $\partial r_1 = 0$. Then the two contributions to the action cancel for D3, but they add for $\overline{D3}$, to give

$$S = -2\tau_3 \left(\frac{r_1}{R}\right)^4 \int d^4x = -2\tau_3 a^4(r_1) \int d^4x \tag{2.39}$$

Notice that $\tau_3 a^4$ is the warped brane tension, and $V = \tau_3 a^4$ is the 4D potential energy associated with this tension. Because of the warp factor, $V$ is minimized at the bottom of the throat. This is why the antibrane sinks to the bottom of the throat, whereas the D3 is neutrally buoyant—it will stay wherever one puts it.

To be consistent, we should recall the words of warning issued earlier in these lectures: since branes carry charge, one is not allowed to simply insert them at will into a compact space. The background must be adjusted to compensate any extra brane charges. This results in a *tadpole* condition

$$\frac{\chi}{24} = N_{D3} - N_{\overline{D3}} + \frac{1}{\kappa_{10}^2 T_3} \int_{C-Y} H_3 \wedge F_3 \tag{2.40}$$

where $\chi$ is the Euler number of the C-Y, and $N_X$ is the number of branes of type $X$. This relation says that for a fixed topology of the C-Y, any change in the net D3 brane charge has to be compensated by a corresponding change in the fluxes. The flux contribution is positive, so there is a limit on the net brane charge which can be accommodated, and this limit is determined by the Euler number, which can be as large as $\sim 10^4$ in some currently known C-Y's.

### 2.3. Warped brane-antibrane inflation

With this background we are ready to think about brane-antibrane inflation in the KS throat, following KKLMMT [28]. Now we want to add not just a D3 or $\overline{D3}$ to the throat, but both together. To find the interaction energy requires a little more work relative to the calculation of (2.35). Following [28], we will consider how the presence of a D3 at $r_1$ perturbs the background geometry and $C_4$ field. Using the perturbed background in the action for the $\overline{D3}$ will then reveal the interaction energy. We can write the perturbation to the background as

$$h(r) \rightarrow h(r) + \delta h(r) \tag{2.41}$$
$$C_4(r) \rightarrow C_4(r) + \delta C_4(r) \tag{2.42}$$

where $\delta h$ is determined by the Poisson equation in 6D,

$$\nabla^2 \delta h = C \delta^{(6)}(\vec{r} - \vec{r}_1) \tag{2.43}$$

in analogy to the gravitational potential for a point mass in 4D; $C$ is a constant which turns out to have the value $C = R^4/N$ in terms of the AdS curvature scale and the product of flux quantum numbers. The solution is

$$\delta h = \frac{R^4}{N r_1^4}, \quad \delta C_4 = -\frac{\delta h}{h^2} \tag{2.44}$$

(see eq.(2.38)). Using the perturbed background fields, we can evaluate the Lagrangian for an antibrane at the bottom of the throat $r = r_0$, and expand to leading order in $(\partial r_1)^2$, to find the result

$$\mathcal{L} = \frac{1}{2}\tau_3(\partial r_1)^2 - 2\tau_3\left(\frac{r_0}{R}\right)^4\left(1 - \left(\frac{r_0}{R}\right)^4 \frac{R^4}{Nr_1^4}\right) \tag{2.45}$$

$$= \frac{1}{2}(\partial\phi)^2 - 2\tau_3 a_0^4\left(1 - a_0^4 \frac{R^4\tau_3^2}{N\phi^4}\right) \tag{2.46}$$

where we introduced the canonically normalized inflaton $\phi = \sqrt{\tau_3}r_1$ and $a_0 = a(r_0)$.

When we compute the slow roll parameters in the new theory with warping, and compare them to our previous calculation without the warp factors, we find that

$$\epsilon = a_0^8 \,\epsilon_{\text{unwarped}}$$

$$\eta = a_0^4 \,\eta_{\text{unwarped}} \tag{2.47}$$

Previously we had the problem that $\eta \sim 1$ unless the branes were separated by more than the size of the extra dimensions. With the new powers of $a_0$, we can easily make $\epsilon \ll \eta \ll 1$ since naturally $a \ll 1$. It looks like warping has beautifully solved the problem of obtaining slow roll in brane-antibrane inflation.

## 2.4. The $\eta$ problem

The story is that KKLLMMT had happily arrived at this result, but then realized the following problem. In fact $\eta$ is not small because one has neglected the effect of stabilizing the Kähler modulus (the overall size of the Calabi-Yau), $T$. The inflaton couples to $T$ in such a way that when a mass is given to $T$, it also inevitably gives a new source of curvature to the inflaton potential which makes $\eta$ of order unity. This is the well-known $\eta$ problem of supergravity inflation models.

To appreciate this, it is useful to consider the low-energy effective SUGRA Lagrangian for the Kähler modulus,

$$\mathcal{L} = G_{T\bar{T}}\partial_\mu T\partial^\mu\bar{T} - e^K\left(D_T W\overline{D_T W} - 3|W|^2\right)$$

$$\equiv G_{T\bar{T}}\partial_\mu T\partial^\mu\bar{T} - V_F \tag{2.48}$$

where $T = \rho + i\chi$ is the complex Kähler modulus, related to the size $L$ of the Calabi-Yau through

$$\rho \equiv e^{4u} \sim L^4, \tag{2.49}$$

$\chi$ is the associated axion,

$$G_{T\overline{T}} = \partial_T \partial_{\overline{T}} K \tag{2.50}$$

is the Kähler metric,

$$K = -3\ln(T + \overline{T}) \tag{2.51}$$

is the Kähler potential, $W(T)$ is the superpotential,

$$D_T W = \partial_T W + W \partial_T K \tag{2.52}$$

is the covariant derivative of $W$, and $V_F$ is the F-term potential.

In the GKP-KS flux compactification, $T$ remains unstabilized because the fluxes generate a superpotential which only depends on the dilaton and complex structure modulus, not on $T$. Once these heavy fields are integrated out, $W$ can be treated as a constant, $W_0$. Then

$$D_T W = W \partial_T K = -3 \frac{W}{T + \overline{T}}, \tag{2.53}$$

$$G_{T\overline{T}} = \frac{3}{(T + \overline{T})^2}, \quad G^{T\overline{T}} = \frac{(T + \overline{T})^2}{3}, \tag{2.54}$$

$$V_F = 3|W|^2 - 3|W|^2 = 0 \tag{2.55}$$

A SUGRA theory where this kind of cancellation occurs is called a *no-scale* model.

Of course having a massless Kähler modulus is unacceptable, for all the reasons mentioned in section 1.1. We need a nontrivial superpotential $W(T)$. We will come back to this. First we focus on the coupling which the inflaton—the position of the mobile D3-brane—has to $T$, which will lead to problems when we introduce $W(T)$. Since Kähler manifolds are complex, we can regard the position of the D3-brane in the 6 extra dimensions as being specified by 3 complex coordinates, which become three complex scalar fields,

$$\phi^i(x^\mu), \quad i = 1, 2, 3 \tag{2.56}$$

Previously we ignored the dependence on the angular directions in $T_{1,1}$ and only kept track of the radial position of the brane in the throat, $r$, but (2.56) is the more precise specification. It can be shown by several arguments (of which we will give one shortly) that when there is a D3 brane in the throat, it changes the Kähler potential by

$$K \to K\left(T + \overline{T} - k(\phi^i, \bar{\phi}^i)\right) \equiv K(2\sigma) \tag{2.57}$$

where $k(\phi^i, \bar{\phi}^i)$ is a real-valued function, known as the Kähler potential for the Calabi-Yau (not to be confused with $K$ which is the Kähler potential for the field space). Now it is the combination $2\sigma$ and not $T + \overline{T}$ which is related to the physical size of the Calabi-Yau,

$$2\sigma = L^4 = e^{4u} = V^{2/3} \tag{2.58}$$

where $V$ is the volume of the extra dimensions. It can be shown (see for example [29]) that

$$k = \sum_i \phi^i \bar{\phi}^i + O(\phi^4) \tag{2.59}$$

in the vicinity of the bottom of the throat, labeled by $\phi^i = 0$. (KKLMMT did not make this identification of $\phi^i = 0$ with the bottom of the throat.)

Next we can generalize the SUGRA action to take account of the additional brane moduli fields, writing

$$\mathcal{L} = G_{I\bar{J}}\partial_\mu \Phi^I \partial^\mu \overline{\Phi}^J - e^K \left(G^{I\bar{J}} D_I W \overline{D_J W} - 3|W|^2\right) \tag{2.60}$$

where $\Phi^I = (T, \phi^i)$. One can show that the new definition of $K$ still leads to a no-scale model if $W$ is constant, so $V_F$ remains zero in the presence of the D3 brane. However the kinetic term is modified; the Kähler metric for the brane moduli is

$$
\begin{aligned}
G_{i\bar{j}} &= \frac{\partial^2 K}{\partial \phi^i \partial \bar{\phi}^j} = \frac{\partial}{\partial \bar{\phi}^j}\left(\frac{3\bar{\phi}^i}{T + \overline{T} - k}\right) \\
&= \frac{3\delta_{ij}}{2\sigma} + \frac{3\phi^j \bar{\phi}^i}{2\sigma}
\end{aligned}
\tag{2.61}
$$

where the last term is considered to be small because we want $\sigma$ to be large to justify integrating out the extra dimensions (so that the corrections from higher dimension operators involving the curvature are small), and this requires that $\phi^2 \ll T + \overline{T}$. Then the kinetic term in the Lagrangian is

$$\mathcal{L}_{\text{kin}} \cong \frac{3}{2\sigma}\left|\partial_\mu \phi^i\right|^2 \tag{2.62}$$

To justify the replacement (2.57), we can derive the kinetic term of the brane modulus by a different method, using the DBI action for the brane [30]. For this, we need the form of the 10D metric that includes not only warping, but also the dependence on the Kähler modulus $\text{Re}(T) = e^{4u}$:

$$ds^2 = \frac{e^{-6u}}{\sqrt{h}}dx^2 + e^{2u}\sqrt{h}\,\tilde{g}^{(6)}_{ab}dy^a dy^b, \tag{2.63}$$

where $\tilde{g}_{ab}^{(6)}$ is the metric with some fiducial volume on the C-Y. The factor $e^{2u}$ in the 6D part of the metric is understandable since $e^u \sim L$; the corresponding factor $e^{-6u}$ in the 4D part is there to put the metric in the Einstein frame, where there is no mixing of the Kähler modulus with the 4D graviton. We can now compute the induced metric in the D3 world-volume,

$$G_{\mu\nu} = G_{AB}\frac{\partial X^A}{\partial\sigma^\mu}\frac{\partial X^B}{\partial\sigma^\nu} = \frac{e^{-6u}}{\sqrt{h}}\eta_{\mu\nu} - e^{2u}\sqrt{h}\,\tilde{g}_{ab}^{(6)}\frac{\partial X^a}{\partial\sigma^\mu}\frac{\partial X^b}{\partial\sigma^\nu} \tag{2.64}$$

Using this in the DBI Lagrangian and Taylor-expanding in the transverse fluctuations, we find that

$$-\tau_3\sqrt{|G_{\mu\nu}|} = -a^4\tau_3 + \frac{3\tau_3}{4\sigma}(\partial_\mu X^a)^2 + O\big((\partial X)^4\big) \tag{2.65}$$

where $a$ is the value of the warp factor at the position of the brane. This reveals the presence of the extra factor of $1/\sigma$ in the brane kinetic term which we did not see previously when we were only considering the effect of warping, and it justifies the SUGRA description given above. We see that the SUGRA and DBI fields are identified through

$$\phi^1 = \sqrt{\frac{3\tau_3}{2\sigma}}(X_1 + iX_2), \; etc. \tag{2.66}$$

With this background we can now present the $\eta$ problem. As long there is no potential for $T$, no new potential is introduced for the inflaton $\phi^i$; however we know that a stabilizing potential for $T$ must be present. Several ways of non-perturbatively generating such a potential were suggested by KKLT [31], which yield an extra contribution to the superpotential,

$$W = W_0 + Ae^{-aT} \tag{2.67}$$

By itself, this new contribution indeed stabilizes $T$ at a nontrivial value, but at the minimum the potential $V_0$ is negative, as shown in fig. 10, which would give an AdS background in 4D, rather than Minkowski or de Sitter space. To lift $V_0$ to a nonnegative value, one needs an additional, supersymmetry-breaking contribution, which thus appears directly as a new contribution to the potential rather than to the superpotential. In fact, placing a $\overline{D3}$ antibrane in the throat does precisely what is needed. As we have already seen, a $\overline{D3}$ has positive energy, twice the warped tension, due to the failure of cancellation of the gravitational and RR potentials. It is minimized at a nonzero value when the $\overline{D3}$ sinks to the bottom of

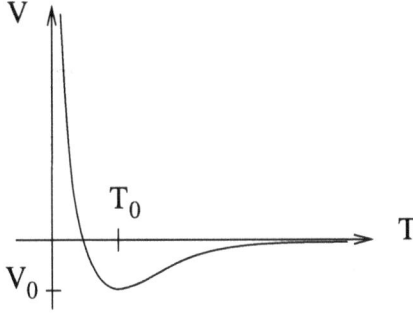

Fig. 10. The KKLT potential for the Kähler modulus, before uplifting by $\overline{D3}$ antibrane.

the throat. The important point is that in the Einstein frame, taking into account the Kähler modulus, the $\overline{D3}$ contribution to the potential takes the form

$$\delta V = \frac{2a_0^4 \tau 3}{(2\text{Re}T)^2} \tag{2.68}$$

However we have seen that when a D3 brane is in the throat, it modifies the volume of the C-Y, and we must replace $2\text{Re}T \rightarrow 2\sigma = T + \overline{T} - |\phi|^2$. In addition, we get the Coulombic interaction energy between the brane and the antibrane so the total inflaton potential becomes

$$\delta V \rightarrow \frac{V(\phi)}{(2\sigma)^2} \cong 2a_0^4 \tau 3 \frac{1 - c|\phi|^{-4}}{(T + \overline{T} - |\phi|^2)^2} \tag{2.69}$$

The $\eta$ problem is now apparent, since even if the Coulomb interaction is arbitrarily small ($c \rightarrow 0$), the inflaton gets a large mass from the term in the denominator:

$$\mathcal{L} \sim \frac{3}{2\sigma} (\partial\phi)^2 - \frac{a_0^4 \tau 3}{2\sigma^2} \left( 1 + \frac{|\phi|^2}{\sigma} \right) \tag{2.70}$$

We see that the inflaton mass is of order the Hubble parameter,

$$m^2 = \frac{2}{3}V = 2H^2 \tag{2.71}$$

(working in units where $M_p = 1$), which implies that $\eta = V''/V = m^2/(3H^2) = 2/3$, whereas inflation requires that $\eta \ll 1$. This is the $\eta$ problem.

### 2.5. Solutions (?) to the η problem

A number of possible remedies to the flatness problem of brane-antibrane inflation have been suggested. We will see that many of these have certain problems
of their own.

### 2.5.1. Superpotential corrections

In an appendix of [28], the possibility of having $\phi$-dependent corrections to the
superpotential was discussed,

$$W = W_0 + Ae^{-aT}\left(1 + \delta|\phi|^2\right) \tag{2.72}$$

which is just an ansatz for the kind of corrections which might arise within string
theory. The inflaton mass from this modified superpotential was computed to be

$$m^2 = 2H^2\left(1 - \frac{|V_{AdS}|}{V_{dS}}(\beta - 2\beta^2)\right) \tag{2.73}$$

where $\beta = -\delta/a$, and $V_{AdS}$ is the value of the potential at the AdS minimum,
before uplifting with the antibrane. By fine-tuning $\beta$, one can make $m^2 \ll H^2$
and thus get acceptable inflation. (It was argued in [32] that the tuning is only
moderate, but this seems to be contingent upon taking $3\sigma$ rather than $1\sigma$ error
bars from the experimental constraints.) The suggestion of using superpotential
corrections was made before any concrete calculations of the actual corrections
had been carried out. Since then, it has been shown that they have a form which
drive the brane more quickly to the bottom of the throat rather than slowing it
down; the corrections have the wrong sign [33, 34].

### 2.5.2. Tuning the length of the throat

In KKLMMT, the impression was given that the function $k(\phi, \bar{\phi}) = |\phi|^2 + O(\phi^4)$
was expanded around some point in the C-Y which did not necessarily coincide
with the bottom of the throat; in their calculation it did not matter where this
point was because they assumed the Coulomb part of the potential was negligible
compared to the dependence via $T + \bar{T} - k(\phi, \bar{\phi})$. However, if the origin of
$\phi^i$ was assumed to be at a position other than the bottom of the throat, thus
located at some position $\phi_0$, then there could be a competing effect between the
two sources of $\phi$-dependence, which when properly tuned could lead to a flat
potential without invoking superpotential corrections. The full potential then had
the form [35]

$$\mathcal{L} = \frac{3T}{(2T - \phi^2)^2}(\partial\phi)^2 - \frac{2\tau_3 a_0^4}{(2T - \phi^2)^2}\left(1 + \frac{a_0^4}{N(\phi - \phi_0)^4}\right)^{-1} \tag{2.74}$$

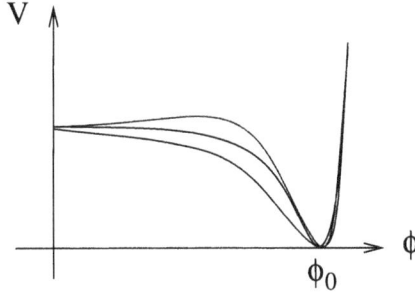

Fig. 11. The brane-antibrane potential for different values of $\phi_0$ relative to other parameters in the potential.

where (recall that) $N = JK$ is the product of the flux quantum numbers, and we have displayed the potential in a form which makes sense even as $\phi \to \phi_0$ (whereas keeping only the first term of the Taylor expansion does not). By tuning $\phi_0$ we can adjust the flatness of the potential in the region between $\phi = 0$, assumed to be somewhere near the top of the throat, and $\phi = \phi_0$, the bottom. Figure 11 shows the effect of changing the value of $\phi_0$ relative to other dimensionful combinations of parameters in the potential.

Unfortunately, the same computations [33, 34] which clarified the nature of the superpotential corrections also made it clear that $\phi = 0$ indeed corresponds to the bottom of the throat, so $\phi_0$ is not a free parameter which can be tuned. However there is a related idea that appears to be viable. Suppose there are two throats, related to each other by a $Z_2$ symmetry. A brane in between them would not be able to decide which throat to fall into, so the potential must be flat at this point in the middle [36].

### 2.5.3. Multibrane inflation
It was also suggested that instead of using a single D3-$\overline{\text{D3}}$ pair, one can use stacks of $M$ D3-branes and $M$ $\overline{\text{D3}}$-branes. If the branes in each stack are coincident, the Lagrangian becomes

$$\mathcal{L} = M \frac{3T}{(2T - M\phi^2)^2}(\partial\phi)^2 - M \frac{\frac{2\tau_3 a_0^4}{(2T-M\phi^2)^2}}{1 + M \frac{a_0^4}{N(\phi-\phi_0)^4}} \tag{2.75}$$

In [37] it was shown that the shape of the potential can vary with $M$ in such a way that for large $M$ there is a metastable minimum in which the brane stacks remain separated, while for small $M$ the minimum becomes a maximum (fig. 12). For an intermediate value of $M$, the potential is nearly flat. The interesting feature

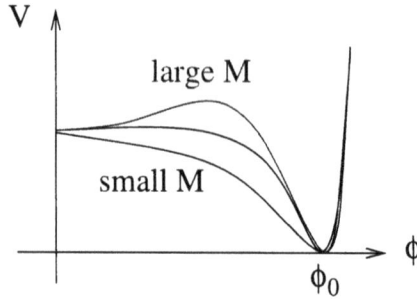

Fig. 12. The brane-antibrane potential for different values of $M$, the number of branes or antibranes in each stack.

is that $M$ can change dynamically by tunneling of branes from the metastable minimum through the barrier to the stack of antibranes. In this way one could eventually pass through a nearly flat potential suitable for inflation. Unfortunately this mechanism also relied on the parameter $\phi_0$ which was shown to be zero.

One might have hoped that even without the possibility of tuning, having $M$ branes in the stack could allow for assisted inflation [38], where the slow-roll parameters get reduced by a factor of $M$ if all inflatons are rolling in the same way. Unfortunately this mechanism only works if the potential for each field by itself is independent of $M$. In the potential (2.75), this is not the case: the $V''/V$ also increases with $M$, undoing the assistance which would have come from renormalizing the field to get a standard kinetic term.[1] On the other hand, ref. [39] finds that assisted inflation does work for multiple M5 branes moving along the 11th dimension of M-theory.

### 2.5.4. DBI inflation (D-celleration)
References [18, 19] explored a different region of the brane-antibrane parameter space and exposed a qualitatively new possibility, in which *fast* roll of the D3 brane could still lead to inflation. In this regime, one does not expand the DBI action in powers of $\dot{\phi}^2$ since it is no longer considered to be small. The Lagrangian has the form

$$\mathcal{L} = -a^3(t)\left(\frac{\tau_3}{h(r)}\sqrt{1 - h(r)\dot{r}^2} - V(r)\right) \qquad (2.76)$$

where $a(t)$ is the scale factor and $r$ is the radial position of the D3-brane in the throat. To be in the qualitatively new regime, one wants the potential to be so steep that the brane is rolling nearly as fast as the local speed limit in the throat,

---

[1] I thank Andrew Frey for this observation.

beyond which the argument of the square root changes sign: $h(r)\dot{r}^2 \sim 1$. One finds that this results in power-law inflation,

$$a(t) \propto t^{1/\epsilon}, \quad \epsilon \cong \frac{M_p}{m} \sqrt{\frac{3g_s}{R^4 M_s^4}} \tag{2.77}$$

where we recall that $R$ is the AdS curvature scale of the throat, $M_s$ is the string scale, and $m$ is the inflaton mass. One sees that inflation works best when $m$ is as large as possible, which is quite different from inflation with a conventional kinetic term. Even if $m \ll M_p$, we can compenstate by taking $R \gg M_s^{-1}$ to get $\epsilon \ll 1$.

However, this takes a huge amount of flux in the original KS model. Recall that $R^4 M_s^4 = \frac{27\pi}{4} g_s N$ where $N = JK$, so

$$\epsilon = \frac{M_p}{m} \sqrt{\frac{4}{9\pi N}} \tag{2.78}$$

If $m$ is of order the GUT scale, then we need $J \sim K \sim 10^3$. In fact the problem is even worse, because the COBE normalization of the CMB power spectrum gives

$$P = \frac{1}{4\pi^2} \frac{g_s}{\epsilon^4 \lambda} \tag{2.79}$$

where $\lambda = R^4 M_s^4$ is identified as the 't Hooft coupling in the context of the AdS-CFT correspondence. The COBE normalization implies that $\lambda \sim 10^{14}$, which requires an Euler number for the C-Y of the same order, according to the tadpole condition (2.40). This exceeds by many orders of magnitude the largest known example. Ref. [19] suggests some ways to overcome this difficulty.

One of the most interesting consequences of DBI inflation is the prediction of large nongaussianity, which is not possible in conventional models of single-field inflation. To see why it occurs in the DBI model, one should consider the form of the Lagrangian for fluctuations $\delta\phi$ of the inflaton. The unperturbed kinetic term has the form

$$\mathcal{L}_{\text{kin}} = -\frac{1}{h}\sqrt{1 - h\dot{\phi}^2} \equiv \frac{1}{h\gamma} \tag{2.80}$$

where $\gamma$ is analogous to its counterpart in special relativity; in the DBI inflation regime, we have $\gamma \gg 1$ since the field is rolling close to its maximum speed. The first variation of this term is

$$\delta\mathcal{L}_{\text{kin}} = \frac{1}{h}\frac{h\dot{\phi}}{\sqrt{1 - h\dot{\phi}^2}}\delta\dot{\phi} = \gamma\dot{\phi}\delta\dot{\phi} \cong \frac{\gamma}{\sqrt{h}}\delta\dot{\phi} \tag{2.81}$$

The fluctuation Lagrangian is enhanced relative to the zeroth order Lagrangian by powers of $\gamma$. The higher the order in $\delta\phi$, the more powers of $\gamma$. Nongaussian features start appearing at order $\delta\phi^3$, through the bispectrum (3-point function) of the fluctuations. DBI inflation predicts that the nonlinearity parameter, which is the conventional measure of nongaussianity, is of order

$$f_{NL} \cong 0.06\gamma^2 \tag{2.82}$$

which is $\gamma^2$ times the prediction of ordinary inflation models. The current experimental limit is $|f_{NL}| \lesssim 100$ [26], (hence $\gamma < 40$), with a future limit of $|f_{NL}| \lesssim 5$ projected for the PLANCK experiment. Thus DBI inflation has a chance of producing observable levels of nongaussianity, which conventional models do not.

Moreover, DBI inflation predicts a tensor component of the CMB with tensor-to-scalar ratio

$$r = 16\frac{\epsilon}{\gamma} \tag{2.83}$$

so that an upper bound on nongaussianity (hence on $\gamma$) implies a lower bound on tensors—a kind of no-lose theorem.

The original work on DBI inflation focused on inflation far from the tip of the throat, but recent work has pointed out that generically (for the KS throat background) one tends to get the last 60 e-foldings of inflation at the bottom the throat, and having larger-than-observed levels of nongaussianity [40].

### 2.5.5. Shift symmetry; D3-D7 inflation

It was suggested [41] that there could be a symmetry of the Lagrangian

$$\phi_i \rightarrow \phi_i + c_i \tag{2.84}$$

$$T \rightarrow \bar{c}_i\phi_i + \frac{1}{2}\sum c_i\bar{c}_i \tag{2.85}$$

which leaves a flat direction despite the combination $T + \bar{T} - |\phi|^2$ in the Kähler potential. In fact, it is only necessary to preserve a remnant of this shift symmetry, $\phi_i \rightarrow \phi_i + \mathrm{Re}(c_i)$ to get a flat inflaton potential; this can be achieved with a superpotential of the form

$$W = W_0 + Ae^{-a(T-\frac{1}{2}\phi^2)+i\beta_i\phi_i} \tag{2.86}$$

Such a symmetry requires an underlying symmetry of the compactification manifold, namely isometries, which would exist for toroidal compactifications. These are not favored for realistic model building.

It was argued that shift symmetry can occur in models where inflation is driven by the interaction between a D3 and a D7 brane [42]. It is interesting to note that the DBI action for D7-branes does not get the factor of $1/\sigma$ which was the harbinger of the $\eta$ problem for D3-$\overline{\text{D3}}$ inflation. If we split the C-Y metric into two factors $\tilde{g}^{(4)}$ for the extra dimensions parallel to D7, and $\tilde{g}^{(2)}$ for those which are transverse, the induced metric for D7 gives

$$\sqrt{G_{ab}} \cong \sqrt{g^{(4)}\tilde{g}^{(2)}} \left(1 - g^{\mu\nu}_{(4)}\, \tilde{g}^{(2)}_{ij}\, \partial_\mu S^i \partial_\nu S^j\right)^{1/2}$$

$$\sim (e^{-6u+2u})^{4/2}\left(1 - \frac{1}{2}e^{6u+2u}(\partial S)^2\right) \tag{2.87}$$

and the factors of $u$ cancel out of the kinetic term for the two transverse fluctuations $S^i$.

D3-D7 inflation can be pictured as a point (the D3 brane fully localized in the extra dimensions) interacting with a 4-brane (the 4 dimensions of the D7 which wrap the compact dimensions), which is like a D3 smeared over 4 extra dimensions. Thus the $1/r^4$ Coulombic potential of D3-$\overline{\text{D3}}$ gets integrated over 4 dimensions to become a logarithmic potential for D3-D7. This potential arises from SUSY-breaking effects, such as turning on fluxes which live in the world-volume of the D7-brane.

The low-energy description of this model is a SUGRA hybrid inflation model, with potential

$$V(S, \phi_+, \phi_-) = 2g^2|S|^2\left(|\phi_+|^2 + |\phi_-|^2\right) + 2g^2\left|\phi_+\phi_- - \frac{1}{2}\zeta_+\right|^2$$

$$+ \frac{1}{2}g^2\left(|\phi_+|^2 - |\phi_-|^2 - \zeta_3\right)^2 \tag{2.88}$$

where $S$ is the brane separation in a $T^2/Z_2$ subspace of the C-Y, $\phi_\pm$ are the lowest modes of the strings stretched between D3 and D7, and $\zeta_3$, $\zeta_+$ are Fayet-Iliopoulos terms, which arise due to a background Kalb-Ramond $H_3$ field strength.

$S$ is the inflaton, while the fields $\phi_\pm$ are "waterfall" fields, whose tachyonic instability brings about the end of inflation, provided that $\zeta_3$ or $\zeta_+ \neq 0$. In that case $m^2_{\phi_\pm}$ changes sign as $S \to 0$. The instability is the dissolution of the D3 in the D7. At large $S$, the waterfall fields' potentials are minimized at $\phi_\pm = 0$, and the potential for $S$ is exactly flat (as demanded by the shift symmetry). However SUSY-breaking generates a potential for $S$ at one-loop,

$$\Delta V = \frac{g^4}{16\pi^2}\zeta_3^2 \ln\left|\frac{S}{S_c}\right|^2 \tag{2.89}$$

as is needed to get inflation to end. However it is still an open question as to whether the shift symmetry gets broken more badly than this, possibly spoiling inflation, when the full quantum corrections are taken into account [43].

### 2.5.6. Racetrack and Kähler moduli inflation

Another alternative is to get rid of the mobile D3-brane entirely and instead use the compactification moduli to get inflation [44]. This requires a slightly more complicated superpotential

$$W = W_0 + Ae^{-aT} + Be^{-bT} \tag{2.90}$$

which for historical reasons is known as a racetrack superpotential. It can come about from gaugino condensation in two strongly interacting gauge groups, *e.g.* SU(N)×SU(M), giving $a = 2\pi/N$, $b = 2\pi/M$. Another possibility is to have two different Kähler moduli [45], with

$$W = W_0 + Ae^{-aT_1} + Be^{-bT_2} \tag{2.91}$$

The latter form arises from a a very specific C-Y compactification [46]; hence this model is rigorously based on string theory. By fine tuning parameters, one can find a saddle point such that the axionic direction Im$T$ (or a linear combination of them in the case where there are two Kähler moduli) is flat and gives rise to inflation. The potential is shown in fig. 13. Interestingly, both models (2.90, 2.91)

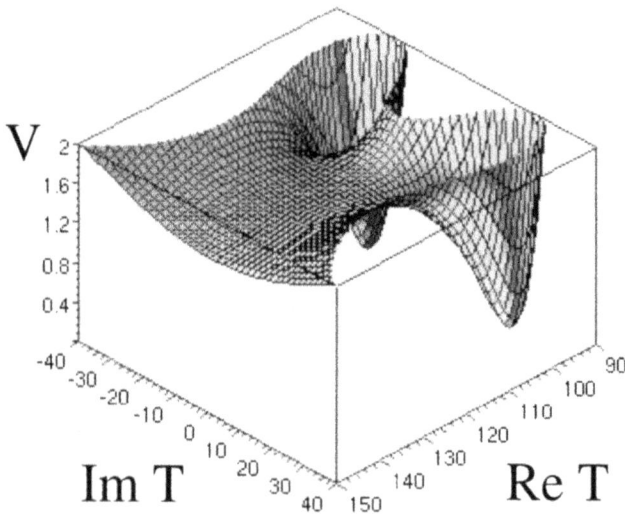

Fig. 13. The potential for the Kähler modulus in a racetrack inflation model.

seem to point to a spectral index of $n_s = 0.95$, in agreement with the recent WMAP measurement. Getting a flat enough spectrum however also seems to require uncomfortably large values for the rank of the gauge group(s). A positive feature of the model is that the existence of degenerate minima leads to inflating domain walls, an example of topological inflation, which solves the initial value problem: it is guaranteed that regions undergoing inflation will always exist in the universe, which are the borderlines between regions existing in the different vacua.

The fine-tuning problem of the simplest racetrack models can be overcome in backgrounds with a larger number of Kähler moduli [47]. In these models, the C-Y volume can naturally be quite large [48], and the potential can be flat since terms proportional to $e^{-aT_i}$ are small corrections in the large-$T_i$ regime, as long as there are some contributions to the potential which are not exponentially suppressed. Ref. [47] finds that this can be realized in models with three or more Kähler moduli, thus alleviating the need for fine tuning.

## 2.6. Confrontation with experiment

Having described a few of the inflationary models which arise from string theory, let's reconsider the experimental constraints which can be used to test them.

### 2.6.1. Power spectrum
The scalar power spectrum of the inflaton is the main connection to observable physics. It is parametrized as

$$P(k) = A_s \left( \frac{k}{k_0} \right)^{n_s - 1} \tag{2.92}$$

where currently $A_s = 4 \times 10^{-10}$ if $k_0 = 7.5a_0 H_0$ (this is the COBE normalization [49]) and $n_s = 0.95 \pm 0.02$. For a single-field inflationary model $P(k)$ is predicted to be

$$P(k) = \frac{1}{50\pi^2} \frac{H^4}{\frac{1}{2}\dot{\phi}^2} \tag{2.93}$$

evaluated at horizon crossing, when $k/a = H$. Using the slow-roll equation of motion for the inflaton, $3H\dot{\phi} = V'$, one can show that [49]

$$P(k) = \frac{V}{150\pi^2 \epsilon M_p^4} \tag{2.94}$$

Many of the inflation models we considered have noncanonically normalized kinetic terms; for example supergravity models typically have $\mathcal{L}_{\text{kin}} \sim (\partial T/T)^2$. It is often useful to be able to evaluate quantities in the original field basis rather than having to change to canonically normalized fields. The correct generalization of (2.93) is simply

$$P(k) = \frac{1}{50\pi^2} \frac{H^4}{\mathcal{L}_{\text{kin}}} \tag{2.95}$$

In addition to the scalar power, there is a tensor contribution to the temperature fluctuations, whose amplitude relative to the scalar contribution is

$$r = \frac{A_t}{A_s} = 16\epsilon \tag{2.96}$$

in terms of the slow-roll parameter $\epsilon$ (but see the different prediction (2.83) from DBI inflation). The current limit is $r < 0.3$ [26].

### 2.6.2. Numerical methods

Moreover in some models, like racetrack inflation, the inflaton is some combination of several fields, and it is useful to have an efficient way of numerically solving the equations of motion. Here is one approach. Let

$$\pi_i = \frac{\partial \mathcal{L}}{\partial \dot{\phi}_i} = f_{ij}(\phi)\dot{\phi}_j \tag{2.97}$$

which can be inverted algebraically to find $\dot{\phi}_i$ as a function of the canonical momenta. The equations of motion for the scalar fields are

$$\frac{d}{dt}\left(a^3 \pi_i\right) = a^3 \frac{\partial}{\partial \phi_i}(\mathcal{L}_{\text{kin}} - V)$$

$$\implies \dot{\pi}_i + 3H\pi_i = \frac{\partial}{\partial \phi_i}(\mathcal{L}_{\text{kin}} - V) \tag{2.98}$$

This together with

$$\dot{\phi}_i = f_{ij}^{-1}\pi_j \tag{2.99}$$

constitute a set of coupled first order equations which can be numerically integrated along with the Friedmann equation which determines $a(t)$. However we are usually not interested in the actual time dependence of the solutions, but rather in the dependence on the number of e-foldings, $N = \int H dt$. It is therefore more efficient to trade $t$ for $N$ using $dN = H dt$, $H \frac{d}{dN} = \frac{d}{dt}$. This makes it

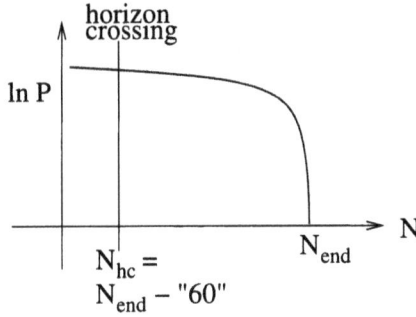

Fig. 14. (a) Scalar power as a function of the number of e-foldings.

unnecessary to also solve the Friedmann equation, and one can simply transform the scalar field equations of motion to

$$\frac{d\pi_i}{dN} + 3\pi_i = \frac{1}{H}\frac{\partial}{\partial \phi_i}(\mathcal{L}_{kin} - V) \tag{2.100}$$

$$\frac{d\phi_i}{dN} = \frac{1}{H}f_{ij}^{-1}\pi_j \tag{2.101}$$

where

$$H = \sqrt{\frac{\mathcal{L}_{kin} + V}{3M_p^2}} \tag{2.102}$$

Knowing $\phi_i(N)$ and $\pi_i(N)$, one can compute the power $P$ as a function of $N$, as shown in figure 14. The tilt at any point can be computed from

$$n_s - 1 = \frac{d\ln P}{d\ln k} = \frac{d\ln P}{d\ln N}\left(1 + H^{-1}\frac{dH}{dN}\right)^{-1} \cong \frac{d\ln P}{d\ln N} \tag{2.103}$$

Here we used the horizon crossing condition $k = aH = e^N H$, hence $\ln k = N + \ln H$, $d\ln k = dN + d\ln H/dN$ to change variables from $k$ to $N$.

We are of course interested in evaluating the properties of the power spectrum at the COBE scale, when perturbations roughly of the size of the present horizon originally crossed the inflationary horizon. This occurs at an $N$ given by $N = N_{hc}$, nominally 60 e-foldings before the end of inflation. However when we say 60, this is an upper bound, determined by the maximum energy scale of inflation $M_{inf}$ consistent with the experimental limit on the tensor contribution to the spectrum. The actual number of e-foldings since horizon crossing is given by

$$N_e = 62 + \ln\left(\frac{M_{inf}}{10^{16}\,\text{GeV}}\right) - \frac{1}{3}\ln\left(\frac{M_{inf}}{T_{rh}}\right) \tag{2.104}$$

where $T_{rh}$ is the reheating temperature at the end of inflation. In the absence of detailed knowledge about reheating, one might assume that $T_{rh} \sim M_{inf}$, meaning that reheating is very fast and efficient, or possibly that $T_{rh} \lesssim 10^{10}$ GeV, since higher reheat temperatures tend to produce too many dangerously heavy and long-lived gravitinos [50].

In many models of inflation, one has the freedom by adjusting parameters to change the overall scale of the inflationary potential without changing its shape:

$$V \to \alpha V \tag{2.105}$$

In the slow-roll regime, $\alpha$ scales out of the inflaton equation of motion, and has no effect on the number of e-foldings or the shape of the spectrum; it only affects the normalization of $P$. In such a case one can always use $\alpha$ to satisfy the COBE normalization. One can then focus attention on tuning parameters such that $n_s = 0.95$ at horizon crossing to fit the observed tilt.

### 2.6.3. New signatures—cosmic strings

One of the great challenges for string cosmology is in producing signals that could distinguish string models of inflation from ordinary field theory ones. We have seen that DBI inflation was distinctive in that respect, since it can give large nongaussianity, in contrast to conventional single-field models (see however [51]). Perhaps the the most dramatic development in this direction has been the possibility of generating cosmologically large superstring remnants in the sky [52]. To see how this can come about, consider the effective action for the complex tachyonic field in the D3-$\overline{\text{D3}}$ system which describes the instability. It resembles the DBI action, but has a multiplicative potential [53],

$$S = -2\tau_3 \int d^4x \, V(|T|)\sqrt{1 - |\partial T|^2} \tag{2.106}$$

where $V = 1/\cosh(|T|/\sqrt{2\alpha'})$. Notice that $V$ has an unstable maximum, indicative of a tachyon. However if the branes are separated by a distance $r$, the potential gets modified at small $T$ such that the curvature can become positive,

$$V(|T|) \cong 1 + \left[\left(\frac{M_s r}{2\pi}\right)^2 - \frac{1}{2}\right]|M_s T|^2 + O\left((M_s T)^4\right) \tag{2.107}$$

This shows that the instability turns on only when D3 and $\overline{\text{D3}}$ come within a critical distance of each other.

The action (2.106) admits classical topologically stable string defect solutions, for example a string oriented along the $x^3$ direction, whose solution in polar coordinates in the $x^1$-$x^2$ plane has the form $T = T(\rho)e^{i\theta}$. Sen has shown that

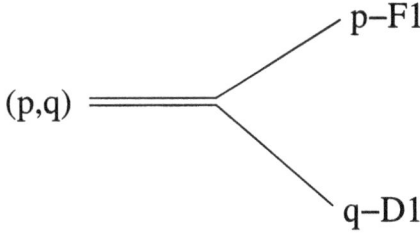

Fig. 15. The junction between a $(p, q)$ string and a bound state of $p$ fundamental strings or $q$ D-strings.

these kinds of solutions are an exact description of D1-branes. Ref. [52] argues that fundamental F1-strings will also be produced, as a consequence of $S$-duality, which exchanges $g_s \leftrightarrow g_s^{-1}$ as well as $F1 \leftrightarrow D1$. The strings remain localized at the bottom of the KS throat because the warp factor provides a gravitational potential barrier.

Again, field theory models can also predict relic cosmic strings, so one would like to find ways of distinguishing them from cosmic superstrings. One possibility is the existence of bound states called $(p, q)$ strings, made of $p$ F1-strings and $p$ D1-branes, whose tension is given by

$$\tau_{p,q} = \frac{M_s^2}{2\pi} \sqrt{(p - C_{(0)}q)^2 + q^2/g_s^2} \tag{2.108}$$

where $C_{(0)}$ is the (possibly vanishing) background value of the RR scalar which gives rise to $F_{(1)}$. (For complications arising from having the strings in a warped throat, see ref. [54].) There exist three-string junctions joining a $(p, 0)$, $(0, q)$ and $(p, q)$ string, as shown in fig. 15. A truly distinctive observation would be to see such a junction in the sky using the gravitationally lensed images of background galaxies to measure the tensions of the three strings, but we would have to be lucky enough to be relatively close to such strings. In the last few years there was some initial evidence for a cosmic string lens (CSL-1) [55], but this has been refuted by Hubble Space Telescope observations.

A network of relic cosmic superstrings could in principle be distinguished from field theoretic strings by the details of their spectrum of emitted gravity waves, which may be observable at LIGO. All relativistic strings produce cusps during their oscillations, which are strong emitters of gravity waves. A major difference between superstrings and field-theory strings is in their interaction probabilities when two strings (or parts of the same string) cross. Field-theory strings almost always intercommute (the ends of the strings at the intersection point swap partners, fig. 16), whereas superstrings can have a smaller intercom-

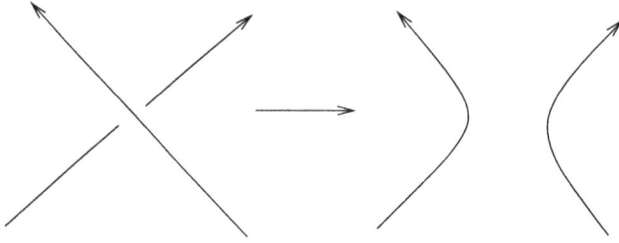

Fig. 16. (a) Intercommutation of two strings after they intersect.

mutation probability, $10^{-3} < P < 1$. If future LIGO observations can eventually measure the spectrum of stochastic gravity waves $dN/dh = Ah^{-B}$, where $h$ is the amplitude of the metric perturbation, the two parameters $A$ and $B$ could be used to determine the intercommutation probability $P$ and the tension of the cosmic strings, $\mu$ [56,57]. In an unwarped model $\mu$ would just be the ordinary string tension $1/\alpha'$, but this can be reduced to much lower values in a warped throat. The current limit on $\mu$ from gravity waves, CMB, and especially pulsar timing is on the order of $G\mu < 10^{-7}$.

### 2.6.4. The reheating problem

One of the interesting consequences of brane-antibrane inflation is its new implications for reheating at the end of inflation. Initially (before the warped models were introduced) it was noticed that Sen's tachyon potential (2.106) has no local minimum around which oscillations and conventional reheating could occur [58], so a new picture of how reheating might take place was suggested, where tachyon condensation following higher dimensional brane annihilation such as D5-$\overline{\text{D5}}$ would give rise to D3-brane defects, including our own universe, which start in a hot, excited state [59]. However, this has the problem that the D3-brane defects could also wrap one of the extra dimensions, appearing as a domain wall to 3D observers, and overclosing the universe [60].

With the warped models came a new twist. While it became understood that the tachyon condensate decays into closed strings [61] as well as lower dimensional branes, there was a potential problem with transferring this energy to the visible sector of the Standard Model (SM), unless we lived in a brane in the same throat as that where inflation was occurring. This was disfavored for two reasons: (1) the mass scale of the throat should be the inflationary scale, which is typically much higher than the scale desired for SUSY breaking to get the SM, and (2) cosmic superstrings left over from brane-antibrane annihilation are unstable to breaking up if there are any D3 branes left in the throat [52]. To overcome these problems, one would like to localize the SM brane in a different throat from the inflationary throat, but in this case, the warp factor acts as a gravitational potential barrier. It was shown that this barrier can be overcome due to the enhanced couplings of Kaluza-Klein (KK) gravitons to the SM brane [62] or if there are resonant effects [63]. Another mechanism for reheating is the oscillations of the SM throat which gets deformed by the large Hubble expansion during inflation [64].

The KS throat can even give rise to unwanted relic KK gravitons during reheating [65]. The reason is that angular momentum in the $T_{1,1}$ directions can be excited within the KS throat. If the throat was not attached to a C-Y which broke the angular isometries, these angular momenta would be exactly conserved, and the excited KK modes carrying them would not be able to decay, only to annihilate, leaving some relic density of superheavy particles to overclose the universe. The breaking effect is suppressed because of the fact that the KK wave functions in the throat are exponentially small in the C-Y region where the isometries are broken. Thus the angular modes are unstable, but with a lifetime that is enhanced by inverse powers of the warp factor in the throat. It is crucial to know exactly what power of $a_0$ arises in the decay rate for determining the observational constraints on the model, but this seems to be quite dependent on the detailed nature of the C-Y to which the throat is attached [66].

## Acknowledgments

I thank N. Barnaby, K. Dasgupta, A. Frey, H. Firouzjahi, and H. Stoica for many helpful comments and answers to questions while I was preparing these lectures. Thanks to L. McAllister for constructive comments on the manuscript.

## References

[1] D. J. Kapner, T. S. Cook, E. G. Adelberger, J. H. Gundlach, B. R. Heckel, C. D. Hoyle and H. E. Swanson, "Tests of the gravitational inverse-square law below the dark-energy length scale," arXiv:hep-ph/0611184.

[2] L. Susskind, "The anthropic landscape of string theory," arXiv:hep-th/0302219.

[3] S. R. Coleman and F. De Luccia, "Gravitational Effects On And Of Vacuum Decay," Phys. Rev. D **21**, 3305 (1980).

[4] A. D. Linde, "Eternal extended inflation and graceful exit from old inflation without Jordan-Brans-Dicke," Phys. Lett. B **249**, 18 (1990).

[5] S. Weinberg, "Anthropic bound on the cosmological constant," Phys. Rev. Lett. **59**, 2607 (1987).

[6] S. Weinberg, "The cosmological constant problem," Rev. Mod. Phys. **61**, 1 (1989).

[7] A. G. Riess *et al.* [Supernova Search Team Collaboration], "Observational Evidence from Supernovae for an Accelerating Universe and a Cosmological Constant," Astron. J. **116**, 1009 (1998) [arXiv:astro-ph/9805201].

S. Perlmutter *et al.* [Supernova Cosmology Project Collaboration], "Measurements of Omega and Lambda from 42 High-Redshift Supernovae," Astrophys. J. **517**, 565 (1999) [arXiv:astro-ph/9812133].

[8] R. Bousso and J. Polchinski, "Quantization of four-form fluxes and dynamical neutralization of the cosmological constant," JHEP **0006**, 006 (2000) [arXiv:hep-th/0004134].

[9] J. D. Brown and C. Teitelboim, 'Neutralization of the cosmological constant by membrane creation," Nucl. Phys. B **297**, 787 (1988); J. D. Brown and C. Teitelboim, "Dynamical neutralization of the cosmological constant," Phys. Lett. B **195**, 177 (1987).

[10] M. R. Douglas, "Basic results in vacuum statistics," Comptes Rendus Physique **5**, 965 (2004) [arXiv:hep-th/0409207].

[11] S. P. de Alwis, "The scales of brane nucleation processes," arXiv:hep-th/0605253.

[12] M. Tegmark and M. J. Rees, "Why is the CMB fluctuation level $10^{-5}$?," Astrophys. J. **499**, 526 (1998) [arXiv:astro-ph/9709058].

M. L. Graesser, S. D. H. Hsu, A. Jenkins and M. B. Wise, "Anthropic distribution for cosmological constant and primordial density perturbations," Phys. Lett. B **600**, 15 (2004) [arXiv:hep-th/0407174].

[13] N. Arkani-Hamed, S. Dimopoulos and S. Kachru, "Predictive landscapes and new physics at a TeV," arXiv:hep-th/0501082.

[14] S. A. Abel, C. S. Chu, J. Jaeckel and V. V. Khoze, "SUSY breaking by a metastable ground state: Why the early universe preferred the non-supersymmetric vacuum," arXiv:hep-th/0610334.

N. J. Craig, P. J. Fox and J. G. Wacker, "Reheating metastable O'Raifeartaigh models," arXiv:hep-th/0611006.

W. Fischler, V. Kaplunovsky, C. Krishnan, L. Mannelli and M. Torres, "Meta-stable supersymmetry breaking in a cooling universe," arXiv:hep-th/0611018.

S. A. Abel, J. Jaeckel and V. V. Khoze, "Why the early universe preferred the non-supersymmetric vacuum. II," arXiv:hep-th/0611130.

L. Pogosian and A. Vilenkin, "Anthropic predictions for vacuum energy and neutrino masses in the light of WMAP-3," arXiv:astro-ph/0611573.

D. Schwartz-Perlov, "Probabilities in the Arkani-Hamed-Dimopolous-Kachru landscape," arXiv:hep-th/0611237.

[15] M. Dine, E. Gorbatov and S. D. Thomas, "Low energy supersymmetry from the landscape," arXiv:hep-th/0407043.

[16] N. Arkani-Hamed and S. Dimopoulos, "Supersymmetric unification without low energy supersymmetry and signatures for fine-tuning at the LHC," JHEP **0506**, 073 (2005) [arXiv:hep-th/0405159].

[17] G. F. Giudice and A. Romanino, "Split supersymmetry," Nucl. Phys. B **699**, 65 (2004) [Erratum-ibid. B **706**, 65 (2005)] [arXiv:hep-ph/0406088].

[18] E. Silverstein and D. Tong, "Scalar speed limits and cosmology: Acceleration from D-cceleration," Phys. Rev. D **70**, 103505 (2004) [arXiv:hep-th/0310221].

[19] M. Alishahiha, E. Silverstein and D. Tong, "DBI in the sky," Phys. Rev. D **70**, 123505 (2004) [arXiv:hep-th/0404084].

[20] I. R. Klebanov and M. J. Strassler, "Supergravity and a confining gauge theory: Duality cascades and $\chi$SB-resolution of naked singularities," JHEP **0008**, 052 (2000) [arXiv:hep-th/0007191].

[21] S. B. Giddings, S. Kachru and J. Polchinski, "Hierarchies from fluxes in string compactifications," Phys. Rev. D **66**, 106006 (2002) [arXiv:hep-th/0105097].

[22] J. Polchinski, "String theory. Vol. 2: Superstring theory and beyond,"

[23] T. Banks and L. Susskind, "Brane - Antibrane Forces," arXiv:hep-th/9511194.

[24] G. R. Dvali and S. H. H. Tye, "Brane inflation," Phys. Lett. B **450**, 72 (1999) [arXiv:hep-ph/9812483].

[25] C. P. Burgess, M. Majumdar, D. Nolte, F. Quevedo, G. Rajesh and R. J. Zhang, "The inflationary brane-antibrane universe," JHEP **0107**, 047 (2001) [arXiv:hep-th/0105204].

[26] D. N. Spergel *et al.*, "Wilkinson Microwave Anisotropy Probe (WMAP) three year results: Implications for cosmology," arXiv:astro-ph/0603449.

[27] L. Randall and R. Sundrum, "A large mass hierarchy from a small extra dimension," Phys. Rev. Lett. **83**, 3370 (1999) [arXiv:hep-ph/9905221]; "An alternative to compactification," Phys. Rev. Lett. **83**, 4690 (1999) [arXiv:hep-th/9906064].

[28] S. Kachru, R. Kallosh, A. Linde, J. M. Maldacena, L. McAllister and S. P. Trivedi, "Towards inflation in string theory," JCAP **0310**, 013 (2003) [arXiv:hep-th/0308055].

[29] P. Candelas and X. C. de la Ossa, "Comments on conifolds," Nucl. Phys. B **342**, 246 (1990).

K. Higashijima, T. Kimura and M. Nitta, "Supersymmetric nonlinear sigma models on Ricci-flat Kaehler manifolds with O(N) symmetry," Phys. Lett. B **515**, 421 (2001) [hep-th/0104184].

[30] O. DeWolfe and S. B. Giddings, "Scales and hierarchies in warped compactifications and brane worlds," Phys. Rev. D **67**, 066008 (2003) [hep-th/0208123].

[31] S. Kachru, R. Kallosh, A. Linde and S. P. Trivedi, "De Sitter vacua in string theory," Phys. Rev. D **68**, 046005 (2003) [arXiv:hep-th/0301240].

[32] H. Firouzjahi and S. H. Tye, "Brane inflation and cosmic string tension in superstring theory," JCAP **0503**, 009 (2005) [arXiv:hep-th/0501099].

[33] D. Baumann, A. Dymarsky, I. R. Klebanov, J. Maldacena, L. McAllister and A. Murugan, "On D3-brane potentials in compactifications with fluxes and wrapped D-branes," JHEP **0611**, 031 (2006) [arXiv:hep-th/0607050].

[34] C. P. Burgess, J. M. Cline, K. Dasgupta and H. Firouzjahi, "Uplifting and inflation with D3 branes," arXiv:hep-th/0610320.

[35] C. P. Burgess, J. M. Cline, H. Stoica and F. Quevedo, "Inflation in realistic D-brane models," JHEP **0409**, 033 (2004) [arXiv:hep-th/0403119].

[36] N. Iizuka and S. P. Trivedi, "An inflationary model in string theory," Phys. Rev. D **70**, 043519 (2004) [arXiv:hep-th/0403203].

[37] J. M. Cline and H. Stoica, "Multibrane inflation and dynamical flattening of the inflaton potential," Phys. Rev. D **72**, 126004 (2005) [arXiv:hep-th/0508029].

[38] A. R. Liddle, A. Mazumdar and F. E. Schunck, "Assisted inflation," Phys. Rev. D **58**, 061301 (1998) [arXiv:astro-ph/9804177].

[39] K. Becker, M. Becker and A. Krause, "M-theory inflation from multi M5-brane dynamics," Nucl. Phys. B **715**, 349 (2005) [arXiv:hep-th/0501130].

[40] S. Kecskemeti, J. Maiden, G. Shiu and B. Underwood, "DBI inflation in the tip region of a warped throat," JHEP **0609**, 076 (2006) [arXiv:hep-th/0605189].

[41] H. Firouzjahi and S. H. H. Tye, "Closer towards inflation in string theory," Phys. Lett. B **584**, 147 (2004) [arXiv:hep-th/0312020].

[42] K. Dasgupta, C. Herdeiro, S. Hirano and R. Kallosh, "D3/D7 inflationary model and M-theory," Phys. Rev. D **65**, 126002 (2002) [arXiv:hep-th/0203019].

J. P. Hsu and R. Kallosh, "Volume stabilization and the origin of the inflaton shift symmetry in string theory," JHEP **0404**, 042 (2004) [arXiv:hep-th/0402047].

[43] M. Berg, M. Haack and B. Kors, "Loop corrections to volume moduli and inflation in string theory," Phys. Rev. D **71**, 026005 (2005) [arXiv:hep-th/0404087]. "String loop corrections to

Kaehler potentials in orientifolds," JHEP **0511**, 030 (2005) [arXiv:hep-th/0508043]; "On volume stabilization by quantum corrections," Phys. Rev. Lett. **96**, 021601 (2006) [arXiv:hep-th/0508171].

[44] J. J. Blanco-Pillado *et al.*, "Racetrack inflation," JHEP **0411**, 063 (2004) [arXiv:hep-th/0406230].

[45] "Inflating in a better racetrack," JHEP **0609**, 002 (2006) [arXiv:hep-th/0603129].

[46] F. Denef, M. R. Douglas and B. Florea, "Building a better racetrack," JHEP **0406**, 034 (2004) [arXiv:hep-th/0404257].

[47] J. P. Conlon and F. Quevedo, "Kaehler moduli inflation," JHEP **0601**, 146 (2006) [arXiv:hep-th/0509012].

[48] V. Balasubramanian and P. Berglund, "Stringy corrections to Kaehler potentials, SUSY breaking, and the cosmological constant problem," JHEP **0411**, 085 (2004) [arXiv:hep-th/0408054].

V. Balasubramanian, P. Berglund, J. P. Conlon and F. Quevedo, "Systematics of moduli stabilisation in Calabi-Yau flux compactifications," JHEP **0503**, 007 (2005) [arXiv:hep-th/0502058].

J. P. Conlon, F. Quevedo and K. Suruliz, "Large-volume flux compactifications: Moduli spectrum and D3/D7 soft supersymmetry breaking," JHEP **0508**, 007 (2005) [arXiv:hep-th/0505076].

[49] D. H. Lyth and A. Riotto, "Particle physics models of inflation and the cosmological density perturbation," Phys. Rept. **314**, 1 (1999) [arXiv:hep-ph/9807278].

[50] J. R. Ellis, J. E. Kim and D. V. Nanopoulos, "Cosmological Gravitino Regeneration And Decay," Phys. Lett. B **145**, 181 (1984).

[51] X. Chen, M. x. Huang, S. Kachru and G. Shiu, "Observational signatures and non-Gaussianities of general single field inflation," arXiv:hep-th/0605045.

[52] E. J. Copeland, R. C. Myers and J. Polchinski, "Cosmic F- and D-strings," JHEP **0406**, 013 (2004) [arXiv:hep-th/0312067].

"Cosmic superstrings II," Comptes Rendus Physique **5**, 1021 (2004).

[53] A. Sen, "Tachyon matter," JHEP **0207**, 065 (2002) [arXiv:hep-th/0203265].

A. Sen, "Field theory of tachyon matter," Mod. Phys. Lett. A **17**, 1797 (2002) [arXiv:hep-th/0204143].

A. Sen, "Dirac-Born-Infeld action on the tachyon kink and vortex," Phys. Rev. D **68**, 066008 (2003) [arXiv:hep-th/0303057].

[54] H. Firouzjahi, L. Leblond and S. H. Henry Tye, "The (p,q) string tension in a warped deformed conifold," JHEP **0605**, 047 (2006) [arXiv:hep-th/0603161].

[55] M. Sazhin *et al.*, "CSL-1: a chance projection effect or serendipitous discovery of a gravitational lens induced by a cosmic string?," Mon. Not. Roy. Astron. Soc. **343**, 353 (2003) [arXiv:astro-ph/0302547].

M. V. Sazhin, M. Capaccioli, G. Longo, M. Paolillo and O. S. Khovanskaya, "The true nature of CSL-1," arXiv:astro-ph/0601494.

[56] T. Damour and A. Vilenkin, "Gravitational radiation from cosmic (super)strings: Bursts, stochastic background, and observational windows," Phys. Rev. D **71**, 063510 (2005) [arXiv:hep-th/0410222].

[57] J. Polchinski, "Introduction to cosmic F- and D-strings," arXiv:hep-th/0412244.

[58] L. Kofman and A. Linde, "Problems with tachyon inflation," JHEP **0207**, 004 (2002) [arXiv:hep-th/0205121].

[59] J. M. Cline, H. Firouzjahi and P. Martineau, "Reheating from tachyon condensation," JHEP **0211**, 041 (2002) [arXiv:hep-th/0207156].

J. M. Cline and H. Firouzjahi, "Real-time D-brane condensation," Phys. Lett. B **564**, 255 (2003) [arXiv:hep-th/0301101].

N. Barnaby and J. M. Cline, "Creating the universe from brane-antibrane annihilation," Phys. Rev. D **70**, 023506 (2004) [arXiv:hep-th/0403223].

[60]  N. Barnaby, A. Berndsen, J. M. Cline and H. Stoica, "Overproduction of cosmic superstrings," JHEP **0506**, 075 (2005) [arXiv:hep-th/0412095].

[61]  N. Lambert, H. Liu and J. M. Maldacena, "Closed strings from decaying D-branes," arXiv:hep-th/0303139.

[62]  N. Barnaby, C. P. Burgess and J. M. Cline, "Warped reheating in brane-antibrane inflation," JCAP **0504**, 007 (2005) [arXiv:hep-th/0412040].

D. Chialva, G. Shiu and B. Underwood, "Warped reheating in multi-throat brane inflation," JHEP **0601**, 014 (2006) [arXiv:hep-th/0508229].

[63]  H. Firouzjahi and S. H. Tye, "The shape of gravity in a warped deformed conifold," JHEP **0601**, 136 (2006) [arXiv:hep-th/0512076].

[64]  A. R. Frey, A. Mazumdar and R. Myers, "Stringy effects during inflation and reheating," Phys. Rev. D **73**, 026003 (2006) [arXiv:hep-th/0508139].

[65]  L. Kofman and P. Yi, "Reheating the universe after string theory inflation," Phys. Rev. D **72**, 106001 (2005) [arXiv:hep-th/0507257].

[66]  A. Berndsen, J.M. Cline and H. Stoica, work in progress.

Course 4

# PHYSICS OF THE EARLY UNIVERSE AND INFLATION

## Lev Kofman

*Canadian Institute for Theoretical Astrophysics*
*University of Toronto, Canada*

*F. Bernardeau, C. Grojean and J. Dalibard, eds.*
*Les Houches, Session LXXXVI, 2006*
*Particle Physcis and Cosmology: The Fabric of Spacetime*
© *2007 Published by Elsevier B.V.*

# Contents

# 1. Preamble

My five lectures will be devoted to the physics of the very early universe, mostly to the physics of inflation. There are different possible perspectives on inflation, from the perspective of practical cosmologist, to the perspective of a string theorist. While I will try to bring elements of both, the focus will be on the theoretical physics side of inflation.

The material is divided into several chapters: Concise History of the Early Universe is devoted to the Big Bang, its answered and unanswered questions. We introduce inflationary solution of the Big Bang problems and briefly discuss attempts at alternative solutions. Then we discuss models of inflation, global geometry of the universe with inflation, and cosmological dynamics with a scalar field. We describe the theory of generation of cosmological fluctuations, preheating after inflation, and finally observational tests of inflation.

I will try to maintain an introductory level adjusted to non-cosmology graduate students. One guru mathematician told pupils about the secret of writing texts: to read them easily you should write them easily. Apologies in advance, it was not easy to put the lectures in writing.

# 2. Concise history of the early universe

The history of the Universe has many common features with human history: in both there were known and unknown events, most mysterious beginning, kings and heroes of the scene of history and heroes and kings of science; and from time to time attempts of to re-interpret the driving forces behind the scene. We will discuss one such re-interpretation in cosmology, the paradigm shift from the Big Bang picture to the picture of primordial Inflation in the early universe.

## 2.1. Big Bang: answered questions

In this section we recall the thermal history of the expanding Universe. As far as we adopt an expanding branch of the Friedmann geometry of the universe

$$ds^2 = dt^2 - a^2 \left( \frac{dr^2}{1 - Kr^2} + r^2 d\Omega_2^2 \right), \tag{2.1}$$

the geometrical degrees of freedom are encoded in the scalar factor $a(t)$. Additionally there is a single number, the "topological" charge of the universe $K = 0, +1, -1$. Why is the universe expanding? Isolated systems with gravity can be in static or stationary equilibrium. A uniform universe cannot be static, it shall collapse or expand depending on the initial conditions. This can be seen in the following way. The scale (conformal) factor is a "moduli" field. Recall that if there is a compact inner space, say, with the portion of the metric $b^2 dy_a dy^a$, its radius $b$ in the four-dimensional theory behaves as the moduli scalar field $\phi = \log b$. What make the moduli $a(t)$ different is the "wrong" sign of its kinetic term. Indeed, for the metric (2.1) $\int d^4 x \sqrt{-g} R = -6 \int d^4 x \, a_{;\mu} a^{;\mu}$. As far as there are no particles associated with $a$, its negative kinetic term is not pathological. However, the rolling modulus $a(t)$ signals about the expansion or contraction of the universe. Because the sign of the kinetic term of the conformal scale factor $a(t)$ differs from that of other moduli, it cannot be stabilized, contrary to other moduli like $b$. We will return to this interpretation in section 3.1 where we introduce an example with the conformal and anisotropy moduli.

Next, we have to describe the matter content of the universe. Together with the background geometry, we assume the matter distribution to be homogeneous and isotropic. From the General Relativity (GR) point of view the matter is characterized phenomenologically by its energy density $\epsilon$ and pressure $p$. Different matter components are described by different equations of state (EOS) which relate $\epsilon$ and $p$. Most important will be several cases: the radiation equation of state

$$p_r \approx \frac{1}{3}\epsilon_r, \tag{2.2}$$

matter equation of state

$$p_m \approx 0, \tag{2.3}$$

and vacuum-like equation of state

$$p_v \approx -\epsilon_v. \tag{2.4}$$

The radiation EOS applies not only to ordinary electro-magnetic radiation (gas of photons), but also to any gas of ultra-relativistic particles (this is why we use $\approx$ instead of $=$ which is valid in the limit of massless particles with negligible interactions). It also describes a coherently oscillating scalar field with a quartic potential. The matter EOS covers any non-relativistic particles as well as coherently oscillating massive scalar fields. The vacuum-like EOS is valid for a cosmological constant, branes, quantum vacuum polarization in de Sitter geometry, the constant parts of the effective potentials, etc.

The conservation of the energy-momentum tensor $T^{\mu}_{\nu;\mu} = 0$ in the Friedmann geometry reads as

$$\dot{\epsilon} + 3\frac{\dot{a}}{a}(\epsilon + p) = 0. \tag{2.5}$$

Combining with the equation of state, this allows us to find how the expansion of the universe dilutes the physical energy density of any particular matter component with time. Substituting (2.3) into (2.5) we find

$$\epsilon_m = \frac{\epsilon_{m0}}{a(t)^3}. \tag{2.6}$$

This law has a simple meaning. The energy density of the non-relativistic gas can be thought of as the number density (abundance) of particles at rest, $n$, multiplied by their rest mass energy $mc^2$. Conservation of particles in an expanding universe results in the constancy of $na^3$, which again leads to equation (2.6).

A similar exercise with the radiation (2.2) + (2.5) leads to the law

$$\epsilon_r = \frac{\epsilon_{r0}}{a(t)^4}. \tag{2.7}$$

Again, one can understand it in the following way. The energy of the gas of photons is their frequency $\omega$ multiplied by their abundance. The last is diluted as $1/a^3$, while the frequency is red-shifted as $1/a$. All together the energy density of photons is diluted as $1/a^4$.

For the vacuum energy-density (2.4) + (2.5) gives

$$\epsilon_v = const. \tag{2.8}$$

For instance, for the universe dominated by a cosmological constant, its curvature is also constant, $R^{\mu}_{\nu} = \delta^{\mu}_{\nu} R$, $R = const$, which is related to the de Sitter metric, one of the realizations of FRW geometry.

Now it is time to include all interesting forms of matter in the Friedmann dynamical equation for the scale factor

$$\left(\frac{\dot{a}}{a}\right)^2 \equiv H^2 = \frac{8\pi}{3M_p^2}\left(\frac{\epsilon_{m0}}{a^4} + \frac{\epsilon_{m0}}{a^3} + \epsilon_v\right) - \frac{K}{a^2}, \tag{2.9}$$

for completeness we add the curvature term $\frac{K}{a^2}$. The universe evolves from early times when $a \to 0$ to the late times when $a$ is big. We immediately see from (2.9) that, if the co-moving values $\epsilon_{v0}, \epsilon_{m0}, \epsilon_{r0}$ are not changing, early times shall be dominated by the relativistic matter and radiation. Correspondingly, at some intermediate time energy density of matter and radiation are equalized, and since then the matter component dominates. Later curvature and ultimately cosmo-logical constant shall dominate the dynamics of an expanding universe. This is close to the actual dynamics of the universe (with $K = 0$). However, if the co-moving value $\epsilon_{v0}$ is changing, say, due to spontaneous symmetry breaking (SSB) in the scalar sector, or $\epsilon_{m0}$, $\epsilon_{r0}$ are changing due to out-of-equilibrium decays or freeze out of particles, the sequence of stages where each component dominates, in principle, can be more complicated.

Let us focus at the early stages of the evolution of the universe dominated by radiation and ultra-relativistic particles. Is it possible to conclude how different species contribute to the dynamics of the universe? So far we did not specify whether radiation is in a thermal equilibrium or out-of-equilibrium state. Assume the radiation is in a thermal equilibrium. The theory of radiation teaches us that

$$\epsilon_r = \frac{\pi^2}{30} g_* T^4, \tag{2.10}$$

where $g_* = g_B + \frac{7}{8} g_F$ and $g_B$ and $g_F$ are the number of bosonic and fermionic inner degrees of freedom.

From (2.10) and (2.8) we find that the temperature is diluted as $1/a(t)$. Recall that for the radiation dominated stage in the Friedmann equation (2.9) we can leave only $\epsilon_r$. It gives $3(\dot{a}/a)^2 \sim 1/a^4$. The solution of this equation is $a(t) = a_0 \sqrt{t}$. Thus, the temperature is diluted with time as $T \sim 1/\sqrt{t}$. Collecting all the coefficients and dimensions, we get formula

$$T = \frac{1.55}{g_*^{1/4}} \frac{Mev}{\sqrt{t(sec)}}. \tag{2.11}$$

This very convenient formula tells us the timing of physically significant epochs of the universe history. For instance the electro-weak phase transition at $\sim 100$ GeV takes place when the universe was as young as $10^{-10}$ sec; the QCD phase transition between free quark-gluon plasma and hadrons at $\sim 250$ MeV oc-curs at $t \sim 10^{-6}$ sec; nucleosynthesis at 0.5 MeV occurs when there were about 3 min on the cosmic clock; etc., see Figure 1.

From those examples we immediately notice that the formula (2.11) cannot be exact. Indeed, its derivation is valid for the ultra-relativistic particles, i.e. for the species with the masses smaller that the temperature. As temperature increase with decreasing time, more and more of what use to be non-relativistic massive particles become relativistic particles. In other words, the factor $g_*$ in (2.11) is a (slowly varying) function of temperature, or function of time in the for-mula (2.10). To know the exact dependence $g_*(T)$, we need to borrow more from high energy particle physics.

observable universe     reciprocal universe

time sec

WIMPs   n,p,ν,e-,γ

temperature/momentum k= 1/λ

$10^8$

present

$10^{-4}$ eV   3.36

LSS

$10^{12}$

$10^2$

CMB

0 1eV

$10^{-6}$

BBN

1 MeV

e-   γ

CDM

QCD transition

250 MeV

16

51.25

$10^{-10}$

EW SSB

100 GeV

$10^{-22}$

baryogenesis

$10^8$ GeV

106.75

q   $A^a_\mu$   q

$g_*$

$10^{-35}$

particlegenesis

$10^{15}$ GeV

$10^{-37}$

GUT

$10^{-43}$

inflation

$10^{19}$ GeV

scale λ

D3   D3

$\alpha_2^{-1}$

$\alpha_1^{-1}$

$\alpha_3^{-1}$

n   p

W-   e-

ν

3   log E   15

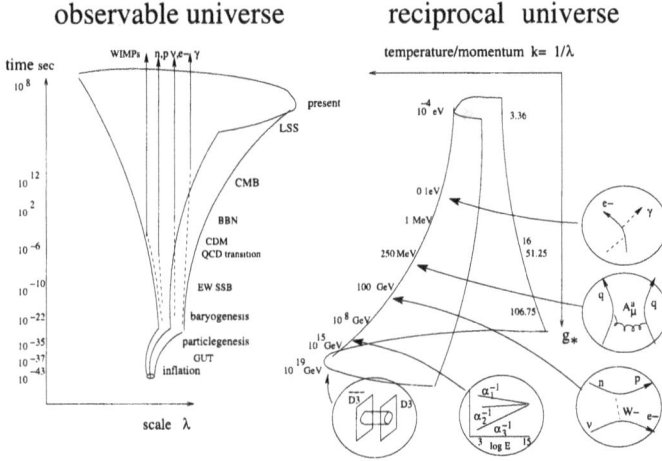

Fig. 1. Left: composition of the expanding universe is changing with time as it cools down. Right: reciprocal universe it terms of physical momenta of particles expands backwards in time to open particle physics at higher and higher temperatures. Icons illustrate physics at different energies.

Although an adiabatically expanding plasma in thermal equilibrium is the basis of the hot Big Bang model, the most interesting stages of evolution and the most important consequences are associated with the relatively short out-of-equilibrium epochs. $g_*(T)$ is notably changing during these epochs. Figure 1 shows some important stages in the evolution of the physical universe. Today the universe is composed of radiation of relic photons $\gamma$ and neutrinos $\nu_e$, $\nu_\mu$, $\nu_\tau$, plus electrons $e^-$ plus protons $p$ and neutrons $p$. The ratio of photon to baryon abundances $n_\gamma/n_B$, which characterizes the entropy of the universe, is enormously high, $10^9$. Compare it with that in the hot interiors of stars $10^{-2}$, or cores of supernovae $\sim \mathcal{O}(1)$. Where is the relative abundance of photons per baryon encoded in thermal description of adiabatically expanding plasma? The answer is in the chemical potential $\mu$ which is associated with the baryon number $n_B$ (defined by the theory of baryogenesis). Since $\frac{n_{\bar{B}}-n_B}{n_B} \sim 10^{-10}$, $\mu$ does not affect radiation but only residual abundance of baryons. At present and up to temperature $\sim 0.1$ MeV, $g_*(T) \approx 3.3626$. The universe becomes transparent for photons $\gamma$ at the epoch of recombination when the age of the universe is $10^{12}$ sec and temperature is about 0.1 eV. There their mean free path due to Thompson scattering on free electrons exceeds the Hubble radius and photons drop out of equilibrium with electrons and protons. A little bit earlier the universe became transparent for $\nu_e$. The epoch when 0.1 Mev $< T <$ 1 MeV corresponds to the very important process of Big Bang Nucleosynthesis (BBN), where the light elements were assembled. BBN is the earliest stage of the universe which is convincingly probed observationally. Any new physics which enters cosmology is filtered by stringent constraints for successful BBN: any extra ultra-relativistic particles beyond the SM are only allowed at the epoch of BBN at the level less than 10% of the SM radiation. Gravitons are allowed at the same level. No heavy particles (say of mass about $m_X \sim 100$ GeV, like gravitino or moduli mass) are allowed to decay during or after BBN unless their abundance is smaller than $n_x/n_\gamma < 10^{-15}$.

Further backwards in time $g_*(T)$ slightly increases due to annihilation of $e^+, e^-$ and $\mu^+, \mu^-$ lepton pairs.

$T_c \simeq 250$ MeV corresponds to the temperature of the QCD phase transition, when the de-confined quark-gluon plasma of gluons and free quarks turns into the quarks confined inside hadrons. Around the QCD phase transition, $g_*(T)$ drops from the value $205/4$ to $69/4$. The larger value of $g_*(T)$ above $T_c$ is mostly due to the gluons. The next important period is the electro-weak phase transition at $\sim 100$ GeV. Above $T \sim 174$ GeV, within the SM, $g_*(T) = 427/4$. At higher energies, above the TeV scale, $g_*(T)$ depends on the extension of the SM; in SUSY theories $g_*(T)$ is bigger by about a factor of two.

The electro-weak phase transition is accompanied by spontaneous symmetry breaking (SSB). Alongside other SSB transitions, in the cosmological context it deserves further discussion. The crucial element of SSB is the scalar Higgs field $\phi$ with the potential

$$V(\phi) = V_0 - \frac{1}{2}m^2\phi^2 + \lambda\phi^4. \tag{2.12}$$

When the temperature is above $\sim 100$ GeV, thermal corrections to the potential restore the symmetry and the Higgs field is in the symmetric phase at $\phi = 0$. The total energy density at that moment is a superposition of the vacuum energy due to $V_0$ and the energy of radiation given by (2.10). After SSB the Higgs field acquires the VeV $\phi = m^2/2\sqrt{\lambda}$. We tune $V_0$ to have zero vacuum energy at present, so that $V_0 = m^4/4\lambda \sim (100 \text{ GeV})^4$. However, around the EW phase transition the contribution of $V(\phi)$ to $\epsilon_v$ is non-zero, and we have $\epsilon \approx (100 \text{ GeV})^4 + \frac{\pi^2}{15}g_*(100 \text{ GeV})^4$. We shall take two lessons from here. The first is that the vacuum energy associated with the scalar potential is subdominant to radiation only by a numerical factor of order $1 : g_*$. If one were to change the initial conditions and assume no hot EW plasma, the energy density would be dominated by the vacuum energy density of the Higgs field. This is exactly what we do to obtain inflation. Second, we tune $V_0$ to have $V(\phi)$ negligible at present. The choice of $V_0$ does not affect particle physics but can significantly affect gravity and the cosmological solution. From the point of view of a hypothetical observer at the beginning of the universe it would be natural to choose $V_0$ to be of the same order of magnitude as $\pm(100 \text{ GeV})^4$. This leads us to one of the three possibilities today: positive vacuum energy of order of $+(100 \text{ GeV})^4$, negative vacuum energy of the same order, and vanishing vacuum energy. In the first case very soon after the EW phase transition the universe will be dominated by positive $\epsilon_v$, and will have no large scale structure, galaxies and planets. In the second case, as we will see below, the universe will recollapse and crunch soon after $10^{-10}$ sec. The only case suitable for our universe with galaxies and planets is the tuning case of negligible $\epsilon_v$. This is one of the many faces of the cosmological constant problem. One can think that the choice of $\epsilon_v$ is conditioned by our existence (anthropic argument), or one can hope to find a mechanism to nullify all contributions to $\epsilon_v$ (for instance by flattening of a three-dimensional brane if SM lives on the brane in a higher dimensional theory).

### 2.2. Big Bang: unanswered questions

The background Friedmann cosmological solution plus small fluctuations around it, together with the Standard Model particles, we will call the Big Bang model. We have to go beyond the SM to account for the dark matter and dark energy necessary to explain the observable universe. With these additions cosmology has been tested backwards in time till Big Bang Nucleosynthesis. From the bottom-up perspective, cosmology is well described by the background (2.1) plus small scalar metric perturbations around it, and $\Lambda$ Cold Dark Matter composition of (2.9) with $K = 0$. We will use density parameters $\Omega$ for corresponding matter components; $\Omega = \epsilon/\epsilon_c$ where the critical energy density is $\epsilon_c = 3H^2 M_p^2/8\pi$. The set of parameters which are found with good accuracy from the wealth of cosmological data includes the background parameters

$$\Omega_m + \Omega_v = \Omega_{tot}, \quad H_0, \quad \Omega_b, \quad \Omega_\gamma. \tag{2.13}$$

The set of parameters from the primordial scalar and tensor perturbations includes

$$P_s = A_s k^{n_s - 1}, \quad r = A_t/A_s, \tag{2.14}$$

where the $A$'s are the amplitudes of the scalar and tensor modes, and $n_s$ is the spectral index, which also may be accompanied by its running. The values of parameters which best fit to the observations are collected in Table 1.

Notice that $\Omega_{tot}$ is almost exactly one, and $n_s$ is almost one, i.e. the spectrum of scalar perturbations is close to the scale free (Zel'dovich) spectrum of primordial fluctuations.

The simple and elegant Big Bang model is remarkably successfully. Yet, this successful Big Bang model is based on the choice of initial conditions which cannot be explained within it. Moreover, extrapolation of the background Friedmann

Table 1

Results for the basic parameter set from [1]. The runs all assumed flat cosmologies, uniform and broad priors on each of the basic six parameters and a weak prior on the Hubble constant ($45 < H_0 < 90$ km s$^{-1}$ Mpc$^{-1}$) and the age ($>10$ Gyr). All runs included the effect of weak gravitational lensing on the spectra which is significant for the $\ell > 1000$ scales probed by ACBAR. The results are nearly identical if we include an SZ template, with the largest, but still relatively minor, impact on $\sigma_8$ and $\tau$.

| | WMAP3 | WMAP3 + ACBAR | CMBall | CMBall +LSS |
|---|---|---|---|---|
| $\Omega_b h^2$ | $0.0226^{+0.0008}_{-0.0007}$ | $0.0225^{+0.0007}_{-0.0007}$ | $0.0226^{+0.0006}_{-0.0006}$ | $0.0226^{+0.0006}_{-0.0006}$ |
| $\Omega_c h^2$ | $0.108^{+0.008}_{-0.009}$ | $0.108^{+0.008}_{-0.007}$ | $0.110^{+0.006}_{-0.006}$ | $0.115^{+0.005}_{-0.005}$ |
| $\theta$ | $1.042^{+0.004}_{-0.004}$ | $1.042^{+0.004}_{-0.003}$ | $1.042^{+0.003}_{-0.003}$ | $1.042^{+0.003}_{-0.003}$ |
| $\tau$ | $0.097^{+0.012}_{-0.014}$ | $0.092^{+0.014}_{-0.014}$ | $0.092^{+0.013}_{-0.014}$ | $0.090^{+0.013}_{-0.013}$ |
| $n_s$ | $0.966^{+0.017}_{-0.017}$ | $0.964^{+0.016}_{-0.015}$ | $0.963^{+0.016}_{-0.014}$ | $0.960^{+0.015}_{-0.014}$ |
| $\log[10^{10}A_s]$ | $3.05^{+0.08}_{-0.06}$ | $3.05^{+0.06}_{-0.07}$ | $3.05^{+0.06}_{-0.06}$ | $3.07^{+0.05}_{-0.06}$ |
| $\Omega_\Lambda$ | $0.75^{+0.03}_{-0.03}$ | $0.76^{+0.03}_{-0.04}$ | $0.75^{+0.03}_{-0.03}$ | $0.72^{+0.03}_{-0.03}$ |
| Age [$Gyrs$] | $13.7^{+0.2}_{-0.1}$ | $13.7^{+0.2}_{-0.1}$ | $13.7^{+0.1}_{-0.1}$ | $13.7^{+0.1}_{-0.1}$ |
| $\Omega_m$ | $0.25^{+0.03}_{-0.03}$ | $0.24^{+0.04}_{-0.03}$ | $0.25^{+0.03}_{-0.03}$ | $0.28^{+0.03}_{-0.03}$ |
| $\sigma_8$ | $0.78^{+0.06}_{-0.05}$ | $0.78^{+0.05}_{-0.05}$ | $0.79^{+0.04}_{-0.04}$ | $0.81^{+0.03}_{-0.03}$ |
| $z_{re}$ | $11.8^{+2.7}_{-2.6}$ | $11.5^{+2.3}_{-2.6}$ | $11.6^{+2.2}_{-2.6}$ | $11.5^{+2.2}_{-2.4}$ |
| $H_0$ | $72.7^{+3.0}_{-2.3}$ | $72.9^{+3.1}_{-3.1}$ | $72.3^{+2.8}_{-2.8}$ | $70.4^{+2.3}_{-2.2}$ |

cosmological solution together with the Standard Model particles with ever increasing temperature, back to the singularity at $t = 0$ results in severe problems which were crystallized by the end of 1970's. Most of the problems are resolved in a new paradigm – the Inflation.

In short, inflationary theory explains the choice of initial conditions (Table 1) for the successfully Big Bang!

At present, the inflationary paradigm is widely accepted, although a few non-believers are trying to find alternatives. Because of that and because my lectures are about Inflation, we will discuss the problems with the Big Bang without Inflation.

The first group of problems is related to General relativity and gravity. The spatial curvature of the observable universe with very high accuracy is close to zero.

• This constitutes the flatness problem:

$K = 0$ means that positive kinetic energy of the expanding universe is exactly balanced by the negative gravitational potential energy. Close to the singularity each of them are enormously big, yet one is canceled by the other. From the point of view of a hypothetical observer at the beginning of the Big Bang the choice $K = 0$ is enormously fine tuned. A good indicator of this attitude is 1960's argument of Dicke: if initial $\Omega$ was less than one it should be much much less than one today. Order of magnitude proximity of present (1960s) $\Omega$ to one forces one to think that it is in fact one. So why is the observable universe flat? The flatness problem can be restated as the oldness problem. Indeed, our hypothetical observer around the singularity has a natural time scale, the Planckian time $10^{-43}$ sec. It would be natural for him to see the universe recollapse within the Planckian time or become curvature dominated very early on. In reality our universe is $10^{17}$ sec or $\sim 10^{60}$ Planckian times old.

The background geometry and the large scale matter distribution up to the observable horizon scales $\sim 1/H \sim 10^{30}$ cm are compatible with the smooth FRW metric. In the Big Bang picture, the present-day horizon encompasses a huge number of patches with practically identical geometry and matter which were not in causal contact in the past.

• This is the horizon problem.

The most straightforward manifestation of the problem is the CBM sky map, as sketched on the left panel of Figure 2. Indeed, the uniform flux of relic photons which come to us from different directions across the sky in fact were emitted at the moment of recombination when the size of the particle horizon (in the Big Bang picture) was about $180h^{-1}$ Mpc (in the present-day units). Two such regions of particle-horizon size at the moment of recombination are shown in Figure 2. At present the size of the horizon is about $12,000h^{-1}$ Mpc, and embraces as many as $(12,000/180)^3 \sim 10^6$ Hubble patches at recombination. Why, near singularity in the geometry, do the physical conditions of matter in $10^6$ physically disconnected regions happen to be synchronized? The number of causally disconnected patches at the moment of BBN (which looks uniform observationally) is $10^{25}$, and $10^{90}$ at the Planckian moment.

In the Big Bang picture:

• the choice of the isotropic and homogeneous FRW solution of the Einstein equations is enormously particular,

which by itself is an issue. In other words, the smoothness of the geometry depicted in the Figure 2 is as stunning as the smoothness of the smooth wave front of EM radiation emitted by a number of uncorrelated sources would be!

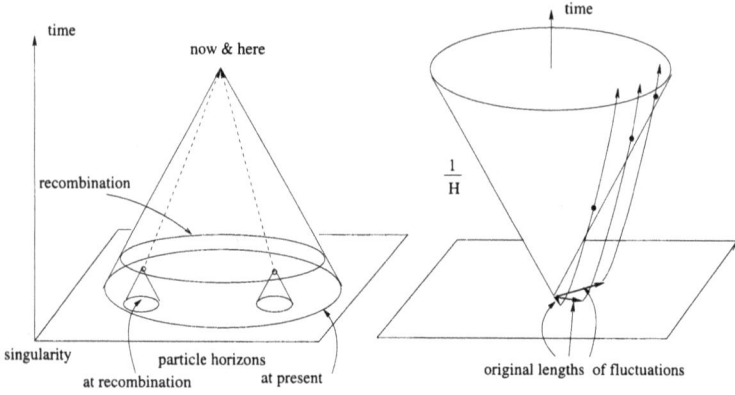

Fig. 2. Left: horizon problem. Colored patches are physically disconnected particle horizons at the moment two photons are emitted from the last scattering surface of recombination. Right: problem of primordial fluctuations. Fluctuations with different present-day wavelengths enter the Hubble radius $1/H(t)$ at different times but their original wavelengths at the time of singularity significantly exceed the original causal patch.

Indeed, we have to seek for the generic cosmological time-dependent solution of the Einstein equations. A more generic ansatz for the cosmological solution in the presence of radiation is a quasi-homogeneous metric

$$ds^2 = dt^2 - \gamma_{ij}(x^\mu)dx^i dx^j, \tag{2.15}$$

where, around the singularity,

$$\gamma(x^\mu)_{ij} = t\alpha_{ij}(\vec{x}) + t^2\beta_{ij}(\vec{x}) + \dots. \tag{2.16}$$

The asymptotic solution around $t = 0$ defines $\beta_{ij}$ and the coefficients of the higher powers of $t$ to be functions of $\alpha_{ij}(\vec{x})$. There are three arbitrary physical functions of the spatial coordinates among six $\alpha_{ij}(\vec{x})$ (the others three are irrelevant due to the residual freedom of coordinate transformations in the synchronous gauge). Thus, the Friedmann solution which corresponds to the particular choice $\gamma(t)_{ij} = t\delta_{ij}$ is a very special solution. Beyond this quasi-homogeneous solution, there is even broader class of anisotropic solutions, when $\gamma(t)_{ij} \sim t^{2p_i}\delta_{ij}$. Within the Big Bang cosmology the question is why among all possible cosmological solution our universe is described by the very degenerate, very symmetric FRW isotropic and homogeneous metric whose measure in the phase space of solutions (with radiation) is negligible?

One of the beyond-the Big bang attempt is the Pre-Big Bang scenario [2]. Here the universe originally contracts and then bounce from singularity. The choice of the uniform solution in this context is even more problematic: generic collapse solution in Kasner-like anisotropic solution.

In fact, the actual geometry of the universe is a composition of the smooth flat FRW geometry plus very small fluctuations around it. The leading scalar metric fluctuations in most cases can be described by a single scalar function $\Phi(t, \vec{x})$. In the longitudinal gauge, metric fluctuations read as

$$ds^2 = a^2 \left( (1 + 2\Phi)d\eta^2 - (1 - 2\Phi)d\vec{x}^2 \right). \tag{2.17}$$

$\Phi(t, \vec{x})$ is a relativistic generalization of the Newtonian gravitational potential. Indeed, recall the the Schwarzschild metric generated by the mass $M$

$$ds^2 = \left( 1 - \frac{2GM}{r} \right) dt^2 - \frac{dr^2}{(1 - \frac{2GM}{r})} - r^2 d\Omega^2, \tag{2.18}$$

and compare (2.17) with frozen expansion with (2.18) in the weak-field limit $\frac{2GM}{r} \ll 1$ to find the familiar formula for gravitational potential around the body of mass $M$, $\Phi = -\frac{GM}{r}$. Notice that in astronomy the potential $\Phi$ is negative and associated with the mass source $M$ (or velocity dispersion $\sigma_v$) like in (4.4). In cosmology $\Phi$ can be negative (like around clusters and galaxies) or positive (like inside giant voids), and smoothly distributed across the universe. In cosmology the amplitude $\Phi \sim 10^{-5}$ is small, so the fluctuations can be treated with linear perturbation theory. Notice that the universe is much smoother than, say, a perfect-looking billiard ball, whose scratches' amplitude I would estimate as $\sim 10^{-3}$. The smallness of cosmological $\Phi$ is relative, say, inside or galaxy $\sigma_v \sim 10^2$ km/sec, $\Phi \sim \sigma_v^2/c^2 \sim 10^{-6}$. Linear metric fluctuations (2.17) are conveniently decomposed into Fourier modes

$$\Phi(t, \vec{x}) = \frac{1}{(2\pi)^{3/2}} \int d^3k \, e^{i\vec{k}\vec{x}} \Phi_k(t). \tag{2.19}$$

The function $\Phi(t, \vec{x})$ is a realization of a random field, which is often modeled by the isotropic uniform random Gaussian field, i.e. superposition of plane waves $e^{i\vec{k}\vec{x}}$ with random phases (example of which is ocean waves). In cosmology its power spectrum $|\Phi_k(t)|^2$ carries a lot of information. Typically at large scales $\Phi_k(t)$ is not significantly varying in time. A useful characteristic of this function is its power per decade of modes $\int d^3k |\Phi_k(t)|^2$. Let us conjecture the power-law form of the spectrum

$$|\Phi_k(t)|^2 \approx \frac{A_s}{k^3} \left( \frac{k}{k_*} \right)^{n_s - 1}, \tag{2.20}$$

where $k_*$ is some pivot point. The power per decade of modes is scale free for a particular choice $n_s = 1$ when $\int d^3k |\Phi_k(t)|^2 = \int dk/k$. This "flat" spectrum is compatible with the initial conditions of smooth background + small fluctuations. Indeed, a blue or red tilt of the spectrum results in non-linear value of $\Phi$ at small or large scales, respectively, with the presence of black holes. Indeed, for tilted spectra (2.20) at some scales $\Phi$ becomes non-linear and proper metric inside the horizon will be closer to (2.18). A particular scale-free form of the spectrum, $n_s = 1$, was conjectured by Zel'dovich long before inflation was proposed. In terms of matter density fluctuations $\delta_k = k^2 \Phi_k$, $|\delta_k|^2 \sim A_s \left(\frac{k}{k_*}\right)^{n_s}$. This allows a consistent setting of initial conditions for Big Bang. Observations favor the value $n_s \approx 0.96$, very close to the scale free spectrum (as we will see, this slight tilt, extrapolated to super-large scale modes, will result in the black holes in the distant future!)

In the Big Bang picture, however,

• there is no known mechanism to explain the causal origin of (almost) scale free primordial cosmological fluctuations.

Indeed, in the Big Bang model, the physical wavelength of fluctuations $\lambda = \frac{2\pi}{k} a(t)$ increases with time slower than the Hubble radius $1/H$. We have $a(t) \sim t^{1/2}$ for radiation and $a(t) \sim t^{2/3}$ for matter equation of states, while $1/H \sim t$, see the right panel of Figure 2. This means that if we trace backwards in time the present-day wavelength of cosmological fluctuations on the scales of Large Scale Structure 10 Kpc–12,000 Mpc, and if we assume $\Phi_k(t)$ is not changing much in time, then at the beginning, around the singularity, these fluctuations are well outside the causal patch, as illustrated in the right panel of Figure 2. On the other hand we expect a causal physical mechanism to generate the primordial fluctuations.

There is a known mechanism to generate fluctuations causally in the Big Bang picture, related to the dynamics of cosmic strings which form in the early universe. In this case $\Phi_k(t)$ is generated in response to the cosmic string dynamics and is significantly time-dependent. Furthermore, $n_s = 4$, fluctuations are not scale free. This was an interesting theoretical alternative until compelling evidence showed that fluctuations in the observed universe are almost scale-free.

Beyond the Big Bang, yet without inflation, there have been several attempts to construct alternative mechanisms for the generation of cosmological fluctuations, one way or another invoking a Pre-Big Bang scenario which contains an initial contracting phase. In the original Pre-Big Bang and the Ekpyrotic scenarios [3], even closing our eyes to the shortcomings of the model, the spectral indices of scalar fluctuations are $n_s = 4$ and $n_s = 3$, correspondingly, which is not acceptable for our observable universe. Another proposal uses thermal fluctuations of the string gas in the Hagedorn pre-big-bang phase [4], but this model yields $n_s = 5$ [5].

The second group of problems is related to the energy particle physics. In the Big Bang picture the initial conditions of the universe are formulated around the singularity, where radiation is maximally hot, $T \sim M_p \simeq 10^{18}$ GeV. To describe particles at energies above the EW phase transition, we have to go beyond the standard model. Possible directions include GUT models, SUSY, SUGRA, and finally Superstring theory. Extensions of the Standard Model contain new particles, many of which are very heavy. It turns out that one way or another embedding of the beyond the Standard Model theory in the Big Bang picture results in emergence of

- the dangerous relics problem.

This problem has specific a face in each specific extension of the SM. GUT models contain monopoles and other topological defects. The abundance of heavy monopoles produced in the Big Bang is so high that it overclose the universe with $\Omega_{monop} \sim 10^{15}$. In SUGRA theories gravitinos play the role of dangerous relics. Gravitinos with the mass $m_{3/2} \sim 100$ GeV decay by the time of BBN and destroy the light-element abundances. To avoid this, we have to keep the gravitino abundance at the level $n_x/n_\gamma < 10^{-15}$, while from the Big Bang we have $n_x/n_\gamma \sim \mathcal{O}(1)$. In SUSY, moduli fields cause similar problems.

## 3. Inflation

### 3.1. Inflation: answered questions

According to the inflationary scenario, the universe at early times, before the standard radiation-dominated phase begins, expands quasi-exponentially in a vacuum-like state without entropy or particles, with the EOS $p \approx -\epsilon$. Indeed, the Friedmann equation gives us

$$\frac{\dot{a}}{a} = \sqrt{\frac{8\pi\epsilon_v}{3M_p^2}} \approx const, \quad a(t) = a_0 e^{\int_{t_0}^{t} dt\, H}. \tag{3.1}$$

For slowly varying $H(t)$, the scale factor is expanding exponentially fast, $a \sim e^{Ht}$. We also assume this EOS terminates after a finite time to provide the graceful exit from inflation. This temporary vacuum-like EOS dominating the expansion of the universe, prior to the hot radiation dominated universe, allows us to resolve the problems of the Big Bang cosmology. Even without knowledge of the microscopic model of inflation we will be able to proceed quite far to overcome the shortcomings of the Big Bang and make new predictions. This is because

GR/gravity solutions are defined by the phenomenological matter equation of state. Similarly we are able to successfully develop the theory of gravitational clustering of Cold Dark Matter (CDM) into the large scale structure of the universe without knowing the microscopic nature of the CDM particles.

The flatness problem is resolved by the acceleration of the universe during inflation due to the negative pressure $p_v = -\epsilon_v$

$$\ddot{a} = -\frac{4\pi^2}{3M_p^2} (\epsilon_v + 3p_v) > 0. \tag{3.2}$$

Indeed, the total energy (spatial curvature) of the universe

$$\Omega_{tot} - 1 = \frac{K}{(Ha)^2} = \frac{K}{\dot{a}^2}. \tag{3.3}$$

Acceleration of the universe drives $\dot{a}$ to a huge value to bring the *l.h.s* of the last equation to a very small number. Another way to explain it is to say that exponentially large $a(t)$ makes very small three-dimensional spatial curvature $\sim 1/a^2$.

Resolution of the horizon problem: a small causal uniform patch of the (quasi)de Sitter geometry is stretched by the exponential expansion to a very large patch $\sim e^N$, where $N$ is a number of inflationary efoldings. Needed number of $N$ is

$$N = 62 - \ln \frac{10^{16} \, GeV}{V^{1/4}} - \Delta, \tag{3.4}$$

where $\Delta$ depends on the details of (p)reheating and will be discussed in Section 6.

We will consider in details global space-time geometry with inflation in Section 4.

Origin of primordial fluctuations. During inflation the Hubble radius is very slowly varying with time, $1/H \sim const$. Physical wavelength of fluctuations exponentially increases $\lambda \sim e^{Ht}$, Therefore fluctuations which initially where well within the causal patch inside the horizon will be send away by superluminal stretch of the space. What is the origin of these fluctuations? Zero vacuum fluctuations of any light scalar field are amplified in (quasi)-de Sitter geometry and their amplitude is frozen with the sufficiently large value $\sim H$. Small inhomogeneities in the background scalar field include small inhomogeneities in geometry. Generation of cosmological fluctuations from inflation will be considered in section 5 in details.

Resolution of Dangerous relics problem. Abundance of undesirable heavy particles depends on time as $n_X \sim 1/a^3$. If there were dangerous relics prior inflation, their abundance after inflation is negligible. During inflation the particle abundance is diluted as

$$n_X \sim e^{-3N}. \tag{3.5}$$

Even for the least lasting inflation original abundance is diluted by a factor $10^{48}$. In many theories there was no hot stage before inflation. In all inflationary models matter of the universe is created in the process of (p)reheating after inflation, which we will consider in details in Section 6. It is assumed that the highest temperature (2.11) of the hot plasma after inflation, reheat temperature $T_r$, is safely low to avoid (re)production of dangerous relics. Monitoring $T_r$ gives an additional constrain on the high energy model of inflation.

The last in the line of this subsection is the origin of smooth uniform background geometry, which we will treat in some detail. The presence of the effective cosmological constant "isotropizes" and "homogenenizes" expanding cosmological solution. Non-gravitational analogy of this will be stretching of brane with the vacuum-like EOS. To illustrate how $\Lambda$ works [25], consider an example of four-dimensional anisotropic cosmological model supported by cosmological constant $\Lambda$, and described by the Nariai metric

$$ds^2 = dt^2 - \cosh^2(H_1 t)dx^2 - b_0^2 d\Omega_2^2, \tag{3.6}$$

where $b_0 = 1/\sqrt{\Lambda}$ is the radius of $S^2$, $H_1 = \sqrt{\Lambda}$ is the Hubble constant of two-dimensional expanding portion. Euclidean version of the metric (3.6) has $S^2 \times S^2$ symmetry. Geometry (3.6) is classically gravitationally unstable. One can conjecture that the instability is evolving in the broader class of metrics

$$ds^2 = dt^2 - a(t)^2 dx^2 - b(t)^2 d\Omega_2^2, \tag{3.7}$$

with two scalar factors $a(t)$ and $b(t)$, which includes (3.6) as a special case. The Einstein equations read as

$$\frac{2\ddot{b}}{b} + \frac{\dot{b}^2}{b^2} + \frac{1}{b^2} = \Lambda, \quad \frac{\ddot{a}}{a} = \frac{\ddot{b}}{b}. \tag{3.8}$$

Let me take an opportunity to introduce the method of two dimensional phase portrait of the dynamical systems which contains all phase trajectories (solutions) of the second order ODE like eqs (3.8). This method is very convenient to analyze cosmological solutions in unified way, and we will apply it to scalar fields/gravity dynamics with inflation. The phase portrait method was invented originally for the celestial mechanics, is further extended to the whole universe! The first equation of (3.8) can be re-written as

$$\frac{dx}{dt} = y = P(y, x), \quad \frac{dy}{dt} = \frac{\Lambda x}{2} - \frac{y^2}{2x} - \frac{1}{2x} = Q(y, x), \tag{3.9}$$

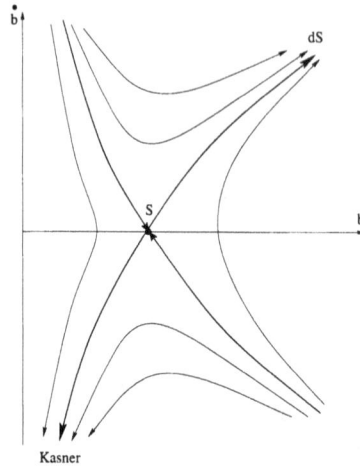

Fig. 3. Phase portrait of eq. (3.8). S is a saddle point which corresponds to the metric (3.6), All phase trajectories in the future approach either expanding dS or contracting Kasner asymptotic.

where $x = b$ and for generality we introduce the r.h.s. functions $P(y, x)$ and $Q(y, x)$. We shall search for the critical points $(y, x)$ where $P(y, x) = 0$ and $Q(y, x) = 0$, and check if they are focus, saddle or attractor/repulsor points. The special points then are connected by non-intersecting trajectories. A phase portrait for the system (3.9) (specified by particular choice of $P(y, x)$ and $Q(y, x)$) is shown in the Figure 3. As you can see the $S^2 \times S^2$ symmetric solution (3.6) corresponding to the saddle point $S = (0, \sqrt{\Lambda})$ is unstable and the expanding solution dynamically evolves towards the asymptotic $a \sim e^{Ht}$, $b \sim e^{Ht}$ with $H = \sqrt{\Lambda/3}$, which corresponds to de Sitter solution. Therefore $\Lambda$ works as "isotropizator". Notice that contracting branch of unstable solutions evolves towards the Kasner asymptotic.

Isotropization by $\Lambda$ can be demonstrated for the metrics (2.15) similar to analysis of Section 2.2 [24]. Remember that in de Sitter geometry scalar factor $a = \frac{1}{-H\eta}$ where asymptotic future is where conformal time $\eta \to 0$ In the context where we may expect that de Sitter geometry asymptotically emerge, it is convenient to consider generic metric $ds^2 = a^2(d\eta^2 - \tilde{\gamma}_{ij}dx^i dx^j)$ where $\tilde{\gamma}_{ij} = \tilde{\gamma}(x^\mu)_{ij}$ is given as a power-law series of small $\eta$

$$\tilde{\gamma}(x^\mu)_{ij} = \alpha_{ij}(\vec{x}) + \eta\beta_{ij}(\vec{x}) + \eta^2\delta_{ij}(\vec{x}) + \eta^3\kappa_{ij}(\vec{x})\ldots. \tag{3.10}$$

The metric (3.10) contains four arbitrary functions, two in $\alpha_{ij}$ and two in $\delta_{ij}$ ($\beta_{ij} = 0$), and the anisotropy Weyl tensor is exponentially diluted.

We conclude that pre-inflation anisotropy and inhomogeneities are erased at the horizon scale (no-hair theorem for de Sitter geometry). As we will see in Section 5, residual classical fluctuations and inevitable quantum fluctuations freeze out at super-horizon scales.

### 3.2. Models of inflation

A simple and natural realization of the vacuum-like equation of state is naturally achievable with the homogeneous scalar field $\phi(t)$ minimally coupled to gravity. Indeed, energy-momentum tensor $T_\nu^\mu = \phi^\mu \phi_\nu - \delta_\nu^\mu \left( \frac{1}{2} \phi^\sigma \phi_\sigma - V \right)$ of a classical moving homogeneous scalar field with the potential $V(\phi)$ is simply $T_\nu^\mu = diag(\epsilon, -p, -p, -p)$ where the pressure and energy are given by

$$p = \frac{1}{2} \dot{\phi}^2 - V, \quad \epsilon = \frac{1}{2} \dot{\phi}^2 + V. \tag{3.11}$$

When potential energy dominates the kinetic energy, we have inflation with $p \approx -\epsilon$. Beginning with this simple idea, the complicated +25 year history of inflation has been about the microscopic nature of $\phi$ and the origin of its potential $V(\phi)$. In Fig. 1 I draw a chronologically ordered broad brush sketch of inflationary models for $V(\phi)$.

Before we will discuss the models of inflation, let us stop at the theory of classical scalar field. The scalar field $\phi(t, \vec{x})$ is s very simple concept. For instance, vector field $A_\mu$ obeys the equation

$$\nabla_\mu \nabla^\nu A_\nu + R_\nu^\mu A_\mu = 0. \tag{3.12}$$

Scalar field obey much simpler relativistic equation

$$\nabla_\mu \nabla^\nu \phi + \xi R \phi + V_{,\phi} = 0, \tag{3.13}$$

where $\xi R \phi$ term describes coupling to gravity, $V_\phi$ is self-interaction potential of the scalar field. the simplest example of the scalar field is the quantum mechanics non-relativistic Schrödinger equation for the scalar function $\Psi(t, \vec{x})$. With this respect, eq. (3.13) is a relativistic generalization of the Schrödinger equation. It is hard to have inflationary solution with the vector field. Indeed, condensate of the uniform EM field anisotropic $C_{\mu\nu\rho\sigma}$ in non-zero, solutions of the Einstein-Maxwell equations have the structure $R_\mu^\nu \sim diag(+1, +1, -1, -1)$. There is no condensate of the spinor field (while $\langle \bar{\psi} \psi \rangle$ works as an effective scalar field). Three-from field in four-dimensions can gives us cosmological constant, but not much dynamics (graceful exit) with it. Next candidate is the QFT vacuum polarization $\langle T_\nu^\mu \rangle$ in the strong gravitational field.

In fact, the Starobinsky 1980 model [22] with the vacuum polarization was the first reliable model of inflation (esoteric style and motives of the paper and science marketing of the cold war era kept it away from the full credit). Con-

**Inflation in the context of ever changing fundamental theory**

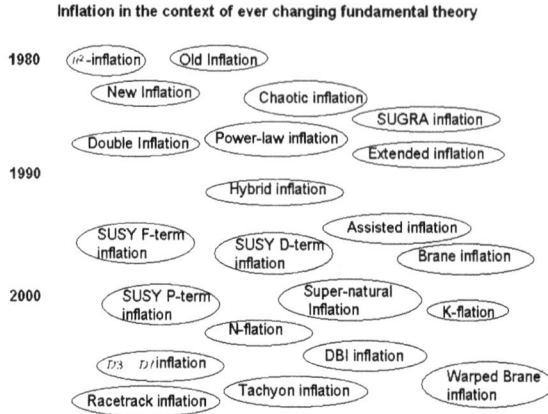

Fig. 4. Sparce selection of models of Inflation.

formal fields in the curved geometry acquire conformal anomaly which depends on the curvature. Self-consistent solutions are extracted from the Einstein equations with the conformal anomaly $\langle T_\nu^\mu \rangle$ constructed from the bi-linear combination of the curvature tensor. Obviously, there is a symmetric de Sitter solution with $\langle T_\nu^\mu \rangle \sim \delta_\nu^\mu$, $H^2 = \frac{360}{\pi k_2} M_p^2$, $k_2$ is the coefficient of the conformal anomaly roughly proportional to the number of conformal fields $N$. Successfully solution with conformal anomaly requires many fields $N \sim 10^6$, which is quite a stretch. The part of conformal anomaly with $\nabla_\mu \nabla^\mu R$ plays crucial role in the dynamics of the model. Although full $\langle T_\nu^\mu \rangle$ cannot be derived from the variation of the local action, $\nabla_\mu \nabla^\mu R$ can. Therefore the essence of the Starobinsky model is catched by its more transparent variation [23, 45]

$$\mathcal{L} = -\frac{M_p^2}{16\pi} \left( R - \frac{1}{6M} R^2 \right), \tag{3.14}$$

This theory can be further simplified by the conformal mapping

$$\tilde{g}_{\mu\nu} = \left( 1 - \frac{R}{3M} \right) g_{\mu\nu}, \quad \phi = \sqrt{\frac{3M_p}{16\pi}} \ln\left( 1 - \frac{R}{3M} \right). \tag{3.15}$$

into the theory with the scalar field in the metric $\tilde{g}_{\mu\nu}$ and the potential

$$V(\phi) = \frac{3M^2 m_p^2}{4} \left[ 1 - e^{-\sqrt{\frac{2}{3}} \frac{\phi}{M_p}} \right]^2, \tag{3.16}$$

which gives $\frac{1}{2} M^2 \phi^2$ around the minimum.

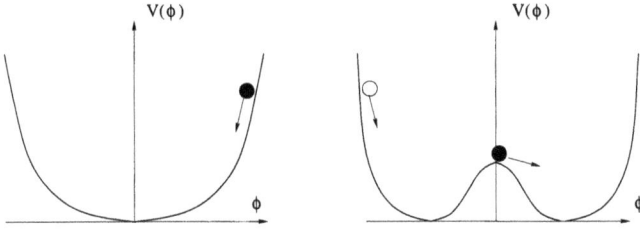

Fig. 5. Left: scalar field potential of the chaotic inflation. Right: double-well scalar field potential.

Vintage 1981 original Guth's paper [11] conceptualizes inflation, directly addresses the Big Bang problems. It is based on the supercooling scenario of SSB with the potential $V(\phi)$ associated with GUT which contains metastable minimum at origin. The potential is sketched on the right panel of Figure 5. Contrary to what we happens with the SSB of the EW theory Higgs field with the potential (2.12), the system stuck for a while in a metastable minimum allowing radiation to be diluted and $V(\phi)$ to dominate and realize inflation. As we saw in section 2.1, it takes only growth of $a$ by factor $g_*^{1/4} \sim 5$ to have domination of $\epsilon_v$. Inflation, however, shall have enough e-foldings given by (3.4). It suppose to end by decay of the metastable vacuum via the nucleation and percolation of the bubbles. Original inflation suffers the graceful exit problem.

Qualitatively new way to realize inflation with slow roll scalar was found by Linde first for the Coleman-Weinberg $SU(5)$ GUT model (new inflation [8], also [9] ) which still bear common features with the old inflation (right panel of Figure 5), and then with chaotic inflation [10] (right panel of Figure 5). One does not need any more hot pre-inflationary stage, as far as initially out-of-equilibrium VeV of $\phi$ is in the right place for prolonged slow roll, inflation is realized. Indeed, consider uniform scalar field $\phi(t)$ with equation of motion in an expanding universe

$$\ddot{\phi} + 3H\dot{\phi} + \xi R\phi + V_{,\phi} = 0, \tag{3.17}$$

where $\xi$ describes coupling to gravity $\frac{\xi}{2}R\phi^2$. This can be achieved with small conformal coupling $\xi \ll 1$ (ideally for minimal coupling) and for shallow $V(\phi)$, when $V_{,\phi} \ll 3H\dot{\phi}$. For instance for the quadratic potential $V = m\phi^2/2$ this simply means $m^2 \ll H^2$ (by factor of tens). For the slow roll scalar, if $\dot{\phi}^2 \ll V$ in (3.11), $p \approx -\epsilon$, and inflation is realized. As we will show in the next section, slow roll inflation is a generic intermediate asymptotic for a broad class of potentials $V(\phi)$. Often we use elegant quadratic $\frac{m^2}{2}\phi^2$ or quartic potentials $\frac{\lambda}{4}\phi^4$ with only a single parameter $m$ or $\lambda$, and the range of $\phi$ to be fixed by cosmology. Inflation of the massive inflaton model described by simple formulas

$$\phi(t) \approx \phi_0 - \sqrt{\frac{2}{3}} mt, \quad a(t) \approx a_0 \exp\left[\frac{2\pi}{M_p^2}(\phi_0^2 - \phi(t)^2)\right]. \tag{3.18}$$

To have 62 efolding, we need large initial inflaton value $\phi_0 \sim 3.5 M_p$. Inflaton mass $m = 10^{13}$ GeV is fixed by the amplitude of scalar fluctuations in the model (as will be derived in Section 5).

More flexibility of parameters, by the price to introduce more of them, can be achieved in the hybrid inflationary models [12]. Hybrid inflation operates with two scalars, inflaton $\phi$ and the Higgs scalar $\sigma$, with the potential

$$V(\phi, \sigma) = \frac{\lambda}{4}(\sigma^2 - v^2)^2 + \frac{g^2}{2}\phi^2 + \sigma^2 + V_{roll}(\phi), \tag{3.19}$$

When inflaton field is larger than critical value $\phi_c = \frac{\sqrt{\lambda}v}{g}$, it rolls in the uplift potential $V_{roll}(\phi)$ while the Higgs field is stuck in the minimum $\sigma = 0$. Inflation is driven by the vacuum energy $\frac{\lambda}{4}v^4$ but terminated when $\phi$ reaches the bifurcation point $\phi_c$. Further field dynamics is that of the SSB, which eventually brings $\sigma$ to the global minimum at $|\sigma| = v$ and $\phi = 0$. Successfully inflation can be achieved for different range of parameters $\lambda$, $g$, $v$ and for VeVs of inflaton $\phi \ll M_p$.

Here we proceed further with inflationary models. The glimpse of the zoo of inflationary models is reflected by the Figure 4. Candidates for inflaton were searched in the scalar sector of SUSY and SUGRA theories and among axions. In the effective four dimensional theory shapes and distances of the higher dimensional geometry manifested as the moduli scalar fields. Some of them suggested at the phenomenological level as the inflaton candidates, e.g. distance between interacting branes [13]. Remarkably, brane and SUSY models are variations of the hybrid model [12].

Among a number of inflationary scenarios, simple scenarios which come in the package with the high energy physics draw an especial attention (to my taste new inflation based on SU(5) GUT was the closest to this criteria) Recently there is significant interest to package up inflation with the string theory. There are several models including $D3/D7$ brane inflation, racetrack inflation with Kahler moduli/axion fields, large volume compactification inflation with Kahler moduli/axion field, etc., see recent review [15] Typical problem for any inflation, including the string theory inflation is the $\eta$-problem, manifested in heavy inflaton mass preventing slow roll of inflaton. Another words, one needs to get a small number from the parameters of the potential to obtain enough efoldings of inflation and right level of scalar fluctuations from inflation. The purpose is to explain the small numbers rather than put them by hand how it is done in the phenomenological models. An example of this problem is warped brane inflation where the inflaton is conformally coupled to the four-dimensional gravity [14]. The scalar fields $\phi$ which is conformally coupled to gravity, acquire effective mass term with $\xi = 1/6$, see equation (3.17). This gives the effective mass $m^2 = 6\xi H^2$ which does not result in the slow-roll. Bi-product of the conformally coupling is the absence of generation of the long-wavelength freezed out scalar field fluctuations in the (quasi) deSitter geometry. This is because equation of motion of the conformally coupled field can be mapped to that in the flat space-time.

For curiosity, I finish this Section with discussion of the unusual model of inflation where in fact conformal coupling is not a problem. This model has new features which make it very different from other models. First, inflaton with $\xi = 1/6$ (including the warped brane-antibrane inflation) can realize the fast-roll inflation, contrary to the customary slow roll inflation. The only feature needed for conformal inflation is the form of its potential: very shallow for the most part of the inflaton rolling and changing sharply at the end point of inflation (exactly as in the warped brane-antibrane inflation or the hybrid models)

Second, fast-roll conformal inflation is a low energy inflation (close to the least energy inflation at 1 Tev scale). As a result it requires significantly less efoldings $N$ than the figure 62 typical for the GUT scale chaotic inflation. In the context of the warp geometry, as we will show, $N$ is directly related to the warp factor of the throat geometry of the inner manifold. Coincidentally, the same number $N \approx 37$ satisfies two different pieces of physics, one for homogeneous and isotropic universe from the low energy 1*Tev* inflation, and another for the warped geometry [16] mass hierarchy $M_p e^{-N} \simeq Tev$.

### 3.3. Inflaton family portrait

Inflationary theory is connected with the dynamics of the universe driven by a scalar field. We can study scalar field/gravity dynamics in great details using the very powerful phase portrait method, which shows the character of *all* solutions for $a(t)$ and $\phi(t)$ as trajectories in the three-dimensional phase space $\mathcal{M}^3$ with the coordinates $(H, \dot{\phi}, \phi)$ [45,51].[1] How big is the fraction of inflationary solutions? For this questions we need the tools of Hamiltonian dynamics to construct an invariant measure on the phase space [26]. There are complications [27] which make the subject controversial [29, 30]. The physics of prior probabilities is needed to see that inflationary trajectories are typical [31].

We consider chaotic inflation with a smooth scalar field potential $V(\phi)$ with a minimum. Here we assume the classical field $\phi(t)$ is homogeneous and neglect

---

[1]Both papers with phase portraits in chaotic inflation appeared simultaneously. The paper [45] was shortened for the Phys Lett B, sacrificing the phase portrait for the scalar field. In anticipation of its merits, we put the phase portrait on the cover page of the preprint version of the paper.

fluctuations of the inflaton. For definiteness we take $V(\phi) = \frac{1}{2}m^2\phi^2 + V_0$. To treat the simplest case first, we start with $V_0 = 0$.

Rigorous treatment of the scalar/gravity dynamics involves the four-dimensional phase space of the canonical variables $(a, p_a, \phi, p_\phi)$ and Hamiltonian form of the Einstein equations. Since the Hamiltonian is zero, $\mathcal{H} = 0$, the phase trajectories actually reside in a three-dimensional space $\mathcal{M}^3$. The scalar field/gravity dynamical system can be described by the familiar equations

$$\ddot{\phi} + 3H\dot{\phi} + m^2\phi = 0, \tag{3.20}$$

$$H^2 + \frac{K}{a^2} = \frac{8\pi}{3M_p^2}\left(\frac{1}{2}\dot{\phi}^2 + \frac{1}{2}m^2\phi^2\right). \tag{3.21}$$

All solutions of these equations can be compactly depicted in the three dimensional phase space $\mathcal{M}^3$ of this dynamical system. It is also convenient to use the Poincare map to bring infinities of the phase space $\mathcal{M}^3$ in to a finite distance. Then all the solution of the system can be put in a box as in the Fig. 6. All phase trajectories begin at the repulsors at infinity $R_1$, $R_2$ and end either at the focus point $F$ at the center or at the attractors at infinity $A_1$, $A_2$. There are also four saddle points at infinity $S_1$, $S_2$, $S_3$, $S_4$ and four separatrices (located at the cone) which connect the saddle points with the focus $F$. For detailed description of the phase portraits of the gravity/scalar dynamical systems and properties of the solutions see [38].

Consider a flat universe with $K = 0$. In this case phase trajectories (solutions) are located at the two-dimensional hyper-surface defined by the constraint (3.21) with $K = 0$. For $V_0 = 0$ this two dimensional hyper-surface consists of two cones which touch each other by their apex in a single point $F$. The upper cone corresponds to a pure expansion branch $H > 0$, while the lower cone corresponds to pure contraction $H < 0$. Consider an expanding universe (contraction is just the time reverse). All trajectories begin at the repulsion points $R_1$ or $R_2$; most of the trajectories approach the separatrices $S_1 - F$ or $S_2 - F$ and then spiral around the focus $F$ where they end.

Around the repulsion points the expansion is dominated by the scalar field kinetic term $\dot{\phi}^2$. The universe begins from a physical singularity and initially expands with the scale factor $a(t) \sim t^{1/3}$. Therefore the asymptotic past of the homogeneous cosmological solutions based on the scalar field dynamics is similar to that of the FRW plain cosmology with initial singularity as in the left panel of Fig. 11. Most of the trajectories then converge to the separatrices. The evolution of the scalar field around the separatrices corresponds to slow rolling, $\ddot{\phi} \ll H\dot{\phi}$, $\dot{\phi}^2 \ll V$. This corresponds to the inflationary expansion $a(t) \sim e^{\int dt\, H}$. In this picture inflation is a very typical stage of the scalar field evolution

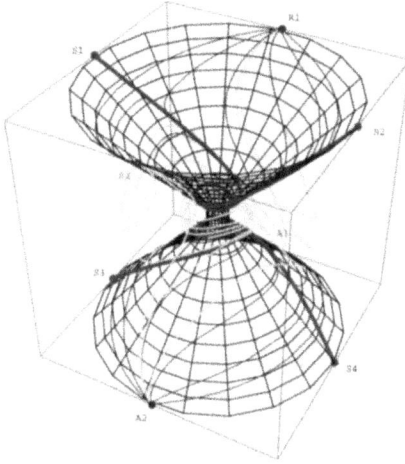

Fig. 6. Phase portrait for the theory $V(\phi) = \frac{1}{2}m^2\phi^2 + V_0$ for $V_0 = 0$. Upper and lower cones correspond to the stages of expansion and contraction of $K = 0$ universe. Trajectories of the universes with $K = +1$ are ouside, $K = -1$ are inside the cones.

but it is an intermediate stage. Eventually the slow-roll regime ends and the field begins to oscillate around the focus point. For a perfectly homogeneous universe this occurs everywhere simultaneously. Scalar field oscillations correspond to a matter equation of state with $a(t) \sim t^{2/3}$.

For $K = -1$, the trajectories propagate in the three-dimensions of $\mathcal{M}^3$ inside the cones of the Figure 6. Their start and end points are the same as for the flat case. Moreover, the trajectories converge towards the inflationary separatrices lying at the cones. Similarly, $K = +1$ trajectories are three-dimensional but outside of the cones, again converging to the separatrices. There are a lower measure fraction of trajectories describing a bouncing closed universe, not shown here.

For all $K$, inflationary separatrix acts as an attractor. To see it better, let us map the upper half of the cone ($K = 0$ case) in the original $\mathcal{M}^3$ space to the two-dimensional plane $(\dot{\phi}, \phi)$, as shown in Figure 7.

For illustrative purposes, we include the case when a flat universe is also filled with radiation. Its presence qualitatively does not change the picture. As we see, most of the trajectories starting at the Planck density approach the inflationary attractors $\dot{\phi} = \pm\sqrt{\frac{2}{3}}m$. In the Figure we have shown the process for the unrealistically large mass $m = 1$. The situation becomes much more impressive in the realistic case $m \sim 10^{-6}$.

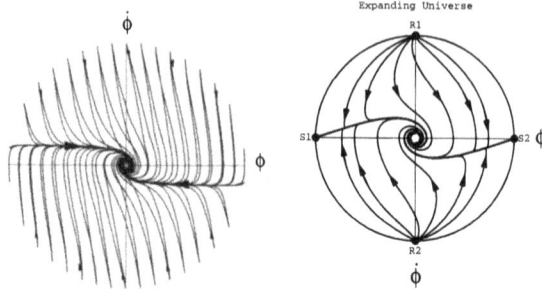

Fig. 7. Left: Phase portrait for the theory $V(\phi) = \frac{1}{2}m^2\phi^2$. The blue (thick) lines show trajectories describing the universe without radiation. The scalar field has half Planck density at the beginning of the simulations. The red (thin) lines show trajectories where an equal amount of energy in radiation was added to the system. In both cases the velocity of the scalar field rapidly decreases, which usually leads to the onset of inflation. The inflationary separatrices (the horizontal lines to which most of the trajectories converge) are attractors when we move forwards in time, but they will be repulsors if we reverse direction of time and move from the center back to large $|\phi|$ and $|\dot\phi|$. Right:Poincare mapping of the phase portrait from the left.

One can introduce a canonical measure in the space of trajectories, $d\mu$, which is invariant under the flow of trajectories

$$d\mu = -a^2 \frac{\dot\phi}{H}d\dot\phi \wedge da + a^2(\dot\phi + V_{,\phi}/H)d\phi \wedge da + d\dot\phi \wedge d\phi. \qquad (3.22)$$

This is a local measure. If we prescribe uniform prior probability of trajectories, then we just have to compare the measure $\int_{inf} d\mu$ trajectories with inflation vs $\int_{non-inf} d\mu$ trajectories without inflation. It turns out that both integrals, $\int_{inf} d\mu$ and $\int_{non-inf} d\mu$, diverge (because the scale factor $a$ is unbounded) and no conclusion with this prescription can be made [27]! Papers [28, 30] argued that since at the classical level scalar/gravity dynamics is reversible, one can try to judge about the measure of trajectories of interest not at the earliest stages of evolution but rather at the latest stages. In particular, if one truncate the two-dimensional integral $\int d\mu$ at some late-time value of $a$ to make it finite, the integral $\int d\mu = \int d\dot\phi \wedge d\phi$ is reduced to a one-dimensional integral $\int d\phi \, p_\phi \equiv J$. At the stage of oscillation $J$ is an adiabatic invariant, and together with the phase forms a pair of canonical variables $(J, \theta)$. They suggested measuring the priors of trajectories at the stage of scalar field oscillations, in terms of a uniform distribution of the phase $\theta$ of the oscillations. The separatrices correspond to a specific choice of the phase $\theta_*(t)$, and inflationary trajectories which converge to the separatrices have their phases crowded towards $\theta_*$.

In contrast, our prescription is to assign priors to trajectories at the very beginning around the repulsors $R_1$, $R_2$. Recent investigation [65] shows that if we repeat the trick with truncation $\int d\mu$ there, we obtain $\int d\phi \, p_\phi = J_0$ which is also conserved! (But it is not conserved in the middle of the trajectories). We can use a uniform distribution of $J_0$ around singularity, and it will not map into a uniform distribution of $J$ at the stage of the oscillations. In contrast, it gives a distribution highly peaked towards the inflationary separatric $J_*(t)$. Moreover, the modern theory of reheating after inflation does not have a long-oscillation post-inflationary stage, see Section 6, so that the prescription based on aposteriory distributions of the phase $\theta$ of oscillations after inflaton is not relevant! The formal solution of the problem can be given in terms of physical probabilities $P(\dot\phi, \phi, )$ for the realization of trajectories around the singularity (see discussion in Section 4.6). Apparently, $P(\dot\phi, \phi, H)$ cannot be given in terms of the phase flow invariants, but that is OK. It shall be calculable in quantum cosmology or in terms of landscape probabilities, but this is another story. In particular, the prescription of uniform oscillation phase $\theta$ would give the highest probabilities for trajectories with initially very high velocity $\dot\phi$. In contrary, physically motivated $P(\dot\phi, \phi)$ is suppressed on those trajectories.

To wrap up the phase portrait story, we finish this section with the case where $V_0$ can become negative. This is a potentially interesting case for present-day inflation (dark energy). The hyper-surface representing a flat universe is still defined by

$$6\frac{M_p^2}{8\pi} H^2 - \dot\phi^2 - m^2\phi^2 = 2V_0, . \tag{3.23}$$

But with $V_0$ negative, this surface is a hyperboloid of one sheet.

Figure 8 shows the phase space for this model and sample trajectories for a flat universe. The phase space is two dimensional, but its topology is very different from that for non-negative potentials. The infinite critical points are unchanged because the finite term $V_0$ has no effect at infinity, but there are no finite critical points. Thus all trajectories begin at infinity with $H > 0$ and end at infinity with $H < 0$. This is possible because the regions corresponding to expansion and contraction are now connected. This property is valid for all types of curvature $K$, i.e. for open, flat or closed universes.

Aside from this "wormhole" connecting the expanding and contracting branches this phase portrait looks a lot like the one for $V_0 > 0$ shown in Figure 8. Note, however, that in this case the separatrices emanating from the saddle points $S_1$ and $S_2$ no longer spiral in to the center, but rather end up reaching the points $A_1$ and $A_2$. Likewise there are separatrices that begin at $R_1$ and $R_2$ and end on $S_3$ and $S_4$. In the expanding phase their segments, and segments of nearby trajectories, represent the rare cases that manage to avoid inflation. In the contracting phase they become the marginal trajectories separating those that end

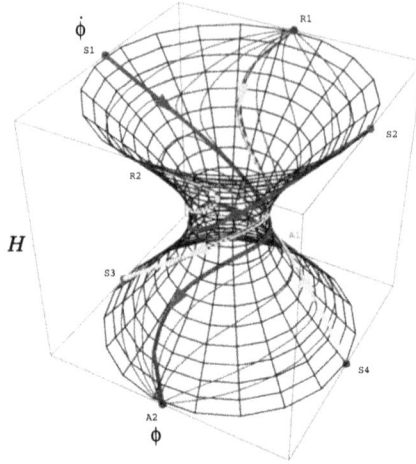

Fig. 8. Phase portrait for the theory $V(\phi) = \frac{1}{2}m^2\phi^2 + V_0$ for $V_0 < 0$ and $K = 0$. The branches describing stages of expansion and contraction (upper and lower parts of the hyperboloid) are connected by a throat.

at positive and negative $\phi$. An important (and obvious) feature of the 3D phase portraits in Figs. 6–8 is that the separatrices, as well as other trajectories, never intersect in 3D. This is a trivial consequence of the fact that we are solving a system of 3 first order equations for 3 variables, $\phi$, $\dot{\phi}$ and $H$. One of the implications of this fact is that a bunch of trajectories in the immediate vicinity of the (green) lines going from $R1$ to $S3$ and from $R2$ to $S4$ never reach the inflationary regime described by the (red) inflationary separatrices going from $S1$ to $A2$ and from $S2$ to $A1$. Only the trajectories that are sufficiently far away from the green lines going from $R1$ to $S3$ and from $R2$ to $S4$ can enter a stage of inflation.

The evolution of this system in an open or closed universe is not very different from the flat universe evolution, although the phase space is three dimensional. Because of the structure of the trajectory flow between their ends at the infinite critical points, all trajectories pass from expansion to contraction, even for an open universe. As with $V_0 > 0$ the trajectories for the open and closed cases will tend to asymptotically approach the flat universe hyper-surface, and more specifically will tend to approach the inflationary separatrices. As before, however, the closed universe will include some trajectories that quickly collapse before experiencing inflation.

## 4. Global geometry

Why are we interested in the properties of the universe on super-horizon scales which we will never observe? There are several reasons.

## 4.1. Generalities

An interesting indicator of the progress in cosmology that has been made is prediction about the the distant future of the universe (alongside with ideas about its beginning). As a graduate student I was impressed by how dramatically these ideas had changed. Compare Eddington's book from the 1920s where he conjectured that the universe in the future would be very hot because it is static and would be gradually filled with the photons emitted from all the dying stars; and the opposite prediction of 1980s cosmology that the universe would cool down to zero temperature. Today we anticipate different possibilities: in the future the universe may stop cooling down and instead be filled up by the radiation from the de Sitter horizon generated by the non-zero cosmological constant; or re-collapse if dark energy is quintessence with a potential which somewhere becomes negative, or will be filled up with the black holes, etc. Early universe inflation is the basis of the modern cosmological paradigm. In particular, it predicts that our patch of the universe is described by a flat FRW geometry with small scalar perturbations around it. How may our patch be embedded in the bigger picture? The question of the initial conditions for our universe can be reduced, in this picture, to the issue of probabilities in the global geometry. The right way to address questions about global geometry is to use the very powerful method of Carter-Penrose diagrams.

Significant information about cosmological solutions of the Einstein equations, in the class of FRW metrics with a scalar field and other matter content, can be extracted from the phase portraits of the gravity/scalar field dynamical system. Trajectories in the phase space correspond to solutions of the dynamical equations. In principle, for each trajectory one can construct the corresponding Penrose diagram. However, as we will see, the Penrose diagrams will be identical for several classes of trajectories. The method of dynamical systems in terms of phase space portraits, and the method of Penrose diagrams, in some sense are complementary to each other.

Another aspect of the problem of the global geometry is how much of it is unique to the problem under consideration, and how much of it comes from simple composition of several more basic geometries. An example of composite geometry is the geometry of several black holes or a single black hole in an FRW universe, which is the composite geometry of a single black hole and a single FRW universe. As we will see, some aspects of the global geometry of the inflationary universe are unique to inflation (such as eternal inflation), while some aspects come from composition of more basic elements (such as domains of FRW universes). However, all together these domains form a fractal structure which is unique to inflation.

Investigation of the space-time geometry requires not only finding solution of the local Einstein equation but also finding a global manifold which contains the inflationary universe with all relevant physics. This means that we deal not with a pure de Sitter geometry but with a quasi-de Sitter geometry; and we deal not with a perfectly homogeneous universe but with inhomogeneities generated by the instability of the vacuum fluctuations of the inflaton field. Even more, we may deal not only with the classical Einstein equations, but with the significant quantum gravity effects beyond the classical equations, as in the picture of the self-reproducing universe. We deal not with the deterministic evolution of geometry but with quantum (stochastic) processes. In general, the multiverse with self-reproducing inflation on super-horizon scales is not spherically symmetric. However, we can continue to use the tool of Penrose diagrams, if we mean two-dimensional sections of four-dimensional space-time, ignoring two spatial dimensions.

To lay down the story, we begin with the geometry of pure de Sitter space-time and the geometry of pure FRW space time, and then add new levels of physics step by step.

### 4.2. De Sitter and FRW geometries

De Sitter geometry, supported by a positive cosmological constant $\Lambda$, has constant scalar curvature $R = -12H^2 = 4\Lambda$, isotropic Ricci curvature tensor $R_{\mu\nu} = \frac{1}{4}Rg_{\mu\nu}$, and vanishing Weyl tensor $C_{\mu\nu\rho\sigma} = 0$. De Sitter geometry corresponds to the 4d hyperboloid of constant curvature embedded in an auxiliary 5d Minkowski space, $-v_1^2 + v_2^2 + v_3^2 + v_4^2 + v_5^2 = \frac{1}{H^2}$. The hyperboloid has the symmetry group $SO(1, 4)$. It can be covered by various different coordinates. The time-dependent coordinate systems include all three FRW cases: closed, flat and open slicings. The closed universe map covers the whole hyperboloid

$$ds^2 = dt^2 - \frac{1}{H^2}\cosh^2(Ht)\left(d\chi^2 + \sin^2\chi\, d\Omega_2^2\right), \tag{4.1}$$

where the line element on $S^2$ is $d\Omega_2^2 = d\theta^2 + \sin^2\theta\, d\phi^2$, $0 \le \chi \le \pi$, see Fig. 9.

Two flat universe maps, corresponding to expanding or contracting universes

$$ds^2 = dt^2 - e^{\pm 2Ht}\left(dr^2 + r^2 d\Omega_2^2\right), \tag{4.2}$$

each cover upper or lower half of the hyperboloid, $-\infty < t < \infty$ for expansion, $-\infty < t < \infty$ for contraction, see Figure 10.

By conformal transformation $ds^2 \to \Omega^2(x^\mu)ds^2$ (with a smooth local function $\Omega$) the global geometry of the space-time and its causal structure can be represented by a compact Penrose diagram. The Penrose diagram of the full dS space-time is shown on the left panel of Fig. 9. It is free of physical singularities.

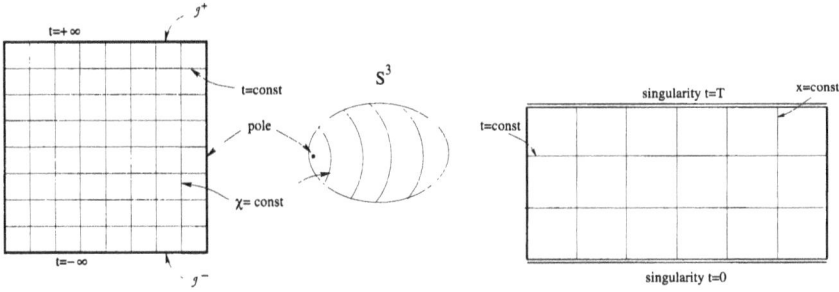

Fig. 9. Left: Penrose diagram of full de Sitter space-time. Right: Penrose diagram of closed FRW universe with matter and radiation.

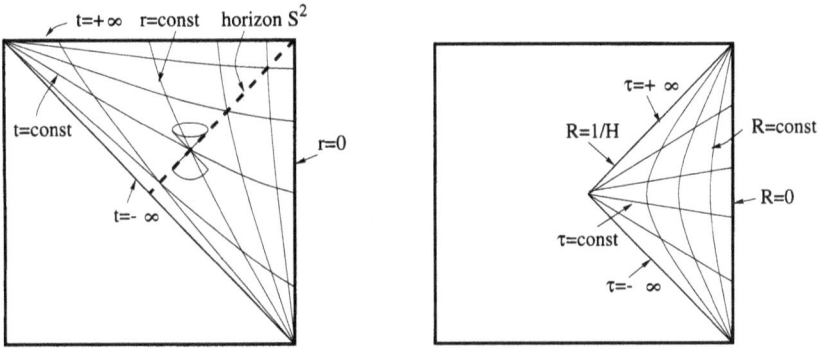

Fig. 10. Penrose diagram of de Sitter space-time in the flat FRW coordinates (left) and the static coordinates (right). Each point represent a sphere $S^2$. Its radius at the horizon (dashed line on the left, edge of diamond on the right) is equal to $\frac{1}{H}$.

Penrose diagrams of the flat FRW de Sitter metric are shown on the left panel of Fig. 10.

Next, we write down a static form of the metric on de Sitter space time

$$ds^2 = (1 - H^2 R^2)d\tau^2 - \frac{dR^2}{(1 - H^2 R^2)} - R^2 d\Omega^2 \tag{4.3}$$

The Penrose diagram of de Sitter space-time in static coordinates is plotted on the right panel of Fig. 10. The diagrams of Fig. 10 can be used to construct the quantum field theory in de Sitter space time in different coordinates.

In de Sitter space-time an observer at the origin $r = 0$, $R = 0$ (right vertical edges of both panels of Fig. 10) is surrounded by an event horizon, shown as the dashed line in time-dependent coordinates at the left panel of Fig. 10, or the edge of the diamond at the right panel of Fig. 10. The radius of each sphere $S^2$ along of the event horizon in both pictures is constant, $R = \frac{1}{H}$.

### 4.3. FRW geometry

In this subsection we recall the FRW geometry (2.1) with matter and radiation. We also consider the models with a small effective cosmological constant $\Lambda \simeq (10^{-3} \, eV)^4$.

A closed universe with radiation and matter begins expanding from a physical singularity, then recollapses and ends by contracting to another physical singularity. The Penrose diagram for a closed FRW universe filled with matter and radiation is plotted in the right panel of Fig. 9. Just as in the full de Sitter map in the left panel of this Figure, the vertical edges of the closed FRW Penrose diagram correspond to coordinate singularities at $\chi = 0, \pi$.

A flat FRW universe with matter and radiation begins from an initial physical singularity. Its Penrose diagram is shown in the left panel of Fig. 11. This is an asymptotically flat metric without a future event horizon. Although formally the de Sitter metric in the form (4.2) also belongs to the class of metrics (2.1) with $K = 0$, globally the dS and FRW geometries are different. The map (4.2) covers only part of the full de Sitter geometry, as shown in Fig. 10, so that the manifold covered by the coordinates (2.1) can be extended. In contrast, the manifold of the generic flat FRW geometry covered by the coordinates (2.1) with $K = 0$ and shown in the left panel of Fig. 11 cannot be extended to the manifold of the closed FRW geometry with $K = +1$ which is shown in Fig. 9. Therefore the cases of FRW geometries with $K = 0$ and $K = +1$, as different solutions, in general shall be considered independently. The Penrose diagram of the open FRW geometry with $K = -1$ is similar to that of the left panel of Fig. 11.

An FRW cosmology dominated by the late time (effective) cosmological constant is observationally favored model. Dark energy will not affect the asymptotic

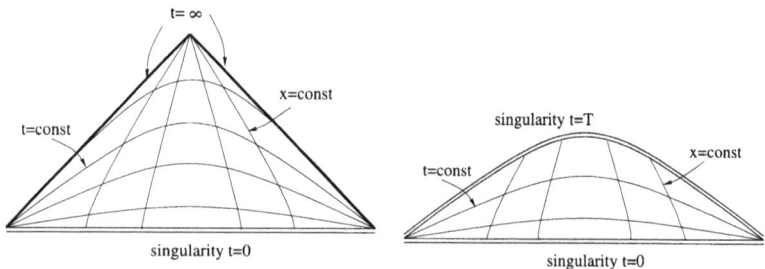

Fig. 11. Left: Penrose diagram of flat (or open) FRW universe with matter or radiation. Right: Penrose diagram of flat recollapsing FRW universe with negative scalar field potential.

past of FRW universes, but will change their asymptotic future. The result, however, depends on the dark energy model. In the simplest case of a pure constant, $\Lambda = const$, the asymptotic future shall be the same as that of the pure de Sitter map in the left panel of Fig. 9. The Penrose diagram of the FRW geometry with asymptotic future de Sitter space looks like a "tall" rectangle with the (horizontal bottom) space-like initial singularity and the (horizontal top) space-like asymptotic future $i^+$. The $K = +1$ case can be constructed by putting left panel diagram of Fig. 9 (without the bottom strip) on top of the right panel diagram of this Figure (without the upper singularity).

Next, consider more exotic model of the effective cosmological constant. Suppose the late time cosmological constant is provided by a scalar field, rolling from the top of its effective potential $V(\phi)$, where it is positive. However, suppose that this potential gradually decreases and at some value of $\phi$ becomes negative. This type of scalar field potential can be encounter in some models of extended supergravity [32]. What will be important for us is that the scalar field evolution in models with potentials which can become negative can lead to contraction of the universe in the future [32, 38]. The Penrose diagram of such a cosmological model is shown in the right panel of Fig. 11. Here the space-time is sandwiched between two physical singularities, in the past and in the future.

The Penrose diagrams of Figs. 9–11 for idealized de Sitter geometry and FRW universe with singularity in the past will be used as building blocks to construct the global geometry of the inflationary universe.

### 4.4. Single FRW domain from inflation

Before constructing the global geometry of more realistic inflationary universes with fluctuations, in this subsection we will study the global geometry of a homogeneous universe with primordial inflation which is followed by an FRW stage.

It turns out that the geometries of globally homogeneous universes, for all relevant cases of the asymptotic future which we discussed in the previous section, can be related to the global geometry of a homogeneous universe filled with a scalar field with potential $V(\phi) + V_0$, where $V(\phi)$ is a monotonic potential for the inflation with minimum at $\phi = 0$, and $V_0$ is a constant which can be negative, zero or positive. Indeed, the case of zero or positive $V_0$ corresponds to the late time universe being filled with an oscillating scalar field. The asymptotic future of this cosmology is similar to that of a FRW model with radiation and matter. The case $V_0 > 0$ corresponds to a cosmological-constant-dominated late-time universe. Finally, negative $V_0$ corresponds to a recollapsing universe.

Therefore, in order to construct the global geometry of the homogeneous FRW cosmologies with inflation in the past, we can use the very simple dynamical system of a scalar field with potential $V(\phi) + V_0$. All qualitatively different solutions of the Einstein equations with a scalar field were reviewed in the previous section with dynamical phase portraits. Thus these two methods (dynamics in phase space and Penrose diagrams) for the same gravity/scalar systems will give a unified picture of the global geometry and the global evolution. We restrict our discussion with inflationary models of Fig. 5.

In this Subsectuin we address a question about the geometry of an inflationary universe which as *a whole* evolves into an FRW universe. One encounters this question in the idealized situation of a perfectly homogeneous universe with a homogeneous scalar field $\phi(t)$ with the potential $V(\phi)$ with a minimum, without considering inflaton field fluctuations. This is a familiar situation from our study of the dynamics of the expanding universe in terms of the scale factor $a(t)$ and the scalar field $\phi(t)$. This problem can be well studied by the phase portrait method.

The asymptotic future of a matter dominated universe has the structure of the future of the Penrose diagram in the left panel of Fig. 11. Although, in a realistic inflationary scenario, inflaton oscillations decay in the process of preheating and set up the hot FRW universe, its global geometry in terms of a Penrose diagram is the same as in the simple case of the evolving scalar field. Furthermore, for rare trajectories which miss inflation, the Penrose diagram will still be the same. Thus, for all solutions corresponding to the the dynamical system (3.20)–(3.21) with $K = 0$, the global geometry is the same as for usual FRW $K = 0$ cosmology shown in the left panel of Fig. 11!

What is the impact of inflation on the global space-time geometry? In short, under the assumption that the universe is globally homogeneous, an intermediate inflationary stage *does not* impact the global geometry.

The local geometry during the slow roll regime is well approximated by a quasi-de Sitter geometry. In the left panel of Fig. 12 we re-plot the Penrose diagram of the flat FRW geometry with an intermediate inflationary stage. Time hyper-surface $t_i$ indicates the beginning of inflation (for a particular phase trajectory), and hyper-surface $t_p$ indicates the end of inflation and the beginning of the preheating stage. We show the past event horizon of the observer at the origin at the moment $t_p$. It is defined by the condition $r = \int^{t_p} \frac{dt}{a}$. The radius of each sphere $S^2$ along this horizon between $t_i$ and $t_p$ is slowly varying and close to $\frac{1}{H}$, where $H$ is the Hubble parameter during inflation. This is similar to what we had for a de Sitter geometry, see left panel Fig. 10. Also, the local geometrical properties between $t_i$ and $t_p$ are similar to that of de Sitter space-time.

Let us consider the apparent horizon in a perfectly homogeneous universe with an intermediate period of inflation. Consider the physical radius $R = a(t)r$. The apparent horizon is defined by the condition $g^{\mu\nu}\nabla_\mu R\nabla_\nu R = 0$. From here we find the evolution of the apparent horizon $r = \pm\frac{1}{a}$. Each sphere along the apparent horizon has the physical radius $R = 1/H$ which is slowly increasing

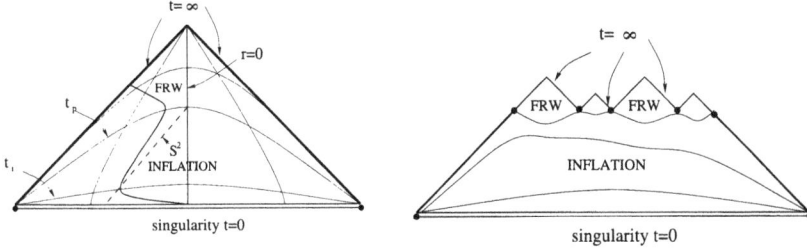

Fig. 12. Left: Penrose diagram of homogeneous flat FRW universe with intermediate inflationary stage. Dashed line is the past event horizon for an observer at the origin at the end of inflation $t_p$. the solid zig-zag line is the apparent horizon. Right: the same with large scale quantum fluctuations taken into account. The asymptotic future is a fractal of FRW domains.

with time. In the left panel of Fig. 12 we plot the apparent horizon in an FRW universe with an intermediate period inflation. The apparent horizon is quite different from that in an FRW universe without inflation (which isa monotonic function), but (locally) is similar to that of a dS geometry. Thus, inflation has an impact on the structure of the apparent horizon.

Consider the case of a closed universe $K = +1$. In this case all phase trajectories are located outside of the cones (see Fig. 6); they begin at the repulsors $R_1$ or $R_2$, and end at the attractors $A_1$ or $A_2$. This means that the universe initially expands and finally contracts. At the stage of expansion the curvature term $\frac{K}{a^2}$ in the constraint equation (3.21) decreases. Correspondingly, most of the phase trajectories converge towards the limiting cone surface. Many trajectories have an intermediate inflationary stage which end by scalar field oscillations. In realistic inflationary models, inflaton oscillations transform into radiation in a closed universe, which eventually will recollapse. In our model with a scalar field, the late time evolution after the oscillations will be different from the closed FRW cosmology in details, but the global geometry will be similar, and the universe will recollapse into a singularity. The Penrose diagram of the global geometry of a closed universe with a period of intermediate chaotic inflation is similar to that of the closed FRW universe shown in the right panel of Fig. 9. Again, an intermediate inflationary stage does not change the global geometry.

To conclude this subsection, we add the cosmological constant term $V_0$ to mimic present day de Sitter asymptotic behavior. We are interested in an expanding solution. Expansion begins with a singularity, around which the scalar field kinetic energy dominates. Then we have an intermediate inflationary stage. After inflation, the scalar field oscillates and sets up the FRW expansion. FRW evolution ends with an asymptotically de Sitter stage. Thus, the Penrose diagram of a closed universe with an intermediate inflationary stage is a "tall" rectangle with an initial singularity and space-like asymptotic future $\rangle^+$. Again, in the idealized closed cosmological model where the geometry is assumed to be homogeneous, an intermediate inflationary stage does not influence the Penrose diagram of the global manifold.

In principle, different classes of inflaton potentials (like those shown in the Figure 5) can have different details of the global geometry, such as the percolation properties of inflating and non-inflating regions. Another setting of the problem is related to the potential of the type depicted on the right of Fig. 5. The scalar field at the top of the potential generates a de Sitter solution with the large symmetry group $S(1, 4)$. This has no preferred time slicing. Suppose that eventually quantum fluctuations of scalar field will displace it from the top of the potential, but only locally. Hence, in a final spatial patch the scalar field is rolling towards of its minimum, while in the rest of space scalar field is still staying at the top of the potential. Locally the rolling scalar field defines time slices, which can be taken as the hyper-surfaces of constant $\phi$. Te geometry in the local domain with the rolling scalar field has less symmetry than de Sitter space-time. The spontaneous origin of a local FRW domain shares many features in common with dynamical symmetry breaking, where a scalar field acts as an order parameter. In a sense, we encounter symmetry breaking of the space-time itself, where the arrow of time arises.

### 4.5.  *Scalar field fluctuations in de Sitter geometry*

Now we introduce the next level of physics, related to QFT in de Sitter space-time. As we will see in the next subsection, it drastically changes the character of the Penrose diagram of a universe with inflation.

Consider a quantum scalar field in a de Sitter geometry. There are two schools of thought about this.

One school, which I refer to as the "ideal dS" approach, originated in late 1960s and included the classic 1977–78 [39] and [40] papers, and has recently drawn significant attention in the theoretical physics and superstring community with respect to holography and the thermodynamics associated with the de Sitter horizon and the "hot tin can" picture. In this context, the static form of the dS metric (4.3) is utilized. The "hot tin can" Penrose diagram is plotted on the right panel of Fig. 10. The key result is that an observer at the origin detects thermal radiation from the horizon at $R = \frac{1}{H}$ with the temperature $T = \frac{H}{2\pi}$, and the horizon area $A = \frac{4\pi}{H^2}$ is associated with the huge (geometrical) entropy $S = \frac{A}{4G}$. This effect can be seen after the Wick rotation $\tau \to iT$ in the metric (4.3), which then turns to the Euclidean form

$$ds_E^2 = (1 - H^2 R^2)dT^2 + \frac{dR^2}{(1 - H^2 R^2)} + R^2 d\Omega^2. \qquad (4.4)$$

This metric has a conical singularity at $R = 1/H$, which implies periodicity of euclidean time $T$ with period

$$T_H = \frac{2\pi}{H}. \qquad (4.5)$$

Now recall that the period of imaginary time in thermal QFT corresponds to the temperature. Thus, the temperature of dS horizon is given by (4.5). An observer at the origin detects thermal excitations of a detector. Notice, however, that this temperature does not result in the radiation equation of state of scalar field excitations. On the contrary, $\langle T_\nu^\mu \rangle \propto H^4 \delta_\nu^\mu$, which corresponds to vacuum polarization in dS geometry. Also, $T_H$ is universal for different fields (not only scalar) of different masses and different couplings $\xi R\phi^2$. The thermal vacuum in the causal patch ("hot tin can") corresponds to the Bunch-Davies vacuum of the metric (4.2).

Consider Bunch-Davies vacuum fluctuations of a scalar field $\langle \delta\phi^2 \rangle_{reg}$ in a dS background. After familiar regularization there will be a finite result which depends on $H$, $m$ and $\xi$. Among quantum scalar fields with mass $m$ and coupling $\xi$ in de Sitter geometry, the case of minimal coupling $\xi = 0$ and very small mass $m \ll H$ plays an especially important role. Indeed, the regularized vacuum expectation value is

$$\langle \delta\phi^2 \rangle_{reg} = \frac{3H^4}{8\pi^2 m^2}. \qquad (4.6)$$

Formally, this is an odd case since its eigen-spectrum contains an infrared divergent term: $\langle \delta\phi^2 \rangle \to \infty$ as $m \to 0$. On the other hand, this is the most interesting case for application to inflation, since the theory of inflaton fluctuations (as well as tensor fluctuations) is reduced exactly to this case. This divergence is due to infrared modes associated with the global dS geometry. We will give a derivation of (4.6) which manifests this infrared effect. Let us make a Wick rotation in the metric (4.1) to bring it to that of $S^4$ and consider euclidean QFT on $S^4$. The Green function of the scalar field is $G(x, x') = \sum_\lambda \frac{\Phi_\lambda(x)\Phi_\lambda^*(x')}{\lambda}$, where $\Phi_\lambda(x)$ is the eigen-function of the Laplace operator on $S^4$ with eigenvalue $\lambda$, $(\nabla_\mu \nabla^\mu + m^2)\Phi_\lambda = \lambda\Phi_\lambda$. For small $m$, the lowest eigenvalue is just $\lambda = m^2$. It corresponds to the homogeneous mode $\Phi_\lambda$, whose amplitude we will find from the normalization $\int d^4x \sqrt{g}\Phi_\lambda(x)\Phi_{\lambda'}^*(x) = \delta_{\lambda\lambda'}$. The area of $S^4$ which envelops an $S^5$ sphere of radius $1/H$ is $\int d^4x \sqrt{g} = \frac{8\pi^2}{3H^4}$, thus for the mode homogeneous at $S^4$, $\Phi_\lambda(x)\Phi_\lambda^*(x) = \frac{3H^4}{8\pi^2}$. In the limit of small $m^2$, the leading contribution to $G(x, x')$ comes from the lowest eigenmode (which is intact by regularization). The VeV of the vacuum fluctuations is just

$\lim_{x' \to x} G(x, x') = \Phi_\lambda(x)\Phi_\lambda^*(x)/m^2 = \frac{3H^4}{8\pi^2 m^2}$, so we derived the Bunch-Davies formula (4.6).

Other non-trivial conjectures in this picture include spontaneous black hole formation from the thermal fluctuations inside dS horizon [41], emergency of "the Boltzmann brain" [42] etc.

Another school, which I refer to as the "time-dependent" school, originated in the early 1980's and treats the QFT of scalar fields in (quasi) de Sitter geometry which (locally) lasts for a finite time interval. This situation corresponds to the inflationary geometry represented by the Penrose diagram 11. Following the time evolution of individual fluctuations, it was found that the infrared divergence can be interpreted as the instability of quantum fluctuations of a very light scalar field, which are accumulated with time [18,43,44]

$$\langle \delta\phi(t)^2 \rangle = \frac{3H^4}{8\pi^2 m^2} \left(1 - e^{-\frac{2m^2}{3H^2}t}\right), \quad \to \frac{H^3}{4\pi}t \quad \text{for} \quad t \ll 1/m, \tag{4.7}$$

and in the limit of infinite inflation $t \gg 1/m$ saturate to the Bunch-Davies expression (4.6). This explains the origin of the infrared divergence. Here we briefly derive (4.7).

The time-dependent form of the metric (4.2) is very convenient for investigating the QFT of a probe scalar field, $\delta\phi$, in the time-dependent background. The field operator in the Heisenberg representation is

$$\delta\hat{\phi}(t, \vec{x}) = \frac{1}{(2\pi)^{3/2}} \int d^3k \left(\hat{a}_{\vec{k}}\phi_{\vec{k}}(t)\, e^{-i\vec{k}\vec{x}} + \hat{a}_{\vec{k}}^+ \phi_{\vec{k}}^*(t)\, e^{i\vec{k}\vec{x}}\right), \tag{4.8}$$

where $\hat{a}_{\vec{k}}$ and $\hat{a}_{\vec{k}}^+$ are the annihilation and creation Bose-operators. The temporal part of the eigenfunction with momentum $\vec{k}$ obeys the equation

$$\ddot{\phi}_k + 3\frac{\dot{a}}{a}\dot{\phi}_k + \left[\left(\frac{k}{a}\right)^2 + m^2 - \xi R\right]\phi_k = 0. \tag{4.9}$$

Let us use the conformal time as a time variable, and introduce the new function $f_k(\eta) = a(\eta)\phi_k(\eta)$. Then instead of (4.9) we have

$$f_k'' + \Omega_k^2 f_k = 0, \quad \Omega_k^2 = (ma)^2 + k^2 - (1 - 6\xi)\frac{a''}{a}, \tag{4.10}$$

where a prime stands for $\frac{d}{d\eta}$. In a de Sitter stage, $H = const$, $a(\eta) = 1/H|\eta|$, $-\infty < \eta < 0$, and a normalized solution of eq. (4.10) is

$$f_k(\eta) = a(\eta)\frac{\sqrt{\pi}}{2} H\, |\eta|^{3/2}\, \mathcal{H}_\nu^{(1)}(|k\eta|)\, e^{i\frac{(2\nu+1)\pi}{4}}, \tag{4.11}$$

where $\nu = \left(9/4 - m^2/H^2 - 12\xi\right)^{1/2}$ is the index of the Hankel function $\mathcal{H}_\nu^{(1)}$. In the short-wavelength limit $|k\eta| \gg 1$ we have the positive-frequency solution $f_k(\eta) \simeq \frac{e^{-ik\eta}}{\sqrt{2k}}$ corresponding to the vacuum fluctuations in the far past. In terms of the decomposition (4.8), the formal answer for the VeV is

$$\langle \delta\phi(t)^2 \rangle = \frac{1}{2\pi^2 a^2} \int_0^\infty dk\, k^2 |f_k(\eta)|^2. \tag{4.12}$$

Now we substitute solution (4.11) in to (4.12), regularize the ultra-violet divergence, and obtain the finite time-dependent answer given by (4.7).

The difference between the "time-dependent" approach in comparison with the "ideal dS" approach is that the answer for the massless scalar field is finite and time-dependent for the first, while it is divergent for the last. Moreover, the "time-dependent" approach allows us to see how $\langle \delta\phi(t)^2 \rangle$ is building up with time. Consider $m^2 \ll H$ so that (4.11) is simply $f_k(\eta) = \frac{iH}{\sqrt{2k^{3/2}}} e^{-ik\eta} (1 - ik\eta)$. After regularization, the residual contribution to the VeV (4.12) comes from the infrared modes $\langle \delta\phi(t)^2 \rangle_{reg} \approx \frac{H^2}{2(2\pi)^2} \int \frac{d^3k}{k^3}$. We have to include in the integration modes which by time $t$ are redshifted outside the horizon $H/a(t) < k < H$. It gives us

$$\langle \delta\phi(t)^2 \rangle = \frac{H^2}{4\pi^2} \int_{H/a}^H \frac{dk}{k} = \frac{H^3}{4\pi^2} t. \tag{4.13}$$

What happens to the fluctuations $\delta\phi_k$ initially oscillating with redshifting wavelengths? As soon as the wavelength exceeds the horizon, the amplitude of the mode ceases to oscillate, and freezes out with a random phase, modulus $k^{3/2}|\delta\phi_k| \sim H$, and scale free spectrum. As time goes on, a more and mode modes contribute to this random gaussian field $\delta\phi(t, \vec{x})$. The field $\delta\phi$ is complex, the phase of each mode is random, and total dispersion is the same as the dispersion of the random walks of a drunk sailor. After $n$ steps the dispersion is proportional to $n$. Think about each step within efold $\delta t \sim 1/H$ as contributing $H^2$ to the dispersion; after $Ht$ steps the dispersion shall be $H^2 \times (Ht)$, in agreement with formula (4.13). The drunk sailor (who probably belongs to the "time-dependent" school) helps us to understand the Bunch-Davies divergence.

Fluctuations of $\delta\phi$ induce scalar metric perturbations [17–21]. This picture is the basis of the inflationary paradigm so successfully confirmed observationally. Further, the backreaction of fluctuations $\delta\phi$ leads to a picture of stochastic evolution of quasi-de Sitter geometry [34, 35], and at large values of $H$ even to eternal self-reproduction of the inflationary universe [37]. A scalar field in the eternal inflationary universe is described naturally in terms of a probability distribution function $P(\phi, t)$ [35, 36]. All of this is critical for drawing a realistic Penrose diagram of the multiverse. For this we have to take with us the formula for scalar metric fluctuations $\Phi$ induced by $\delta\phi$, which we will derive in Section 5

$$\Phi_k = C \, \frac{V^{3/2}}{M_p^3 V_{,\phi}} \, \frac{1}{k^{3/2}}, \quad C = \frac{16}{15} (3\pi)^{3/2}. \tag{4.14}$$

For definiteness, we will bear in mind the case of a quadratic potential which give us

$$\Phi_k \approx 0.4 \frac{m}{M_p} \frac{\log k}{k^{3/2}}. \tag{4.15}$$

Similar formula takes place for the potential $V_0 - \frac{m^2}{2}\phi^2$ due to the degeneracy of the spectrum of scalar fluctuations, as we will see in Section (5.4). Therefore this formula covers both potential of Fig. 5. This formula gives a prediction for the observable spectrum of scalar perturbations in the model, as well as predicting its cousin: super-large scale fluctuations well outside of the observable horizon. If we accept the first, we have to accept the last. As we will see in the next section, they will have a drastic effect on the global Penrose diagram.

Some final remark about the two ways to think about QFT in dS. While it is expected that the approaches based on the time-dependent form of the de Sitter metric with unstable fluctuations, and the static form of the de Sitter metric with thermal fluctuations, should give us complementary insights, their languages are apparently different. This is partly due to the difference between quasi-de Sitter and pure de Sitter geometries, and partly because different questions are addressed. However, we have to understand how these two different approaches to (quasi-)de Sitter geometry with a scalar field are compatible with each other with respect to such important issues as the generation of fluctuations, entropy and global geometry

It is not clear to me, however, how quantum fluctuations in the "hot tin can" picture correspond to the instability of quantum inflaton fluctuations $\delta\phi$ and generation of metric perturbations. While the heuristic picture of the Hawking temperature $\delta\phi = \frac{H}{2\pi}$ is common for both pictures, the subtleties are not. Notice that his temperature does not result in a radiation equation of state for scalar field excitations. On the contrary, $\langle T_\nu^\mu \rangle \propto H^4 \delta_\nu^\mu$, which corresponds to the vacuum polarization in dS geometry. Also, $T_H$ is universal for different fields (not only scalar) of different masses and different conformal couplings $\xi$. However, freeze out of quantum fluctuations occurs for light minimally coupled fields. On the other hand, it is not clear how to understand the quantum gravity effects of black hole emanation from thermal fluctuations inside the hot dS cave, or emergency of the "Boltzmann brain" in the "time-dependent" picture.

## 4.6. Multiple FRW domains from inflation and fractal geometry in the future

Classical perturbations in a (quasi) de Sitter geometry die out in accordance with the "no-hair" theorem. However, one of the features of the quasi-de Sitter inflationary stage is that the quantum fluctuations of a light scalar field $\delta\phi$ (minimally coupled to gravity) are amplified and become classical inhomogeneities $\delta\phi(t, \vec{x})$. Those inevitably generate scalar metric perturbations $\Phi(t, \vec{x})$, so that the metric becomes perturbed (2.17). Although these fluctuations are small within our observable horizon, in many inflationary models their *r.m.s.* value is gradually increasing on larger and larger scales, well beyond the observable horizon, see formula (4.15) above.

The fact that the universe is not homogeneous on superhorizon scales has many important consequences for the global description of the universe. In particular it means that globally the inflationary universe is not homogeneous. Time slices can be naturally chosen as slices of constant values of scalar field. Inhomogeneous scalar field means wiggling of constant time hyper-surfaces. Wiggling of constant $\phi$ hyper-surfaces, taking fractal shape.

In the presence of the accumulating fluctuations, strictly speaking, we cannot anymore use the whole phase space trajectories of the previous section. But this picture is still valid for a domain of the universe where fluctuations are sufficiently small (i.e. for the portion of trajectories on the left of some critical trajectory which marks the regime of self-reproducing universe.)

In general, due to the initial conditions plus accumulating super-horizon fluctuations, the inflationary epoch will be dynamically ended at different moments of time in different spatial domains. At the moment $t_e$, the scalar field inside the domain ceases to roll slowly and begins to oscillate around a minimum. This domain has the rest of the universe in its past. Thus, $t_e$ is a space-like hyper-surface which separates the FRW domain from the rest of the universe. In the multiverse, there are infinite numbers of FRW domains which originated from the ever-inflating universe. They form a fractal structure in the asymptotic future, as shown in Fig. 13. The details of the geometry inside an FRW domain depends on the equation of state in the future. The matter domination in the FRW domains is shown in the left panel of Fig. 13, and the domination of a late-time cosmological constant in the right panel of Fig. 13.

The fractal structure means that the asymptotic future is divided into an infinite number of segments. The segments correspond to an infinite number of causally disconnected FRW domains. The sizes of the domains in the Penrose diagrams are irrelevant, since by conformal transformations (which keep intact the causal structure) one can make the apparently large domains smaller and vice verse. For the late-time cosmological-constant-dominated FRW domains, for the same reason it is also irrelevant exactly how the future infinity is

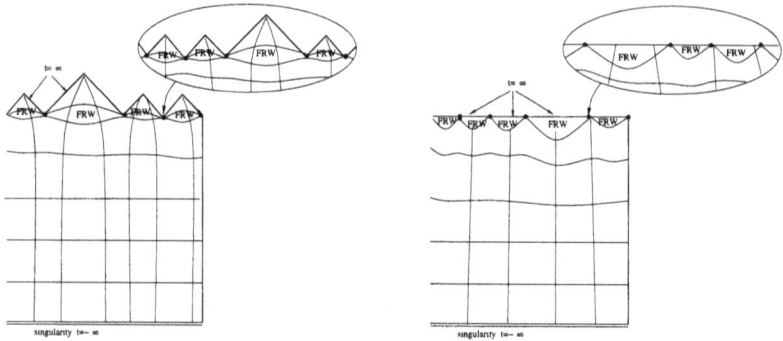

Fig. 13. Left: Penrose diagram for FRW domains originating from inflation. The asymptotic future in each domain is matter dominated. Right: Same as in the left panel but the asymptotic future in each domain is dominated by a late time cosmological constant. The asymptotic future has a fractal structure.

drawn, it can be either horizontal in each domain, or just be a space-like segment.

As time goes on, more and more FRW domains form from the ever inflating space between them. Although the co-moving volume is more and more consumed by FRW domains, the physical volume of the inflating phase between them is exponentially increasing. The time flow in the Penrose diagrams in Fig. 13 is manifested in the production of more and more elements of the fractal structure.

The picture of Fig. 13 works for the globally closed universe. For $K = 0, -1$ we have cut the tip of the FRW triangle and ued it to make the fractal structure at the top. The resulting Penrose diagram is shown in the right panel of Fig. 12. Thus, the impact of chaotic inflation on the global geometry of the universe is that the global homogeneity is broken and the asymptotic future has a fractal structure, consisting of an infinite number of causally disconnected FRW patches.

The details of the fractal structure depend on the inflationary model. For the potential of the new inflationary model (right panel of Fig. 5), the scalar field fluctuations can send some patches of the universe to the left while other patches go to the right. However, there will always still be inflating patches. As a result, there will be a number of disconnected FRW domains associated either with the right or with the left minimum of the double-well potential. The asymptotic future has a distinct fractal structure.

Thus, the Penrose diagram corresponding to inflation in the double-well potential is very different from that for the simple homogeneous universe rolling from the top to the bottom. When implementing the dS/CFT correspondence

FRW FRACTALS

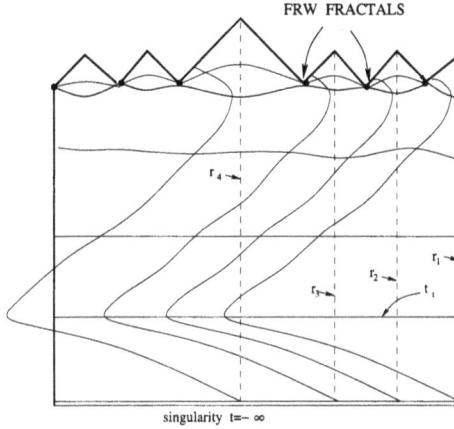

singularity t=− ∞

Fig. 14. Different observers $r_1$, $r_2$, $r_3$, $r_4$ etc. will see different apparent horizons and different radii of $S^2$ spheres alone them.

in a universe with inflation, one has yo bear in mind the fractal structure of the asymptotic future.

In the perturbed space-time, the physical radius is $R = (1 - \Phi)\, e^{\int dt\, H}\, r$. The radius of the apparent horizon is

$$r_h = H^{-1}\left(1 + \Phi + \frac{\dot{\Phi}}{H} - \frac{\Phi_{,r}}{\dot{a}}\right). \tag{4.16}$$

Consider the area of the apparent horizon $A = 4\pi (r_h)^2$. For frozen long-wavelength fluctuations, we can drop the derivatives in the formula (4.16) for $r_h$. Then the leading contribution of the frozen classical fluctuations to the area is

$$A = \frac{4\pi}{H^2}\, (1 + 2\Phi). \tag{4.17}$$

Thus, the quantum fluctuations of the inflaton field generate wiggles of the quasi-de Sitter apparent horizon area [33], as described by (4.17) and sketched in Fig. 14.

Locally (within the Hubble patch), an almost spherical metric perturbation $\Phi$ can be viewed as a variation of the Hubble parameter $H$, $\tilde{H} = (1 - \Phi)H$. As expected, the area of the horizon is the same

$$A = \frac{4\pi}{H^2}\, (1 + 2\Phi) = \frac{4\pi}{\tilde{H}^2}. \tag{4.18}$$

Therefore an observer within a Hubble patch sees the local Hubble parameter $\tilde{H}$ simultaneously slow rolling and wiggling due to the generation of quantum fluctuations of the inflaton field, in the spirit of the stochastic picture of inflation.

This can be illustrated by the apparent horizon for different observers, as shown in Fig. 14. The wiggles of the apparent horizon can be removed from the Penrose diagram by a conformal transformation, but they are real effects for the physical radius at the apparent horizon. To stress this point, we have left the wiggles in the apparent horizons of different observers.

Let us discuss the formula (4.17). Scalar metric fluctuations $\Phi(t, r)$ have zero mean value but their dispersion increases linearly with time. Thus, the area of the apparent horizon looks like a random (Wiener) process with the mean value $\bar{A} = \frac{4\pi}{H(t)^2}$, and with the dispersion $\langle A^2 \rangle = \frac{8\pi}{H^2} \langle \Phi^2 \rangle$ increasing linearly with time. Imagine an ensemble of inflationary Hubble patches with apparent horizon areas given by the formula (4.17). Then their area is a statistical value which is governed by the Gaussian distribution $P(A, t) \sim e^{-\frac{(A-\bar{A})^2}{2\langle A^2 \rangle}}$. Now, if we take into account variation of background values like $H(t)$ and $V(\phi(t))$ with time, the time-dependence and statistical properties of the horizon area given by (4.17) or equivalently (4.18) will be more complicated.

A lesson from our calculations is that in the quasi-de Sitter geometry with a slowly rolling scalar field, the horizon area, or geometrical entropy, are perturbed by the scalar metric fluctuations according to the formulas (4.17), (4.18). This means that the area of the horizon for an ensemble of different Hubble patches is a statistical variable by itself. Equivalently, we can talk about statistical properties of the local Hubble values $\tilde{H}$ [35,36]. Notably, this picture is converging with the stochastic description of inflation in terms of the probability distribution of the inflaton field [34–36]. The probability to have a particular value of the inflaton field $\phi$ in the quasi-static regime is $P(\phi) \sim \exp\left(-\frac{3M_p^4}{8V(\phi)}\right)$, and can be interpreted in terms of the entropy $S = \log P(\phi)$ [37]. Remarkably, this entropy is identical to the geometrical entropy $\frac{A}{4G}$. During inflation and the rolling of $\phi$, the Wiener process of accumulating fluctuations $\delta\phi$ (or similarly perturbations of $\Phi$) changes the distribution $P(\phi)$. The distribution of the horizon areas of different Hubble patches, defined by the local values of the Hubble parameters, depends on the background slow roll regime and the regime of accumulation of fluctuations, both of which depend on the model of inflation [34–36].

It would be interesting to understand further the correspondence between the stochastic approach to inflation and the geometrical entropy (4.17), (4.18).

Let us summarize the results. As far as inflation is an intermediate stage, it affects only the asymptotic future of the global geometry. Independent of the model of late-time cosmology, the asymptotic future will have a fractal structure. The details of the fractal geometry can depend on the late-time cosmology.

As we mentioned before, metric fluctuations (2.17) have the spectrum (4.15) logarithmically increasing with the growth of the wavelegths of fluctuations. Although this increase is logarithmically slow, in the context of an infinitely large multiverse it leads to drastic conclusions. We can compare the impact of logarithmic corrections to the the logarithmic corrections to the running couplings in a gauge QFT. Indeed, sooner or later at the super-large scales $\Phi$ will be of order unity, which signals the formation of a giant black hole in the asymptic future within our observable universe. Thus (independent of dark matter or dark energy domination), our future is to find ourselves inside a giant black hole! There may be a number of such black holes inside our FRW domain, and they will coalesce eventually.

## 5. Generation of fluctuations from inflation

Before the development of the inflationary universe scenario the origin of primordial density perturbations remained obscure. Inflation provides universal powerful mechanism for the generation of scalar field fluctuations. We first consider a toy model of a "cosmic piano" which illustrates what happens to the amplitude and wavelength of fluctuations, and then turn to a realistic model for the coupling of metric and field fluctuations. We include a discussion of primordial gravity wave production from inflation, and an alternative mechanism to generate scalar metric perturbations, based on the modulated (inhomogeneous reheating) fluctuations.

### 5.1. Cosmic piano

We begin our discussion of cosmological fluctuations with a simple example of a "cosmic piano" assembled from straight cosmic strings with the fixed end points. The evolution of a string's fluctuations provides an one-object illustration of what happens to the amplitude and fluctuations of oscillators during inflation. Imagine a cosmic string of co-moving length $L$ stretched in the $x$-direction. Consider the vibration of this string when placed in a de Sitter geometry. Small transverse string oscillations $Y(t, x)$ during inflation obey an oscillator-like equation of motion in an expanding universe, which is simpler in terms of the co-moving variable $y = Y/a$ and the conformal time $\eta$

$$\partial_\eta^2 y + 2\frac{\dot{a}}{a}\partial_\eta y - \partial_x^2 y = 0. \tag{5.1}$$

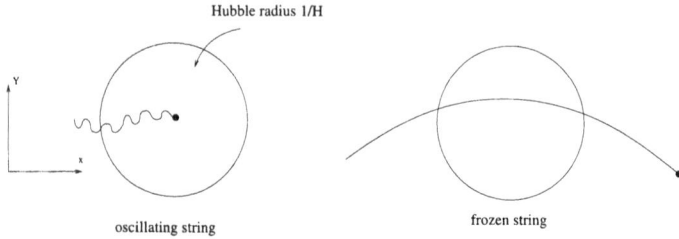

Fig. 15. Evolution of cosmic string during inflation.

The string oscillations are described by a superposition of Fourier harmonics $y_k(\eta)sin(kx)$, where $k = n\pi/L$ is the co-moving wave-number of the string excitation. For the time-dependent eigenfunction $y_k(t)$ we have

$$\partial_\eta^2 y_k + 2\frac{\dot{a}}{a}\partial_\eta y_k + k^2 y_k = 0. \tag{5.2}$$

Suppose the initial conditions for the string vibrations correspond to their vacuum fluctuations. Then the normalized positive-frequency solution of this equation is

$$y_k(\eta) = H\sqrt{\frac{\pi}{2L}}\,(|\eta|)^{3/2}\,\mathcal{H}_{3/2}^{(1)}(|k\eta|). \tag{5.3}$$

where $\mathcal{H}_{3/2}^{(1)}$ is a Hankel function. Let us analyze this solution. Initially the physical wavelength of the string (vacuum) oscillations $\lambda(t) = (2\pi/k)a(t) = \frac{2L}{n}\frac{1}{|H\eta|}$ is smaller than the Hubble radius $1/H$; and the physical amplitude $Y_k$ of the string's oscillations is constant, as shown in the left panel of Fig. 15. At some moment of time $t_k$, the physical wavelength $\lambda(t)$ exceeds the size of the causal patch $1/H$, and the solution (5.3) reaches its asymptotic value, so that the co-moving displacement $y$ stops oscillating and freezes at a constant amplitude. The physical displacement $Y$ is still growing with time as the scalar factor $a(t)$, see the right panel of Fig. 15.

All together the "cosmic piano" works like that: initially during inflation all strings are vibrating due to zero-point vacuum fluctuations. Then one by one, in order from lower to higher co-moving frequencies, waves on the strings are stretched outside the horizon and simultaneously cease to oscillate so that their co-moving amplitude become frozen. After inflation, the physical wavelengths of the string fluctuations frozen outside of the horizon still increases with time as $a(t)$, but slower than the linear growth of the Hubble radius with time. One has to wait until a string fluctuation with $\lambda(t)$ re-enters the horizon, de-frosts, and starts oscillating again! In a sense, we do hear primordial oscillations, although not from strings, but in the form of acoustic oscillations of the photon-baryon plasma, as part of the manifestation of the primordial cosmological fluctuations.

## 5.2. Generation of the scalar metrics fluctuations

Above in section 4.5 we considered the QFT of a scalar field placed in de Sitter geometry. Since the inflaton field is already part of the inflationary model, we can consider the QFT of the inflaton field itself.

The basic mechanism for the generation of the density fluctuations is connected with the production of large-scale fluctuations $\delta\phi$ of the inflaton scalar field. These fluctuations obey the oscillator-like equation (4.9) involving the friction term $3H\delta\dot\phi$. Inflaton fluctuations correspond to a light ($V_{,\phi,\phi} \equiv m^2 \ll H^2$) minimally coupled field with $\xi = 0$. Formula (4.11) readily describes the evolution of the eigen-modes of the inflaton quantum fluctuations. For short waves $k > aH$, the friction term can be neglected and the fluctuations oscillate, whereas long waves $k < aH$ do not oscillate due to the friction term and behave like a long-wave classical scalar field. Fluctuations $\delta\phi$ result in fluctuations of the inflaton energy-momentum tensor. Therefore there should be linear gravitational response (2.17) of the metric to the long-wave classical perturbations of $\delta\phi$. Notice that not just any fluctuations will do that. For instance, thermal fluctuations around a minimum of the potential $\delta\phi_k \sim e^{ik_\mu x^\mu}$ lead to uniform $\delta T_\nu^\mu$ with small-scale temperature fluctuations. A specific feature of inflaton fluctuations is that $\delta T_\nu^\mu$ is constructed from a bi-linear combination of $\phi$, and the VeV of the rolling inflaton is non-zero, so its fluctuations are linear w.r.t. $\delta\phi$, $\delta T_\nu^\mu \sim \phi\delta\phi$.

There are several subtleties in the QFT of fluctuations from inflation, which made the road to a clear picture quite bumpy. First, gravity is a gauge theory, and the formalism of linear cosmological fluctuations looks different in different gauge. Second, scalar metric fluctuations are not freely propagating waves (contrary to, say, tensor metric perturbations – gravitational waves) but are coupled to the matter fluctuations. Thus one has to quantize a proper canonical degree of freedom, which is a composite object. Third, the final answer should be written in terms of an observable variable.

Here we list the main results. Working in the longitudinal gauge (2.17), the classical theory of scalar/gravity fluctuations governed by the linearized Einstein equations is in two equations

$$\frac{1}{a^2}\nabla^2\Phi - 3H\dot\Phi - 3H^2\Phi = \frac{4\pi}{M_p^2}\left(\dot\phi\delta\phi - \dot\phi^2\Phi + V_{,\phi}\delta\phi\right),$$

$$\dot\Phi + H\Phi = \frac{4\pi}{M_p^2}\dot\phi\delta\phi. \tag{5.4}$$

Inside the horizon the first equation gives us the Poisson equation for the gravitational potential $\Phi$. Outside the horizon, energy density fluctuations obey $\delta\epsilon/\epsilon \approx -2\Phi$. During slow roll inflation $\delta\epsilon/\epsilon \sim V_{,\phi}\delta\phi/V$. After inflation $\delta\epsilon/\epsilon = C(\dot{\phi}, \phi)\delta\phi$. The main purpose of the theory of scalar perturbation is to determine the function $C(\dot{\phi}, \phi)$.

As usual for the linear fluctuations, we turn to the Fourier modes $\Phi_k$, $\delta\phi_k$. A proper variable for canonical quantization of fluctuations is $v = a(\delta\phi + z\Phi)$, where we are using conformal time. The equation for $v$ is an oscillator-like equation

$$v_k'' + \left(k^2 - \frac{z''}{z}\right)v_k = 0, \tag{5.5}$$

where $z = a\phi'/H$. Initial conditions correspond to vacuum fluctuations of "quasi-particles" $v_k = e^{-ik\eta}/\sqrt{2k}$ at $\eta \to -\infty$. In the long wave limit $v_k = C_1 z + C_2 \int d\eta/z^2$. The gravitational potential can be expressed through $v$ as $k^2\Phi_k = 4\pi\dot{\phi}z = (v_k/z)'$. In the long wave limit, the solution of eqs. (5.4)+(5.5) with vacuum initial fluctuations is

$$\Phi_k = \frac{16}{15}\frac{(3\pi V)^{3/2}}{M_p^3 V_{,\phi}}\bigg|_{\phi=\phi(t_k)}. \tag{5.6}$$

In particular, for the potential $V = \frac{m^2\phi^2}{2}$ we obtain formula (4.15).

### 5.3. Generation of the tensor metrics fluctuations

The theory of cosmological fluctuations also includes tensor perturbations around FRW geometry

$$ds^2 = dt^2 - a^2\left(\delta_{ij} + h^{ij}\right)dx^i dx^j, \quad i, j = 1, 2, 3. \tag{5.7}$$

Tensor modes are traceless and transverse, $h_i^i = 0$, $h^{ij}_{\cdot\cdot j} = 0$. The physical degrees of freedom are described by two independent components, which can be attributed to two polarizations $h_+$, $h_\times$, so that

$$h_{\mu\nu} = h_+ e_{\mu\nu}^+ + h_\times e_{\mu\nu}^\times. \tag{5.8}$$

At second order, tensor perturbations are governed by the action

$$\delta S = \frac{M_p^2}{2}\int d^4x\sqrt{-g}\left(\frac{1}{2}\nabla_\sigma h_{\mu\nu}\nabla^\sigma h^{\mu\nu} + \ldots\right). \tag{5.9}$$

From here (or the linearized Einstein equations) we obtain the equation of motion for gravity waves $\nabla_\sigma \nabla^\sigma h^{\mu\nu} = 0$. After Fourier decomposition, we get a free wave equation for the Fourier amplitudes $h_+, h_\times$

$$h_k'' + \left( k^2 - \frac{a''}{a} \right) h_k = 0. \tag{5.10}$$

Compare this with (4.10) to find that the equation for gravitational waves is identical to that for a scalar field with minimal coupling $\xi = 0$. Therefore all the results for massless scalar fields are also valid for tensor modes: they are produced from the vacuum fluctuations [46]. To quantize the tensor fluctuations, notice that the action (5.9) is reduced to a canonical field $\chi$ if we define $\chi = \frac{M_p}{\sqrt{2}} h$. After we quantize $\chi$, we find its amplitude and spectrum $k^3 |\chi_k|^2 = \left( \frac{H}{2\pi} \right)^2$. Returning to the tensor mode, we have

$$k^3 |h_k|^2 = \frac{H^2}{2\pi^2 M_p^2} \Big|_{t=t_k}. \tag{5.11}$$

The amplitude of the tensor mode depends on the energy scale of inflation. For a chaotic inflation with a large field value and large Hubble value $H \sim 10^{13}$ GeV, the amplitude of the tensor mode is of order $h \sim 10^{-6}$. For other models of inflation, like the family of hybrid inflation, $h$ is much lower.

### 5.4. Probing models with fluctuations

Scalar metric fluctuations produced from the slow roll inflation acquire almost scale free spectrum of fluctuations (5.6). This small deviation from the perfect Zel'dovich spectrum $n_s = 1$ in fact is the big window of opportunity to probe models of inflation observationally.

To illustrate this, consider an example of monomial potential $V(\phi) \sim \phi^n$. Formula (5.6) gives us

$$k^3 |\Phi_k|^2 = A_s \left( \ln \frac{k_0}{k} \right)^{\frac{n+2}{2}}, \tag{5.12}$$

where $k_0$ is the last mode leaving inflation, $1/k_0 \sim$ mm. We observe fluctuations at the astronomically large scales, let the pivot point $k_*$ be at the median of the these scales, $1/k_* \gg 1/k_0$. Around pivot point the log factor in (5.12) can be simplified $\left( \ln \frac{k_0}{k} \right)^{\frac{n+2}{2}} = \left( \ln \frac{k_*}{k} + \ln \frac{k_0}{k_*} \right)^{\frac{n+2}{2}} \approx N^{\frac{n+2}{2}} \left( 1 - \frac{n+2}{2N} \ln \frac{k_*}{k} \right)$. Here $N$ is the number of efolding $N = \ln \frac{k_0}{k_*}$. Now recall the power-law fit to the spectrum (2.20), which one rewrites as $\left( \frac{k}{k_*} \right)^{(n_s - 1)} = e^{(n_s - 1) \ln \frac{k}{k_*}} \approx 1 + (n_s -$

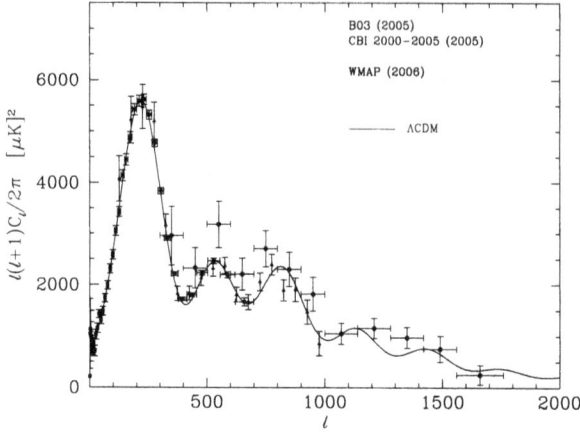

Fig. 16. Power spectrum of angular CMB temperature anisotropiesest: best fit ΛCDM model vs data.

1) $\ln \frac{k}{k_*}$. Compare this with decomposition of ln above to find formula for the index of the scalar power spectrum in the model with monomial inflaton potential

$$n_s - 1 \approx -\frac{n+2}{2N}. \tag{5.13}$$

For quadratic inflaton potential $n = 2$, and $N = 55$ we have $n_s \approx 0.96$. For quartic potential $n = 4$, $n_s \approx 0.94$. Observational table of section 2.2 prefers quadratic potential and rejects quartic potential.

Scalar metric perturbations induce CMB temperature anisotropies $\frac{\Delta T}{T}$ on the sky. The amplitude and angular power spectrum $C_l = (\Delta T)_l^2$ of linear temperature anisotropies is controlable tool box to extract a lot of information about primordial perurbations [47]. For the favored ΛCDM model best fit power spectrum with $n_s \approx 0.96$ leads to the $C_l$-s plotted in Figure 16 against the recent CMB experimental data [6]. In plane words, these results confirm inflation.

Formula (5.13) can be generalized for arbitrary slow roll potential

$$n_s - 1 \approx \frac{M_p^2}{4\pi} \left( \frac{V_{,\phi\phi}}{V} - \frac{3}{2} \frac{V_{,\phi}^2}{V^2} \right). \tag{5.14}$$

For aficionados of mathematical physics we notice interesting structure of this formula. In has the form of the Schwarz derivative (which emerges in the theory of ODE)

$$n_s - 1 \approx \frac{M_p^2}{4\pi} \{F, \phi\}, \quad F = \int d\phi V. \tag{5.15}$$

where by definition $\{y, x\} = \frac{y'''}{y'} - \frac{3}{2} \left( \frac{y''}{y'} \right)^2$. Its origin here is not clear to me.

There are several effective tools to estimate observables of the slow-roll inflation, detouring detailed treatment of the problem. I will mention method of the slow-roll parameters [7]. It will be convenient to switch to the Hamilton-Jacobi formulation of the scalar field dynamics. Consider the Hubble parameter as the function of the field $H = H(\phi)$. Field momentum is $\dot{\phi} = -\frac{M_p^2}{4\pi} H'$, where $H' = \partial H/\partial \phi$. Constrain (Friedmann) equation takes the form of the Hamilton-Jacobi equation

$$V(\phi) = \frac{M_p^4}{32\pi^2} \left( \frac{12\pi}{M_p^2} H^2 - H'^2 \right). \tag{5.16}$$

Scalar metric fluctuations, instead of functions $V(\phi)$ as in (5.6), can be expressed as functions of $H(\phi)$

$$P_s(k) = \frac{4\pi}{M_p^2} \frac{1}{(2\pi)^2} \frac{H^2}{\epsilon} = \frac{8\pi H^4}{M_p^4 H'^2}, \tag{5.17}$$

the value of $k$ is taken to be $k = aH$. Formulas (5.6), (5.17) are derived in the uniform acceleration approximation.

Define the slow roll parameters

$$\epsilon = \frac{M_p^2}{4\pi} \left( \frac{H'}{H} \right)^2 \quad \eta = \frac{M_p^2}{4\pi} \frac{H''}{H} \quad \xi^2 = \left( \frac{M_p^2}{4\pi} \right)^2 \frac{H'H'''}{H^2}, \dots \tag{5.18}$$

Assumed that during slow roll inflation slow roll parameters gradually decrease with the derivatives of $H$. Similar to how we derived expression (5.14), we can rewrite it in terms of the slow roll parameters

$$n_s - 1 \approx -4\epsilon + 2\eta + \mathcal{O}(\epsilon^2, \eta^2, \xi). \tag{5.19}$$

The ratio of tensor (5.11) and scalar (5.17) amplitudes is $r \equiv A_t/A_s = 16\epsilon$.

How unambiguous is connection between power spectrum of scalar fluctuations and the underlying inflaton potential? We end up this section to demonstrate that unfortunately, it is not, unless we observe gravity waves! We show that the reconstruction of the scalar field potential $V(\phi)$ from the known spectrum of the scalar perturbations $P_s(k)$ alone is strongly degenerate. Namely, it turns out that a given shape of $P_s(k)$ corresponds to a whole one-parametric family of potentials $V(\phi)$, which have very different appearance, varying from convey to convex potentials within the same family. In other words, potentials from the new inflationary type to the chaotic type and all in between can give the same spectrum of scalar perturbations. An example of ambiguity between a pair of potentials $V(\phi) = \frac{1}{4}\lambda\phi^4$ and $V(\phi) = V_0 - \frac{1}{4}\lambda\phi^4$ was given in [52], other examples are in [53].

We seek to express the potential $V(\phi)$ through the known function $P_s(k)$. In equation (5.16) we shall substitute $H'$ from equation (5.17) to obtain

$$V(\phi) = \frac{M_p^2 H^2}{8\pi} \left( 3 - \pi \frac{H^2}{P_s} \right). \tag{5.20}$$

To complete the derivation, we have to express the Hubble parameter $H$ through the function $P_s$. From (5.17) we have

$$\frac{dH}{d\phi} = \frac{\sqrt{8\pi} H^2}{M_p^2 \sqrt{P_s}}. \tag{5.21}$$

We want to define $H$ as a function of the momenta $\ln k$. For this we need connection between $d\phi$ and $d\ln k$. We have

$$d\phi = \sqrt{\frac{M_p^2}{4\pi}} \frac{\sqrt{\epsilon}}{\epsilon - 1} d\ln k. \tag{5.22}$$

After we substitute (5.22) into (5.21), we obtain the first order equation for the function $H(k)$:

$$\frac{dH}{d\ln k} = \frac{H^3}{H^2 - \pi M_p^2 P_s(k)}. \tag{5.23}$$

Thus, the given function $P_s(k)$ defines one parametric family of functions $H(k)$. The parameter is given by the constant of integration which is the initial value $H_0$ corresponding to the value of the Hubble parameter at the moment where the field value is $\phi_0$. Since the Hubble parameter is related to the power spectrum of gravitational waves the value of $H_0$ is related to the power of the tensor mode at a chosen moment corresponding to $\phi_0$.

Now substitute the solution $H(H_0, k)$ in the eq. (5.20), which defines the shape of the potential $V(\phi)$ for a given scalar power spectrum $P_s(k)$. The reconstruction of $V(\phi)$ is degenerate because it involves the continuous parameter $H_0$. So far the potential $V$ reconstructed via the function $P_s(k)$, so that the potential is a function of $k$. To finish, we have to map the argument $k$ into $\phi$. This mapping again will involve the function $P_s$ and $H(k)$, and will depend on the parameter $H_0$. We already have a relation between $d\phi$ and $d\ln k$. After integration of this relation and simple manipulations with equations (5.21) and (5.23), we get

$$\phi - \phi_0 = \frac{M_p^2}{2} \int_{\ln k_0}^{\ln k} d\ln k' \frac{\sqrt{P_s(k')}}{H^2(k')} \frac{dH}{d\ln k'}. \tag{5.24}$$

We observe $C_l$, which depend not only on $P_s(k)$ but potentially on $P_t(k)$. The derivation above can be extended to show that for given $C_l$, reconstruction of $V(\phi)$ is degenerate.

## 6. Preheating after inflation

The last chapetr of inflation is about what happens at the end point of inflation. The list of topics I will touch includes: preheating after inflation treated with the Quantum Field Theory (QFT), reheating after string theory inflation, potential observables which may be associated with (p)reheating, which are (modulated) cosmological fluctuations generated from preheating; and generation of gravitational waves after preheating.

### 6.1. Heating the universe after inflation

Theory of inflation is accompanying with the theory of origin of particle after inflation. Details of this theory somehow depend on the model of inflation. Let us briefly review the "particlegenethis" after inflation, in parallel with the structure of Figure 4.

In the Big Bang picture where the universe starts with the singularity, it was assumed that the very hot matter from the very beginning was in the state of the thermal equilibrium, with its temperature gradually decreases as the universe expands. The very early versions of inflation were embedded in the Big Bang picture and evoked supercooling of matter which reheats once again after inflation ends. Therefore the theory of particle creation and thermalization after inflation was dubbed as reheating after inflation – an anachronism of +25 history. In the modern version of inflation it is not necessary to postulate the hot pre-inflationary stage.

Let us return to Eq. (3.11). During the stage of inflation, all energy is contained in a classical slowly moving inflaton field $\phi$. Eventually the inflaton field decays and transfers all of its energy to relativistic particles, to start the thermal history of the hot Friedmann universe.

QFT of (p)reheating, or theory of particle creation from inflaton field in an expanding universe is a process where quantum effects of particle creation is not small but a spectacular process where all the particles of universe are created from the rolling classical inflaton. The theory of particle creation and their subsequent thermalization after inflation has a record of theoretical developments within the QFT. The four-dimensional QFT Lagrangian $\mathcal{L}(\phi, \chi, \psi, A_i, h_{ik}, \ldots)$ contains the inflaton part and other fields which give subdominant contributions to gravity. Due to the interactions of other fields with the inflaton in $\mathcal{L}$, the inflaton field decays and transfers all of its energy to relativistic particles. If the creation of particles is sufficiently slow (for instance, if the inflaton is coupled only gravitationally to the matter fields) the decay products simultaneously interact with each other and come to a state of thermal equilibrium at the reheating temperature $T_r$. This temperature will be the largest one in formula (2.11). This gradual reheating can be treated with the perturbative theory of particle creation and thermalization [55, 56]. However, typically the particle production from a inflaton occurs in the non-perturbative regime. In chaotic inflationary models, soon after the end of inflation the almost homogeneous inflaton field $\phi(t)$ coherently oscillates with a very large amplitude of the order of the Planck mass $M_p$ around the minimum of its potential. The particle production from a coherently oscillating inflaton occurs in the regime of parametric excitation [48, 57]. This picture, with variation in its details, is extended to other inflationary models. For instance, in hybrid inflation (including $D$-term inflation) inflaton decay proceeds via a tachyonic instability of the inhomogeneous modes which accompany the symmetry breaking [58, 59]. One consistent feature of preheating – non-perturbative copious particle production immediately after inflation – is that the process occurs far away from thermal equilibrium.

Recent development in string theory inflation, while in the early stages, has to address the issue of end point of inflation, (p)reheating immediately after inflation. As it was in QFT theory of reheating, String Theory reheating shall be compatible with the thermal history. Yet, we are especially interested in the specific String Theory effects of reheating [60].

Sooner or later in an expanding universe the stage of (p)reheating dynamically evolves into the stage of thermal equilibrium of particles, where a lot of information about initial state will be lost. What is then an output of (p)reheatng for cosmology? It is important to figure out the character of particlegenethis and thermalization in a specific model of inflation (say from the Fig. 4) and reheat temperature $T_r$, Short out-of-equilibrium stages in the thermal history of the universe are responsible for the variety of the observable forms of matter in the universe. Strong out-of-equilibrium character of preheating opens the possibility for crucial phenomena associated with non-equilibrium physics, including cosmological baryo/leptogenesis phase transitions, non-thermal production of heavy particles, etc. We need to know evolution of EOS $w(t)$ to find how connection $N-\log k$ of number of needed e-folds $N$ of inflation and wavelength of comolgical fluctuations. Most interesting output of preheating will be potential observables of it, like gravitational waves, and modulated cosmological fluctuations.

In the rest of this chapter I briefly review the basics and new developments in the theory of different types of preheating and reheating, and in the light of of our current understanding, discuss the outputs of (p)reheating for cosmology.

## 6.2. Pair creation by an electric field

Before we turn to the theory of particle creation from oscillating inflaton, it is instructive to consider a simple prototype problem of particles creation in a scalar electrodynamics. Suppose $E$ is the strength of the constant electric field aligned with $z$-direction. Let $A_\mu$ be the four-potentials of the external classical EM field, and $\chi$ is the quantum field describing massless scalar particles (of the charge $e$) with the equation of motion

$$D_\mu D^\mu \chi = 0, \tag{6.1}$$

where $D_\mu = \partial_\mu - ieA_\mu$.

Next we have to choose the gauge for $A_\mu$. It is convenient to deal with the time-dependent problem of particle creation. We can put $A_\mu = (0, 0, 0, -Et)$ Then the time-dependent part of the eigenmodes $X_k(t)e^{i\vec{k}\vec{x}}$ obey the equation

$$\frac{d^2 X_k}{d\tau^2} + (\kappa^2 + \tau^2)X_k = 0, \tag{6.2}$$

where we introduced dimensionless time $\tau = eEt$ and dimensionless momentum $\kappa = k/eE$. Initial condition of the equation (6.2) shall correspond to the positive frequency vacuum fluctuation $X_k(t) = \frac{1}{\sqrt{2k}}e^{i\kappa\tau}$. Then equation (6.2) can be interpreted as scattering of the incoming wave on the inverse parabolic potential. Far away from the apex of the potential at $\tau = 0$ where $A_\mu$ crosses zero value, solution can be written in the WKB form

$$X_k(t) = \frac{\alpha_k(t)}{\sqrt{2k}} e^{-ikt} + \frac{\beta_k(t)}{\sqrt{2k}} e^{+ikt}, \tag{6.3}$$

Initially $\beta_k = 0, \alpha_k = 1$. After scattering $\beta_k$ is not zero anymore and corresponds to created particles $n_k = |\beta_k|^2$. Analytic solution of equation (6.2) with the asymptotic (6.3) is well known and gives

$$n_k = \exp\left(-\frac{\pi k^2}{eE}\right). \tag{6.4}$$

This formula is a variance of the celebrated Schwinger formula for the rate of electron-positron pair creation from the constant electric field. We will take with us several lessons from the scalar ED example. Despite external electric field is constant, the problem of the pair creation can be reduced to the time-dependent problem. The instance of particles creation is associated with the time when $A_\mu$ amplitude crosses zero value. Of course the final answer (6.4) does not depend on this moment. Pair creation is a non-perturbative effect. The answer (6.4) is non-analytic function of the coupling $eE$. As we will see, very similar physics will take place in the process of pair creation from oscillating inflaton field.

## 6.3. Linear resonant preheating

In this section we follow theory of preheating developed in [50]. Consider simple chaotic inflation with the potential $V(\phi) = \frac{m^2}{2}\phi^2$. Soon after the end of inflation, an almost homogeneous inflaton field $\phi(t)$ coherently oscillates as $\phi(t) \approx \Phi(t)\sin(m_\phi t)$, with a very large initial amplitude of the order of the Planck mass $\simeq 0.1 M_p$, $\Phi(t) = \frac{M_p}{\sqrt{3\pi}} \cdot \frac{1}{m_\phi t}$. A toy model describing the interaction between inflatons and radiation, i.e. other massless Bose particles $\chi$ will be $\mathcal{L} = -\frac{1}{2}g^2\phi^2\chi^2$. QFT of particles creation can be constructed in the following way. Consider the Heisenberg representation of the quantum scalar field $\hat{\chi}$, with the eigenfunctions $\chi_k(t)\, e^{-i\mathbf{kx}}$ where $\mathbf{k}$ is a co-moving momenta. The temporal part of the eigenfunction obeys the equation

$$\ddot{\chi}_k + 3\frac{\dot{a}}{a}\dot{\chi}_k + \left(\frac{\mathbf{k}^2}{a^2} + g^2\phi^2\right)\chi_k = 0, \tag{6.5}$$

with the vacuum-like initial condition $\chi_k \simeq \frac{e^{-ikt}}{\sqrt{2k}}$ in the far past. Let us seek the solutions of Eq. (6.5) in the adiabatic WKB approximation form

$$a^{3/2}\chi_k(t) \equiv X_k(t) = \frac{\alpha_k(t)}{\sqrt{2\omega}}\, e^{-i\int^t \omega dt} + \frac{\beta_k(t)}{\sqrt{2\omega}}\, e^{+i\int^t \omega dt}, \tag{6.6}$$

where the time-dependent frequency is $\omega_k^2(t) = \frac{\mathbf{k}^2}{a^2} + g^2\phi^2$ (neglecting small corrections $\sim H^2, \dot{H}$); initially $\beta_k(t) = 0$. The goal is to calculate particle occupation number $n_k = |\beta_k|^2$.

Eq. (6.5) describes parametric oscillator in an expanding universe. Introduce new time variable $z = mt$ and the essential dimensionless coupling parameter $q = \frac{g^2\Phi^2}{4m^2}$. Since $\frac{m}{M_p} \simeq 10^{-6}$, it is expected that $q \simeq 10^{10}g^2 \gg 1$, the parametric resonance is broad. For the large value of $q$ the eigenfunction $\chi_k(t)$ is changing adiabatically between the moments $t_j$, $j = 1, 2, 3, \ldots$, where the inflaton field is equal to zero $\phi(t_j) = 0$. The non-adiabatic changes of $\chi_k(t)$ occur only in the vicinity of $t_j$. Therefore, the semi-classical solution (6.6) is valid everywhere but around $t_j$. Thus zeros of inflaton oscillations play a crucial role, very similar to zeros of $A_\mu$ in the scalar ED example of the previous section. Let the wave $\chi_k(t)$ have the form of the WKB solution with the pair of coefficients $(\alpha_k^j, \beta_k^j)$ before the scattering at the point $t_j$; and the pair $(\alpha_k^{j+1}, \beta_k^{j+1})$ after the scattering at $t_j$. The interaction term around all the points $t_j$ is parabolic $g^2\phi^2(t) \approx g^2\Phi^2 m^2(t - t_j)^2$, and the eigenmode equation around zero point is reduced to the equation (6.2) of the previous section. Therefore, the outgoing amplitudes $(\alpha_k^{j+1}, \beta_k^{j+1})$ can be expressed through the incoming amplitudes

$(\alpha_k^j, \beta_k^j)$ with help of the well-known reflection and transmission amplitudes for scattering from a parabolic potential at $t_j$: The net result in terms of number of $\chi$-particles $n_k^{j+1} = |\beta_k^{(j+1)}|$ created after $t_j$ is

$$n_k^{j+1} = e^{-\pi\kappa^2} + \left(1 + 2e^{-\pi\kappa^2}\right) n_k^j$$

$$- 2e^{-\pi\kappa^2/2}\sqrt{1 + e^{-\pi\kappa^2}}\sqrt{n_k^j(1 + n_k^j)}\sin\theta^j. \tag{6.7}$$

Here $\kappa = \frac{k^2}{a\sqrt{gm\Phi}}$ and $\theta^j$ is the phase accumulated between zeros of $\phi(t)$.

The first term of the formula (6.7) is nothing but the variance of the Schwinger formula for the spontaneous pair creation from an external field, similar to the formula (6.4) of the previous section. The last term in (6.7) corresponds to the induced pair creation in the presence of particles created from the previous cycle of the zero crossing by inflaton. There is no analogy of this part in the scalar ED with the constant electric field. Finally, the second term in (6.7) in an interference between induced and spontaneous particles creation.

Formula (6.7) describes parametric resonant particle creation from oscillating inflaton. Very quickly $n_k^j$ becomes large and formula (6.7) can be approximated by much simpler formula

$$n_k^{j+1} = e^{2\pi\mu_k} n_k^j \quad \rightarrow \quad n_k(t) \simeq e^{\mu_k t}. \tag{6.8}$$

Here $\mu_k$ is a complex exponent, which is real for the resonant modes (positive interference) and complex for stable modes (negative interference). Expanding of the universe makes the phase $\theta_k$ a random and resonant process stochastic, but broad band of modes exponentially amplified.

### 6.4. Non-linear dynamics of resonant preheating

Due to the rapid growth of its occupation numbers the field $\chi(t, \vec{x})$ can be treated as a classical scalar field. Its appearance is described by the realization of the random gaussian field i.e. as a superposition of standing waves with random phases and Rayleigh-distributed amplitudes One can use many different quantities to characterize a random field, such as the spatial density of its peaks of a given height, etc. The scale of the peaks and their density depend on the characteristic scale $R$ of the spectrum, which in our case is related to the leading resonant momentum $k_* \simeq \sqrt{gm\phi_0}a^{1/4}$ [50]. At the linear stage the phases are constant, so that the structure of the random field $\chi$ stays almost the same.

Once one field is amplified in this way, other fields that are coupled to it are themselves amplified [61, 62], so within a short time of linear preheating (of order dozens of inflaton oscillations) fluctuations of $\chi$ generate inhomogeneous fluctuations of the field $\phi$. It is easy to see that fluctuations of $\phi$ will have a non-linear, non-gaussian character. From the equation of motion for $\phi$

$$\nabla_\mu \nabla^\mu \phi + m^2 \phi^2 + g^2 \chi^2 \phi = 0, \tag{6.9}$$

we have in Fourier-space

$$\ddot{\phi}_k + 3H\dot{\phi}_k + \left((k^2/a^2) + m^2\right)\phi_k = g^2\phi_0(t)\int d^3q\,\chi_q\chi^*_{k-q}, \tag{6.10}$$

where we neglect the term that is third order with respect to fluctuations; $\phi_0(t)$ is the background oscillation. The solution of this equation with Green's func-tions [50] shows that $\phi$ fluctuations grow with twice the exponent of $\chi$ fluctua-tions. It also shows that the fluctuations of $\phi$ are non-gaussian. Sometimes this solution is interpreted as re-scattering of the particle $\chi_q$ against the condensate particle $\phi_0$ at rest producing $\chi_{k-q}$ and $\phi_k$, $\chi\phi_0 \to \chi\delta\phi$. However, this interpre-tation has significant limitations.

When the amplitudes of $\chi$ and $\phi$ become sufficiently large we have to deal with the fully non-linear problem. The field evolution can be well approximated using the classical equation of motion (6.9) supplemented by another equation for $\chi$

$$\nabla_\mu \nabla^\mu \chi + g^2 \phi^2 \chi = 0. \tag{6.11}$$

Eqs. (6.9) and (6.11) of the non-linear preheating can be solved numerically using the LATTICEEASY program. For chaotic inflation, these results were presented in terms of the time evolution of occupation numbers $n_k(t)$ or total number den-sity of particles $N(t)$. Figures 17–18 show the results of our simulations in these familiar terms of $n_k(t)$ (in combination $k^3\omega_k n_k$) and $N(t)$, as well as showing the evolution of the field statistics (departures from gaussianity). Here all sim-ulation results are for model with $m = 10^{-6}M_p$ (fixed by CMB normalization) and $g^2 = 2.5 \times 10^{-7}$. The size of the box was $L = 10$ m$^{-1}$ and the grid contained $256^3$ points. We also tried other values of $g^2$ and found qualitatively similar results.

Figure 17 shows the evolution of the spectra. The spectra show rapid growth of the occupation numbers of both fields, with a resonant peak that develops first in the infrared ($k \simeq k_*$) and then moves towards the ultraviolet as a result of re-scattering. In left panel of figure 18 you can see clearly that the occupation number of $\chi$ initially grows exponentially fast due to parametric resonance, fol-lowed by even faster growth of the $\phi$ field due to the interaction, in accordance with the solution of eq. (6.10).

The gaussianity of classical fields can be measured in different ways. Right panel of Figure 18 shows the evolution of the ratio $\langle f^2\rangle^2/\langle f^4\rangle$ (kurtosis), which

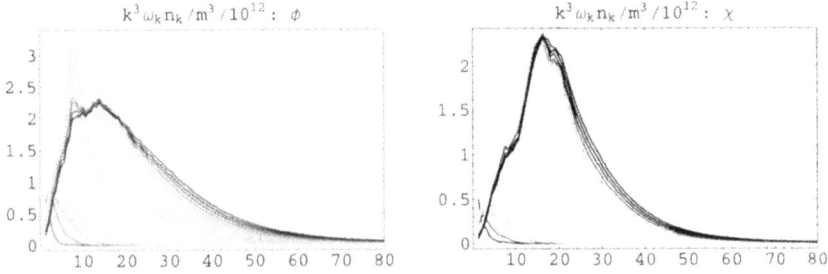

$k^3 \omega_k n_k /m^3 /10^{12}: \phi$

$k^3 \omega_k n_k /m^3 /10^{12}: \chi$

Fig. 17. Evolution of spectra of created particles.

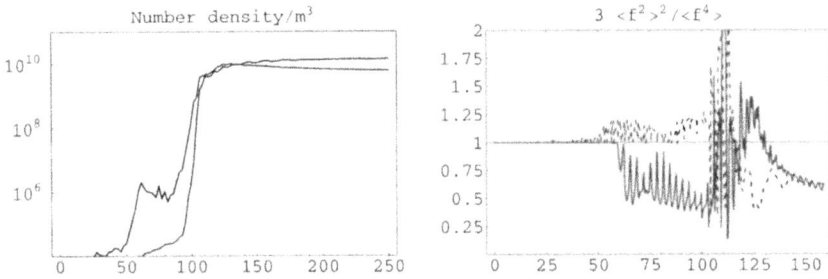

Number density/$m^3$

$3 \ <f^2>^2 /<f^4>$

Fig. 18. Left: evolution of total number of particles. Right: evolution of non-gaussianity.

is equal to unity for a gaussian field. During the linear stage of preheating, the field fluctuations are a random gaussian field, reflecting the initial quantum fluctuations that seeded them. The inhomogeneous field $\phi$ is generated as a non-gaussian field, in agreement with the solution of eq. (6.10). When the fluctuation amplitude begins to get large, both fields are non-gaussian. During the later turbulent stage both fields begin to return to gaussianity.

Another known feature of preheating is the onset of chaos, when small differences in the initial conditions for the fields lead to exponentially divergent solutions: $D(t) \simeq e^{\lambda t}$, where $D$ is the distance in phase space between the solutions and $\lambda$ is the Lyapunov exponent (see [62] for details). The distance $D$ begins to diverge exponentially exactly after the violent transition to the turbulent stage.

Let us summarize the picture which emerges when we study preheating, turbulence and thermalization in momentum space with the occupation numbers $n_k$. There is initial exponential amplification of the field $\chi$, peaked around the mode $k_*$. At this stage the $\chi$ fluctuations form a squeezed state, which is a superposition of standing waves that make up a realization of a random gaussian field. Interactions of the two fields lead to very rapid excitation of fluctuations of $\phi$, with its energy spectrum also sharply peaked around $k_*$. To describe generation of $\phi$ inhomogeneities, people use the terminology of "re-scattering" of waves. However, there is a short violent stage when occupation numbers have a sharply peaked and rapidly changing spectrum. The field at this stage is non-gaussian, which signals that the waves phases are correlated. In some sense, the concept of "particles" is not very useful around that time. In the later turbulent stage when $n_k(t)$ gradually evolves and gaussianity is restored (due to the loss of phase coherency) the picture of re-scattering particles becomes proper. As we will see in the next section, gaussianity is not restored for some time after the end of preheating. To understand this violent, intermediate stage, however, it is useful to turn to the reciprocal picture of field dynamics in position space [63].

### 6.5. Inflaton fragmentation

The features in the occupation number spectra $n_k(t)$, namely, sharp time variations, peaks at $k \sim k_*$, and strong non-gaussianity of the fields around the time of transition between preheating and turbulence suggest that we are dealing with distinct spatial features of the fields in the position space. This prompted to study the dynamics of the fields in position space.

The evolution of the fields in position space is shown in Figure 19. Each frame shows the spatial profile of the fields $\phi$ and $\chi$ along a two-dimensional slice of the 3D lattice.

The initial evolution of the fields ($t < 100$) is characterized by linear growth of fluctuations of $\chi$. During this stage the fluctuations have the form of a superposition of standing waves with random phases, which make up a random gaussian field. The eye captures positive and negative peaks that correspond to the peaks of the initial gaussian random field $\chi$. The peaks in this early stage correspond to the peaks of the initial gaussian random field $\chi$. Next the oscillations of $\chi$ excite oscillations of $\phi$. The first panel of Figure 19 shows a typical profile near the end of this period, just as the oscillations are becoming nonlinear and $\phi$ is becoming excited. The amplitude of these $\phi$ oscillations grows much faster than the initial $\chi$ oscillations (see our discussion of eq. (6.10)) and the oscillations have different (and changing) frequencies. The peaks of the $\phi$ oscillations occur in the same places as the peaks of the $\chi$ oscillations, however, as can be seen in the bottom three panels on the left side of Fig. 19.

The profile of $\phi(t, \mathbf{x})$ is a superposition of the still oscillating homogeneous part plus inhomogeneities induced by the Yukawa-type interaction $(g^2\phi_0)\phi\chi^2$ in the Lagrangian. Since the Yukawa interaction is a short-range interaction (defined by the length scale $1/m$ for a space-like interval $G(r) \sim e^{-mr}$), induced inhomogeneities of $\phi$ appear in the vicinity of those in $\chi$.

In the next stage ($t \sim 110$) the peaks reach their maximum amplitude, comparable to the initial value of the homogeneous field $\phi$, and begin to spread. The

t=102

t=106

t=111

t=122

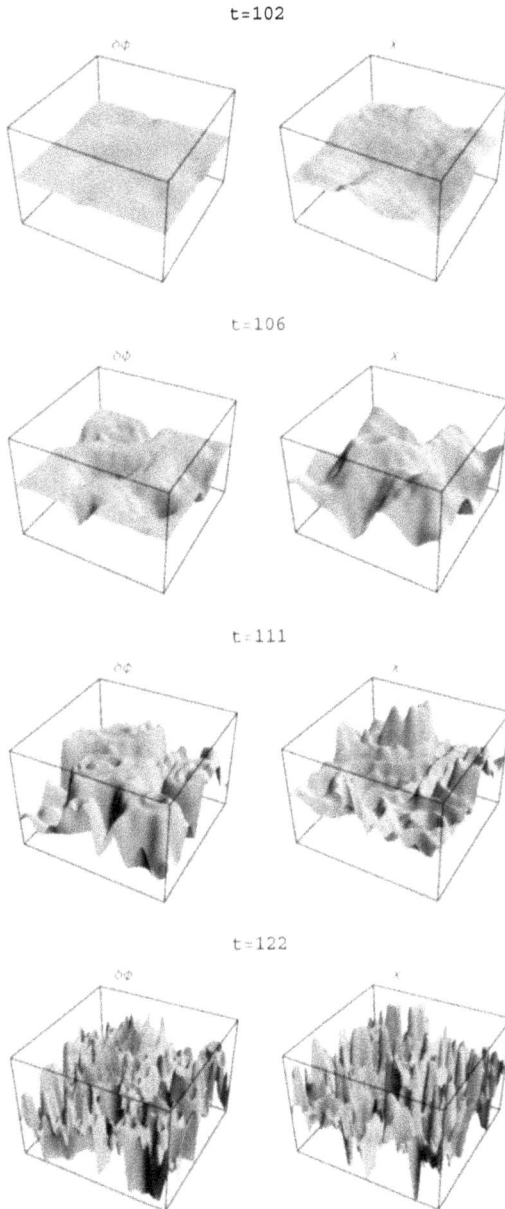

Fig. 19. Inflaton fragmentation via parametric resonance preheating. Values of the $\phi$ and $\chi$ fields in a two-dimensional slice through the lattice.

two lower left panels of Fig. 19 shows the peaks expanding and colliding. In the panels on the right you can see the standing wave pattern lose coherence as the peaks send out ripples that collide and interfere. By $t = 124$ the fluctuations have spread throughout the lattice, but you can still see waves spreading from the original locations of the peaks. Shortly after that time all coherence is lost and the field positions appear to be like random turbulence.

The bubble-like structure of the fields is reflected in their statistics. Perhaps most surprisingly, the statistics of both fields remain non-gaussian for a long time after preheating. At the end of our simulation, at $t = 300$, the fields were still noticeably non-gaussian. During all this time the random phase approximation of interacting scalars is not well justified.

### 6.6. Tachyonic preheating

Hybrid inflation is another very important class of inflationary models. At first glance preheating in hybrid inflation, which contains a symmetry breaking mechanism in the Higgs field sector, has a very different character than in chaotic inflation. Preheating in hybrid inflation occurs via tachyonic preheating [58], in which a tachyonic instability of the homogeneous modes drives the production of field fluctuations. In hybrid inflation, the decay of the homogeneous fields leads to fast non-linear growth of scalar field lumps associated with the peaks of the initial (quantum) fluctuations. The lumps then build up, expand and superpose in a random manner to form turbulent, interacting scalar waves.

Like parametric resonance, tachyonic preheating can be interpreted via the reciprocal picture of copious particle production far away from thermal equilibrium, and consequent cascades of energy through interacting, excited modes.

Fig. 20 illustrates tachyonic instability with the results of numerical simulations in the model with the Higgs part of the potential $V(\phi) = \frac{1}{2}m^2\phi^2 + \frac{1}{3}\sigma\phi^3 + \frac{1}{4}\lambda\phi^4$. Because of the nonlinear dependence of the tachyonic mass on $\phi$, initial linear gaussian fluctuations quickly turn into non-linear non-gaussian field, with pronounced bubble structures.

### 6.7. Gravitational waves from preheating

Eventually after preheating the fields reach thermal equilibrium characterized only by the temperature. Does that mean that all traces of inflaton fragmentation history are erased? For instance, people have discussed realizations of baryogenesis at the electroweak scale via tachyonic preheating after hybrid inflation, and this process is ultimately related to the bubble-like lumps of the Higgs field that form during tachyonic preheating. Since we now see that fragmentation through bubbles can also occur in chaotic inflation, baryogenesis via out-of-equilibrium bubbles can also be extended to these models.

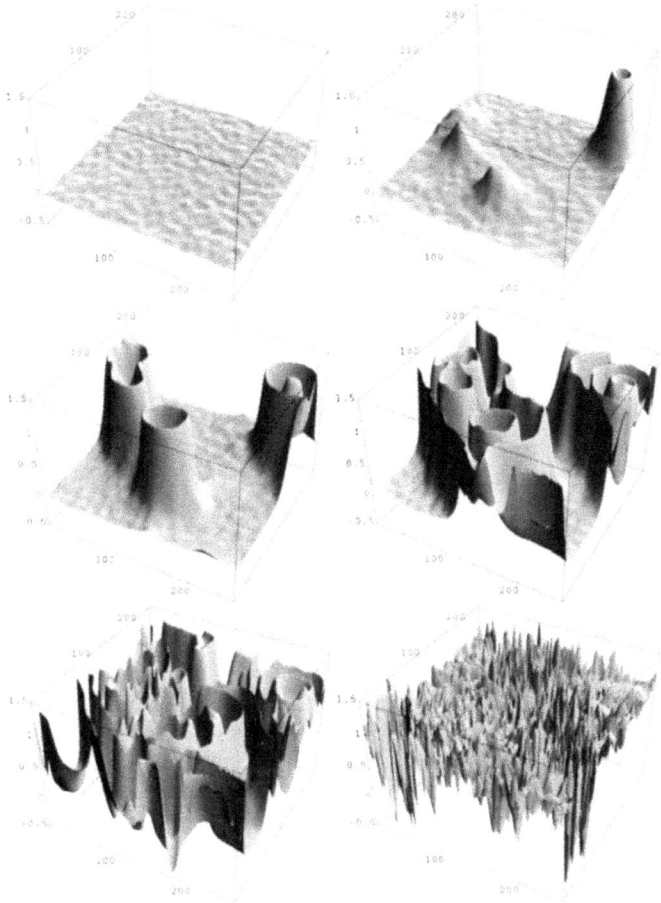

Fig. 20. Inflaton fragmentation via tachyonic preheating. Evolution of the field is shown at slices through lattic.

There is another, potentially observable consequence of the non-linear "bubble" stage of inflaton fragmentation. Lumps of the scalar fields correspond to large (order of unity) energy density inhomogeneities at the scale of those bubbles, $R$. Collisions of bubbles generate gravitational waves. The fraction of the total energy at the time of preheating converted into gravitational waves is significant. We estimate it is of the order of

$$\frac{\rho_{gw}}{\rho_{rad}} \simeq 0.05(RH)^2, \tag{6.12}$$

where $1/H$ is the Hubble radius. This corresponds to a present-day fraction of energy density $\Omega_{GW} \sim 10^{-5}(RH)^2$. The way to understand formula (6.12) is the following: The energy converted into gravitational waves from the collision of two black holes is of the order of the black hole masses. If the mass of lumps of size $R$ is a fraction $f$ of a black hole of the same size, then the fraction of energy converted to gravitational waves from two lumps colliding is $f$. Scalar field lumps at the Hubble scale would form black holes, so in our case $f = (RH)^2$.

The present-day frequency of this gravitational radiation is

$$f \simeq \frac{M}{10^7 \, GeV} Hz, \qquad (6.13)$$

where $M = V^{1/4}$ is the energy scale of inflation with the potential $V$.

For the chaotic inflation model considered in this paper the size of the bubbles is $R \sim few/m$ and at the time they begin colliding $H \sim m/100$, so that the fraction of energy converted into gravitational waves is of the order $10^{-3}$–$10^{-4}$. This figure is in agreement with the numerical calculations of gravitational wave radiation from preheating after chaotic inflation.

For chaotic inflation with $M$ at the GUT scale the frequency (6.13) is too short and not observable. Gravitational waves continue to be generated during the turbulent stage and even during equilibrium due to thermal fluctuations, but with a smaller amplitude. It is a subject of further investigation if they can be observed. The most promising possibility for observations is, however, generation of gravity waves from low energy hybrid inflation, where $f$ can much much smaller.

## References

[1] C. Kuo *et al.* arXiv:astro-ph/0611198.
[2] M. Gasperino and G. Veneziano, Astropart Phys **1** (1993) 317.
[3] J. Khoury, B. Ovrut, N. Turok and P. Steinhardt, Phys Rev D **64** (2001) 123522.
[4] A. Nayeri, R. Brandenberger and C. Vafa, Phys Rev Lett **97** (2006) 021302.
[5] N. Kaloper, L. Kofman, A. Linde and V. Mukhanov, JCAP **0610** (2006) 0006.
[6] Compiled by C. Contaldi from the data of arXiv:astro-ph/0611198, astro-ph/0507494, astro-ph/0409569 and astro-ph/060345.
[7] E. Stewart and D. Lyth, Phys. Lett. B **302** (1993) 171.
[8] A. Linde, Phys Lett B **108** (1982) 384.
[9] A. Albrecht and P. Steinhardt, Phys Rev Lett **48** (1982) 1220.
[10] A. Linde, Phys Lett B **129** (1983) 177.
[11] A. Guth, Phys Rev D **23** (1981) 347.
[12] A. Linde, Phys Rev D **49** (1994) 748.
[13] G. Dvali and H. Tye, Phys Lett B **450** (1999) 72.
[14] S. Kachru *et al*, JCAP **0310** (2003) 013.

[15] R. Kallosh, arXiv:hep-th/0702059.

[16] L. Randall and R. Sundrum, Phys Rev Lett **83** (1999) 3370.

[17] A. Guth, and S. Pi, Phys Rev Lett **49** (1982) 1110.

[18] A. Starobinsky, Phys. Lett. B **117** (1982) 175.

[19] S. Hawking, Phys. Lett. B **115** (1982) 295.

[20] J. Bardeen, P. Steinhardt and M. Turner, Phys.Rev.D **28** (1983) 679.

[21] V. Mukhanov and G. Chibisov, Sov Phys JETsP Letter **33** (1981) 54, and references therein.

[22] A. Starobinsky, Phys. Lett. B **91** (1980) 99.

[23] A. Starobinsky, Sov Astron Lett **9** (1983) 579.

[24] A. Starobinsky, Sov JETP Lett **37** (1985) 66.

[25] V. Sahni and L. Kofman, Phys Lett B **117** (1986) 275.

[26] G. Gibbons, S. Hawking and J. Stewart, Nucl Phys B **281** (1987) 609.

[27] S. Hawking and D. Page, Nucl Phys B **298** (1988) 789.

[28] S. Hollands and R. Wald, Gen Rel Grav **34** (1992) 2043.

[29] L. Kofman, A. Linde, and V. Mukhanov, JHEP **0210** (1992) 057.

[30] G. Gibbons and N. Turok, arXiv:hep-th/0609095.

[31] L. Kofman and A. Linde, In preparation.

[32] R. Kallosh, A. Linde, S. Prokushkin, and M. Shmakova, Phys Rev D **65** (2002) 105016.

[33] A. Frolov and L. Kofman, JCAP **0305** (2003) 009.

[34] A. Vilenkin, Phys Rev D **27** (1983) 2848.

[35] A. Starobinsky, Lect Notes in Phys **246** (1986) 107.

[36] A. Linde, Phys Lett B **175** (1986) 395.

[37] A. Linde, Phys Rev D **58** (1998) 083514.

[38] G. Felder, A. Frolov, L. Kofman and A. Linde, Phys Rev D **66** (2002) 023507.

[39] G. Gibbons and S. Hawking, Phys Rev D **15** (1977) 2738.

[40] T. Bunch and P. Davies, Proc Roy Soc A **360** (1978) 117.

[41] R. Bousso and S. Hawking, Phys Rev D **54** (1996) 6312.

[42] L. Dyson, M. Kleban and L. Susskind, JHEP **0210** (2002) 011.

[43] A. Vilenkin and L. Ford, Phys Rev D **26** (1982) 1231.

[44] A. Linde, Phys Lett B **116** (1982) 335.

[45] L. Kofman, A. Linde and A. Starobinsky, Phys. Lett. B **157** (1985) 361.

[46] A. Starobinsky, Sov JETP Lett **30** (1979) 682.

[47] J.R. Bond, Les Houches **Session LX 1993** (1996) Elsvier Sci.

[48] L. Kofman, A. Linde and A. Starobinsky, Phys Rev Lett **73** (1994) 3195.

[49] R. Kallosh, L. Kofman and A. Linde, Phys Rev D **64** (2001) 123523.

[50] L. Kofman, A. Linde and A. Starobinsky, Phys. Rev D **56** (1997) 3258.

[51] V. Belinsky, L. Grishuk, I. Khalatnikov and Ya. Zel'dovich, Sov. Phys. JETP **89** (1985) 346.

[52] A. Linde, *Particle Physics and Inflationary Cosmology* (Academic Publisher, Chur, Switzerland 1990).

[53] A. Mezhlumian and A. Starobinsky, arXiv:astro-ph/9406045.

[54] A. Linde, Phys Lett B **175** (1986) 395.

[55] A. Dolgov and A. Linde, Phys Lett B **116** (1982) 329.

[56] L. Abbott, E. Fahri and M. Wise, Phys Lett B **117** (1982) 29.

[57]  J. Traschen, and R. Brandenberger, Phys Rev D **42** (1990) 2491.

[58]  G. Felder *et al*, Phys Rev Lett **87** (2001) 011601.

[59]  G. Felder, L. Kofman and A. Linde, Phys Rev D **64** (2001) 123517.

[60]  L. Kofman and P. Yi, Phys Rev D **72** (2005) 106001, and references therein.

[61]  A. Khlebnikov and I. Tkachev, Phys Rev Lett **77** (1996) 219.

[62]  G. Felder and L. Kofman, Phys Rev D **63** (2001) 103503.

[63]  G. Felder and L. Kofman, Phys Rev D **75** (2007) 047518.

[64]  L. Kofman, arXiv:astro-ph/0303614.

[65]  L. Kofman and A. Linde, in preparation.

Course 5

# COSMIC MICROWAVE BACKGROUND ANISOTROPIES
# UP TO SECOND ORDER

N. Bartolo[1], S. Matarrese[1] and A. Riotto[2]

[1] *Dipartimento di Fisica "G. Galilei", Università di Padova, and INFN, Sezione di Padova, via Marzolo 8, I-35131 Padova, Italy*
[2] *CERN, Theory Division, CH-1211 Geneva 23, Switzerland*

*F. Bernardeau, C. Grojean and J. Dalibard, eds.*
*Les Houches, Session LXXXVI, 2006*
*Particle Physcis and Cosmology: The Fabric of Spacetime*
© *2007 Published by Elsevier B.V.*

# Contents

## 1. Preamble

These lecture notes present the computation of the full system of Boltzmann equations describing the evolution of the photon, baryon and cold dark matter fluids up to second order in perturbation theory, as recently studied in Refs. [1, 2]. These equations allow to follow the time evolution of the cosmic microwave background anisotropies at all angular scales from the early epoch, when the cosmological perturbations were generated, to the present, through the recombination era. The inclusion of second-order contributions is mandatory when one is interested in studying possible deviations from Gaussianity of cosmological perturbations, either of primordial (e.g. inflationary) origin or due to their subsequent evolution. Most of the emphasis in these lectures notes will be given to the derivation of the relevant equations for the study of cosmic microwave background anisotropies and to their analytical solutions.

## 2. Introduction

Cosmic Microwave Background (CMB) anisotropies play a special role in cosmology, as they allow an accurate determination of cosmological parameters and may provide a unique probe of the physics of the early universe and in particular of the processes that gave origin to the primordial perturbations.

Cosmological inflation [3] is nowadays considered the dominant paradigm for the generation of the initial seeds for structure formation. In the inflationary picture, the primordial cosmological perturbations are created from quantum fluctuations "redshifted" out of the horizon during an early period of accelerated expansion of the universe, where they remain "frozen". They are observable through CMB temperature anisotropies (and polarization) and the large-scale clustering properties of the matter distribution in the Universe.

This picture has recently received further spectacular confirmation from the results of the Wilkinson Microwave Anisotropy Probe (WMAP) three year dataset [4]. Since the observed cosmological perturbations are of the order of $10^{-5}$, one might think that first-order perturbation theory will be adequate for all comparisons with observations. This might not be the case, though. Present [4] and future experiments [5] may be sensitive to the non-linearities of the cosmological perturbations at the level of second- or higher-order perturbation theory. The detection of these non-linearities through the non-Gaussianity (NG) in the CMB [6] has become one of the primary experimental targets.

There is one fundamental reason why a positive detection of NG is so relevant: it might help in discriminating among the various mechanisms for the generation of the cosmological perturbations. Indeed, various models of inflation, firmly rooted in modern particle physics theory, predict a significant amount of primordial NG generated either during or immediately after inflation when the comoving curvature perturbation becomes constant on super-horizon scales [6]. While single-field [7] and two(multi)-field [8] models of inflation predict a tiny level of NG, "curvaton"-type models, in which a significant contribution to the curvature perturbation is generated after the end of slow-roll inflation by the perturbation in a field which has a negligible effect on inflation, may predict a high level of NG [9]. Alternatives to the curvaton model are models where a curvature perturbation mode is generated by an inhomogeneity in the decay rate [10, 11], the mass [12] or the interaction rate [13] of the particles responsible for the reheating after inflation. Other opportunities for generating the curvature perturbations occur at the end of inflation [14], during preheating [15], and at a phase-transition producing cosmic strings [16].

Statistics like the bispectrum and the trispectrum of the CMB can then be used to assess the level of primordial NG on various cosmological scales and to discriminate it from the one induced by secondary anisotropies and systematic effects [6, 17–19]. A positive detection of a primordial NG in the CMB at some level might therefore confirm and/or rule out a whole class of mechanisms by which the cosmological perturbations have been generated.

Despite the importance of NG in CMB anisotropies, little effort has ben made so far to provide accurate theoretical predictions of it. On the contrary, the vast majority of the literature has been devoted to the computation of the bispectrum of either the comovig curvature perturbation or the gravitational potential on large scales within given inflationary models. These, however, are not the physical quantities which are observed. One should instead provide a full prediction for the second-order radiation transfer function. A preliminary step towards this goal has been taken in Ref. [20] (see also [21]) where the full second-order radiation transfer function for the CMB anisotropies on large angular scales in a flat universe filled with matter and cosmological constant was computed, including the second-order generalization of the Sachs-Wolfe effect, both the early and late Integrated Sachs-Wolfe (ISW) effects and the contribution of the second-order tensor modes.

There are many sources of NG in CMB anisotropies, beyond the primordial ones. The most relevant sources are the so-called secondary anisotropies, which arise after the last scattering epoch. These anisotropies can be divided into two categories: scattering secondaries, when the CMB photons scatter with electrons along the line of sight, and gravitational secondaries when effects are mediated by gravity [22]. Among the scattering secondaries we may list the thermal Sunyaev-Zeldovich effect, where hot electrons in clusters transfer energy to the CMB photons, the kinetic Sunyaev-Zeldovich effect produced by the bulk motion of the electrons in clusters, the Ostriker-Vishniac effect, produced by bulk motions modulated by linear density perturbations, and effects due to reionization processes. The scattering secondaries are most significant on small angular scales as density inhomogeneities, bulk and thermal motions grow and become sizeable on small length-scales when structure formation proceeds.

Gravitational secondaries arise from the change in energy of photons when the gravitational potential is time-dependent, the ISW effect, and gravitational lensing. At late times, when the Universe becomes dominated by the dark energy, the gravitational potential on linear scales starts to decay, causing the ISW effect mainly on large angular scales. Other secondaries that result from a time dependent potential are the Rees-Sciama effect, produced during the matter-dominated epoch by the time evolution of the potential on non-linear scales.

The fact that the potential never grows appreciably means that most second order effects created by gravitational secondaries are generically small compared to those created by scattering ones. However, when a photon propagates from the last scattering to us, its path may be deflected because of the gravitational lensing. This effect does not create anisotropies, but only modifies existing ones. Since photons with large wavenumbers $k$ are lensed over many regions ($\sim k/H$, where $H$ is the Hubble rate) along the line of sight, the corresponding second-order effect may be sizeable. The three-point function arising from the correlation of the gravitational lensing and ISW effects generated by the matter distribution along the line of sight [23, 24] and the Sunyaev-Zeldovich effect [25] are large and detectable by Planck [26].

Another relevant source of NG comes from the physics operating at the recombination. A naive estimate would tell that these non-linearities are tiny being suppressed by an extra power of the gravitational potential. However, the dynamics at recombination is quite involved because all the non-linearities in the evolution of the baryon-photon fluid at recombination and the ones coming from general relativity should be accounted for. This complicated dynamics might lead to unexpected suppressions or enhancements of the NG at recombination. A step towards the evaluation of the three-point correlation function has been taken in Ref. [27] where some effects were taken into account in the in so-called squeezed triangle limit, corresponding to the case when one wavenumber is much smaller than the other two and was outside the horizon at recombination.

These notes, which are based on Refs. [1, 2], present the computation of the full system of Boltzmann equations, describing the evolution of the photon, baryon and Cold Dark Matter (CDM) fluids, at second order and neglecting polarization, These equations allow to follow the time evolution of the CMB anisotropies at second order on all angular scales from the early epochs, when the cosmological perturbations were generated, to the present time, through the recombination era. These calculations set the stage for the computation of the full second-order radiation transfer function at all scales and for a a generic set of initial conditions specifying the level of primordial non-Gaussianity. Of course on small angular scales, fully non-linear calculations of specific effects like Sunyaev-Zel'dovich, gravitational lensing, etc. would provide a more accurate estimate of the resulting CMB anisotropy, however, as long as the leading contribution to second-order statistics like the bispectrum is concerned, second-order perturbation theory suffices.

These notes are organized as follows. In Section 3 we provide the second-order metric and corresponding Einstein equations. In Section 4 the left-hand-side of the Boltzmann equation for the photon distribution function is derived at second order. The collision term is computed in Section 5. In Section 6 we present the second-order Boltzmann equation for the photon brightness function, its formal solution with the method of the integration along the line of sight and the corresponding hierarchy equations for the multipole moments. Section 7 contains the derivation of the Boltzmann equations at second order for baryons and CDM. Section 8 contains the approximate solution of the Boltzmann equations up to first order. Section 9 contains a brief summary of the results. In Appendix A we give the explicit form of Einstein's equations up to second order, while in Appendix B we provide the first-order solutions of Einstein's equations in various cosmological eras.

In performing the computation presented in these lecture notes, we have mainly followed Ref. [28] (in particular chapter 4), where an excellent derivation of the Boltzmann equations for the baryon-photon fluid at first order is given, and Refs. [1, 2] for their second-order extension. Since the derivation at second order is straightforward, but lenghty, the reader might benefit from reading the appropriate sections of Ref. [28]. In the Conclusions (Section 9) we have also provided a Table which summarizes the many symbols appearing throughout these notes.

## 3. Perturbing gravity

Before tackling the problem of interest – the computation of the Boltzmann equations for the baryon-photon and CDM fluids – we first provide the necessary tools to deal with perturbed gravity, giving the expressions for the Einstein tensor perturbed up to second order around a flat Friedmann-Robertson-Walker background, and the relevant Einstein equations. In the following we will adopt the Poisson gauge which eliminates one scalar degree of freedom from the $g_{0i}$ component of the metric and one scalar and two vector degrees of freedom from $g_{ij}$. We will use a metric of the form

$$ds^2 = a^2(\eta) \left[ -e^{2\Phi} d\eta^2 + 2\omega_i dx^i d\eta + \left( e^{-2\Psi} \delta_{ij} + \chi_{ij} \right) dx^i dx^j \right], \qquad (3.1)$$

where $a(\eta)$ is the scale factor as a function of the conformal time $\eta$, and $\omega_i$ and $\chi_{ij}$ are the vector and tensor peturbation modes respectively. Each metric perturbation can be expanded into a linear (first-order) and a second-order part, as for example, the gravitational potential $\Phi = \Phi^{(1)} + \Phi^{(2)}/2$. However in the metric (3.1) the choice of the exponentials greatly helps in computing the relevant expressions, and thus we will always keep them where it is convenient. From Eq. (3.1) one recovers at linear order the well-known longitudinal gauge while at second order, one finds $\Phi^{(2)} = \phi^{(2)} - 2(\phi^{(1)})^2$ and $\Psi^{(2)} = \psi^{(2)} + 2(\psi^{(1)})^2$ where $\phi^{(1)}$, $\psi^{(1)}$ and $\phi^{(2)}$, $\psi^{(2)}$ (with $\phi^{(1)} = \Phi^{(1)}$ and $\psi^{(1)} = \Psi^{(1)}$) are the first and second-order gravitational potentials in the longitudinal (Poisson) gauge adopted in Refs. [6, 29] as far as scalar perturbations are concerned. For the vector and tensor perturbations, we will neglect linear vector modes since they are not produced in standard mechanisms for the generation of cosmological perturbations (as inflation), and we also neglect tensor modes at linear order, since they give a negligible contribution to second order perturbations. Therefore we take $\omega_i$ and $\chi_{ij}$ to be second-order vector and tensor perturbations of the metric.

Let us now give our definitions for the connection coefficients and their expressions for the metric (3.1). The number of spatial dimensions is $n = 3$. Greek indices $(\alpha, \beta, \ldots, \mu, \nu, \ldots)$ run from 0 to 3, while latin indices $(a, b, \ldots, i, j, k, \ldots, m, n, \ldots)$ run from 1 to 3. The space-time metric $g_{\mu\nu}$ has signature $(-, +, +, +)$. The connection coefficients are defined as

$$\Gamma^\alpha_{\beta\gamma} = \frac{1}{2} g^{\alpha\rho} \left( \frac{\partial g_{\rho\gamma}}{\partial x^\beta} + \frac{\partial g_{\beta\rho}}{\partial x^\gamma} - \frac{\partial g_{\beta\gamma}}{\partial x^\rho} \right). \qquad (3.2)$$

The Riemann tensor is defined as

$$R^\alpha_{\beta\mu\nu} = \Gamma^\alpha_{\beta\nu,\mu} - \Gamma^\alpha_{\beta\mu,\nu} + \Gamma^\alpha_{\lambda\mu}\Gamma^\lambda_{\beta\nu} - \Gamma^\alpha_{\lambda\nu}\Gamma^\lambda_{\beta\mu}. \qquad (3.3)$$

The Ricci tensor is a contraction of the Riemann tensor, $R_{\mu\nu} = R^\alpha_{\mu\alpha\nu}$ and in terms of the connection coefficient it is given by

$$R_{\mu\nu} = \partial_\alpha \Gamma^\alpha_{\mu\nu} - \partial_\mu \Gamma^\alpha_{\nu\alpha} + \Gamma^\alpha_{\sigma\alpha} \Gamma^\sigma_{\mu\nu} - \Gamma^\alpha_{\sigma\nu} \Gamma^\sigma_{\mu\alpha}. \qquad (3.4)$$

The Ricci scalar is the trace of the Ricci tensor, $R = R^{\mu}_{\mu}$. The Einstein tensor is defined as $G_{\mu\nu} = R_{\mu\nu} - \frac{1}{2}g_{\mu\nu}R$.

The Einstein equations are written as $G_{\mu\nu} = \kappa^2 T_{\mu\nu}$, so that $\kappa^2 = 8\pi G_N$, where $G_N$ is the usual Newtonian gravitational constant.

## 4. The collisionless Boltzmann equation for photons

We are interested in the anisotropies in the cosmic distribution of photons and inhomogeneities in the matter. Photons are affected by gravity and by Compton scattering with free electrons. The latter are tightly coupled to protons. Both are, of course, affected by gravity. The metric which determines the gravitational forces is influenced by all these components plus CDM (and neutrinos). Our plan is to write down Boltzmann equations for the phase-space distributions of each species in the Universe.

The phase-space distribution of particles $g(x^i, P^{\mu}, \eta)$ is a function of spatial coordinates $x^i$, conformal time $\eta$, and momentum of the particle $P^{\mu} = dx^{\mu}/d\lambda$ where $\lambda$ parametrizes the particle path. Through the constraint $P^2 \equiv g_{\mu\nu}P^{\mu}P^{\nu} = -m^2$, where $m$ is the mass of the particle one can eliminate $P^0$ and $g(x^i, P^j, \eta)$ gives the number of particles in the differential phase-space volume $dx^1 dx^2 dx^3 dP^1 dP^2 dP^3$. In the following we will denote the distribution function for photons with $f$.

The photons' distribution evolves according to the Boltzmann equation

$$\frac{df}{d\eta} = \overline{C}[f], \tag{4.1}$$

where the collision term is due to the scattering of photons off free electrons. In the following we will derive the left-hand side of Eq. (4.1) while in the next section we will compute the collision term.

For photons we can impose $P^2 \equiv g_{\mu\nu}P^{\mu}P^{\nu} = 0$ and using the metric (3.1) in the conformal time $\eta$ we find

$$P^2 = a^2\left[-e^{2\Phi}(P^0)^2 + \frac{p^2}{a^2} + 2\omega_i P^0 P^i\right] = 0, \tag{4.2}$$

where we define

$$p^2 = g_{ij}P^i P^j. \tag{4.3}$$

Using the constraint (4.2), we find

$$P^0 = e^{-\Phi}\left(\frac{p^2}{a^2} + 2\omega_i P^0 P^i\right)^{1/2}. \tag{4.4}$$

Notice that we immediately recover the usual zero and first-order relations $P^0 = p/a$ and $P^0 = p(1 - \Phi^{(1)})/a$.

The components $P^i$ are proportional to $pn^i$, where $n^i$ is a unit vector with $n^i n_i = \delta_{ij} n^i n^j = 1$. We can write $P^i = Cn^i$, where $C$ is determined by

$$g_{ij} P^i P^j = C^2 a^2 \left(e^{-2\Psi} + \chi_{ij} n^i n^j\right) = p^2, \tag{4.5}$$

so that

$$P^i = \frac{p}{a} n^i \left(e^{-2\Psi} + \chi_{km} n^k n^m\right)^{-1/2} = \frac{p}{a} n^i e^\Psi \left(1 - \frac{1}{2} \chi_{km} n^k n^m\right), \tag{4.6}$$

where the last equality holds up to second order in the perturbations. Again we recover the zero and first-order relations $P^i = pn^i/a$ and $P^i = pn^i(1 + \Psi^{(1)})/a$ respectively. Thus, up to second order, we can write

$$P^0 = e^{-\Phi} \frac{p}{a} \left(1 + \omega_i n^i\right). \tag{4.7}$$

Eq. (4.6) and (4.7) allow us to replace $P^0$ and $P^i$ in terms of the variables $p$ and $n^i$. Therefore, as it is standard in the literature, from now on we will consider the phase-space distribution $f$ as a function of the momentum $\mathbf{p} = pn^i$ with magnitude $p$ and angular direction $n^i$, $f \equiv f(x^i, p, n^i, \eta)$.

Thus, in terms of these variables, the total time derivative of the distribution function reads

$$\frac{df}{d\eta} = \frac{\partial f}{\partial \eta} + \frac{\partial f}{\partial x^i} \frac{dx^i}{d\eta} + \frac{\partial f}{\partial p} \frac{dp}{d\eta} + \frac{\partial f}{\partial n^i} \frac{dn^i}{d\eta}. \tag{4.8}$$

In the following we will compute $dx^i/d\eta$, $dp/d\eta$ and $dn^i/d\eta$.

a) $dx^i/d\eta$:

Starting from

$$P^i = \frac{dx^i}{d\lambda} = \frac{dx^i}{d\eta} \frac{d\eta}{d\lambda} = \frac{dx^i}{d\eta} P^0 \tag{4.9}$$

and from Eq. (4.6) and (4.7), we find

$$\frac{dx^i}{d\eta} = n^i e^{\Phi+\Psi} \left(1 - \omega_j n^j - \frac{1}{2} \chi_{km} n^k n^m\right). \tag{4.10}$$

b) $dp/d\eta$:

For $dp/d\eta$ we make use of the time component of the geodesic equation $dP^0/d\lambda = -\Gamma^0_{\alpha\beta} P^\alpha P^\beta$, where $d/d\lambda = (d\eta/d\lambda) d/d\eta = P^0 d/d\eta$, and

$$\frac{dP^0}{d\eta} = -\Gamma^0_{\alpha\beta} \frac{P^\alpha P^\beta}{P^0}, \tag{4.11}$$

Using the metric (3.1) we find

$$2\Gamma^0_{\alpha\beta} P^\alpha P^\beta = g^{0\nu} \left[2\frac{\partial g_{\nu\alpha}}{\partial x^\beta} - \frac{\partial_{\alpha\beta}}{\partial x^\nu}\right] P^\alpha P^\beta$$

$$= 2(\mathcal{H} + \Phi')(P^0)^2 + 4\Phi_{,i} P^0 P^i + 4\mathcal{H}\omega_i P^0 P^i$$

$$+ 2e^{-2\Phi}\left[(\mathcal{H} - \Psi')e^{-2\Psi}\delta_{ij} - \omega_{i,j} + \frac{1}{2}\chi'_{ij} + \mathcal{H}\chi_{ij}\right] P^i P^j. \tag{4.12}$$

On the other hand the expression (4.7) of $P^0$ in terms of $p$ and $n^i$ gives

$$\frac{dP^0}{d\eta} = -\frac{p}{a}\frac{d\Phi}{d\eta} e^{-\Phi}\left(1 + \omega_i n^i\right) + e^{-\Phi}\left(1 + \omega_i n^i\right) \frac{d(p/a)}{d\eta}$$

$$+ \frac{p}{a} e^{-\Phi}\frac{d(\omega_i n^i)}{d\eta}. \tag{4.13}$$

Thus Eq. (4.11) allows us express $dp/d\eta$ as

$$\frac{1}{p}\frac{dp}{d\eta} = -\mathcal{H} + \Psi' - \Phi_{,i} n^i e^{\Phi+\Psi} - \omega'_i n^i - \frac{1}{2}\chi'_{ij} n^i n^j, \tag{4.14}$$

where in Eq. (4.12) we have replaced $P^0$ and $P^i$ by Eqs. (4.7) and (4.6). Notice that in order to obtain Eq. (4.14) we have used the following expressions for the total time derivatives of the metric perturbations

$$\frac{d\Phi}{d\eta} = \frac{\partial\Phi}{\partial\eta} + \frac{\partial\Phi}{\partial x^i}\frac{dx^i}{d\eta}$$

$$= \frac{\partial\Phi}{\partial\eta} + \frac{\partial\Phi}{\partial x^i} n^i e^{\Phi+\Psi}\left(1 - \omega_j n^j - \frac{1}{2}\chi_{km}n^k n^m\right) \tag{4.15}$$

and

$$\frac{d(\omega_i n^i)}{d\eta} = n^i\left(\frac{\partial\omega_i}{\partial\eta} + \frac{\partial\omega_i}{\partial x^j}\frac{dx^j}{d\eta}\right) = \frac{\partial\omega_i}{\partial\eta} n^i + \frac{\partial\omega_i}{\partial x^j} n^i n^j, \tag{4.16}$$

where we have taken into account that $\omega_i$ is already a second-order perturbation so that we can neglect $dn^i/d\eta$ which is at least a first order quantity, and we can take the zero-order expression in Eq. (4.10), $dx^i/d\eta = n^i$. In fact there is also an alternative expression for $dp/d\eta$ which turns out to be useful later and which can be obtained by applying once more Eq. (4.15)

$$\frac{1}{p}\frac{dp}{d\eta} = -\mathcal{H} - \frac{d\Phi}{d\eta} + \Phi' + \Psi' - \omega'_i\, n^i - \frac{1}{2}\chi'_{ij} n^i n^j.$$  (4.17)

c) $dn^i/d\eta$:

We can proceed in a similar way to compute $dn^i/d\eta$. Notice that since in Eq. (4.8) it multiplies $\partial f/\partial n^i$ which is first order, we need only the first order perturbation of $dn^i/d\eta$. We use the spatial components of the geodesic equations $dP^i/d\lambda = -\Gamma^i_{\alpha\beta} P^\alpha P^\beta$ written as

$$\frac{dP^i}{d\eta} = -\Gamma^i_{\alpha\beta}\frac{P^\alpha P^\beta}{P^0}.$$  (4.18)

For the right-hand side we find, up to second order,

$$
\begin{aligned}
2\Gamma^i_{\alpha\beta} P^\alpha P^\beta &= g^{iv}\left[\frac{\partial g_{\alpha v}}{\partial x^\beta} + \frac{\partial g_{\beta v}}{\partial x^\alpha} - \frac{\partial g_{\alpha\beta}}{\partial x^v}\right] P^\alpha P^\beta\\
&= 4(\mathcal{H} - \Psi')P^i P^0 + 2\left(\chi^i_k + \omega^i_{,k} - \omega_k^{,i}\right)P^0 P^k\\
&\quad + \left(2\frac{\partial\Phi}{\partial x^i}e^{2\Phi+2\Psi} + 2\omega^{i\prime} + 2\mathcal{H}\omega^i\right)(P^0)^2\\
&\quad - 4\frac{\partial\Psi}{\partial x^k}P^i P^k + 2\frac{\partial\Psi}{\partial x^i}\delta_{km} P^k P^m\\
&\quad - \left[2\mathcal{H}\omega^i\delta_{jk} - \left(\frac{\partial\chi^i_{\ j}}{\partial x^k} + \frac{\partial\chi^i_{\ k}}{\partial x^j} - \frac{\partial\chi_{jk}}{\partial x_i}\right)\right]P^j P^k,
\end{aligned}
$$  (4.19)

while the expression (4.6) of $P^i$ in terms of our variables $p$ and $n^i$ in the left-hand side of Eq. (4.18) brings

$$
\begin{aligned}
\frac{dP^i}{d\eta} &= \frac{p}{a}e^\Psi\left[\frac{d\Psi}{d\eta}n^i + \frac{a}{p}\frac{d(p/a)}{d\eta}n^i + \frac{dn^i}{d\eta}\right]\left(1 - \frac{1}{2}\chi_{km}n^k n^m\right)\\
&\quad - \frac{p}{a}n^i e^\Psi\frac{1}{2}\frac{d(\chi_{km}n^k n^m)}{d\eta}.
\end{aligned}
$$  (4.20)

Thus, using the expression (4.6) for $P^i$ and (4.4) for $P^0$ in Eq. (4.19), together with the previous result (4.14), the geodesic equation (4.18) gives the following expression $dn^i/d\eta$ (valid up to first order)

$$\frac{dn^i}{d\eta} = (\Phi_{,k} + \Psi_{,k})n^k n^i - \Phi^{,i} - \Psi^{,i}.$$  (4.21)

To proceed further we now expand the distribution function for photons around the zero-order value $f^{(0)}$ which is that of a Bose-Einstein distribution

$$f^{(0)}(p, \eta) = 2 \frac{1}{\exp\{\frac{p}{T(\eta)}\} - 1}, \tag{4.22}$$

where $T(\eta)$ is the average (zero-order) temperature and the factor 2 comes from the spin degrees of photons. The perturbed distribution of photons will depend also on $x^i$ and on the propagation direction $n^i$ so as to account for inhomogeneities and anisotropies

$$f(x^i, p, n^i, \eta) = f^{(0)}(p, \eta) + f^{(1)}(x^i, p, n^i, \eta) + \frac{1}{2}f^{(2)}(x^i, p, n^i, \eta), \tag{4.23}$$

where we split the perturbation of the distribution function into a first and a second-order part. The Boltzmann equation up to second order can be written in a straightforward way by recalling that the total time derivative of a given $i$-th perturbation, as e.g. $df^{(i)}/d\eta$ is at least a quantity of the $i$-th order. Thus it is easy to realize, looking at Eq. (4.8), that the left-hand side of Boltzmann equation can be written up to second order as

$$\begin{aligned} \frac{df}{d\eta} &= \frac{df^{(1)}}{d\eta} + \frac{1}{2}\frac{df^{(2)}}{d\eta} - p\frac{\partial f^{(0)}}{\partial p}\frac{d}{d\eta}\left(\Phi^{(1)} + \frac{1}{2}\Phi^{(2)}\right) \\ &\quad + p\frac{\partial f^{(0)}}{\partial p}\frac{\partial}{\partial \eta}\left(\Phi^{(1)} + \Psi^{(1)} + \frac{1}{2}\Phi^{(2)} + \frac{1}{2}\Psi^{(2)}\right) \\ &\quad - p\frac{\partial f^{(0)}}{\partial p}\frac{\partial \omega_i}{\partial \eta}n^i - \frac{1}{2}p\frac{\partial f^{(0)}}{\partial p}\frac{\partial \chi_{ij}}{\partial \eta}n^i n^j, \end{aligned} \tag{4.24}$$

where for simplicity in Eq. (4.24) we have already used the background Boltzmann equation $(df/d\eta)|^{(0)} = 0$. In Eq. (4.24) there are all the terms which will give rise to the integrated Sachs-Wolfe effects (corresponding to the terms which explicitly depend on the gravitational perturbations), while other effects, such as the gravitational lensing, are still hidden in the (second-order part) of the first term. In fact in order to obtain Eq. (4.24) we just need for the time being to know the expression for $dp/d\eta$, Eq. (4.17).

## 5. Collision term

### 5.1. The collision integral

In this section we focus on the collision term due to Compton scattering

$$e(\mathbf{q})\gamma(\mathbf{p}) \longleftrightarrow e(\mathbf{q}')\gamma(\mathbf{p}'), \tag{5.1}$$

where we have indicated the momentum of the photons and electrons involved in the collisions. The collision term will be important for small scale anisotropies and spectral distortions. The important point to compute the collision term is that for the epoch of interest very little energy is transferred. Therefore one can proceed by expanding the right hand side of Eq. (4.1) both in the small perturbation, Eq. (4.23), and in the small energy transfer. Part of the computation up to second order has been done in Refs. [30–32] (see also [33]). In particular Refs. [30, 31] are focused on the effects of reionization on the CMB anisotropies thus keeping in the collision term those contributions which are relevant for the small-scale effects due to reionization and neglecting the effects of the metric perturbations on the left-hand side of Eq. (4.1). We will mainly follow the formalism of Ref. [31] and we will keep all the terms arising from the expansion of the collision term up to second order.

The collision term is given (up to second order) by

$$\overline{C}(\mathbf{p}) = C(\mathbf{p})ae^{\Phi}, \tag{5.2}$$

where $a$ is the scale factor and[1]

$$
\begin{aligned}
C(\mathbf{p}) = \frac{1}{E(\mathbf{p})} \int &\frac{d\mathbf{q}}{(2\pi)^3 2E(\mathbf{q})} \frac{d\mathbf{q}'}{(2\pi)^3 2E(\mathbf{q}')} \frac{d\mathbf{p}'}{(2\pi)^3 2E(\mathbf{p}')} \\
&\times (2\pi)^4 \delta^4(q + p - q' - p')|M|^2 \\
&\times \left\{ g(\mathbf{q}')f(\mathbf{p}')[1 + f(\mathbf{p})] - g(\mathbf{q})f(\mathbf{p})[1 + f(\mathbf{p}')] \right\}
\end{aligned}
\tag{5.3}
$$

where $E(\mathbf{q}) = (q^2 + m_e^2)^{1/2}$, $M$ is the amplitude of the scattering process, $\delta^4(q + p - q' - p') = \delta^3(\mathbf{q} + \mathbf{p} - \mathbf{q}' - \mathbf{p}')\delta(E(\mathbf{q}) + p - E(\mathbf{q}') - p')$ ensures the energy-momentum conservation and $g$ is the distribution function for electrons. The Pauli suppression factors $(1 - g)$ have been dropped since for the epoch of interest the density of electrons $n_e$ is low. The electrons are kept in thermal equilibrium by Coulomb interactions with protons and they are non-relativistic, thus we can take a Maxwell-Boltzmann distribution around some bulk velocity $\mathbf{v}$

$$g(\mathbf{q}) = n_e \left(\frac{2\pi}{m_e T_e}\right)^{3/2} \exp\left\{-\frac{(\mathbf{q} - m_e \mathbf{v})^2}{2m_e T_e}\right\} \tag{5.4}$$

By using the three dimensional delta function the energy transfer is given by $E(\mathbf{q}) - E(\mathbf{q} + \mathbf{p} - \mathbf{p}')$ and it turns out to be small compared to the typical thermal

---

[1] The reason why we write the collision term as in Eq. (5.2) is that the starting point of the Boltzmann equation requires differentiation with respect to an affine parameter $\lambda$, $df/d\lambda = C'$. In moving to the conformal time $\eta$ one rewrites the Boltzmann equation as $df/d\eta = C'(P^0)^{-1}$, with $P^0 = d\eta/d\lambda$ given by Eq. (4.7). Taking into account that the collision term is at least of first order, Eq. (5.2) then follows.

energies

$$E(\mathbf{q}) - E(\mathbf{q} + \mathbf{p} - \mathbf{p}') \simeq \frac{(\mathbf{p} - \mathbf{p}') \cdot \mathbf{q}}{m_e} = \mathcal{O}(Tq/m_e). \tag{5.5}$$

In Eq. (5.5) we have used $E(\mathbf{q}) = m_e + q^2/2m_e$ and the fact that, since the scattering is almost elastic ($p \simeq p'$), $(\mathbf{p} - \mathbf{p}')$ is of order $p \sim T$, with $q$ much bigger than $(\mathbf{p} - \mathbf{p}')$. In general, the electron momentum has two contributions, the bulk velocity ($q = m_e v$) and the thermal motion ($q \sim (m_e T)^{1/2}$) and thus the parameter expansion $q/m_e$ includes the small bulk velocity $\mathbf{v}$ and the ratio $(T/m_e)^{1/2}$ which is small because the electrons are non-relativistic.

The expansion of all the quantities entering the collision term in the energy transfer parameter and the integration over the momenta $\mathbf{q}$ and $\mathbf{q}'$ is described in details in Ref. [31]. It is easy to realize that we just need the scattering amplitude up to first order since at zero order $g(\mathbf{q}') = g(\mathbf{q} + \mathbf{p} - \mathbf{p}') = g(\mathbf{q})$ and $\delta(E(\mathbf{q}) + p - E(\mathbf{q}') - p') = \delta(p - p')$ so that all the zero-order quantities multiplying $|M|^2$ vanish. To first order

$$|M|^2 = 6\pi\sigma_T m_e^2 [(1 + \cos^2\theta) - 2\cos\theta(1 - \cos\theta)\mathbf{q} \cdot (\hat{\mathbf{p}} + \hat{\mathbf{p}}')/m_e], \tag{5.6}$$

where $\cos\theta = \mathbf{n} \cdot \mathbf{n}'$ is the scattering angle and $\sigma_T$ the Thompson cross-section. The resulting collision term up to second order is given by [31]

$$C(\mathbf{p}) = \frac{3n_e\sigma_T}{4p} \int dp' p' \frac{d\Omega'}{4\pi} \Big[ c^{(1)}(\mathbf{p}, \mathbf{p}') + c_\Delta^{(2)}(\mathbf{p}, \mathbf{p}') + c_v^{(2)}(\mathbf{p}, \mathbf{p}')$$
$$+ c_{\Delta v}^{(2)}(\mathbf{p}, \mathbf{p}') + c_{vv}^{(2)}(\mathbf{p}, \mathbf{p}') + c_K^{(2)}(\mathbf{p}, \mathbf{p}') \Big], \tag{5.7}$$

where we arrange the different contributions following Ref. [31]. The first order term reads

$$c^{(1)}(\mathbf{p}, \mathbf{p}') = (1 + \cos^2\theta) \Big[ \delta(p - p')\big(f^{(1)}(\mathbf{p}') - f^{(1)}(\mathbf{p})\big)$$
$$+ \big(f^{(0)}(p') - f^{(0)}(p)\big)(\mathbf{p} - \mathbf{p}') \cdot \mathbf{v}\frac{\partial\delta(p - p')}{\partial p'} \Big], \tag{5.8}$$

while the second-order terms have been separated into four parts. There is the so-called anisotropy suppression term

$$c_\Delta^{(2)}(\mathbf{p}, \mathbf{p}') = \frac{1}{2}(1 + \cos^2\theta)\delta(p - p')\big(f^{(2)}(\mathbf{p}') - f^{(2)}(\mathbf{p})\big); \tag{5.9}$$

a term which depends on the second-order velocity perturbation defined by the expansion of the bulk flow as $\mathbf{v} = \mathbf{v}^{(1)} + \mathbf{v}^{(2)}/2$

$$c_v^{(2)}(\mathbf{p}, \mathbf{p}') = \frac{1}{2}(1 + \cos^2\theta)\left(f^{(0)}(p') - f^{(0)}(p)\right)(\mathbf{p} - \mathbf{p}') \cdot \mathbf{v}^{(2)} \frac{\partial \delta(p - p')}{\partial p'};$$

a set of terms coupling the photon perturbation to the velocity

$$c_{\Delta v}^{(2)}(\mathbf{p}, \mathbf{p}') = \left(f^{(1)}(\mathbf{p}') - f^{(1)}(\mathbf{p})\right)\left[(1 + \cos^2\theta)(\mathbf{p} - \mathbf{p}') \cdot \mathbf{v}\right.$$
$$\times \frac{\partial \delta(p - p')}{\partial p'} - 2\cos\theta(1 - \cos\theta)\delta(p - p')(\mathbf{n} + \mathbf{n}') \cdot \mathbf{v}\Big],$$

and a set of source terms quadratic in the velocity

$$c_{vv}^{(2)}(\mathbf{p}, \mathbf{p}') = \left(f^{(0)}(p') - f^{(0)}(p)\right)(\mathbf{p} - \mathbf{p}') \cdot \mathbf{v}\left[(1 + \cos^2\theta)\right.$$
$$\times \frac{(\mathbf{p} - \mathbf{p}') \cdot \mathbf{v}}{2} \frac{\partial^2 \delta(p - p')}{\partial p'^2}$$
$$- 2\cos\theta(1 - \cos\theta)(\mathbf{n} + \mathbf{n}') \cdot \mathbf{v}\frac{\partial \delta(p - p')}{\partial p'}\Big]. \tag{5.10}$$

The last contribution are the Kompaneets terms describing spectral distortions to the CMB

$$c_K^{(2)}(\mathbf{p}, \mathbf{p}') = (1 + \cos^2\theta)\frac{(\mathbf{p} - \mathbf{p}')^2}{2m_e}\left[\left(f^{(0)}(p') - f^{(0)}(p)\right)T_e\right.$$
$$\times \frac{\partial^2 \delta(p - p')}{\partial p'^2} - \left(f^{(0)}(p') + f^{(0)}(p) + 2f^{(0)}(p')f^{(0)}(p)\right)$$
$$\times \frac{\partial \delta(p - p')}{\partial p'}\right] + \frac{2(p - p')\cos\theta(1 - \cos^2\theta)}{m_e}\left[\delta(p - p')\right.$$
$$\times f^{(0)}(p')\left(1 + f^{(0)}(p)\right)\left(f^{(0)}(p') - f^{(0)}(p)\right)\frac{\partial \delta(p - p')}{\partial p'}\Big].$$
$$\tag{5.11}$$

Let us make a couple of comments about the various contributions to the collision term. First, notice the term $c_v^{(2)}(\mathbf{p}, \mathbf{p}')$ due to second-order perturbations in the velocity of electrons which is absent in Ref. [31]. In standard cosmological scenarios (like inflation) vector perturbations are not generated at linear order, so that linear velocities are irrotational $v^{(1)i} = \partial^i v^{(1)}$. However at second order vector perturbations are generated after horizon crossing as non-linear combinations

of primordial scalar modes. Thus we must take into account also a transverse
(divergence-free) component, $v^{(2)i} = \partial^i v^{(2)} + v_T^{(2)i}$ with $\partial_i v_T^{(2)i} = 0$. As we
will see such vector perturbations will break azimuthal symmetry of the collision
term with respect to a given mode $\mathbf{k}$, which instead usually holds at linear order.
Secondly, notice that the number density of electrons appearing in Eq. (5.7) must
be expanded as $n_e = \bar{n}_e(1 + \delta_e)$ and then

$$\delta_e^{(1)} c^{(1)}(\mathbf{p}, \mathbf{p}') \tag{5.12}$$

gives rise to second-order contributions in addition to the list above, where we
split $\delta_e = \delta_e^{(1)} + \delta_e^{(2)}/2$ into a first- and second-order part. In particular the
combination with the term proportional to $\mathbf{v}$ in $c^{(1)}(\mathbf{p}, \mathbf{p}')$ gives rise to the so-
called Vishniac effect, as discussed in Ref. [31].

### 5.2. *Computation of different contributions to the collision term*

In the integral (5.7) over the momentum $\mathbf{p}'$ the first-order term gives the usual
collision term

$$C^{(1)}(\mathbf{p}) = n_e \sigma_T \left[ f_0^{(1)}(p) + \frac{1}{2} f_2^{(1)} P_2(\hat{\mathbf{v}} \cdot \mathbf{n}) - f^{(1)} - p \frac{\partial f^{(0)}}{\partial p} \mathbf{v} \cdot \mathbf{n} \right], \tag{5.13}$$

where one uses the decomposition in Legendre polynomials

$$f^{(1)}(\mathbf{x}, p, \mathbf{n}) = \sum_\ell (2\ell + 1) f_\ell^{(1)}(p) P_\ell(\cos\vartheta), \tag{5.14}$$

where $\vartheta$ is the polar angle of $\mathbf{n}$, $\cos\vartheta = \mathbf{n} \cdot \hat{\mathbf{v}}$.

In the following we compute the second-order collision term separately for
the different contributions, using the notation $C(\mathbf{p}) = C^{(1)}(\mathbf{p}) + C^{(2)}(\mathbf{p})/2$. We
have not reported the details of the calculation of the first-order term because for
its second-order analog, $c_\Delta^{(2)}(\mathbf{p}, \mathbf{p}') + c_v^{(2)}(\mathbf{p}, \mathbf{p}')$, the procedure is the same. The
important difference is that the second-order velocity term includes a vector part,
and this leads to a generic angular decomposition of the distribution function (for
simplicity drop the time dependence)

$$f^{(i)}(\mathbf{x}, p, \mathbf{n}) = \sum_\ell \sum_{m=-\ell}^{\ell} f_{\ell m}^{(i)}(\mathbf{x}, p)(-i)^\ell \sqrt{\frac{4\pi}{2\ell + 1}} Y_{\ell m}(\mathbf{n}), \tag{5.15}$$

such that

$$f_{\ell m}^{(i)} = (-i)^{-\ell} \sqrt{\frac{2\ell + 1}{4\pi}} \int d\Omega f^{(i)} Y_{\ell m}^*(\mathbf{n}). \tag{5.16}$$

Such a decomposition holds also in Fourier space [34]. The notation at this stage is a bit confusing, so let us restate it: superscripts denote the order of the perturbation; the subscripts refer to the moments of the distribution. Indeed at first order one can drop the dependence on $m$ setting $m = 0$ using the fact that the distribution function does not depend on the azimuthal angle $\phi$. In this case the relation with $f_l^{(1)}$ is

$$f_{\ell m}^{(1)} = (-i)^{-\ell}(2\ell + 1)\delta_{m0} f_\ell^{(1)}. \tag{5.17}$$

a) $c_\Delta^{(2)}(\mathbf{p}, \mathbf{p}')$:

The integral over $\mathbf{p}'$ yields

$$C_\Delta^{(2)}(\mathbf{p}) = \frac{3n_e\sigma_T}{4p} \int dp'\, p' \frac{d\Omega'}{4\pi} c_\Delta^{(2)}(\mathbf{p}, \mathbf{p}') = \frac{3n_e\sigma_T}{4p} \int dp'\, p'\delta(p - p')$$

$$\times \int \frac{d\Omega'}{4\pi}[1 + (\mathbf{n} \cdot \mathbf{n}')^2][f^{(2)}(\mathbf{p}') - f^{(2)}(\mathbf{p})]. \tag{5.18}$$

To perform the angular integral we write the angular dependence on the scattering angle $\cos\theta = \mathbf{n} \cdot \mathbf{n}'$ in terms of the Legendre polynomials

$$[1 + (\mathbf{n} \cdot \mathbf{n}')^2] = \frac{4}{3}\left[1 + \frac{1}{2}P_2(\mathbf{n} \cdot \mathbf{n}')\right]$$

$$= \left[1 + \frac{1}{2}\sum_{m=-2}^{2} Y_{2m}(\mathbf{n})Y_{2m}^*(\mathbf{n}')\frac{4\pi}{2\ell + 1}\right], \tag{5.19}$$

where in the last step we used the addition theorem for spherical harmonics

$$P_\ell = \frac{4\pi}{2\ell + 1}\sum_{m=-2}^{2} Y_{\ell m}(\mathbf{n})Y_{\ell m}^*(\mathbf{n}'). \tag{5.20}$$

Using the decomposition (5.16) and the orthonormality of the spherical harmonics we find

$$C_\Delta^{(2)}(\mathbf{p}) = n_e\sigma_T\left[f_{00}^{(2)}(p) - f^{(2)}(\mathbf{p}) - \frac{1}{2}\sum_{m=-2}^{2} \frac{\sqrt{4\pi}}{5^{3/2}} f_{2m}^{(2)}(p)\, Y_{2m}(\mathbf{n})\right].$$

It is easy to recover the result for the corresponding first-order contribution in Eq. (5.13) by using Eq. (5.17).

b) $c_v^{(2)}(\mathbf{p}, \mathbf{p}')$:

Let us fix for simplicity our coordinates such that the polar angle of $\mathbf{n}'$ is defined by $\mu' = \hat{\mathbf{v}}^{(2)} \cdot \mathbf{n}'$ with $\phi'$ the corresponding azimuthal angle. The contribution of $c_v^{(2)}(\mathbf{p}, \mathbf{p}')$ to the collision term is then

$$
C_v^{(2)}(\mathbf{p}) = \frac{3 n_e \sigma_T}{4p} v^{(2)} \int dp' p' \big[ f^{(0)}(p') - f^{(0)}(p) \big] \frac{\partial \delta(p - p')}{\partial p'}
$$
$$
\times \int_{-1}^{1} \frac{d\mu'}{2} (p\mu - p'\mu') \int_{0}^{2\pi} \frac{d\phi'}{2\pi} \big[ 1 + (\mathbf{p} \cdot \mathbf{p}')^2 \big].   \tag{5.21}
$$

We can use Eq. (5.19) which in our coordinate system reads

$$
\frac{4}{3} \left[ 1 + \frac{1}{2} \sum_{m=-2}^{m} \frac{(2 - m)!}{(2 + m)!} P_2^m(\mathbf{n} \cdot \hat{\mathbf{v}}^{(2)}) P_2^m(\mathbf{n}' \cdot \hat{\mathbf{v}}^{(2)}) e^{im(\phi' - \phi)} \right],   \tag{5.22}
$$

so that

$$
\int \frac{d\phi'}{2\pi} P_2(\mathbf{n} \cdot \mathbf{n}') = P_2(\mathbf{n} \cdot \hat{\mathbf{v}}^{(2)}) P_2(\mathbf{n}' \cdot \hat{\mathbf{v}}^{(2)}) = P_2(\mu) P_2(\mu').   \tag{5.23}
$$

By using the orthonormality of the Legendre polynomials and integrating by parts over $p'$ we find

$$
C_v^{(2)}(\mathbf{p}) = -n_e \sigma_T \, p \frac{\partial f^{(0)}}{\partial p} v^{(2)} \cdot \mathbf{n}.   \tag{5.24}
$$

As it is clear by the presence of the scalar product $v^{(2)} \cdot \mathbf{p}$ the final result is independent of the coordinates chosen.

c) $c_{\Delta v}^{(2)}(\mathbf{p}, \mathbf{p}')$:

Let us consider the contribution from the first term

$$
c_{\Delta v(I)}^{(2)}(\mathbf{p}, \mathbf{p}') = (1 + \cos^2\theta) \big( f^{(1)}(\mathbf{p}') - f^{(1)}(\mathbf{p}) \big)(\mathbf{p} - \mathbf{p}') \cdot \mathbf{v} \frac{\partial \delta(p - p')}{\partial p'},
$$

where the velocity has to be considered at first order. In the integral (5.7) it brings

$$
\frac{1}{2} C_{\Delta v(I)}^{(2)} = \frac{3 n_e \sigma_T v}{4p} \int dp' p' \frac{\partial \delta(p - p')}{\partial p'} \int_{-1}^{1} \frac{d\mu'}{2} \big[ f^{(1)}(\mathbf{p}') - f^{(1)}(\mathbf{p}) \big]
$$
$$
\times (p\mu - p'\mu') \int_{0}^{2\pi} \frac{d\phi'}{2\pi} (1 + \cos^2\theta),   \tag{5.25}
$$

The procedure to do the integral is the same as above. We use the same relations as in Eqs. (5.22) and (5.23) where now the angles are those taken with respect to the first-order velocity. This eliminates the integral over $\phi'$, and integrating by parts over $p'$ yields

$$\frac{1}{2}C^{(2)}_{\Delta v(I)}(\mathbf{p}) = -\frac{3n_e\sigma_T v}{4p}\int_{-1}^1 \frac{d\mu'}{2}\left[\frac{4}{3}+\frac{2}{3}P_2(\mu)P_2(\mu')\right] \qquad (5.26)$$

$$\times\left[p(\mu-2\mu')\left(f^{(1)}(p,\mu')-f^{(1)}(p,\mu)\right)+p^2(\mu-\mu')\frac{\partial f^{(1)}(p,\mu')}{\partial p}\right].$$

We now use the decomposition (5.14) and the orthonormality of the Legendre polynomials to find

$$\int \frac{d\mu'}{2}\mu' f^{(1)}(p,\mu')P_2(\mu') = \sum_\ell \int \frac{d\mu'}{2}\mu' P_2(\mu')P_l(\mu')f^{(1)}_\ell(p)$$

$$= \sum_\ell \int \frac{d\mu'}{2}\left[\frac{2}{5}P_1(\mu')+\frac{3}{5}P_3(\mu')\right]P_\ell(\mu')f^{(1)}_\ell(p)$$

$$= \frac{2}{5}f^{(1)}_1(p)+\frac{3}{5}f^{(1)}_3(p), \qquad (5.27)$$

where we have used $\mu' P_2(\mu')P_l(\mu') = \frac{2}{5}P_1(\mu')+\frac{3}{5}P_3(\mu')$, with $P_1(\mu') = \mu'$. Thus from Eq. (5.26) we get

$$\frac{1}{2}C^{(2)}_{\Delta v(I)}(\mathbf{p}) = n_e\sigma_T\left\{\mathbf{v}\cdot\mathbf{n}\left[f^{(1)}(\mathbf{p})-f^{(1)}_0(p)-p\frac{\partial f^{(1)}_0(p)}{\partial p}\right.\right.$$

$$-\frac{1}{2}P_2(\hat{\mathbf{v}}\cdot\mathbf{n})\left(f^{(1)}_2(p)+p\frac{\partial f^{(1)}_2(p)}{\partial p}\right)\right]$$

$$+v\left[2f^{(1)}_1(p)+p\frac{\partial f^{(1)}_1(p)}{\partial p}+\frac{1}{5}P_2(\hat{\mathbf{v}}\cdot\mathbf{n})\left(2f^{(1)}_1(p)\right.\right.$$

$$\left.\left.\left.+p\frac{\partial f^{(1)}_1(p)}{\partial p}+3f^{(1)}_3(p)+\frac{3}{2}p\frac{\partial f^{(1)}_3(p)}{\partial p}\right)\right]\right\}. \qquad (5.28)$$

In $c^{(2)}(\mathbf{p},\mathbf{p}')$ there is a second term

$$c^{(2)}_{\Delta v(II)} = -2\cos\theta(1-\cos\theta)\left(f^{(1)}(\mathbf{p}')-f^{(1)}(\mathbf{p})\right)\delta(p-p')(\mathbf{n}+\mathbf{n}')\cdot\mathbf{v},$$

whose contribution to the collision term is

$$\frac{1}{2}C^{(2)}_{\Delta v(II)}(\mathbf{p}) = -\frac{3n_e\sigma_T v}{2p}\int dp'\,p'\delta(p-p')\int_{-1}^1\frac{d\mu'}{2}(f^{(1)}(\mathbf{p}')$$

$$-f^{(1)}(\mathbf{p}))(\mu+\mu')\int_0^{2\pi}\frac{d\phi'}{2\pi}\cos\theta(1-\cos\theta). \qquad (5.29)$$

This integration proceeds through the same steps as for $C^{(2)}_{\Delta v(I)}(\mathbf{p})$. In particular by noting that $\cos\theta(1 - \cos\theta) = -1/3 + P_1(\cos\theta) - 2P_3(\cos\theta)/3$, Eqs. (5.22) and (5.23) allows to compute

$$\int \frac{d\phi'}{2\pi} \cos\theta(1 - \cos\theta) = -\frac{1}{3} + P_1(\mu)P_1(\mu') - \frac{2}{3}P_2(\mu)P_2(\mu'), \qquad (5.30)$$

and using the decomposition (5.14) we arrive at

$$\frac{1}{2}C^{(2)}_{\Delta v(II)}(\mathbf{p}) = -n_e\sigma_T \left\{ \mathbf{v}\cdot\mathbf{n}\, f^{(1)}_2(p)(1 - P_2(\hat{\mathbf{v}}\cdot\mathbf{n})) \right.$$
$$\left. + v\left[\frac{1}{5}P_2(\hat{\mathbf{v}}\cdot\mathbf{n})\left(3f^{(1)}_1(p) - 3f^{(1)}_3(p)\right)\right]\right\}. \qquad (5.31)$$

We then obtain

$$\frac{1}{2}C^{(2)}_{\Delta v}(\mathbf{p}) = n_e\sigma_T \left\{ \mathbf{v}\cdot\mathbf{n}\left[f^{(1)}(\mathbf{p}) - f^{(1)}_0(p) - p\frac{\partial f^{(1)}_0(p)}{\partial p} - f^{(1)}_2(p)\right.\right.$$
$$\left. + \frac{1}{2}P_2(\hat{\mathbf{v}}\cdot\mathbf{n})\left(f^{(1)}_2(p) - p\frac{\partial f^{(1)}_2(p)}{\partial p}\right)\right] + v\left[2f^{(1)}_1(p)\right.$$
$$+ p\frac{\partial f^{(1)}_1(p)}{\partial p} + \frac{1}{5}P_2(\hat{\mathbf{v}}\cdot\mathbf{n})\left(-f^{(1)}_1(p)\right.$$
$$\left.\left.\left. + p\frac{\partial f^{(1)}_1(p)}{\partial p} + 6f^{(1)}_3(p) + \frac{3}{2}p\frac{\partial f^{(1)}_3(p)}{\partial p}\right)\right]\right\}. \qquad (5.32)$$

As far as the remaining terms, these have already been computed in Ref. [31] (see also Ref. [30]) and here we just report them.

d) $c^{(2)}_{vv}(\mathbf{p}, \mathbf{p}')$:

The term proportional to the velocity squared yield a contribution to the collision term

$$\frac{1}{2}C^{(2)}_{vv}(\mathbf{p}) = n_e\sigma_T \left\{ (\mathbf{v}\cdot\mathbf{n})^2\left[p\frac{\partial f^{(0)}}{\partial p} + \frac{11}{20}p^2\frac{\partial^2 f^{(0)}}{\partial p^2}\right]\right.$$
$$\left. + v^2\left[p\frac{\partial f^{(0)}}{\partial p} + \frac{3}{20}p^2\frac{\partial^2 f^{(0)}}{\partial p^2}\right]\right\}. \qquad (5.33)$$

e) $c_K^{(2)}(\mathbf{p}, \mathbf{p}')$:

The terms responsible for the spectral distortions give

$$\frac{1}{2}C_K^{(2)}(\mathbf{p}) = \frac{1}{m_e^2}\frac{\partial}{\partial p}\left\{p^4\left[T_e\frac{\partial f^{(0)}}{\partial p} + f^{(0)}(1 + f^{(0)})\right]\right\}. \tag{5.34}$$

Finally, we write also the part of the collision term coming from Eq. (5.12)

$$\delta_e^{(1)}c^{(1)}(\mathbf{p}, \mathbf{p}') \rightarrow \delta_e^{(1)}C^{(1)}(\mathbf{p}) = n_e\sigma_T\,\delta_e^{(1)}\left[f_0^{(1)}(p)\right.$$

$$\left. + \frac{1}{2}f_2^{(1)}P_2(\hat{\mathbf{v}}\cdot\mathbf{n}) - f^{(1)} - p\frac{\partial f^{(0)}}{\partial p}\mathbf{v}\cdot\mathbf{n}\right]. \tag{5.35}$$

### 5.3. Final expression for the collision term

Summing all the terms we find the final expression for the collision term (5.7) up to second order

$$C(\mathbf{p}) = C^{(1)}(\mathbf{p}) + \frac{1}{2}C^{(2)}(\mathbf{p}) \tag{5.36}$$

with

$$C^{(1)}(\mathbf{p}) = n_e\sigma_T\left[f_0^{(1)}(p) + \frac{1}{2}f_2^{(1)}P_2(\hat{\mathbf{v}}\cdot\mathbf{n}) - f^{(1)} - p\frac{\partial f^{(0)}}{\partial p}\mathbf{v}\cdot\mathbf{n}\right] \tag{5.37}$$

and

$$\frac{1}{2}C^{(2)}(\mathbf{p}) = n_e\sigma_T\left\{\frac{1}{2}f_{00}^{(2)}(p) - \frac{1}{4}\sum_{m=-2}^{2}\frac{\sqrt{4\pi}}{5^{3/2}}f_{2m}^{(2)}(p)Y_{2m}(\mathbf{n})\right.$$

$$-\frac{1}{2}f^{(2)}(\mathbf{p}) + \delta_e^{(1)}\left[f_0^{(1)}(p) + \frac{1}{2}f_2^{(1)}P_2(\hat{\mathbf{v}}\cdot\mathbf{n}) - f^{(1)}\right.$$

$$\left. - p\frac{\partial f^{(0)}}{\partial p}\mathbf{v}\cdot\mathbf{n}\right] - \frac{1}{2}p\frac{\partial f^{(0)}}{\partial p}\mathbf{v}^{(2)}\cdot\mathbf{n} + \mathbf{v}\cdot\mathbf{n}\left[f^{(1)}(\mathbf{p})\right.$$

$$- f_0^{(1)}(p) - p\frac{\partial f_0^{(1)}(p)}{\partial p} - f_2^{(1)}(p) + \frac{1}{2}P_2(\hat{\mathbf{v}}\cdot\mathbf{n})$$

$$\times\left(f_2^{(1)}(p) - p\frac{\partial f_2^{(1)}(p)}{\partial p}\right)\right] + v\left[2f_1^{(1)}(p) + p\frac{\partial f_1^{(1)}(p)}{\partial p}\right.$$

$$+ \frac{1}{5}P_2(\hat{\mathbf{v}}\cdot\mathbf{n})\left(-f_1^{(1)}(p) + p\frac{\partial f_1^{(1)}(p)}{\partial p} + 6f_3^{(1)}(p)\right)$$

$$+ \frac{3}{2} p \frac{\partial f_3^{(1)}(p)}{\partial p} \Big) \Big] + (\mathbf{v} \cdot \mathbf{n})^2 \Big[ p \frac{\partial f^{(0)}}{\partial p} + \frac{11}{20} p^2 \frac{\partial^2 f^{(0)}}{\partial p^2} \Big]$$

$$+ v^2 \Big[ p \frac{\partial f^{(0)}}{\partial p} + \frac{3}{20} p^2 \frac{\partial^2 f^{(0)}}{\partial p^2} \Big]$$

$$+ \frac{1}{m_e^2} \frac{\partial}{\partial p} \Big[ p^4 \Big( T_e \frac{\partial f^{(0)}}{\partial p} + f^{(0)}(1 + f^{(0)}) \Big) \Big] \Big\}. \tag{5.38}$$

Notice that there is an internal hierarchy, with terms which do not depend on the baryon velocity $\mathbf{v}$, terms proportional to $\mathbf{v} \cdot \mathbf{n}$ and then to $(\mathbf{v} \cdot \mathbf{n})^2$, $v$ and $v^2$ (apart from the Kompaneets terms). In particular notice the term proportional to $\delta_e^{(1)} \mathbf{v} \cdot \mathbf{n}$ is the one corresponding to the Vishniac effect. We point out that we have kept all the terms up to second order in the collision term. In Refs. [30, 31] many terms coming from $c_{\Delta v}^{(2)}$ have been dropped mainly because these terms are proportional to the photon distribution function $f^{(1)}$ which on very small scales (those of interest for reionization) is suppressed by the diffusion damping. Here we want to be completely general and we have to keep them.

## 6. The Brightness equation

### 6.1. First order

The Boltzmann equation for photons is obtained by combining Eq. (4.24) with Eqs. (5.37)–(5.38). At first order the left-hand side reads

$$\frac{df}{d\eta} = \frac{df^{(1)}}{d\eta} - p \frac{\partial f^{(0)}}{\partial p} \frac{\partial \Phi^{(1)}}{\partial x^i} \frac{dx^i}{d\eta} + p \frac{\partial f^{(0)}}{\partial p} \frac{\partial \Psi^{(1)}}{\partial \eta}. \tag{6.1}$$

At first order it is useful to characterize the perturbations to the Bose-Einstein distribution function (4.22) in terms of a perturbation to the temperature as

$$f(x^i, p, n^i, \eta) = 2 \Big[ \exp \Big\{ \frac{p}{T(\eta)(1 + \Theta^{(1)})} \Big\} - 1 \Big]^{-1}. \tag{6.2}$$

Thus it turns out that

$$f^{(1)} = -p \frac{\partial f^{(0)}}{\partial p} \Theta^{(1)}, \tag{6.3}$$

where we have used the fact that $\partial f / \partial \Theta|_{\Theta=0} = -p \partial f^{(0)} / \partial p$. In terms of this variable $\Theta^{(1)}$ the linear collision term (5.37) will now become proportional to $-p \partial f^{(0)} / \partial p$ which contains the only explicit dependence on $p$, and the same happens for the left-hand side, Eq. (6.1). This is telling us that at first order $\Theta^{(1)}$ does not depend on $p$ but only on $x^i$, $n^i$, $\eta$, $\Theta^{(1)} = \Theta^{(1)}(x^i, n^i, \tau)$. This is well known and the physical reason is that at linear order there is no energy transfer in Compton collisions between photons and electrons. Therefore, the Boltzmann equation for $\Theta^{(1)}$ reads

$$
\frac{\partial \Theta^{(1)}}{\partial \eta} + n^i \frac{\partial \Theta^{(1)}}{\partial x^i} + \frac{\partial \Phi^{(1)}}{\partial x^i} n^i - \frac{\partial \Psi^{(1)}}{\partial \eta}
$$

$$
= n_e \sigma_T a \left[ \Theta_0^{(1)} + \frac{1}{2} \Theta_2^{(1)} P_2(\hat{\mathbf{v}} \cdot \mathbf{n}) - \Theta^{(1)} + \mathbf{v} \cdot \mathbf{n} \right],
\tag{6.4}
$$

where we made us of $f_\ell^{(1)} = -p \partial f^{(0)} / \partial p \Theta_\ell^{(1)}$, according to the decomposition of Eq. (5.14), and we have taken the zero-order expressions for $dx^i/d\eta$, dropping the contribution from $dn^i/d\eta$ in Eq. (4.8) since it is already first-order.

Notice that, since $\Theta^{(1)}$ is independent of $p$, it is equivalent to consider the quantity

$$
\Delta^{(1)}(x^i, n^i, \tau) = \frac{\int dp p^3 f^{(1)}}{\int dp p^3 f^{(0)}},
\tag{6.5}
$$

being $\Delta^{(1)} = 4\Theta^{(1)}$ at this order. The physical meaning of $\Delta^{(1)}$ is that of a fractional energy perturbation (in a given direction). From Eq. (4.24) another way to write an equation for $\Delta^{(1)}$ – the so-called brightness equation – is

$$
\frac{d}{d\eta} [\Delta^{(1)} + 4\Phi^{(1)}] - 4\frac{\partial}{\partial \eta}(\Phi^{(1)} + \Psi^{(1)})
$$

$$
= n_e \sigma_T a \left[ \Delta_0^{(1)} + \frac{1}{2} \Delta_2^{(1)} P_2(\hat{\mathbf{v}} \cdot \mathbf{n}) - \Delta^{(1)} + 4\mathbf{v} \cdot \mathbf{n} \right].
\tag{6.6}
$$

## 6.2. Second order

The previous results show that at linear order the photon distribution function has a Planck spectrum with the temperature that at any point depends on the photon direction. At second order one could characterize the perturbed photon distribution function in a similar way as in Eq. (6.2)

$$
f(x^i, p, n^i, \eta) = 2 \left[ \exp \left\{ \frac{p}{T(\eta) e^\Theta} - 1 \right\} \right]^{-1},
\tag{6.7}
$$

where by expanding $\Theta = \Theta^{(1)} + \Theta^{(2)}/2 + \ldots$ as usual one recovers the first-order expression. For example, in terms of $\Theta$, the perturbation of $f^{(1)}$ is given by Eq. (6.3), while at second order

$$\frac{f^{(2)}}{2} = -\frac{p}{2}\frac{\partial f^{(0)}}{\partial p}\Theta^{(2)} + \frac{1}{2}\left(p^2\frac{\partial^2 f^{(0)}}{\partial p^2} + p\frac{\partial f^{(0)}}{\partial p}\right)(\Theta^{(1)})^2. \tag{6.8}$$

However, as discussed in details in Refs. [30, 31], now the second-order perturbation $\Theta^{(2)}$ will not be momentum independent because the collision term in the equation for $\Theta^{(2)}$ does depend explicitly on $p$ (defining the combination $-(p\partial f^{(0)}/\partial p)^{-1}f^{(2)}$ does not lead to a second-order momentum independent equation as above). Such dependence is evident, for example, in the terms of $C^{(2)}(\mathbf{p})$, Eq. (5.38), proportional to $v$ or $v^2$, and in the Kompaneets terms. The physical reason is that at the non-linear level photons and electrons do exchange energy during Compton collisions. As a consequence spectral distortions are generated. For example, in the isotropic limit, only the Kompaneets terms survive giving rise to the Sunyaev-Zeldovich distortions. As discussed in Ref. [30], the Sunyaev-Zeldovich distortions can also be obtained with the correct coefficients by replacing the average over the direction electron $\langle v^2 \rangle$ with the mean squared thermal velocity $\langle v_{th}^2 \rangle = 3T_e/m_e$ in Eq. (5.38). This is due simply to the fact that the distinction between thermal and bulk velocity of the electrons is just for convenience. This fact also shows that spectral distortions due to the bulk flow (kinetic Sunyaev-Zeldovich) has the same form as the thermal effect. Thus spectral distortions can be in general described by a global Compton $y$-parameter (see Ref. [30] for a full discussion of spectral distortions). However in the following we will not be interested in the frequency dependence but only in the anisotropies of the radiation distribution. Therefore we can integrate over the momentum $p$ and define [30, 31]

$$\Delta^{(2)}(x^i, n^i, \tau) = \frac{\int dp\, p^3 f^{(2)}}{\int dp\, p^3 f^{(0)}}, \tag{6.9}$$

as in Eq. (6.5).

Integration over $p$ of Eqs. (4.24)–(5.38) is straightforward using the following relations

$$\int dp\, p^3 p\frac{\partial f^{(0)}}{\partial p} = -4\,N; \qquad \int dp\, p^3 p^2\frac{\partial^2 f^{(0)}}{\partial p^2} = 20\,N;$$

$$\int dp\, p^3 f^{(1)} = N\Delta^{(1)}; \qquad \int dp\, p^3 p\frac{\partial f^{(1)}}{\partial p} = -4\,N\Delta^{(1)}. \tag{6.10}$$

Here $N = \int dp\, p^3 f^{(0)}$ is the normalization factor (it is just proportional the background energy density of photons $\bar{\rho}_\gamma$). At first order one recovers Eq. (6.6).

At second order we find

$$
\frac{1}{2}\frac{d}{d\eta}\left[\Delta^{(2)}+4\Phi^{(2)}\right]+\frac{d}{d\eta}\left[\Delta^{(1)}+4\Phi^{(1)}\right]-4\Delta^{(1)}\left(\Psi^{(1)\prime}-\Phi^{(1)}_{,i}n^i\right)
$$
$$
-2\frac{\partial}{\partial\eta}\left(\Psi^{(2)}+\Phi^{(2)}\right)+4\frac{\partial\omega_i}{\partial\eta}n^i+2\frac{\partial\chi_{ij}}{\partial\eta}n^in^j
$$
$$
=-\frac{\tau'}{2}\left[\Delta^{(2)}_{00}-\Delta^{(2)}-\frac{1}{2}\sum_{m=-2}^{2}\frac{\sqrt{4\pi}}{5^{3/2}}\Delta^{(2)}_{2m}Y_{2m}(\mathbf{n})+2\left(\delta_e^{(1)}+\Phi^{(1)}\right)\right.
$$
$$
\left(\Delta^{(1)}_0+\frac{1}{2}\Delta^{(1)}_2 P_2(\hat{\mathbf{v}}\cdot\mathbf{n})-\Delta^{(1)}+4\mathbf{v}\cdot\mathbf{n}\right)+4\mathbf{v}^{(2)}\cdot\mathbf{n}
$$
$$
+2(\mathbf{v}\cdot\mathbf{n})\left[\Delta^{(1)}+3\Delta^{(1)}_0-\Delta^{(1)}_2\left(1-\frac{5}{2}P_2(\hat{\mathbf{v}}\cdot\mathbf{n})\right)\right]
$$
$$
\left.-v\Delta^{(1)}_1\left(4+2P_2(\hat{\mathbf{v}}\cdot\mathbf{n})\right)+14(\mathbf{v}\cdot\mathbf{n})^2-2v^2\right],\tag{6.11}
$$

where we have expanded the angular dependence of $\Delta$ as in Eq. (5.15)

$$
\Delta^{(i)}(\mathbf{x},\mathbf{n})=\sum_{\ell}\sum_{m=-\ell}^{\ell}\Delta^{(i)}_{\ell m}(\mathbf{x})(-i)^\ell\sqrt{\frac{4\pi}{2\ell+1}}Y_{\ell m}(\mathbf{n}),\tag{6.12}
$$

with

$$
\Delta^{(i)}_{\ell m}=(-i)^{-\ell}\sqrt{\frac{2\ell+1}{4\pi}}\int d\Omega\,\Delta^{(i)}Y^*_{\ell m}(\mathbf{n}),\tag{6.13}
$$

where we recall that the superscript stands by the order of the perturbation. At first order one can drop the dependence on $m$ setting $m=0$ so that $\Delta^{(1)}_{\ell m}=(-i)^{-\ell}(2\ell+1)\delta_{m0}\,\Delta^{(1)}_{\ell}$. In Eq. (6.11) we have introduced the differential optical depth

$$
\tau'=-\bar{n}_e\sigma_T a.\tag{6.14}
$$

It is understood that on the left-hand side of Eq. (6.11) one has to pick up for the total time derivatives only those terms which contribute to second order. Thus we have to take

$$
\frac{1}{2}\frac{d}{d\eta}\left[\Delta^{(2)}+4\Phi^{(2)}\right]+\frac{d}{d\eta}\left[\Delta^{(1)}+4\Phi^{(1)}\right]\Big|^{(2)}\tag{6.15}
$$
$$
=\frac{1}{2}\left(\frac{\partial}{\partial\eta}+n^i\frac{\partial}{\partial x^i}\right)\left(\Delta^{(2)}+4\Phi^{(2)}\right)+n^i\left(\Phi^{(1)}+\Psi^{(1)}\right)
$$
$$
\times\partial_i\left(\Delta^{(1)}+4\Phi^{(1)}\right)+\left[\left(\Phi^{(1)}_{,j}+\Psi^{(1)}_{,j}\right)n^in^j-\left(\Phi^{,i}+\Psi^{,i}\right)\right]\frac{\partial\Delta^{(1)}}{\partial n^i},
$$

where we used Eqs. (4.10) and (4.21). Notice that we can write $\partial \Delta^{(1)}/\partial n^i = (\partial \Delta^{(1)}/\partial x^i)(\partial x^i/\partial n^i) = (\partial \Delta^{(1)}/\partial x^i)(\eta - \eta_i)$, from the integration in time of Eq. (4.10) at zero-order when $n^i$ is constant in time.

### 6.3. Hierarchy equations for multipole moments

Let us now move to Fourier space. In the following, for a given wave-vector $\mathbf{k}$ we will choose the coordinate system such that $\mathbf{e}_3 = \hat{\mathbf{k}}$ and the polar angle of the photon momentum is $\vartheta$, with $\mu = \cos\vartheta = \hat{\mathbf{k}} \cdot \mathbf{n}$. Then Eq. (6.11) can be written as

$$\Delta^{(2)\prime} + ik\mu\Delta^{(2)} - \tau'\Delta^{(2)} = S(\mathbf{k}, \mathbf{n}, \eta), \tag{6.16}$$

where $S(\mathbf{k}, \mathbf{n}, \eta)$ can be easily read off Eq. (6.11). We now expand the temperature anisotropy in the multipole moments $\Delta_{\ell m}^{(2)}$ in order to obtain a system of coupled differential equations. By applying the angular integral of Eq. (6.13) to Eq. (6.16) we find

$$\Delta_{\ell m}^{(2)\prime}(\mathbf{k}, \eta) = k\left[\frac{\kappa_{\ell m}}{2\ell - 1}\Delta_{\ell-1,m}^{(2)} - \frac{\kappa_{\ell+1,m}}{2\ell + 3}\Delta_{\ell+1,m}^{(2)}\right] + \tau'\Delta_{\ell m}^{(2)} + S_{\ell m} \tag{6.17}$$

where the expansion coefficients of the source term are given by

$$
\begin{aligned}
S_{\ell m} =\ & \left(4\Psi^{(2)\prime} - \tau'\Delta_{00}^{(2)}\right)\delta_{\ell 0}\delta_{m 0} + 4k\Phi^{(2)}\delta_{\ell 1}\delta_{m 0} - 4\omega'_{\pm 1}\delta_{\ell 1} \\
& - 8\tau' v_m^{(2)}\delta_{\ell 1} - \frac{\tau'}{10}\Delta_{\ell m}^{(2)}\delta_{\ell 2} - 2\chi'_{\pm 2}\delta_{\ell 2} \\
& - 2\tau'\int \frac{d^3 k_1}{(2\pi)^3}\left[v_0^{(1)}(\mathbf{k}_1)v_0^{(1)}(\mathbf{k}_2)\hat{\mathbf{k}}_2 \cdot \hat{\mathbf{k}}_1 + \left(\delta_e^{(1)}(\mathbf{k}_1) + \Phi^{(1)}(\mathbf{k}_1)\right)\right. \\
& \left. \times \Delta_0^{(1)}(\mathbf{k}_2) - i\frac{2}{3}v(\mathbf{k}_1)\Delta_{10}^{(1)}(\mathbf{k}_2)\right]\delta_{\ell 0}\delta_{m 0} \\
& + 16k\int \frac{d^3 k_1}{(2\pi)^3}\left[\Phi^{(1)}(\mathbf{k}_1)\Phi^{(1)}(\mathbf{k}_2)\right]\delta_{\ell 1}\delta_{m 0} - 2\left[(\Psi^{(1)}\nabla\Phi^{(1)})_m\right. \\
& \left. + 8\tau'\left[(\delta_e^{(1)} + \Phi)\mathbf{v}\right]_m + 6\tau'\left(\Delta_0^{(1)}\mathbf{v}\right)_m - 2\tau'\left(\Delta_2^{(1)}\mathbf{v}\right)_m\right]\delta_{\ell 1} + \tau' \\
& \times \int \frac{d^3 k_1}{(2\pi)^3}\left[\left(\delta_e^{(1)}(\mathbf{k}_1) + \Phi^{(1)}(\mathbf{k}_1)\right)\Delta_2^{(1)}(\mathbf{k}_2) - i\frac{2}{3}v(\mathbf{k}_1)\Delta_{10}^{(1)}(\mathbf{k}_2)\right] \\
& \times \delta_{\ell 2}\delta_{m 0} + \int \frac{d^3 k_1}{(2\pi)^3}\left[8\Psi^{(1)}(\mathbf{k}_1) + 2\tau'\left(\delta_e^{(1)}(\mathbf{k}_1) + \Phi^{(1)}(\mathbf{k}_1)\right)\right. \\
& \left. - (\eta - \eta_i)(\Psi^{(1)} + \Phi^{(1)})(\mathbf{k}_1)\,\mathbf{k}_1 \cdot \mathbf{k}_2\right]\Delta_{\ell 0}^{(1)}(\mathbf{k}_2)\delta_{m 0} - i(-i)^{-\ell}
\end{aligned}
$$

$$\times (-1)^{-m}(2\ell+1)\sum_{\ell''}\sum_{m'=-1}^{1}(2\ell''+1)\left[8\Delta_{\ell''}^{(1)}\nabla\Phi^{(1)}-2(\Phi^{(1)}\right.$$

$$+\Psi^{(1)})\nabla\Delta_{\ell''}^{(1)}\big]_{m'}\begin{pmatrix}\ell'' & 1 & \ell \\ 0 & 0 & 0\end{pmatrix}\begin{pmatrix}\ell'' & 1 & \ell \\ 0 & m' & -m\end{pmatrix}$$

$$+\tau'i(-i)^{-\ell}(-1)^{-m}(2\ell+1)\sum_{\ell''}\sum_{m'=-1}^{1}(2\ell''+1)\left[2\Delta_{\ell''}^{(1)}\mathbf{v}\right.$$

$$+5\delta_{\ell''2}\,\Delta_2^{(1)}\mathbf{v}\big]_{m'}\begin{pmatrix}\ell'' & 1 & \ell \\ 0 & 0 & 0\end{pmatrix}\begin{pmatrix}\ell'' & 1 & \ell \\ 0 & m' & -m\end{pmatrix}$$

$$+14\tau'(-i)^{-\ell}(-1)^{-m}\sum_{m',m''=-1}^{1}\int\frac{d^3k_1}{(2\pi)^3}\left[v_0^{(1)}(\mathbf{k}_1)\frac{v_0^{(1)}(\mathbf{k}_2)}{k_2}\right.$$

$$\times\frac{4\pi}{3}Y_{1m'}^*(\hat{\mathbf{k}}_1)\left(kY_{1m''}^*(\hat{\mathbf{k}})-k_1Y_{1m''}^*(\hat{\mathbf{k}}_1)\right)\bigg]$$

$$\times\begin{pmatrix}1 & 1 & \ell \\ 0 & 0 & 0\end{pmatrix}\begin{pmatrix}1 & 1 & \ell \\ m' & m'' & -m\end{pmatrix}+2(\eta-\eta_i)(-i)^{-\ell}$$

$$\times(-1)^{-m}\sqrt{\frac{2\ell+1}{4\pi}}\sum_{L}\sum_{m',m''=-1}^{1}\int\frac{d^3k_1}{(2\pi)^3}\sqrt{\frac{4\pi}{2L+1}}\left(\frac{4\pi}{3}\right)^2$$

$$\times\Delta_L^{(1)}(\mathbf{k}_1)(\Phi^{(1)}+\Psi^{(1)})(\mathbf{k}_2)k_1Y_{1m'}^*(\hat{\mathbf{k}}_1)\left(kY_{1m''}^*(\hat{\mathbf{k}})\right.$$

$$-k_1Y_{1m''}^*(\hat{\mathbf{k}}_1))\int d\Omega Y_{1m'}(\mathbf{n})Y_{1m''}(\mathbf{n})Y_{L0}(\mathbf{n})Y_{\ell-m}(\mathbf{n}),\qquad(6.18)$$

where $\mathbf{k}_2=\mathbf{k}-\mathbf{k}_1$ and $k_2=|\mathbf{k}_2|$. In Eq. (6.18) it is understood that $|m|\le\ell$.

Let us explain the notations we have adopted in writing Eq. (6.18). The baryon velocity at linear order is irrotational, meaning that it is the gradient of a potential, and thus in Fourier space it is parallel to $\hat{\mathbf{k}}$, and following the same notation of Ref. [37], we write

$$\mathbf{v}^{(1)}(\mathbf{k})=-iv_0^{(1)}(\mathbf{k})\hat{\mathbf{k}}.\qquad(6.19)$$

The second-order velocity perturbation will contain a transverse (divergence-free) part whose components are orthogonal to $\hat{\mathbf{k}}=\mathbf{e}_3$, and we can write

$$\mathbf{v}^{(2)}(\mathbf{k})=-iv_0^{(2)}(\mathbf{k})\mathbf{e}_3+\sum_{m=\pm1}v_m^{(2)}\frac{\mathbf{e}_2\mp i\mathbf{e}_1}{\sqrt{2}},\qquad(6.20)$$

where $\mathbf{e}_i$ form an orthonormal basis with $\hat{\mathbf{k}}$. The second-order perturbation $\omega_i$ is decomposed in a similar way, with $\omega_{\pm1}$ the corresponding components (in this case in the Poisson gauge there is no scalar component). Similarly for the tensor perturbation $\chi_{ij}$ we have indicated its amplitudes as $\chi_{\pm 2}$ in the decomposition [36]

$$\chi_{ij} = \sum_{m=\pm 2} -\sqrt{\frac{3}{8}} \chi_m (\mathbf{e}_1 \pm i\mathbf{e}_2)_i (\mathbf{e}_1 \pm i\mathbf{e}_2)_j. \tag{6.21}$$

We have taken into account that in the gravitational part of the Boltzmann equation and in the collsion term there are some terms, like $\delta_e^{(1)} \mathbf{v}$, which still can be decomposed in the scalar and transverse parts in Fourier space as in Eq. (6.20). For a generic quantity $f(\mathbf{x})\mathbf{v}$ we have indicated the corresponding scalar and vortical components with $(f\mathbf{v})_m$ and their explicit expression is easily found by projecting the Fourier modes of $f(\mathbf{x})\mathbf{v}$ along the $\hat{\mathbf{k}} = \mathbf{e}_3$ and $(\mathbf{e}_2 \mp i\mathbf{e}_1)$ directions

$$(f\mathbf{v})_m(\mathbf{k}) = \int \frac{d^3 k_1}{(2\pi)^3} v_0^{(1)}(\mathbf{k}_1) f(\mathbf{k}_2) Y_{1m}^*(\hat{\mathbf{k}}_1) \sqrt{\frac{4\pi}{3}}. \tag{6.22}$$

Similarly for a term like $f(\mathbf{x})\nabla g(\mathbf{x})$ we used the notation

$$(f\nabla g)_m(\mathbf{k}) = -\int \frac{d^3 k_1}{(2\pi)^3} k_1 g(\mathbf{k}_1) f(\mathbf{k}_2) Y_{1m}^*(\hat{\mathbf{k}}_1) \sqrt{\frac{4\pi}{3}}. \tag{6.23}$$

Finally, the first term on the right-hand side of Eq. (6.17) has been obtained by using the relation

$$i\mathbf{k} \cdot \mathbf{n} \Delta^{(2)}(\mathbf{k}) = \sum_{\ell m} \Delta_{\ell m}^{(2)}(\mathbf{k}) \frac{k}{2\ell+1} \left[ \kappa_{\ell m} \tilde{G}_{\ell-1,m} - \kappa_{\ell+1,m} \tilde{G}_{\ell+1,m} \right]$$

$$= k \sum_{\ell m} \left[ \frac{\kappa_{\ell m}}{2\ell-1} \Delta_{\ell-1,m}^{(2)} - \frac{\kappa_{\ell m}}{2\ell+3} \Delta_{\ell+1,m}^{(2)} \right] \tilde{G}_{\ell m}, \tag{6.24}$$

where $\tilde{G}_{\ell m} = (-i)^\ell \sqrt{4\pi/(2\ell+1)} Y_{\ell m}(\mathbf{n})$ is the angular mode for the decomposition (6.12) and $\kappa_{\ell m} = \sqrt{l^2 - m^2}$. This relation has been discussed in Refs. [34, 36] and corresponds to the term $n^i \partial \Delta^{(2)}/\partial x^i$ in Eq. (6.11).

As expected, at second order we recover some intrinsic effects which are characteristic of the linear regime. In Eq. (6.17) the relation (6.24) represents the free streaming effect: when the radiation undergoes free-streaming, the inhomogeneities of the photon distribution are seen by the observer as angular anisotropies. At first order it is responsible for the hierarchy of Boltzmann equations coupling the different $\ell$ modes, and it represents a projection effect of fluctuations on a scale $k$ onto the angular scale $\ell \sim k\eta$. The term $\tau' \Delta_{\ell m}^{(2)}$ causes an exponential suppression of anisotropies in the absence of the source term $S_{\ell m}$. The first line of the source term (6.18) just reproduces the expression of the first order case. Of course the dynamics of the second-order metric and baryon-velocity perturbations which appear will be different and governed by the second-order Einstein equations and continuity equations. The remaining terms in the source are second-order effects generated as non-linear combinations of the primordial (first-order) perturbations. We have ordered them according to the increasing number of $\ell$ modes they contribute to. Notice in particular that they involve the first-order anisotropies $\Delta_\ell^{(1)}$ and as a consequence such terms contribute to generate the hierarchy of equations (apart from the free-streaming effect). The source term contains additional scattering processes and gravitational effects. On large scales (above the horizon at recombination) we can say that the main effects are due to gravity, and they include the Sachs-Wolfe and the (late and early) Sachs-Wolfe effect due to the redshift photons suffer when travelling through the second-order gravitational potentials. These, toghether with the contribution due to the second-order tensor modes, have been already studied in details in Ref. [20]. Another important gravitational effect is that of lensing of photons as they travel from the last scattering surface to us. A contribution of this type is given by the last term of Eq. (6.18).

### 6.4. Integral solution of the second-order Boltzmann equation

As in linear theory, one can derive an integral solution of the Boltzmann equation (6.11) in terms of the source term $S$. Following the standard procedure (see *e.g.* Ref. [28, 35]) for linear perturbations, we write the left-hand side as $\Delta^{(2)\prime} + ik\mu\Delta^{(2)} - \tau'\Delta^{(2)} = e^{-ik\mu\eta+\tau}d[\Delta^{(2)}e^{ik\mu\eta-\tau}]/d\eta$ in order to derive the integral solution

$$\Delta^{(2)}(\mathbf{k}, \mathbf{n}, \eta_0) = \int_0^{\eta_0} d\eta\, S(\mathbf{k}, \mathbf{n}, \eta)e^{ik\mu(\eta-\eta_0)-\tau}, \tag{6.25}$$

where $\eta_0$ stands by the present time. The expression of the photon moments $\Delta_{\ell m}^{(2)}$ can be obtained as usual from Eq. (6.13). In the previous section we have already found the coefficients for the decomposition of source term $S$

$$S(\mathbf{k}, \mathbf{n}, \eta) = \sum_\ell \sum_{m=-\ell}^\ell S_{\ell m}(\mathbf{k}, \eta)(-i)^\ell \sqrt{\frac{4\pi}{2\ell+1}} Y_{\ell m}(\mathbf{n}). \tag{6.26}$$

In Eq. (6.25) there is an additional angular dependence in the exponential. It is easy to take it into account by recalling that

$$e^{i\mathbf{k}\cdot\mathbf{x}} = \sum_\ell (i)^\ell (2\ell+1) j_\ell(kx) P_\ell(\hat{\mathbf{k}} \cdot \hat{\mathbf{x}}). \tag{6.27}$$

Thus the angular integral (6.13) is computed by using the decomposition of the source term (6.26) and Eq. (6.27)

$$
\Delta_{\ell m}^{(2)}(\mathbf{k}, \eta_0) = (-1)^{-m}(-i)^{-\ell}(2\ell + 1) \int_0^{\eta_0} d\eta \, e^{-\tau(\eta)}
$$

$$
\times \sum_{\ell_2} \sum_{m_2=-\ell_2}^{\ell_2} (-i)^{\ell_2} S_{\ell_2 m_2} \sum_{\ell_1} i^{\ell_1} j_{\ell_1}(k(\eta - \eta_0))
$$

$$
\times (2\ell_1 + 1) \begin{pmatrix} \ell_1 & \ell_2 & \ell \\ 0 & 0 & 0 \end{pmatrix} \begin{pmatrix} \ell_1 & \ell_2 & \ell \\ 0 & m_2 & -m \end{pmatrix}, \quad (6.28)
$$

where the Wigner $3 - j$ symbols appear because of the Gaunt integrals

$$
\mathcal{G}_{l_1 l_2 l_3}^{m_1 m_2 m_3} \equiv \int d^2 \hat{\mathbf{n}} Y_{l_1 m_1}(\hat{\mathbf{n}}) Y_{l_2 m_2}(\hat{\mathbf{n}}) Y_{l_3 m_3}(\hat{\mathbf{n}})
$$

$$
= \sqrt{\frac{(2l_1 + 1)(2l_2 + 1)(2l_3 + 1)}{4\pi}}
$$

$$
\times \begin{pmatrix} \ell_1 & \ell_2 & \ell_3 \\ 0 & 0 & 0 \end{pmatrix} \begin{pmatrix} \ell_1 & \ell_2 & \ell_3 \\ m_1 & m_2 & m_3 \end{pmatrix}, \quad (6.29)
$$

Since the second of the Wigner 3-$j$ symbols in Eq. (6.28) is nonzero only if $m = m_2$, our solution can be rewritten to recover the corresponding expression found for linear anisotropies in Refs. [34, 36]

$$
\frac{\Delta_{\ell m}^{(2)}(\mathbf{k}, \eta_0)}{2\ell + 1} = \int_0^{\eta_0} d\eta \, e^{-\tau(\eta)} \sum_{\ell_2} S_{\ell_2 m} \, j_\ell^{(l_2, m)}[k(\eta_0 - \eta)], \quad (6.30)
$$

where $j_\ell^{(l_2, m)}[k(\eta_0 - \eta)]$ are the so called radial functions. Of course the main information at second order is included in the source term containing different effects due to the non-linearity of the perturbations. In the total angular momentum method of Refs. [34, 36] Eq. (6.30) is interpreted just as the intergration over the radial coordinate ($\chi = \eta_0 - \eta$) of the projected source term. Another important comment is that, as in linear theory, the integral solution (6.28) is in fact just a formal solution, since the source term $S$ contains itself the second-order photon moments up to $l = 2$ (see Eq. (6.18)). This means that one has anyway to resort to the hierarchy equations for photons, Eq. (6.17), to solve for these moments. Nevertheless, as in linear theory [35], one expects to need just a few moments beyond $\ell = 2$ in the hierarchy equations, and once the moments entering in the source function are computed the higher moments are obtained from the integral solution. Thus the integral solution should in fact be more advantageous than solving the system of coupled equations (6.17).

## 7. The Boltzmann equation for baryons and cold dark matter

In this section we will derive the Boltzmann equation for massive particles, which is the case of interest for baryons and dark matter. These equations are necessary to find the time evolution of number densities and velocities of the baryon fluid which appear in the brightness equation, thus allowing to close the system of equations. Let us start from the baryon component. Electrons are tightly coupled to protons via Coulomb interactions. This forces the relative energy density contrasts and the velocities to a common value, $\delta_e = \delta_p \equiv \delta_b$ and $\mathbf{v}_e = \mathbf{v}_p \equiv \mathbf{v}$, so that we can identify electrons and protons collectively as "baryonic" matter.

To derive the Boltzmann equation for baryons let us first focus on the collisionless equation and compute therefore $dg/d\eta$, where $g$ is the distribution function for a massive species with mass $m$. One of the differences with respect to photons is just that baryons are non-relativistic for the epochs of interest. Thus the first step is to generalize the formulae in Section 4 up to Eq. (4.21) to the case of a massive particle. In this case one enforces the constraint $Q^2 = g_{\mu\nu} Q^\mu Q^\nu = -m^2$ and it also useful to use the particle energy $E = \sqrt{q^2 + m^2}$, where $q$ is defined as in Eq. (4.3). Moreover in this case it is very convenient to take the distribution function as a function of the variables $q^i = qn^i$, the position $x^i$ and time $\eta$, without using the explicit splitting into the magnitude of the momentum $q$ (or the energy E) and its direction $n^i$. Thus the total time derivative of the distribution functions reads

$$\frac{dg}{d\eta} = \frac{\partial g}{\partial \eta} + \frac{\partial g}{\partial x^i}\frac{dx^i}{d\eta} + \frac{\partial g}{\partial q^i}\frac{dq^i}{d\eta}. \tag{7.1}$$

We will not give the details of the calculation since we just need to replicate the same computation we did for the photons. For the four-momentum of the particle notice that $Q^i$ has the same form as Eq. (4.6), while for $Q^0$ we find

$$Q^0 = \frac{e^{-\Phi}}{a} E\left(1 + \omega_i \frac{q^i}{E}\right). \tag{7.2}$$

In the following we give the expressions for $dx^i/d\eta$ and $dq^i/d\eta$.

a) As in Eq. (4.10) $dx^i/d\eta = Q^i/Q^0$ and it turns out to be

$$\frac{dx^i}{d\eta} = \frac{q}{E}n^i e^{\Phi+\Psi}\left(1 - \omega_i n^i \frac{q}{E}\right)\left(1 - \frac{1}{2}\chi_{km}n^k n^m\right). \tag{7.3}$$

b) For $dq^i/d\eta$ we need the expression of $Q^i$ which is the same as that of Eq. (4.6)

$$Q^i = \frac{q^i}{a}e^\Psi\left(1 - \frac{1}{2}\chi_{km}n^k n^m\right). \tag{7.4}$$

The spatial component of the geodesic equation, up to second order, reads

$$
\frac{dQ^i}{d\eta} = -2(\mathcal{H} - \Psi')\left(1 - \frac{1}{2}\chi_{km}n^k n^m\right)\frac{q}{a}n^i e^{\Psi} + e^{\Phi + 2\Psi}
$$

$$
\times \left(\frac{\partial\Psi}{\partial x^k}\frac{q^2}{aE}(2n^i n^k - \delta^{ik}) - \frac{\partial\Phi}{\partial x^i}\frac{E}{a}\right) - \frac{E}{a}\big[\omega^{i\prime} + \mathcal{H}\omega^i + q^k(\chi_k^{i\prime}
$$

$$
+ \omega_k^i - \omega_k^{i\prime})\big] + \left[\mathcal{H}\omega^i\delta_{jk} - \frac{1}{2}(\chi^i_{j,k} + \chi^i_{k,j} - \chi_{jk}^{\cdot i})\right]\frac{q^j q^k}{Ea}. \quad (7.5)
$$

Proceeding as in the massless case we now take the total time derivative of Eq. (7.4) and using Eq. (7.5) we find

$$
\frac{dq^i}{d\eta} = -(\mathcal{H} - \Psi')q^i + \Psi_{,k}\frac{q^i q^k}{E}e^{\Phi + \Psi} - \Phi^{,i}Ee^{\Phi + \Psi} - \Psi_{,i}\frac{q^2}{E}e^{\Phi + \Psi}
$$

$$
- E(\omega^{i\prime} + \mathcal{H}\omega^i) - (\chi_k^{i\prime} + \omega_k^i - \omega_k^{i\prime})Eq^k
$$

$$
+ \left[\mathcal{H}\omega^i\delta_{jk} - \frac{1}{2}(\chi^i_{j,k} + \chi^i_{k,j} - \chi_{jk}^{\cdot i})\right]\frac{q^j q^k}{E}. \quad (7.6)
$$

We can now write the total time derivative of the distribution function as

$$
\frac{dg}{d\eta} = \frac{\partial g}{\partial\eta} + \frac{q}{E}n^i e^{\Phi + \Psi}\left(1 - \omega_i n^i - \frac{1}{2}\chi_{km}n^k n^m\right)\frac{\partial g}{\partial x^i}
$$

$$
+ \Big[-(\mathcal{H} - \Psi')q^i + \Psi_{,k}\frac{q^i q^k}{E}e^{\Phi + \Psi} - \Phi^{,i}Ee^{\Phi + \Psi} - \Psi_{,i}\frac{q^2}{E}e^{\Phi + \Psi}
$$

$$
- E(\omega^{i\prime} + \mathcal{H}\omega^i) - (\chi_k^{i\prime} + \omega_k^i - \omega_k^{i\prime})Eq^k
$$

$$
+ \left(\mathcal{H}\omega^i\delta_{jk} - \frac{1}{2}(\chi^i_{j,k} + \chi^i_{k,j} - \chi_{jk}^{\cdot i})\right)\frac{q^j q^k}{E}\Big]\frac{\partial g}{\partial q^i}. \quad (7.7)
$$

This equation is completely general since we have just solved for the kinematics of massive particles. As far as the collision terms are concerned, for the system of electrons and protons we consider the Coulomb scattering processes between the electrons and the protons and the Compton scatterings between the photons and the electrons

$$
\frac{dg_e}{d\eta}(\mathbf{x}, \mathbf{q}, \eta) = \langle c_{ep}\rangle_{QQ'q'} + \langle c_{e\gamma}\rangle_{pp'q'} \quad (7.8)
$$

$$
\frac{dg_p}{d\eta}(\mathbf{x}, \mathbf{Q}, \eta) = \langle c_{ep}\rangle_{qq'Q'}, \quad (7.9)
$$

where we have adopted the same formalism of Ref. [28] with **p** and **p'** the initial and final momenta of the photons, **q** and **q'** the corresponding quantities for the electrons and for protons $\mathbf{Q}$ and $\mathbf{Q}'$. The integral over different momenta is indicated by

$$\langle \cdots \rangle_{pp'q'} \equiv \int \frac{d^3p}{(2\pi)^3} \int \frac{d^3p'}{(2\pi)^3} \int \frac{d^3q'}{(2\pi)^3} \cdots, \tag{7.10}$$

and thus one can read $c_{ey}$ as the unintegrated part of Eq. (5.2), and similarly for $c_{ep}$ (with the appropriate amplitude $|M|^2$). In Eq. (7.8) Compton scatterings between protons and photons can be safely neglected because the amplitude of this process has a much smaller amplitude than Compton scatterings with electrons being weighted by the inverse squared mass of the particles.

At this point for the photons we considered the perturbations around the zero-order Bose-Einstein distribution function (which are the unknown quantities). For the electrons (and protons) we can take the thermal distribution described by Eq. (5.4). Moreover we will take the moments of Eqs. (7.8)–(7.9) in order to find the energy-momentum continuity equations.

## 7.1. Energy continuity equations

We now integrate Eq. (7.7) over $d^3q/(2\pi)^3$. Let us recall that in terms of the distribution function the number density $n_e$ and the bulk velocity $\mathbf{v}$ are given by

$$n_e = \int \frac{d^3q}{(2\pi)^3} g, \tag{7.11}$$

and

$$v^i = \frac{1}{n_e} \int \frac{d^3q}{(2\pi)^3} g \frac{qn^i}{E}, \tag{7.12}$$

where one can set $E \simeq m_e$ since we are considering non-relativistic particles. We will also make use of the following relations when integrating over the solid angle $d\Omega$

$$\int d\Omega \, n^i = \int d\Omega \, n^i n^j n^k = 0, \qquad \int \frac{d\Omega}{4\pi} n^i n^j = \frac{1}{3} \delta^{ij}. \tag{7.13}$$

Finally notice that $dE/dq = q/E$ and $\partial g/\partial q = (q/E)\partial g/\partial E$.

Thus the first two integrals just brings $n'_e$ and $(n_e v^i)_{,i}$. Notice that all the terms proportional to the second-order vector and tensor perturbations of the metric give a vanishing contribution at second order since in this case we can take the zero-order distribution functions which depends only on $\eta$ and $E$, integrate over the direction and use the fact that $\delta^{ij} \chi_{ij} = 0$. The trick to solve the remaining integrals is an integration by parts over $q^i$. We have an integral like (the one multiplying $(\Psi' - \mathcal{H})$)

$$\int \frac{d^3q}{(2\pi)^3} q^i \frac{\partial g}{\partial q^i} = -3 \int \frac{d^3q}{(2\pi)^3} g = -3n_e, \qquad (7.14)$$

after an integration by parts over $q^i$. The remaining integrals can be solved still by integrating by parts over $q^i$. The integral proportional to $\Phi^{\cdot i}$ in Eq. (7.7) gives

$$\int \frac{d^3q}{(2\pi)^3} E = -v_i n_e, \qquad (7.15)$$

where we have used the fact that $dE/dq^i = q^i/E$. For the integral

$$\int \frac{d^3q}{(2\pi)^3} \frac{q^i q^k}{E} \frac{\partial g}{\partial q^i}, \qquad (7.16)$$

the integration by parts brings two pieces, one from the derivation of $q^i q^k$ and one from the derivation of the energy $E$

$$-4 \int \frac{d^3q}{(2\pi)^3} g \frac{q^k}{E} + \int \frac{d^3q}{(2\pi)^3} g \frac{q^2}{E} \frac{q^k}{E} = -4v^k n_e + \int \frac{d^3q}{(2\pi)^3} g \frac{q^2}{E^2} \frac{q^k}{E}. \qquad (7.17)$$

The last integral in Eq. (7.17) can indeed be neglected. To check this one makes use of the explicit expression (5.4) for the distribution function $g$ to derive

$$\frac{\partial g}{\partial v^i} = g \frac{q_i}{T_e} - \frac{m_e}{T_e} v_i g, \qquad (7.18)$$

and

$$\int \frac{d^3q}{(2\pi)^3} g q^i q^j = \delta^{ij} n_e m_e T_e + n_e m_e^2 v^i v^j. \qquad (7.19)$$

Thus it is easy to compute

$$\frac{\Psi_{,k}}{m_e^3} \int \frac{d^3q}{(2\pi)^3} g q^2 q^k = -\Psi_{,k} v^2 \frac{T_e}{m_e} + 3\Psi_{,k} v_k n_e \frac{T_e}{m_e} + \Psi_{,k} v_k v^2, \qquad (7.20)$$

which is negligible taking into account that $T_e/m_e$ is of the order of the thermal velocity squared.

With these results we are now able to compute the left-hand side of the Boltzmann equation (7.8) integrated over $d^3q/(2\pi)^3$. The same operation must be done for the collision terms on the right hand side. For example for the first of the equations in (7.8) this brings to the integrals $\langle c_{ep} \rangle_{QQ'qq'} + \langle c_{e\gamma} \rangle_{pp'qq'}$.

However looking at Eq. (5.3) one realizes that $\langle c_{e\gamma}\rangle_{pp'qq'}$ vanishes because the integrand is antisymmetric under the change $\mathbf{q} \leftrightarrow \mathbf{q}'$ and $\mathbf{p} \leftrightarrow \mathbf{p}'$. In fact this is simply a consequence of the fact that the electron number is conserved for this process. The same argument holds for the other term $\langle c_{ep}\rangle_{QQ'qq'}$. Therefore the right-hand side of Eq. (7.8) integrated over $d^3q/(2\pi)^3$ vanishes and we can give the evolution equation for $n_e$. Collecting the results of Eq. (7.14) to (7.20) we find

$$\frac{\partial n_e}{\partial \eta} + e^{\Phi+\Psi}\frac{\partial(v^i n_e)}{\partial x^i} + 3(\mathcal{H} - \Psi')n_e + e^{\Phi+\Psi}v^k n_e(\Phi_{,k} - 2\Psi_{,k}) = 0. \quad (7.21)$$

Similarly, for CDM particles, we find

$$\frac{\partial n_{CDM}}{\partial \eta} + e^{\Phi+\Psi}\frac{\partial(v^i n_{CDM})}{\partial x^i} + 3(\mathcal{H} - \Psi')n_{CDM}$$
$$+ e^{\Phi+\Psi}v^k_{CDM} n_{CDM}(\Phi_{,k} - 2\Psi_{,k}) = 0. \quad (7.22)$$

## 7.2. Momentum continuity equations

Let us now multiply Eq. (7.7) by $(q^i/E)/(2\pi)^3$ and integrate over $d^3q$. In this way we will find the continuity equation for the momentum of baryons. The first term just gives $(n_e v^i)'$. The second integral is of the type

$$\frac{\partial}{\partial x^j}\int \frac{d^3q}{(2\pi)^3}g\frac{qn^j}{E}\frac{qn^i}{E} = \frac{\partial}{\partial x^j}\left(n_e\frac{T_e}{m_e}\delta^{ij} + n_e v^i v^j\right), \quad (7.23)$$

where we have used Eq. (7.19) and $E = m_e$. The third term proportional to $(\mathcal{H} - \Psi')$ is

$$\int \frac{d^3q}{(2\pi)^3}q^k\frac{\partial g}{\partial q_k}\frac{q^i}{E} = 4n_e + \int \frac{d^3q}{(2\pi)^3}g\frac{q^2}{E^2}\frac{q^i}{E}, \quad (7.24)$$

where we have integrated by parts over $q^i$. Notice that the last term in Eq. (7.24) is negligible being the same integral we discussewd above in Eq. (7.20). By the same arguments that lead to neglect the term of Eq. (7.20) it is easy to check that all the remaining integrals proportional to the gravitational potentials are negligible except for

$$-e^{\Phi+\Psi}\Phi_{,k}\int \frac{d^3q}{(2\pi)^3}\frac{\partial g}{\partial q_k}q^i = n_e e^{\Phi+\Psi}\Phi^{,i}. \quad (7.25)$$

The integrals proportional to the second-order vector and tensor perturbations vanish as vector and tensor perturbations are traceless and divergence-free. The only one which survives is the term proportional to $\omega^{i\prime} + \mathcal{H}\omega^i$ in Eq. (7.7).

Therefore for the integral over $d^3qq^i/E$ of the left-hand side of the Boltz-mann equation (7.7) for a massive particle with mass $m_e$ ($m_p$) and distribution function (5.4) we find

$$
\int \frac{d^3q}{(2\pi)^3} \frac{q^i}{E} \frac{dg_e}{d\eta} = \frac{\partial(n_e v^i)}{\partial\eta} + 4(\mathcal{H} - \Psi')n_e v^i + \Phi^{\cdot i} e^{\Phi+\Psi} n_e
$$

$$
+ e^{\Phi+\Psi} \left(n_e \frac{T_e}{m_e}\right)^{\cdot i} + e^{\Phi+\Psi} \frac{\partial}{\partial x^j}(n_e v^j v^i) + \frac{\partial\omega^i}{\partial\eta} n_e + \mathcal{H}\omega^i n_e. \qquad (7.26)
$$

Now, in order to derive the momentum conservation equation for baryons, we take the first moment of both Eq. (7.8) and (7.9) multiplying them by $\mathbf{q}$ and $\mathbf{Q}$ respectively and integrating over the momenta. Since previously we integrated the left-hand side of these equations over $d^3qq^i/E$, we just need to multiply the previous integrals by $m_e$ for the electrons and for the protons by $m_p$ for the protons. Therefore if we sum the first moment of Eqs. (7.8) and (7.9) the dominant contribution on the left-hand side will be that of the protons

$$
\int \frac{d^3Q}{(2\pi)^3} Q^i \frac{dg_p}{d\eta} = \langle c_{ep}(q^i + Q^i) \rangle_{QQ'qq'} + \langle c_{e\gamma} q^i \rangle_{pp'qq'}. \qquad (7.27)
$$

Notice that the integral of the Coulomb collision term $c_{ep}(q^i + Q^i)$ over all momenta vanishes simply because of momentum conservation (due to the Dirac function $\delta^4(q + Q - q' - Q')$). As far as the Compton scattering is concerned we have that, following Ref. [28],

$$
\langle c_{e\gamma} q^i \rangle_{pp'qq'} = -\langle c_{e\gamma} p^i \rangle_{pp'qq'}, \qquad (7.28)
$$

still because of the total momentum conservation. Therefore what we can com-pute now is the integral over all momenta of $c_{e\gamma} p^i$. Notice however that this is equivalent just to multiply the Compton collision term $C(\mathbf{p})$ of Eq. (5.3) by $p^i$ and integrate over $d^3p/(2\pi^3)$

$$
\langle c_{e\gamma} p^i \rangle_{pp'qq'} = ae^\Phi \int \frac{d^3p}{(2\pi)^3} p^i C(\mathbf{p}). \qquad (7.29)
$$

where $C(\mathbf{p})$ has been already computed in Eqs. (5.37) and (5.38).

We will do the integral (7.29) in the following. First let us introduce the defi-nition of the velocity of photons in terms of the distribution function

$$
(\rho_\gamma + p_\gamma)v_\gamma^i = \int \frac{d^3p}{(2\pi)^3} f p^i, \qquad (7.30)
$$

where $p_\gamma = \rho_\gamma/3$ is the photon pressure and $\rho_\gamma$ the energy density. At first order we get

$$\frac{4}{3} v_\gamma^{(1)i} = \int \frac{d\Omega}{4\pi} \Delta^{(1)} n^i,$$  (7.31)

where $\Delta$ is the photon distribution anisotropies defined in Eq. (6.9). At second order we instead find

$$\frac{4}{3} \frac{v_\gamma^{(2)i}}{2} = \frac{1}{2} \int \frac{d\Omega}{4\pi} \Delta^{(2)} n^i - \frac{4}{3} \delta_\gamma^{(1)} v_\gamma^{(1)i}.$$  (7.32)

Therefore the terms in Eqs. (5.37) and (5.38) proportional to $f^{(1)}(\mathbf{p})$ and $f^{(2)}(\mathbf{p})$ will give rise to terms containing the velocity of the photons. On the other hand the terms proportional to $f_0^{(1)}(p)$ and $f_{00}^{(2)}(p)$, once integrated, vanish because of the integral over the momentum direction $n^i$, $\int d\Omega n^i = 0$. Also the integrals involving $P_2(\hat{\mathbf{v}} \cdot \mathbf{n}) = [3(\hat{\mathbf{v}} \cdot \mathbf{n})^2 - 1]/2$ in the first line of Eq. (5.37) and (5.38) vanish since

$$\int d\Omega P_2(\hat{\mathbf{v}} \cdot \mathbf{n}) n^i = \hat{v}^k \hat{v}^j \int d\Omega n_k n_j n^i = 0,$$  (7.33)

where we are using the relations (7.13). Similarly all the terms proportional to $v$, $(\mathbf{v} \cdot \mathbf{n})^2$ and $v^2$ do not give any contribution to Eq. (7.29) and, in the second-order collision term, one can check that $\int d\Omega Y_2(\mathbf{n}) n^i = 0$. Then there are terms proportional to $(\mathbf{v} \cdot \mathbf{n}) f^{(0)}(p)$, $(\mathbf{v} \cdot \mathbf{n}) p \partial f^{(0)}/\partial p$ and $(\mathbf{v} \cdot \mathbf{n}) p \partial f_0^{(1)}/\partial p$ for which we can use the rules (6.10) when integrating over $p$ while the integration over the momentum direction is

$$\int \frac{d\Omega}{4\pi} (\mathbf{v} \cdot \mathbf{n}) n^i = v_k \int \frac{d\Omega}{4\pi} n^k n^i = \frac{1}{3} v^i.$$  (7.34)

Finally from the second line of Eq. (5.38) we get three integrals. One is

$$\int \frac{d^3 p}{(2\pi)^3} p^i (\mathbf{v} \cdot \mathbf{n}) f^{(1)}(\mathbf{p}) = \bar{\rho}_\gamma \int \frac{d\Omega}{4\pi} \Delta^{(1)} (\mathbf{v} \cdot \mathbf{n}) n^i,$$  (7.35)

where $\bar{\rho}_\gamma$ is the background energy density of the photons. The second comes from

$$\frac{1}{2} \int \frac{d^3 p}{(2\pi)^3} p^i (\mathbf{v} \cdot \mathbf{n}) P_2(\hat{\mathbf{v}} \cdot \mathbf{n}) \left( f_2^{(1)}(p) - p \frac{\partial f_2^{(1)}(p)}{\partial p} \right)$$

$$= \frac{5}{4} \bar{\rho}_\gamma \Delta_2^{(1)} \left[ 3 v_j \hat{v}_k \hat{v}_l \int \frac{d\Omega}{4\pi} n^i n^j n^k n^l - v_j \int \frac{d\Omega}{4\pi} n^i n^j \right]$$

$$= \frac{1}{3} \bar{\rho}_\gamma \Delta_2^{(1)} \hat{v}^i,$$  (7.36)

where we have used the rules (6.10), Eq. (7.13) and $\int (d\Omega/4\pi) n^i n^j n^k n^l = (\delta^{ij}\delta^{kl} + \delta^{ik}\delta^{lj} + \delta^{il}\delta^{jk})/15$. In fact the third integral

$$-\int \frac{d^3 p}{(2\pi)^3} p^i (\mathbf{v} \cdot \mathbf{n}) f_2^{(1)}(p), \qquad (7.37)$$

exactly cancels the previous one. Summing the various integrals we find

$$\int \frac{d\mathbf{p}}{(2\pi)^3} C(\mathbf{p})\mathbf{p} = n_e \sigma_T \bar{\rho}_\gamma \left[ \frac{4}{3}(\mathbf{v}^{(1)} - \mathbf{v}_\gamma^{(1)}) - \int \frac{d\Omega}{4\pi} \frac{\Delta^{(2)}}{2} \mathbf{n} + \frac{4}{3} \frac{\mathbf{v}^{(2)}}{2} \right.$$
$$\left. + \frac{4}{3}\delta_e^{(1)}(\mathbf{v}^{(1)} - \mathbf{v}_\gamma^{(1)}) + \int \frac{d\Omega}{4\pi} \Delta^{(1)}(\mathbf{v} \cdot \mathbf{n})\mathbf{n} + \Delta_0^{(1)}\mathbf{v} \right]. \qquad (7.38)$$

Eq. (7.38) can be further simplified. Recalling that $\delta_\gamma^{(1)} = \Delta_0^{(1)}$ we use Eq. (7.32) and notice that

$$\int \frac{d\Omega}{4\pi} \Delta^{(1)} (\mathbf{v} \cdot \mathbf{n})n^i = v_j^{(1)}\Pi_\gamma^{ji} + \frac{1}{3}v^i \Delta_0^{(1)}, \qquad (7.39)$$

where the photon quadrupole $\Pi_\gamma^{ij}$ is defined as

$$\Pi_\gamma^{ij} = \int \frac{d\Omega}{4\pi} \left( n^i n^j - \frac{1}{3}\delta^{ij} \right) \left( \Delta^{(1)} + \frac{\Delta^{(2)}}{2} \right). \qquad (7.40)$$

Thus, our final expression for the integrated collision term (7.29) reads

$$\int \frac{d^3 p}{(2\pi)^3} C(\mathbf{p})p^i = n_e \sigma_T \bar{\rho}_\gamma \left[ \frac{4}{3}(v^{(1)i} - v_\gamma^{(1)i}) + \frac{4}{3}\left( \frac{v^{(2)i}}{2} - \frac{v_\gamma^{(2)i}}{2} \right) \right.$$
$$\left. + \frac{4}{3}(\delta_e^{(1)} + \Delta_0^{(1)})(v^{(1)i} - v_\gamma^{(1)i}) + v_j^{(1)}\Pi_\gamma^{ji} \right]. \qquad (7.41)$$

We are now able to give the momentum continuity equation for baryons by combining $m_p dg_p/d\eta$ from Eq. (7.26) with the collision term (7.29)

$$\frac{\partial(\rho_b v^i)}{\partial \eta} + 4(\mathcal{H} - \Psi')\rho_b v^i + \Phi^{,i} e^{\Phi+\Psi}\rho_b + e^{\Phi+\Psi}\left( \rho_b \frac{T_b}{m_p} \right)^{,i}$$
$$+ e^{\Phi+\Psi}\frac{\partial}{\partial x^j}(\rho_b v^j v^i) + \frac{\partial \omega^i}{\partial \eta}\rho_b + \mathcal{H}\omega^i \rho_b$$
$$= -n_e \sigma_T a \bar{\rho}_\gamma \left[ \frac{4}{3}(v^{(1)i} - v_\gamma^{(1)i}) + \frac{4}{3}\left( \frac{v^{(2)i}}{2} - \frac{v_\gamma^{(2)i}}{2} \right) \right.$$
$$\left. + \frac{4}{3}(\delta_b^{(1)} + \Delta_0^{(1)} + \Phi^{(1)})(v^{(1)i} - v_\gamma^{(1)i}) + v_j^{(1)}\Pi_\gamma^{ji} \right], \qquad (7.42)$$

where $\rho_b$ is the baryon energy density and, as we previously explained, we took into account that to a good approximation the electrons do not contribute to the mass of baryons. In the following we will expand explicitly at first and second-order Eq. (7.42).

### 7.2.1. First-order momentum continuity equation for baryons

At first order we find

$$\frac{\partial v^{(1)i}}{\partial \eta} + \mathcal{H}v^{(1)i} + \Phi^{(1),i} = \frac{4}{3}\tau'\frac{\bar{\rho}_\gamma}{\bar{\rho}_b}\left(v^{(1)i} - v_\gamma^{(1)i}\right). \tag{7.43}$$

### 7.2.2. Second-order momentum continuity equation for baryons

At second order there are various simplifications. In particular notice that the term on the right-hand side of Eq. (7.42) which is proportional to $\delta_b$ vanishes when matched to expansion of the left-hand side by virtue of the first-order equation (7.43). Thus, at the end we find a very simple equation

$$\frac{1}{2}\left(\frac{\partial v^{(2)i}}{\partial \eta} + \mathcal{H}v^{(2)i} + 2\frac{\partial \omega^i}{\partial \eta} + 2\mathcal{H}\omega_i + \Phi^{(2),i}\right) - \frac{\partial \Psi^{(1)}}{\partial \eta}v^{(1)i}$$

$$+ v^{(1)j}\partial_j v^{(1)i} + (\Phi^{(1)} + \Psi^{(1)})\Phi^{(1),i} + \left(\frac{T_b}{m_p}\right)^{,i}$$

$$= \frac{4}{3}\tau'\frac{\bar{\rho}_\gamma}{\bar{\rho}_b}\left[\left(\frac{v^{(2)i}}{2} - \frac{v_\gamma^{(2)i}}{2}\right) + (\Delta_0^{(1)} + \Phi^{(1)})(v^{(1)i} - v_\gamma^{(1)i}) + \frac{3}{4}v_j^{(1)}\Pi_\gamma^{ji}\right], \tag{7.44}$$

with $\tau' = -\bar{n}_e\sigma_T a$.

### 7.2.3. First-order momentum continuity equation for CDM

Since CDM particles are collisionless, at first order we find

$$\frac{\partial v_{\text{CDM}}^{(1)i}}{\partial \eta} + \mathcal{H}v_{\text{CDM}}^{(1)i} + \Phi^{(1),i} = 0. \tag{7.45}$$

### 7.2.4. Second-order momentum continuity equation for CDM

At second order we find

$$\frac{1}{2}\left(\frac{\partial v_{\text{CDM}}^{(2)i}}{\partial \eta} + \mathcal{H}v_{\text{CDM}}^{(2)i} + 2\frac{\partial \omega^i}{\partial \eta} + 2\mathcal{H}\omega_i + \Phi^{(2),i}\right) - \frac{\partial \Psi^{(1)}}{\partial \eta}v_{\text{CDM}}^{(1)i}$$

$$+ v_{\text{CDM}}^{(1)j}\partial_j v_{\text{CDM}}^{(1)i} + (\Phi^{(1)} + \Psi^{(1)})\Phi^{(1),i} + \left(\frac{T_{\text{CDM}}}{m_{\text{CDM}}}\right)^{,i} = 0. \tag{7.46}$$

## 8. Linear solution of the Boltzmann equations

In this section we will solve the Boltzmann equations at first order in perturbation theory. The interested reader will find the extension of these formulae to second order in Ref. [2]. The first two moments of the photon Boltzmann equation are obtained by integrating Eq. (6.6) over $d\Omega_{\mathbf{n}}/4\pi$ and $d\Omega_{\mathbf{n}}n^i/4\pi$ respectively and they lead to the density and velocity continuity equations

$$\Delta_{00}^{(1)'} + \frac{4}{3}\partial_i v_\gamma^{(1)i} - 4\Psi^{(1)'} = 0, \tag{8.1}$$

$$v_\gamma^{(1)i'} + \frac{3}{4}\partial_j \Pi_\gamma^{(1)ji} + \frac{1}{4}\Delta_{00}^{(1),i} + \Phi^{(1),i} = -\tau'\left(v^{(1)i} - v_\gamma^{(1)}\right), \tag{8.2}$$

where $\Pi^{ij}$ is the photon quadrupole moment, defined in Eq. (7.40).

Let us recall here that $\delta_\gamma^{(1)} = \Delta_{00}^{(1)} = \int d\Omega \Delta^{(1)}/4\pi$ and that the photon velocity is given by Eq. (7.31).

The two equations above are complemented by the momentum continuity equation for baryons, which can be conveniently written as

$$v^{(1)i} = v_\gamma^{(1)i} + \frac{R}{\tau'}\left[v^{(1)i'} + \mathcal{H}v^{(1)i} + \Phi^{(1),i}\right], \tag{8.3}$$

where we have introduced the baryon-photon ratio $R \equiv 3\rho_b/(4\rho_\gamma)$.

Eq. (8.3) is in a form ready for a consistent expansion in the small quantity $\tau^{-1}$ which can be performed in the tight-coupling limit. By first taking $v^{(1)i} = v_\gamma^{(1)i}$ at zero order and then using this relation in the left-hand side of Eq. (8.3) one obtains

$$v^{(1)i} - v_\gamma^{(1)i} = \frac{R}{\tau'}\left[v_\gamma^{(1)i'} + \mathcal{H}v_\gamma^{(1)i} + \Phi^{(1),i}\right]. \tag{8.4}$$

Such an expression for the difference of velocities can be used in Eq. (8.2) to give the evolution equation for the photon velocity in the limit of tight coupling

$$v_\gamma^{(1)i'} + \mathcal{H}\frac{R}{1+R}v_\gamma^{(1)i} + \frac{1}{4}\frac{\Delta_{00}^{(1),i}}{1+R} + \Phi^{(1),i} = 0. \tag{8.5}$$

Notice that in Eq. (8.5) we are neglecting the quadrupole of the photon distribution $\Pi^{(1)ij}$ (and all the higher moments) since it is well known that at linear order such moment(s) are suppressed in the tight-coupling limit by (successive powers of) $1/\tau$ with respect to the first two moments, the photon energy density and velocity. Eqs. (8.1) and (8.5) are the master equations which govern the photon-baryon fluid acoustic oscillations before the epoch of recombination when photons and baryons are tightly coupled by Compton scattering.

In fact one can combine these two equations to get a single second-order differential equation for the photon energy density perturbations $\Delta_{00}^{(1)}$. Deriving Eq. (8.1) with respect to conformal time and using Eq. (8.5) to replace $\partial_i v_\gamma^{(1)i}$ yields

$$
\left(\Delta_{00}^{(1)''} - 4\Psi^{(1)''}\right) + \mathcal{H}\frac{R}{1+R}\left(\Delta_{00}^{(1)'} - 4\Psi^{(1)'}\right)
$$
$$
- c_s^2 \nabla^2\left(\Delta_{00}^{(1)} - 4\Psi^{(1)}\right) = \frac{4}{3}\nabla^2\left(\Phi^{(1)} + \frac{\Psi^{(1)}}{1+R}\right),
\tag{8.6}
$$

where $c_s = 1/\sqrt{3(1+R)}$ is the speed of sound of the photon-baryon fluid. Indeed, in order to solve Eq. (8.6) one needs to know the evolution of the gravitational potentials. We will come back later to the discussion of the solution of Eq. (8.6).

A useful relation we will use in the following is obtained by considering the continuity equation for the baryon density perturbation. By perturbing at first order Eq. (7.21) we obtain

$$
\delta_b^{(1)'} + v_{,i}^i - 3\Psi^{(1)'} = 0.
\tag{8.7}
$$

Subtracting Eq. (8.7) form Eq. (8.1) brings

$$
\Delta_{00}^{(1)'} - \frac{4}{3}\delta_b^{(1)'} + \frac{4}{3}\left(v_\gamma^{(1)i} - v^{(1)i}\right)_{,i} = 0,
\tag{8.8}
$$

which implies that at lowest order in the tight-coupling approximation

$$
\Delta_{00}^{(1)} = \frac{4}{3}\delta_b^{(1)},
\tag{8.9}
$$

for adiabatic perturbations.

### 8.1. Linear solutions in the limit of tight coupling

In this section we briefly recall how to obtain at linear order the solutions of the Boltzmann equations (8.6). These correspond to the acoustic oscillations of the photon-baryon fluid for modes which are within the horizon at the time of recombination. It is well known that, in the variable $(\Delta_{00}^{(1)} - 4\Psi^{(1)})$, the solution can be written as [32, 38]

$$
[1 + R(\eta)]^{1/4}\left(\Delta_{00}^{(1)} - 4\Psi^{(1)}\right) = A\cos[kr_s(\eta)] + B\sin[kr_s(\eta)]
$$
$$
- 4\frac{k}{\sqrt{3}}\int_0^\eta d\eta'[1 + R(\eta')]^{3/4}\left(\Phi^{(1)}(\eta') + \frac{\Psi^{(1)}(\eta')}{1+R(\eta')}\right)
$$
$$
\times \sin[k(r_s(\eta) - r_s(\eta'))],
\tag{8.10}
$$

where the sound horizon is given by $r_s(\eta) = \int_0^\eta d\eta' c_s(\eta')$, with $R = 3\rho_b/(4\rho_\gamma)$. The constants $A$ and $B$ in Eq. (8.10) are fixed by the choice of initial conditions.

In order to give an analytical solution that catches most of the physics underlying Eq. (8.10) and which remains at the same time very simple to treat, we will make some simplifications following Ref. [28, 39]. First, for simplicity, we are going to neglect the ratio $R$ wherever it appears, *except* in the arguments of the varying cosines and sines, where we will treat $R = R_*$ as a constant evaluated at the time of recombination. In this way we keep track of a damping of the photon velocity amplitude with respect to the case $R = 0$ which prevents the acoustic peaks in the power-spectrum to disappear. Treating $R$ as a constant is justified by the fact that for modes within the horizon the time scale of the ocillations is much shorter than the time scale on which $R$ varies. If $R$ is a constant the sound speed is just a constant $c_s = 1/\sqrt{3(1 + R_*)}$, and the sound horizon is simply $r_s(\eta) = c_s \eta$.

Second, we are going to solve for the evolutions of the perturbations in two well distinguished limiting regimes. One regime is for those perturbations which enter the Hubble radius when matter is the dominant component, that is at times much bigger than the equality epoch, with $k \ll k_{eq} \sim \eta_{eq}^{-1}$, where $k_{eq}$ is the wavenumber of the Hubble radius at the equality epoch. The other regime is for those perturbations with much smaller wavelenghts which enter the Hubble radius when the universe is still radiation dominated, that is perturbations with wavenumbers $k \gg k_{eq} \sim \eta_{eq}^{-1}$. In fact we are interested in perturbation modes which are within the horizon by the time of recombination $\eta_*$. Therefore we will further suppose that $\eta_* \gg \eta_{eq}$ in order to study such modes in the first regime. Even though $\eta_* \gg \eta_{eq}$ is not the real case, it allows to obtain some analytical expressions.

Before solving for these two regimes let us fix our initial conditions, which are taken on large scales deep in the radiation dominated era (for $\eta \to 0$). During this epoch, for adiabatic perturbations, the gravitational potentials remain constant on large scales (we are neglecting anisotropic stresses so that $\Phi^{(1)} \simeq \Psi^{(1)}$) and from the $(0 - 0)$-component of Einstein equations

$$\Phi^{(1)}(0) = -\frac{1}{2}\Delta_{00}^{(1)}(0). \tag{8.11}$$

On the other hand, from the energy continuity equation (8.1) on large scales

$$\Delta_{00}^{(1)} - 4\Psi^{(1)} = \text{const.}; \tag{8.12}$$

using Eq. (8.11) the constant on the right-hand side of Eq. (8.12) is fixed to be $-6\Phi^{(1)}(0)$; thus we find $B = 0$ and $A = -6\Phi^{(1)}(0)$.

With our simplifications Eq. (8.10) then reads

$$\Delta_{00}^{(1)} - 4\Psi^{(1)} = -6\Phi^{(1)}(0)\cos(\omega_0\eta)$$
$$- \frac{8k}{\sqrt{3}} \int_0^{\eta} d\eta' \Phi^{(1)}(\eta') \sin[\omega_0(\eta - \eta')], \tag{8.13}$$

where $\omega_0 = kc_s$.

## 8.2. Perturbation modes with $k \ll k_{eq}$

This regime corresponds to perturbation modes which enter the Hubble radius when the universe is matter dominated at times $\eta \gg \eta_{eq}$. During matter domination the gravitational potential remains constant (both on super-horizon and sub-horizon scales), as one can see for example from Eq. (B.1), and its value is fixed to $\Phi^{(1)}(\mathbf{k}, \eta) = \frac{9}{10}\Phi^{(1)}(0)$, where $\Phi^{(1)}(0)$ corresponds to the gravitational potential on large scales during the radiation dominated epoch. Since we are interested in the photon anisotropies around the time of recombination, when matter is dominating, we can perform the integral appearing in Eq. (8.10) by taking the gravitational potential equal to its value during matter domination so that it is easily computed

$$2 \int_0^{\eta} d\eta' \Phi^{(1)}(\eta') \sin[\omega_0(\eta - \eta')] = \frac{18}{10}\frac{\Phi^{(1)}(0)}{\omega_0}(1 - \cos(\omega_0\eta)). \tag{8.14}$$

Thus Eq. (8.13) gives

$$\Delta_{00}^{(1)} - 4\Psi^{(1)} = \frac{6}{5}\Phi^{(1)}(0)\cos(\omega_0\eta) - \frac{36}{5}\Phi^{(1)}(0). \tag{8.15}$$

The baryon-photon fluid velocity can then be obtained as $\partial_i v_\gamma^{(1)i} = -3(\Delta_{00}^{(1)} - 4\Psi^{(1)})'/4$ from Eq. (8.1). In Fourier space

$$ik_i v_\gamma^{(1)i} = \frac{9}{10}\Phi^{(1)}(0)\sin(\omega_0\eta)\omega_0, \tag{8.16}$$

where, going to Fourier space, $\partial_i v_\gamma^{(1)i} \to ik_i v_\gamma^{(1)i}(\mathbf{k})$ and

$$v_\gamma^{(1)i} = -i\frac{k^i}{k}\frac{9}{10}\Phi^{(1)}(0)\sin(\omega_0\eta)c_s, \tag{8.17}$$

since the linear velocity is irrotational.

## 8.3. Perturbation modes with $k \gg k_{eq}$

This regime corresponds to perturbation modes which enter the Hubble radius when the universe is still radiation dominated at times $\eta \ll \eta_{eq}$. In this case an approximate analytical solution for the evolution of the perturbations can be obtained by considering the gravitational potential for a pure radiation dominated epoch, given by Eq. (B.8). For the integral in Eq. (8.13) we thus find

$$\int_0^\eta \Phi^{(1)}(\eta') \sin[\omega_0(\eta - \eta')] = -\frac{3}{2\omega_0} \cos(\omega_0 \eta), \tag{8.18}$$

where we have kept only the dominant contribution oscillating in time, while neglecting terms which decay in time. The solution (8.13) becomes

$$\Delta_{00}^{(1)} - 4\Psi^{(1)} = 6\Phi^{(1)}(0) \cos(\omega_0 \eta), \tag{8.19}$$

and the velocity is given by

$$v_\gamma^{(1)i} = -i\frac{k^i}{k}\frac{9}{2}\Phi^{(1)}(0) \sin(\omega_0 \eta)c_s, \tag{8.20}$$

Notice that the solutions (8.19)–(8.20) are actually correct only when radiation dominates. Indeed, between the epoch of equality and recombination, matter starts to dominate. Full account of such a period is given e.g. in Section 7.3 of Ref. [28], while its consequences for the CMB anisotropy evolution can be found e.g. in Ref. [40].

## 9. Conclusions

In these lecture notes we derived the equations which allow to evaluate CMB anisotropies, by computing the Boltzmann equations describing the evolution of the baryon-photon fluid up to second order. This allows to follow the time evolution of CMB anisotropies (up to second order) on all scales, from the early epoch, when the cosmological perturbations were generated, to the present time, through the recombination era. The dynamics at second order is particularly important when dealing with the issue of non-Gaussianity in CMB anisotropies. Indeed, many mechanisms for the generation of the primordial inhomogeneities predict a level of non-Gaussianity in the curvature perturbations which might be detectable by present and future experiments.

| Symbol | Definition | Equation |
|---|---|---|
| $\Phi, \Psi$ | Gravitational potentials in Poisson gauge | (3.1) |
| $\omega_i$ | 2nd-order vector perturbation in Poisson gauge | (3.1) |
| $\chi_{ij}$ | 2nd-order tensor perturbation in Poisson gauge | (3.1) |
| $\eta$ | Conformal time | (3.1) |
| $f$ | Photon distribution function | (4.8) |
| $g$ | Distribution function for massive particles | (5.4) & (7.1) |
| $f^{(i)}$ | $i$-th order perturbation of the photon distribution function | (4.23) |
| $f^{(i)}_{\ell m}$ | Moments of the photon distribution function | (5.16) |
| $C(\mathbf{p})$ | Collision term | (5.3) & (5.7) |
| $p$ | Magnitude of photon momentum ($\mathbf{p} = pn^i$) | (4.3) |
| $n^i$ | Propagation direction | (4.6) |
| $\Delta^{(1)}(x^i, n^i, \eta)$ | First-order fractional energy photon fluctuations | (6.5) |
| $\Delta^{(2)}(x^i, n^i, \eta)$ | Second-order fractional energy photon fluctuations | (6.9) |
| $n_e$ | Electron number density | (7.11) |
| $\delta_e(\delta_b)$ | Electron (baryon) density perturbation | (5.12) |
| $\mathbf{k}$ | Wavenumber | (6.16) |
| $v_m$ | Baryon velocity perturbation | (6.19) & (6.20) |
| $v^{(2)i}_{\text{CDM}}$ | Cold dark matter velocity | (7.46) |
| $v^{(2)i}_{\gamma}$ | Second-order photon velocity | (7.32) |
| $S_{\ell m}$ | Temperature source term | (6.17) |
| $\tau$ | Optical depth | (6.14) |
| $\bar{\rho}_\gamma(\bar{\rho}_b)$ | Background photon (baryon) energy density | (7.43) |

## Acknowledgments

A.R. is on leave of absence from INFN, Sezione di Padova. S.M. is partially supported by INAF.

## Appendix A. Einstein's equations

In this Appendix we provide the necessary expressions to deal with the gravitational part of the problem we are interested in, namely the second-order CMB anisotropies generated at recombination as well as the acoustic oscillations of the baryon-photon fluid. The first part of the Appendix contains the expressions for the metric, connection coefficients and Einstein's tensor perturbed up to second order around a flat Friedmann-Robertson-Walker background, the energy-momentum tensors for massless (photons) and massive (baryons and cold dark matter) particles, and the relevant Einstein's equations. The second part deals with the evolution equations and the solutions for the second-order gravitational potentials in the Poisson gauge.

*Appendix A.1. The metric tensor*

As discussed in Section 3, we write our second-order metric in the Poisson gauge,

$$ds^2 = a^2(\eta)\left[-e^{2\Phi}d\eta^2 + 2\omega_i dx^i d\eta + \left(e^{-2\Psi}\delta_{ij} + \chi_{ij}\right)dx^i dx^j\right], \qquad (A.1)$$

where $a(\eta)$ is the scale factor as a function of conformal time $\eta$, and $\omega_i$ and $\chi_{ij}$ are vector and tensor perturbation modes respectively. Each metric perturbation is expanded into a linear (first-order) and a second-order part, as discussed in Section 3.

*Appendix A.2. The connection coefficients*

For the connection coefficients we find

$$
\begin{aligned}
\Gamma^0_{00} &= \mathcal{H} + \Phi', \\
\Gamma^0_{0i} &= \frac{\partial \Phi}{\partial x^i} + \mathcal{H}\omega_i, \\
\Gamma^i_{00} &= \omega^{i\prime} + \mathcal{H}\omega^i + e^{2\Psi + 2\Phi}\frac{\partial \Phi}{\partial x_i}, \\
\Gamma^0_{ij} &= -\frac{1}{2}\left(\frac{\partial \omega_j}{\partial x^i} + \frac{\partial \omega_i}{\partial x^j}\right) + e^{-2\Psi - 2\Phi}(\mathcal{H} - \Psi')\delta_{ij} + \frac{1}{2}\chi'_{ij} + \mathcal{H}\chi_{ij}, \\
\Gamma^i_{0j} &= (\mathcal{H} - \Psi')\delta_{ij} + \frac{1}{2}\chi'_{ij} + \frac{1}{2}\left(\frac{\partial \omega_i}{\partial x^j} - \frac{\partial \omega_j}{\partial x^i}\right), \\
\Gamma^i_{jk} &= -\mathcal{H}\omega^i\delta_{jk} - \frac{\partial \Psi}{\partial x^k}\delta^i_{\ j} - \frac{\partial \Psi}{\partial x^j}\delta^i_{\ k} + \frac{\partial \Psi}{\partial x_i}\delta_{jk} \\
&\quad + \frac{1}{2}\left(\frac{\partial \chi^i_{\ j}}{\partial x^k} + \frac{\partial \chi^i_{\ k}}{\partial x^j} - \frac{\partial \chi_{jk}}{\partial x_i}\right).
\end{aligned}
\qquad (A.2)
$$

*Appendix A.3. Einstein tensor*

The components of Einstein's tensor read

$$
\begin{aligned}
G^0_{\ 0} &= -\frac{e^{-2\Phi}}{a^2}\Big[3\mathcal{H}^2 - 6\mathcal{H}\Psi' + 3(\Psi')^2 \\
&\quad - e^{2\Phi + 2\Psi}\left(\partial_i\Psi\partial^i\Psi - 2\nabla^2\Psi\right)\Big], \\
\end{aligned}
\qquad (A.3)
$$

$$
\begin{aligned}
G^i_{\ 0} &= 2\frac{e^{2\Psi}}{a^2}\left[\partial^i\Psi' + (\mathcal{H} - \Psi')\partial^i\Phi\right] - \frac{1}{2a^2}\nabla^2\omega^i \\
&\quad + \left(4\mathcal{H}^2 - 2\frac{a''}{a}\right)\frac{\omega^i}{a^2}, \\
\end{aligned}
\qquad (A.4)
$$

$$G^i{}_j = \frac{1}{a^2}\left[ e^{-2\Phi}\left( \mathcal{H}^2 - 2\frac{a''}{a} - 2\Psi'\Phi' - 3(\Psi')^2 + 2\mathcal{H}(\Phi' + 2\Psi') + 2\Psi'' \right) \right.$$

$$\left. + e^{2\Psi}\left( \partial_k \Phi \partial^k \Phi + \nabla^2\Phi - \nabla^2\Psi \right) \right]\delta^i{}_j + \frac{e^{2\Psi}}{a^2}$$

$$\times \left( -\partial^i\Phi\partial_j\Phi - \partial^i\partial_j\Phi + \partial^i\partial_j\Psi - \partial^i\Phi\partial_j\Psi + \partial^i\Psi\partial_j\Psi - \partial^i\Psi\partial_j\Phi \right)$$

$$- \frac{\mathcal{H}}{a^2}\left( \partial^i\omega_j + \partial_j\omega^i \right) - \frac{1}{2a^2}\left( \partial^i\omega'_j + \partial_j\omega^{i'} \right)$$

$$+ \frac{1}{a^2}\left( \mathcal{H}\chi^{i'}_j + \frac{1}{2}\chi^{i''}_j - \frac{1}{2}\nabla^2\chi^i_j \right). \tag{A.5}$$

Taking the traceless part of Eq. (A.5), we find $\Psi - \Phi = \mathcal{Q}$, where $\mathcal{Q}$ is defined by $\nabla^2\mathcal{Q} = -P + 3N$, with $P \equiv P^i{}_i$,

$$P^i{}_j = \partial^i\Phi\partial_j\Psi + \frac{1}{2}(\partial^i\Phi\partial_j\Phi - \partial^i\Psi\partial_j\Psi) + 4\pi G_N a^2 e^{-2\Psi} T^i{}_j \tag{A.6}$$

and $\nabla^2 N = \partial_i \partial^j P^i{}_j$.

The trace of Eq. (A.5) gives

$$e^{-2\Phi}\left( \mathcal{H}^2 - 2\frac{a''}{a} - 2\Phi'\Psi' - 3(\Psi')^2 + 2\mathcal{H}\left(3\Psi' - \mathcal{Q}'\right) + 2\Psi'' \right)$$

$$+ \frac{e^{2\Psi}}{3}\left( 2\partial_k \Phi \partial^k \Phi + \partial_k \Psi \partial^k \Psi - 2\partial_k \Phi \partial^k \Psi + 2(P - 3N) \right)$$

$$= \frac{8\pi G_N}{3} a^2 T^k{}_k. \tag{A.7}$$

Using Eq. (A.4) we may deduce an equation for $\omega^i$

$$-\frac{1}{2}\nabla^2\omega^i + \left( 4\mathcal{H}^2 - 2\frac{a''}{a} \right)\omega^i$$

$$= -\left( \delta^i_j - \frac{\partial^i\partial_j}{\nabla^2} \right)\left( 2(\partial^j\Psi' + (\mathcal{H} - \Psi')\partial^j\Phi) - 8\pi G_N a^2 e^{-2\Psi} T^j_0 \right). \tag{A.8}$$

### Appendix A.4.  Energy-momentum tensor

#### Appendix A.4.1.  Energy-momentum tensor for photons
The energy-momentum tensor for photons is defined as

$$T^\mu_{\gamma\,\nu} = \frac{2}{\sqrt{-g}}\int \frac{d^3P}{(2\pi)^3}\frac{P^\mu P_\nu}{P^0}\, f, \tag{A.9}$$

where $g$ is the determinant of the metric (A.1) and $f$ is the distribution function. We thus obtain

$$T^0_{\gamma\,0} = -\bar{\rho}_\gamma \left( 1 + \Delta^{(1)}_{00} + \frac{\Delta^{(2)}_{00}}{2} \right), \tag{A.10}$$

$$T^i_{\gamma\,0} = -\frac{4}{3} e^{\Psi+\Phi} \bar{\rho}_\gamma \left( v^{(1)i}_\gamma + \frac{1}{2} v^{(2)i}_\gamma + \Delta^{(1)}_{00} v^{(1)i}_\gamma \right) + \frac{1}{3} \bar{\rho}_\gamma e^{\Psi-\Phi} \omega^i \tag{A.11}$$

$$T^i_{\gamma\,j} = \bar{\rho}_\gamma \left( \Pi^i_{\gamma\,j} + \frac{1}{3} \delta^i{}_j \left( 1 + \Delta^{(1)}_{00} + \frac{\Delta^{(2)}_{00}}{2} \right) \right), \tag{A.12}$$

where $\bar{\rho}_\gamma$ is the background energy density of photons and

$$\Pi^{ij}_\gamma = \int \frac{d\Omega}{4\pi} \left( n^i n^j - \frac{1}{3} \delta^{ij} \right) \left( \Delta^{(1)} + \frac{\Delta^{(2)}}{2} \right), \tag{A.13}$$

is the quadrupole moment of the photons.

### Appendix A.4.2. Energy-momentum tensor for massive particles

The energy-momentum tensor for massive particles of mass $m$, number density $n$ and degrees of freedom $g_d$

$$T^\mu_{m\,\nu} = \frac{g_d}{\sqrt{-g}} \int \frac{d^3 Q}{(2\pi)^3} \frac{Q^\mu Q_\nu}{Q^0} g_m, \tag{A.14}$$

where $g_m$ is the distribution function. We obtain

$$T^0_{m\,0} = -\rho_m = -\bar{\rho}_m \left( 1 + \delta^{(1)}_m + \frac{1}{2} \delta^{(2)}_m \right), \tag{A.15}$$

$$T^i_{m\,0} = -e^{\Psi+\Phi} \rho_m v^i_m = -e^{\Phi+\Psi} \bar{\rho}_m \left( v^{(1)i}_m + \frac{1}{2} v^{(2)i}_m + \delta^{(1)}_m v^{(1)i}_m \right) \tag{A.16}$$

$$T^i_{m\,j} = \rho_m \left( \delta^i{}_j \frac{T_m}{m} + v^i_m v_{m\,j} \right) = \bar{\rho}_m \left( \delta^i{}_j \frac{T_m}{m} + v^{(1)i}_m v^{(1)}_{m\,j} \right), \tag{A.17}$$

where $\bar{\rho}_m$ is the background energy density of massive particles and we have included the equilibrium temperature $T_m$.

## Appendix B. First-order solutions of Einstein's equations in various eras

*Appendix B.1. Matter-dominated era*

During the phase in which the CDM is dominating the energy density of the Universe, $a \sim \eta^2$ and we may use Eq. (A.7) to obtain an equation for the gravitational potential at first order in perturbation theory (for which $\Phi^{(1)} = \Psi^{(1)}$)

$$\Phi^{(1)''} + 3\mathcal{H}\Phi^{(1)'} = 0, \tag{B.1}$$

which has two solutions $\Phi_{+}^{(1)} = $ constant and $\Phi_{-}^{(1)} = \mathcal{H}/a^2$. At the same order of perturbation theory, the CDM velocity can be read off from Eq. (A.4)

$$v^{(1)i} = -\frac{2}{3\mathcal{H}}\partial^i \Phi^{(1)}. \tag{B.2}$$

The matter density contrast $\delta^{(1)}$ satisfies the first-order continuity equation

$$\delta^{(1)'} = -\frac{\partial v^{(1)i}}{\partial x^i} = -\frac{2}{3\mathcal{H}}\nabla^2\Phi^{(1)}. \tag{B.3}$$

Going to Fourier space, this implies that

$$\delta_k^{(1)} = \delta_k^{(1)}(0) + \frac{k^2\eta^2}{6}\Phi_k^{(1)}, \tag{B.4}$$

where $\delta_k^{(1)}(0)$ is the initial condition in the matter-dominated period.

*Appendix B.2. Radiation-dominated era*

We consider a universe dominated by photons and massless neutrinos. The energy-momentum tensor for massless neutrinos has the same form as that for photons. During the phase in which radiation is dominating the energy density of the Universe, $a \sim \eta$ and we may combine Eqs. (A.3) and (A.7) to obtain an equation for the gravitational potential $\Psi^{(1)}$ at first order in perturbation theory

$$\Psi^{(1)''} + 4\mathcal{H}\Psi^{(1)'} - \frac{1}{3}\nabla^2\Psi^{(1)} = \mathcal{H}Q^{(1)'} + \frac{1}{3}\nabla^2 Q^{(1)},$$

$$\nabla^2 Q^{(1)} = \frac{9}{2}\mathcal{H}^2\frac{\partial_i\partial^j}{\nabla^2}\Pi_T^{(1)i}{}_j, \tag{B.5}$$

where the total anisotropic stress tensor is

$$\Pi_T^i{}_j = \frac{\bar{\rho}_\gamma}{\bar{\rho}_T}\Pi_\gamma^i{}_j + \frac{\bar{\rho}_\nu}{\bar{\rho}_T}\Pi_\nu^i{}_j. \tag{B.6}$$

We may safely neglect the quadrupole and solve Eq. (B.5) setting $u_\pm = \Phi_\pm^{(1)}\eta$. Then Eq. (B.5), in Fourier space, becomes

$$u'' + \frac{2}{\eta}u' + \left(\frac{k^2}{3} - \frac{2}{\eta^2}\right)u = 0. \tag{B.7}$$

This equation has as independent solutions $u_+ = j_1(k\eta/\sqrt{3})$, the spherical Bessel function of order 1, and $u_- = n_1(k\eta/\sqrt{3})$, the spherical Neumann function of order 1. The latter blows up as $\eta$ gets small and we discard it on the basis of initial conditions. The final solution is therefore

$$\Phi_k^{(1)} = 3\Phi^{(1)}(0)\frac{\sin(k\eta/\sqrt{3}) - (k\eta/\sqrt{3})\cos(k\eta/\sqrt{3})}{(k\eta/\sqrt{3})^3} \tag{B.8}$$

where $\Phi^{(1)}(0)$ represents the initial condition deep in the radiation era.

At the same order in perturbation theory, the radiation velocity can be read off from Eq. (A.4)

$$v_\gamma^{(1)i} = -\frac{1}{2\mathcal{H}^2}\frac{(a\partial^i\Phi^{(1)})'}{a}. \tag{B.9}$$

## References

[1] N. Bartolo, S. Matarrese and A. Riotto, JCAP **0606**, 024 (2006) [arXiv:astro-ph/0604416].

[2] N. Bartolo, S. Matarrese and A. Riotto, arXiv:astro-ph/0610110.

[3] For a review, see D. H. Lyth and A. Riotto, Phys. Rept. **314**, 1 (1999).

[4] D. N. Spergel *et al.*, arXiv:astro-ph/0603449.

[5] See *http://planck.esa.int/*.

[6] N. Bartolo, E. Komatsu, S. Matarrese and A. Riotto, Phys. Rept. **402**, 103 (2004) [arXiv:astro-ph/0406398].

[7] V. Acquaviva, N. Bartolo, S. Matarrese and A. Riotto, Nucl. Phys. B **667**, 119 (2003); J. Maldacena, JHEP **0305**, 013 (2003).

[8] N. Bartolo, S. Matarrese and A. Riotto, Phys. Rev. D **65**, 103505 (2002); F. Bernardeau and J. P. Uzan, Phys. Rev. D **66**, 103506 (2002); F. Vernizzi and D. Wands, arXiv:astro-ph/0603799.

[9] D. H. Lyth, C. Ungarelli and D. Wands, Phys. Rev. D **67** (2003) 023503.

[10] T. Hamazaki and H. Kodama, Prog. Theor. Phys. **96** (1996) 1123.

[11] G. Dvali, A. Gruzinov and M. Zaldarriaga, Phys. Rev. D **69**, 023505 (2004); L. Kofman, arXiv:astro-ph/0303614.

[12] G. Dvali, A. Gruzinov and M. Zaldarriaga, Phys. Rev. D **69** (2004) 083505.

[13] C. W. Bauer, M. L. Graesser and M. P. Salem, Phys. Rev. D **72** (2005) 023512.

[14] D. H. Lyth, JCAP **0511** (2005) 006; M. P. Salem, Phys. Rev. D **72** (2005) 123516.

[15] M. Bastero-Gil, V. Di Clemente and S. F. King, Phys. Rev. D **70**, 023501 (2004); E. W. Kolb, A. Riotto and A. Vallinotto, Phys. Rev. D **71**, 043513 (2005); C. T. Byrnes and D. Wands, Phys. Rev. D **73**, 063509 (2006) [arXiv:astro-ph/0512195].

[16] T. Matsuda, Phys. Rev. D **72** (2005) 123508.

[17] W. Hu, Phys. Rev. D **64**, 083005 (2001); T. Okamoto and W. Hu, Phys. Rev. D **66**, 063008 (2002).

[18] G. De Troia *et al.*, Mon. Not. Roy. Astron. Soc. **343**, 284 (2003).

[19] J. Lesgourgues, M. Liguori, S. Matarrese and A. Riotto, Phys. Rev. D **71**, 103514 (2005).

[20] N. Bartolo, S. Matarrese and A. Riotto, JCAP **0605**, 010 (2006) [arXiv:astro-ph/0512481].

[21] T. Pyne and S. M. Carroll, Phys. Rev. D **53**, 2920 (1996); S. Mollerach and S. Matarrese, Phys. Rev. D **56**, 4494 (1997); S. Matarrese, S. Mollerach and M. Bruni, Phys. Rev. D **58**, 043504 (1998); N. Bartolo, S. Matarrese and A. Riotto, Phys. Rev. Lett. **93**, 231301 (2004); K. Tomita, Phys. Rev. D **71**, 083504 (2005); N. Bartolo, S. Matarrese and A. Riotto, JCAP **0508**, 010 (2005).

[22] W. Hu and S. Dodelson, "Cosmic Microwave Background Anisotropies," Ann. Rev. Astron. Astrophys. **40**, 171 (2002).

[23] U. Seljak and M. Zaldarriaga, "Direct Signature of Evolving Gravitational Potential from Cosmic Microwave Background," Phys. Rev. D **60**, 043504 (1999).

[24] D. M. Goldberg and D. N. Spergel, "Microwave Background Bispectrum. 2. A Probe Of The Low Redshift Universe," Phys. Rev. D **59**, 103002 (1999).

[25] D. M. Goldberg and D. N. Spergel, Phys. Rev. D **59**, 103002 (1999).

[26] E. Komatsu and D. N. Spergel, Phys. Rev. D **63**, 063002 (2001).

[27] P. Creminelli and M. Zaldarriaga, Phys. Rev. D **70**, 083532 (2004).

[28] S. Dodelson, *Modern cosmology*, Amsterdam, Netherlands: Academic Pr. (2003) 440 pp.

[29] S. Matarrese, S. Mollerach and M. Bruni, Phys. Rev. D **58**, 043504 (1998).

[30] W. Hu, D. Scott and J. Silk, Phys. Rev. D **49**, 648 (1994).

[31] S. Dodelson and J. M. Jubas, Astrophys. J. **439**, 503 (1995).

[32] W. T. Hu, PhD thesis *Wandering in the background: A Cosmic microwave background explorer*, arXiv:astro-ph/9508126.

[33] R. Maartens, T. Gebbie and G.R. Ellis, Phys. Rev. D **59**, 083506 (1999).

[34] W. Hu, U. Seljak, M. J. White and M. Zaldarriaga, Phys. Rev. D **57**, 3290 (1998).

[35] U. Seljak and M. Zaldarriaga, Astrophys. J. **469**, 437 (1996).

[36] W. Hu and M. J. White, Phys. Rev. D **56**, 596 (1997).

[37] W. Hu, Astrophys. J. **529**, 12 (2000).

[38] W. Hu and N. Sugiyama, Astrophys. J. **444**, 489 (1995).

[39] M. Zaldarriaga and D. D. Harari, Phys. Rev. D **52**, 3276 (1995).

[40] V. F. Mukhanov, Int. J. Theor. Phys. **43**, 623 (2004).

Course 6

# PHYSICS BEYOND THE STANDARD MODEL
# AND DARK MATTER

Hitoshi Murayama

*Department of Physics, University of California*
*Berkeley, CA 94720, USA*
and
*Theoretical Physcis Group, Lawrence Berkeley National Laboratory*
*Berkeley, CA 94720, USA*

*F. Bernardeau, C. Grojean and J. Dalibard, eds.*
*Les Houches, Session LXXXVI, 2006*
*Particle Physcis and Cosmology: The Fabric of Spacetime*
© *2007 Published by Elsevier B.V.*

# Contents

# 1. Introduction

I'm honored to be invited as a lecturer at a Les Houches summer school which has a great tradition. I remember reading many of the lectures from past summer schools when I was a graduate student and learned a lot from them. I'm also looking forward to have a good time in the midst of beautiful mountains, even though the weather doesn't seem to be cooperating. I'm not sure if I will ever get to see Mont Blanc!

I was asked to give four general lectures on physics beyond the standard model. This is in some sense an ill-defined assignment, because it is a vast subject for which we know pratically nothing about. It is vast because there are so many possibilities and speculations, and a lot of ink and many many pages of paper had been devoted to explore it. On the other hand, we know practically nothing about it by definition, because if we did, it should be a part of the standard model of particle physics already. I will therefore focus more on the motivation why we should consider physics beyond the standard model and discuss a few candidates, and there is no way I can present all the examples exhaustively. In addition, after reviewing the program, I've realized that there are no dedicated lectures on dark matter. Since this is a topic where particle physics and cosmology (I believe) are likely to come together in the near future, it is relevant to the theme of the school "Particle Physics and Cosmology: the Fabric of Spacetime." Therefore I will emphasize this connection in some detail.

Because I try to be pedagogical in lectures, I will probably discuss many points which some of you already know very well. Given the wide spectrum of background you have, I aim at the common denominator. Hopefully I don't end up boring you all!

## 1.1. Particle physics and cosmology

At the first sight, it seems crazy to talk about particle physics and cosmology together. Cosmology is the study of the universe, where the distance scale involved is many Gigaparsecs $\sim 10^{28}$ cm. Particle physics studies the fundamental constituent of matter, now reaching the distance scale of $\sim 10^{-17}$ cm. How can they have anything in common?

The answer is the Big Bang. Discovery of Hubble expansion showed that the visible universe was much smaller in the past, and the study of cosmic microwave background showed the universe was filled with a hot plasma made of photons, electrons, and nuclei in thermal equilibrium. It was *hot*. As we contemplate earlier and earlier epochs of the universe, it was correspondingly smaller and hotter.

On the other hand, the study of small scales $d$ in particle physics translates to large momentum due to the uncertainty principle, $p \sim \hbar/d$. Since large momentum requires relativity, it also means high energy $E \sim cp \sim \hbar c/d$. Physics at higher energies is relevant for the study of higher temperatures $T \sim E/k$, which was the state of the earlier universe.

This way, Big Bang connects microscopic physics to macroscopic physics. And we have already seen two important examples of this connection.

Atomic and molecular spectroscopy is based on quantum physics at the atomic distance $d \sim 10^{-8}$ cm. This spectroscopy is central to astronomy to identify the chemical composition of faraway stars and galaxies which we never hope to get to directly and measure their redshifts to understand their motion including the expansion of the space itself. The cosmic microwave background also originates from the atomic-scale physics when the universe was as hot as $T \sim 4000$ K and hence was in the plasma state. This is the physics which we believe we understand from the laboratory experiments and knowledge of quantum mechanics and hence we expect to be able to extract interesting information about the universe. Ironically, cosmic microwave background also poses a "wall" because the universe was opaque and we cannot "see" with photons the state of the universe before this point. We have to rely on other kinds of "messengers" to extract information about earlier epochs of the universe.

The next example of the micro-macro connection concerns with nuclear physics. The stars are powered by nuclear fusion, obviously a topic in nuclear physics. This notion is now well tested by the recent fantastic development in the study of solar neutrinos, where the core temperature of the Sun is inferred from the helioseismology and solar neutrinos which agree at better than a percent level. Nuclear physics also determines death of a star. Relatively heavy stars even end up with nuclear matter, *i.e.* neutron stars, where the entire star basically becomes a few kilometer-scale nucleus. On the other hand, when the universe was as hot as MeV (ten billion degrees Kelvin), it was too hot for protons and neutrons to be bound in nuclei. One can go through theoretical calculations on how the protons and neutrons became bound in light nuclear species, such as deuterium, $^{3}$He, $^{4}$He, $^{7}$Li, based on the laboratory measurements of nuclear fusion cross sections, as well as number of neutrino species from LEP (Large Electron Positron collider at CERN). This process is called Big-Bang Nucleosynthesis (BBN). There is only one remaining free parameter in this calculation: cosmic

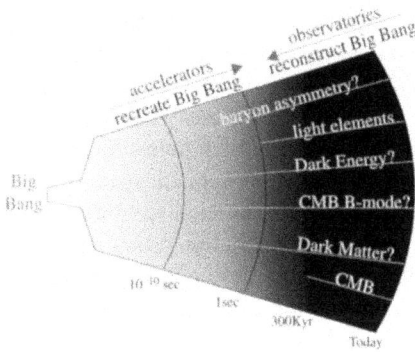

Fig. 1. Possible messengers from early universe.

baryon density. The resulting predictions can be compared to astronomical determinations of light element abundances by carefully selecting the sites which are believed to be not processed by stellar evolutions. There is (in my humble opinion) reasonable agreement between the observation and theoretical predictions (see, *e.g.*, [1]). This agreement gives us confidence that we understand the basic history of the universe since it was as hot as MeV.

We currently do not have messengers from epochs in early universe above the MeV temperature. In other words, our understanding of early universe physics is not tested well for $T \gtrsim$ MeV. Yet many of the topics discussed in this school are *possible* messengers from earlier era: dark matter ($10^3$ GeV?), baryon asymmetry of the universe ($10^{10}$ GeV?), density perturbations (scalar and tensor components) from the inflationary era ($10^{16}$ GeV?). These are the energy scales that laboratory measurements have not reached to reveal the full particle spectrum and their interactions, hence the realm of *physics beyond the standard model*. Understanding of such early stages of the universe requires the development in particle physics, while the universe as a whole may be regarded as a testing ground of hypothesized particle physics at high energies beyond the reach of accelerators. This way, cosmology and particle physics help and require each other.

## 1.2. Next threshold

There is a strong anticipation in the community that we are just about to reveal a new threshold in physics. Let me tell you why from a historical perspective.

We (physicists) do not witness crossing a new threshold very often, but each time it happened, it resulted in a major change in our understanding of Mother Nature.

Around year 1900, we crossed the threshold of atomic scale. It is impressive to recall how much progress chemists have made without knowing the underlying dynamics of atoms and molecules. But the empirical understanding of chemistry had clear limitation. For example, van der Waals equation of state showed there was the distance scale of about $10^{-8}$ cm below which the state-of-art scientific knowledge of the time could not be applied, namely the size of atoms. Once the technology improved to study precision spectroscopy that allowed people to probe physics inside the atoms, a revolution followed. It took about three decades for quantum mechanics to be fully developed but it forever changed our understanding of nature. The revolution went on well into the 40's when the marriage of quantum mechanics and relativity was completed in Quantum ElectroDynamics.

Next important threshold was crossed around 1950 when new hadron resonances and strange particles were discovered, crossing the threshold of the strong interaction scale $\sim 10^{-13}$ cm. Discovery of a zoo of "elementary particles" led to a great deal of confusion for about three decades. It eventually led to the revelation of non-perturbative dynamics of quantum field theory, namely confinement of quarks, dimensional transmutation, and dynamical symmetry breaking of chiral symmetry. More importantly, it showed a new layer in nature where quarks and gluons take over the previous description of subatomic world with protons and neutrons. The experimental verification of this theory, Quantum ChromoDynamics, took well into 90's at numerous accelerators PETRA, PEP, TRISTAN, LEP, and HERA.

One more force that is yet to be fully understood is the weak interaction. Its scale was known from the time of Fermi back in 1933 when he wrote the first theory of nuclear beta decay. The theory contained one dimensionful constant $G_F \approx (300 \text{ GeV})^{-2} \approx (10^{-16} \text{ cm})^2$. Seven decades later, we are just about to reach this energy scale in accelerator experiments, at Tevatron and LHC. We do not really know what Nature has in store for us, but at least we've known all along that this is another important energy scale in physics. If we are not misled, this is the energy scale associated with the cosmic superconductor. Just like the Meissner effect lets magnetic field penetrate into a superconductor only over a finite distance, the cosmic superconductor lets the weak force carried by $W$ and $Z$ bosons go over a tiny distance: a billionth of a nanometer. Right now we are only speculating what revolution may take place at this distance scale. A new layer of matter? New dimensions of space? Quantum dimensions? Maybe string theory? We just don't know yet.

Of course historical perspective does not guarantee that history repeats itself in an equally exciting fashion. But from all what we know, there is a good reason to think that indeed a new threshold is waiting to be discovered at the TeV energy scale, as I will discuss in the next section. Another simple fact is that crossing a new threshold is something like twice-in-a-century experience. I'm excited to think that we are just about to witness one, a historic moment.

An interesting question is what fundamental physics determines these thresholds. The atomic scale, that looked like a fundamental limitation in understanding back in the 19th century, did not turn out to be a fundamental scale at all. It is a derived scale from the mass of the electron and the fundamental constants,

$$a_B = \frac{\hbar^2}{e^2 m_e} \approx 10^{-8} \text{ cm.} \tag{1.1}$$

The strong-interaction scale is also a derived energy scale from the coupling constant

$$a_s = M e^{-8\pi^2/g_s^2(M)b_0} \approx 10^{-13} \text{ cm,} \tag{1.2}$$

where $g_s$ is the strong coupling constant defined at a high-energy scale $M$ and $b_0$ is the beta function coefficient. Because of the asymptotic freedom, the strong coupling constant is weak (what an oxymoron!) at high energies, while it becomes infinitely strong at low energies. The scale of strong interaction is where the strength of the interaction blows up. In other words, the two thresholds we have crossed so far were extremely exciting, yet they turned out to be not fundamental! They point to yet deeper physics that determine these parameters in nature. Maybe the weak-interaction scale is also a derived scale from some deeper physics at yet shorter distances.

## 2. Why beyond the standard model

### 2.1. Empirical reasons

Until about ten years ago, particle physicists lamented that the standard model described every new data that came out from experiments and we didn't have a clue what may lie beyond the standard model. Much of the discussions on physics beyond the standard model therefore were not based on data, but rather on theoretical arguments, primarily philosophical and aesthetic displeasure with the standard model. It all changed the last ten years when empirical evidence appeared that demonstrated that the standard model is incomplete:

- Non-baryonic dark matter,
- Dark energy,
- Neutrino mass,
- Nearly scale-invariant, Gaussian, and apparently acausal density perturbations,
- Baryon asymmetry.

I will discuss strong evidence for non-baryonic dark matter and dark matter later in my lectures. Density fluctuation is covered in many other lectures in this school by Lev Kofman, Sabino Matarrese, Yannick Mellier, Simon Prunet, and Romain Teyssier. Neutrino mass is discussed by Sergio Pastor, and baryon asymmetry by Jim Cline. The bottom line is simple: we already *know* that there must be physics beyond the standard model. However, we don't necessarily know the energy (or distance) scale for this new physics, nor what form it takes. One conservative approach is to try to accommodate all of these established empirical facts into the standard model with minimum particle content: The New Minimal Standard Model [2]. I will discuss some aspects of the model later. But theoretical arguments suggest the true model be much bigger, richer, and more interesting.

## 2.2. Philosophical and aesthetic reasons

What are the theoretical arguments that demand physics beyond the standard model? As I mentioned already, they are based on somewhat philosophical arguments and aesthetic desires and not exactly on firm footing. Nonetheless they are useful and suggestive, especially because nature did solve some of the similar problems in the past by invoking interesting mechanisms. A partial list relevant to my lectures here is

- Hierarchy problem: why $G_F \sim 10^{-5}$ GeV$^{-2} \ll G_N \sim 10^{-38}$ GeV$^{-2}$?
- Why $\theta_{QCD} \lesssim 10^{-10} \ll 1$?
- Why are there three generations of particles?
- Why are the quantum numbers of particles so strange, yet do anomalies cancel so non-trivially?

For an expanded list of the "big questions", see *e.g.*, [3].

To understand what these questions are about, it is useful to remind ourselves how the standard model works. It is a gauge theory based on the $SU(3) \times SU(2) \times U(1)$ gauge group with the Lagrangian

$$
\mathcal{L}_{SM} = -\frac{1}{4g'^2} B_{\mu\nu} B^{\mu\nu} - \frac{1}{2g^2} \text{Tr}(W_{\mu\nu} W^{\mu\nu}) - \frac{1}{2g_s^2} \text{Tr}(G_{\mu\nu} G^{\mu\nu})
$$

$$
+ \bar{Q}_i i \slashed{D} Q_i + \bar{L}_i i \slashed{D} L_i + \bar{u}_i i \slashed{D} u_i + \bar{d}_i i \slashed{D} d_i + \bar{e}_i i \slashed{D} e_i
$$

$$
+ \left( Y_u^{ij} \bar{Q}_i u_j \tilde{H} + Y_d^{ij} \bar{Q}_i d_j H + Y_l^{ij} \bar{L}_i e_j H + \text{h.c.} \right)
$$

$$
+ (D_\mu H)^\dagger (D^\mu H) - \lambda (H^\dagger H)^2 - m^2 H^\dagger H + \frac{\theta}{32\pi^2} \epsilon^{\mu\nu\rho\sigma} \text{Tr}(G_{\mu\nu} G_{\rho\sigma}).
$$

$$
(2.1)
$$

It looks compact enough that it should fit on a T-shirt.[1] Why don't we see such a T-shirt while we see Maxwell equations a lot?

The first two lines describe the gauge interactions. The covariant derivatives $\not{D} = \gamma^\mu D_\mu$ in the second line are determined by the gauge quantum numbers given in this table:

|   | $SU(3)$ | $SU(2)$ | $U(1)$ | chirality |
|---|---------|---------|--------|-----------|
| $Q$ | 3 | 2 | $+1/6$ | left |
| $U$ | 3 | 1 | $+2/3$ | right |
| $D$ | 3 | 1 | $-1/3$ | right |
| $L$ | 1 | 2 | $-1/2$ | left |
| $E$ | 1 | 1 | $-1$ | right |

This part of the Lagrangian is well tested, especially by the LEP/SLC data in the 90's. However, the quantum number assignments (especially $U(1)$ hypercharges) appear very strange and actually hard to remember.[2] Why this peculiar assignment is one of the things people don't like about the standard model. In addition, they are subject to non-trivial anomaly cancellation conditions for $SU(3)^2U(1)$, $SU(2)^2U(1)$, gravity$^2U(1)$, $U(1)^3$, and Witten's $SU(2)$ anomalies. Many of us are left with the feeling that there must be a deep reason for this baroque quantum number assignments which had led to the idea of grand unification.

The third line of the Lagrangian comes with the generation index $i, j = 1, 2, 3$ and is responsible for masses and mixings of quarks and masses of charged leptons. The quark part has been tested precisely in this decade at $B$-factories while there is a glaring omission of neutrino masses and mixings that became established since 1998. In addition, it appears unnecessary for nature to repeat elementary particles three times. The repetition of generations and the origin of mass and mixing patterns remains an unexplained mystery in the standard model.

The last line is completely untested. The first two terms describe the Higgs field and its interaction to the gauge fields and itself. Having not seen the Higgs boson so far, it is far from established. The mere presence of the Higgs field poses an aesthetic problem. It is the only spinless field in the model, but it is introduced

---

[1] It reminds me of an anecdote from when the standard model was just about getting off the ground around 1978. There was a convergence of the data to the standard model and people got very excited about it. Then Tini Veltman gave a talk asking "do you really think this is great model?" and wrote down every single term in the Lagrangian without using a compact notation used here over pages and pages of transparencies. Unfortunately I don't remember who told me this story.

[2] I often told my friends that I chose physics over chemistry or biology because I didn't want to memorize anything, but this kind of table casts serious doubt on my choice!

for the purpose of doing the most important job in the model. In addition, we have not seen *any* elementary spinless particle in nature! Moreover, the potential needs to be chosen with $m^2 < 0$ to cause the cosmic superconductivity which does not give any reason why our universe is in this state. I will discuss more problems about it in a few minutes. Overall, this part of the model looks very artificial.

The last term is the so-called $\theta$-term in QCD and violates $T$ and $CP$. The vacuum angle $\theta$ is periodic under $\theta \rightarrow \theta + 2\pi$, and hence a "natural" value of $\theta$ is believed to be order unity. On the other hand, the most recent experimental upper limit on the neutron electric dipole moment $|d_n| < 2.9 \times 10^{-26} e$ cm (90% CL) [4][3] translates to a stringent upper limit $\theta < (1.2 \pm 0.6) \times 10^{-10}$ using the formula in [5]. Why $\theta$ is so much smaller than the "natural" value is the strong CP problem, and again the standard model does not offer any explanations.

Now we have more to say about the Higgs sector (the third line). Clearly it is very important because (1) this is the only part of the Standard Model which has a dimensionful parameter and hence sets the overall energy scale for the model, and (2) it has the effect of causing cosmic superconductivity without explaining its microscopic mechanism. For the usual superconductors studied in the laboratory, we can use the same Lagrangian, but it is *derived* from the more fundamental theory by Bardeen, Cooper, and Schrieffer. The weak attractive force between electrons by the phonon exchange causes electrons to get bound and condense. The "Higgs" field is the Cooper pair of electrons. And one can show why it has this particular potential. In the standard model, we do not know if Higgs field is elementary or if it is made of something else, nor what mechanism causes it to have this potential.

All the puzzles raised here (and more) cry out for a more fundamental theory underlying the Standard Model. What history suggests is that the fundamental theory lies always at shorter distances than the distance scale of the problem. For instance, the equation of state of the ideal gas was found to be a simple consequence of the statistical mechanics of free molecules. The van der Waals equation, which describes the deviation from the ideal one, was the consequence of the finite size of molecules and their interactions. Mendeleev's periodic table of chemical elements was understood in terms of the bound electronic states, Pauli exclusion principle and spin. The existence of varieties of nuclide was due to the composite nature of nuclei made of protons and neutrons. The list could go on and on. Indeed, seeking answers at more and more fundamental level is the heart of the physical science, namely the reductionist approach.

---

[3]This is an amazing limit. If you blow up the neutron to the size of the Earth, this limit corresponds to a possible displacement of an electron by less than ten microns.

The distance scale of the Standard Model is given by the size of the Higgs boson condensate $v = 250$ GeV. In natural units, it gives the distance scale of $d = \hbar c / v = 0.8 \times 10^{-16}$ cm. We therefore would like to study physics at distance scales shorter than this eventually, and try to answer puzzles whose partial list was given in the previous section.

Then the idea must be that we imagine the Standard Model to be valid down to a distance scale shorter than $d$, and then new physics will appear which will take over the Standard Model. But applying the Standard Model to a distance scale shorter than $d$ poses a serious theoretical problem. In order to make this point clear, we first describe a related problem in the classical electromagnetism, and then discuss the case of the Standard Model later along the same line [9].

### 2.3. Positron analogue

In the classical electromagnetism, the only dynamical degrees of freedom are electrons, electric fields, and magnetic fields. When an electron is present in the vacuum, there is a Coulomb electric field around it, which has the energy of

$$\Delta E_{\text{Coulomb}} = \frac{1}{4\pi\varepsilon_0} \frac{e^2}{r_e}. \tag{2.2}$$

Here, $r_e$ is the "size" of the electron introduced to cutoff the divergent Coulomb self-energy. Since this Coulomb self-energy is there for every electron, it has to be considered to be a part of the electron rest energy. Therefore, the mass of the electron receives an additional contribution due to the Coulomb self-energy:

$$(m_e c^2)_{obs} = (m_e c^2)_{bare} + \Delta E_{\text{Coulomb}}. \tag{2.3}$$

Experimentally, we know that the "size" of the electron is small, $r_e \lesssim 10^{-17}$ cm. This implies that the self-energy $\Delta E$ is greater than 10 GeV or so, and hence the "bare" electron mass must be negative to obtain the observed mass of the electron, with a fine cancellation like[4]

$$0.000511 = (-3.141082 + 3.141593) \text{ GeV}. \tag{2.4}$$

Even setting a conceptual problem with a negative mass electron aside, such a fine-cancellation between the "bare" mass of the electron and the Coulomb self-energy appears ridiculous. In order for such a cancellation to be absent, we conclude that the classical electromagnetism cannot be applied to distance scales shorter than $e^2/(4\pi\varepsilon_0 m_e c^2) = 2.8 \times 10^{-13}$ cm. This is a long distance in the present-day particle physics' standard.

---

[4]Do you recognize $\pi$?

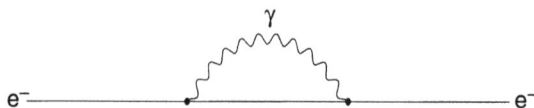

Fig. 2. The Coulomb self-energy of the electron.

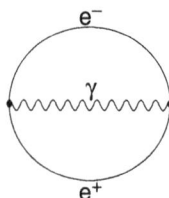

Fig. 3. The bubble diagram which shows the fluctuation of the vacuum.

The resolution to this problem came from the discovery of the anti-particle of the electron, the positron, or in other words by doubling the degrees of freedom in the theory. The Coulomb self-energy discussed above can be depicted by a diagram Fig. 2 where the electron emits the Coulomb field (a virtual photon) which is absorbed later by the electron (the electron "feels" its own Coulomb field).[5] But now that we know that the positron exists (thanks to Anderson back in 1932), and we also know that the world is quantum mechanical, one should think about the fluctuation of the "vacuum" where the vacuum produces a pair of an electron and a positron out of nothing together with a photon, within the time allowed by the energy-time uncertainty principle $\Delta t \sim \hbar/\Delta E \sim \hbar/(2m_e c^2)$ (Fig. 3). This is a new phenomenon which didn't exist in the classical electrodynamics, and modifies physics below the distance scale $d \sim c\Delta t \sim \hbar c/(2m_e c^2) = 200 \times 10^{-13}$ cm. Therefore, the classical electrodynamics actually did have a finite applicability only down to this distance scale, much earlier than $2.8 \times 10^{-13}$ cm as exhibited by the problem of the fine cancellation above. Given this vacuum fluctuation process, one should also consider a process where the electron sitting in the vacuum by chance annihilates with the positron and the photon in the vacuum fluctuation, and the electron which used to be a part of the fluctuation remains instead as a real electron (Fig. 4). V. Weisskopf [10] calculated this contribution to the electron self-energy, and found that it is negative and cancels the leading piece in the

---

[5]The diagrams Figs. 2, 4 are not Feynman diagrams, but diagrams in the old-fashioned perturbation theory with different $T$-orderings shown as separate diagrams. The Feynman diagram for the self-energy is the same as Fig. 2, but represents the *sum* of Figs. 2, 4 and hence the linear divergence is already cancelled within it. That is why we normally do not hear/read about linearly divergent self-energy diagrams in the context of field theory.

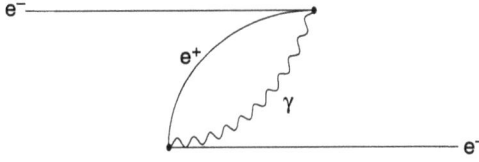

Fig. 4. Another contribution to the electron self-energy due to the fluctuation of the vacuum.

Coulomb self-energy exactly:[6]

$$\Delta E_{\text{pair}} = -\frac{1}{4\pi\varepsilon_0}\frac{e^2}{r_e}. \tag{2.5}$$

After the linearly divergent piece $1/r_e$ is canceled, the leading contribution in the $r_e \to 0$ limit is given by

$$\Delta E = \Delta E_{\text{Coulomb}} + \Delta E_{\text{pair}} = \frac{3\alpha}{4\pi}m_e c^2 \log\frac{\hbar}{m_e c r_e}. \tag{2.6}$$

There are two important things to be said about this formula. First, the correction $\Delta E$ is proportional to the electron mass and hence the total mass is proportional to the "bare" mass of the electron,

$$(m_e c^2)_{obs} = (m_e c^2)_{bare}\left[1 + \frac{3\alpha}{4\pi}\log\frac{\hbar}{m_e c r_e}\right]. \tag{2.7}$$

Therefore, we are talking about the "percentage" of the correction, rather than a huge additive constant. Second, the correction depends only logarithmically on the "size" of the electron. As a result, the correction is only a 9% increase in the mass even for an electron as small as the Planck distance $r_e = 1/M_{Pl} = 1.6 \times 10^{-33}$ cm.

The fact that the correction is proportional to the "bare" mass is a consequence of a new symmetry present in the theory with the antiparticle (the positron): the chiral symmetry. In the limit of the exact chiral symmetry, the electron is massless and the symmetry protects the electron from acquiring a mass from self-energy corrections. The finite mass of the electron breaks the chiral symmetry explicitly, and because the self-energy correction should vanish in the chiral symmetric limit (zero mass electron), the correction is proportional to the electron mass. Therefore, the doubling of the degrees of freedom and the cancellation

---

[6]An earlier paper by Weisskopf actually found two contributions to add up. After Furry pointed out a sign mistake, he published an errata with no linear divergence. I thank Howie Haber for letting me know.

of the power divergences lead to a sensible theory of electron applicable to very short distance scales.

## 2.4. Hierarchy problem

In the Standard Model, the Higgs potential is given by

$$V = m^2|H|^2 + \lambda|H|^4, \tag{2.8}$$

where $v^2 = \langle H \rangle^2 = -m^2/2\lambda = (176 \text{ GeV})^2$. Because perturbative unitarity requires that $\lambda \lesssim 1$, $-m^2$ is of the order of $(100 \text{ GeV})^2$. However, the mass squared parameter $m^2$ of the Higgs doublet receives a quadratically divergent contribution from its self-energy corrections. For instance, the process where the Higgs doublets splits into a pair of top quarks and come back to the Higgs boson gives the self-energy correction

$$\Delta m_{\text{top}}^2 = -6 \frac{h_t^2}{4\pi^2} \frac{1}{r_H^2}, \tag{2.9}$$

where $r_H$ is the "size" of the Higgs boson, and $h_t \approx 1$ is the top quark Yukawa coupling. Based on the same argument in the previous section, this makes the Standard Model not applicable below the distance scale of $10^{-17}$ cm. This is the hierarchy problem. In other words, if we don't solve this problem, we can't even talk about physics at much shorter distances without an excessive fine-tuning in parameters.

It is worth pondering if the mother nature may fine-tune. Now that the cosmological constant appears to be fine-tuned at the level of $10^{-120}$, should we be really worried about the fine-tuning of $v^2/M_{Pl}^2 \approx 10^{-30}$ [6]? In fact, some people argued that the hierarchy exists because intelligent life cannot exist otherwise [7]. On the other hand, a different way of varying the hierarchy does seem to support stellar burning and life [8]. I don't get into this debate here, but I'd like to just point out that a different fine-tuning problem in cosmology, horizon and flatness problems, pointed to the theory of inflation, which in turn appears to be empirically supported by data. I just hope that proper solutions will be found to both of these fine-tuning problems and we will see their manifestations at the relevant energy scale, namely TeV. You have to be an optimist to work on big problems.

## 3. Examples of physics beyond the standard model

Given various problems in the standard model discussed in the previous section, especially the hierarchy problem, many possible directions of physics beyond the standard model have been proposed. I can review only a few of them here given the spacetime constraint. But I especially emphasize the aspect of the models that leads to a (nearly) stable neutral particle as a good dark matter candidate.

## 3.1. Supersymmetry

The motivation for supersymmetry is to make the Standard Model applicable to much shorter distances so that we can hope that the answers to many of the puzzles in the Standard Model can be given by physics at shorter distance scales [11]. In order to do so, supersymmetry repeats what history did with the positron: doubling the degrees of freedom with an explicitly broken new symmetry. Then the top quark would have a superpartner, the stop,[7] whose loop diagram gives another contribution to the Higgs boson self energy

$$\Delta m_{\text{stop}}^2 = +6\frac{h_t^2}{4\pi^2}\frac{1}{r_H^2}.\tag{3.1}$$

The leading pieces in $1/r_H$ cancel between the top and stop contributions, and one obtains the correction to be

$$\Delta m_{\text{top}}^2 + \Delta m_{\text{top}}^2 = -6\frac{h_t^2}{4\pi^2}\left(m_{\tilde{t}}^2 - m_t^2\right)\log\frac{1}{r_H^2 m_{\tilde{t}}^2}.\tag{3.2}$$

One important difference from the positron case, however, is that the mass of the stop, $m_{\tilde{t}}$, is unknown. In order for the $\Delta m^2$ to be of the same order of magnitude as the tree-level value $m^2 = -2\lambda v^2$, we need $m_{\tilde{t}}^2$ to be not too far above the electroweak scale. TeV stop mass is already a fine tuning at the level of a percent. Similar arguments apply to masses of other superpartners that couple directly to the Higgs doublet. This is the so-called naturalness constraint on the superparticle masses (for more quantitative discussions, see papers [12]).

Supersymmetry doubles the number of degrees of freedom in the standard model. For each fermion (quarks and leptons), you introduce a complex scalar field (squarks and sleptons). For each gauge boson, you introduce gaugino, a partner Majorana fermion (a fermion field whose anti-particle is itself). I do not go into technical aspect of how to write a supersymmetric quantum field theory; you should consult some review articles [13, 14].

One important point related to dark matter is the proton longevity. We know from experiments such as SuperKamiokande that proton is *very* long lived (if not immortal). The life time for the decay mode $p \rightarrow e^+\pi^0$ is longer than

---

[7]This is a terrible name, which was originally meant to be "scalar top" or "supersymmetric top." Some other names are even worse: *sup*, *sstrange*, etc. If supersymmetry will be discovered at LHC, we should seriously look for better names for the superparticles, maybe after the names of rich donors.

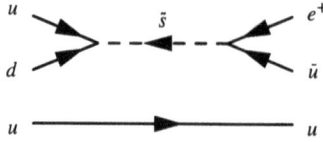

Fig. 5. A possible Feynman diagram with supersymmetric particles that can lead to a too-rapid proton decay $p \to e^+ \pi^0$.

$1.6 \times 10^{33}$ years, at least twenty-three orders of magnitude longer than the age of the universe! On the other hand, if you write the most general renormalizable theory with standard model particle content consistent with supersymmetry, it allows for vertices such as $\epsilon_{ijk} u^i d^j \tilde{s}^k$ and $e u^i \tilde{s}^*_i$ (here $i, j, k$ are color indices). Then one can draw a Feynman diagram like one in Fig. 5. If the couplings are $O(1)$, and superparticles around TeV, one finds the proton lifetime as short as $\tau_p \sim m_{\tilde{s}}^4 / m_p^5 \sim 10^{-12}$ sec; *a little too short!*

Because of this embarrassment, we normally introduce a $\mathbb{Z}_2$ symmetry called "*R*-parity" defined by

$$R_p = (-1)^{3B+L+2s} = (-1)^{\text{matter}} R_{2\pi} \tag{3.3}$$

where $s$ is the spin. What it does is to flip the sign of all matter fields (quarks and leptons) and perform $2\pi$ rotation of space at the same time. In effect, it assigns even parity to all particles in the standard model, and odd parity to their superpartners. Here is a quick check. For the quarks, $B = 1/3$, $L = 0$, and $s = 1/2$, and we find $R_p = +1$, while for squarks the difference lies in $s = 0$ and hence $R_p = -1$. This symmetry forbids both of the bad vertices in Fig. 5.

Once the *R*-parity is imposed,[8] there are no baryon- and lepton-number violating interaction you can write down in a renormalizable Lagrangian with the standard model particle content. This way, the *R*-parity makes sure that proton is long lived. Then the lightest supersymmetric particle (LSP), with odd *R*-parity, cannot decay because there are no other states with the same *R*-parity with smaller mass it can decay into by definition. In most models it also turns out to be electrically neutral. Then one can talk about the possibility that the LSP is the dark matter of the universe.

---

[8] An obvious objection is that imposing *R*-parity appears *ad hoc*. Fortunately there are several ways for it to emerge from a more fundamental theory. Because the *R*-parity is anomaly-free [15], it may come out from string theory. Or $R_p$ can arise as a subgroup of the $SO(10)$ grand unified gauge group because the matter belongs to the spinor representation and Higgs to vector, and hence $2\pi$ rotation in the gauge group leads precisely to $(-1)^{\text{matter}}$. It may also be an accidental symmetry due to other symmetries of the theory [16, 17] so that it is slightly broken and dark matter may eventually decay.

## 3.2. Composite Higgs

Another way the hierarchy problem may be solved is by making the Higgs boson to actually have a finite size. Then the correction in Eq. (2.9) does not require tremendous fine-tuning as long as the physical size of the Higgs boson is about $r_H \approx (\text{TeV})^{-1} \approx 10^{-17}$ cm. This is possible if the Higgs boson is a composite object made of some elementary constituents.

The original idea along this line is called technicolor (see reviews [18, 19]), where a new strong gauge force binds fermions and anti-fermions much like mesons in the real QCD. Again just like in QCD, fermion anti-fermion pair have a condensate $\langle \bar{\psi} \psi \rangle \neq 0$ breaking chiral symmetry. In technicolor theories, this chiral symmetry breaking is nothing but the breaking of the electroweak $SU(2) \times U(1)$ symmetry to the $U(1)$ QED subgroup. Because the Higgs boson is heavy and strongly interacting, it is expected to be too wide to be seen as a particle state.

It is fair to say, however, that the technicolor models suffer from various problems. First of all, it is difficult to find a way of generating sufficient masses for quarks and leptons, especially the top quark, because you have to rely on higher dimension operators of type $\bar{q} q \bar{\psi} \psi / \Lambda^2$. The scale $\Lambda$ must be low enough to generate $m_t$, while high enough to avoid excessive flavor-changing neutral current. In addition, there is tension with precision electroweak observables. These observables are precise enough that they constrain heavy particles coupled to $Z$- and $W$-bosons even though we cannot produce them directly.[9]

Because of this issue, there are various other incarnations of composite Higgs idea, which try to get a relatively light Higgs boson as a bound state [26, 27]. One of the realistic models is called "little Higgs" [24, 25]. Because of the difficulty of achieving Higgs compositeness at the TeV scale, we are better off putting off the compositeness scale to about 10 TeV to avoid various phenomenological constraints. Then you must wonder if the problem with Eq. (2.9) comes back. But there is a way of protecting the scale of Higgs mass much lower than the compositeness scale by using symmetries similar to the reason why a pion is so much lighter than a proton. If you are clever, you can arrange the structure of symmetry such that it eliminates the one-loop correction in Eq. (2.9) and the correction arises only at the two-loop level. Then the compositeness $\sim 10$ TeV is not a problem.

Another attractive idea is to use extra dimensions to generate the Higgs field from a gauge field, called "Higgs-gauge unification" [28–32]. We know the mass of the gauge boson is forbidden by the gauge invariance. If the Higgs field is

---

[9]It is curious that higher dimensional versions of technicolor models called Higgsless models [21] do much better [22]. A supersymmetric version of technicolor also does better than the original technicolor [23].

actually a gauge boson (spin one), but if it is spinning in extra dimensions, we (as observers stuck in four dimensions) perceive it not to spin. Not only this gives us *raison d'être* of (apparently) spinless degrees of freedom, it also provides protection for the Higgs mass and hence solves the hierarchy problem. The best implementation of this line of thinking is probably the holographic Higgs model in Refs. [33, 34] which involves the warped extra dimension I will briefly discuss in the next section. It should also be said that many of the ideas mentioned here are closely related to each other [35].

Similarly to the case of supersymmetry, people often introduce a $\mathbb{Z}_2$ symmetry to avoid certain phenomenological embarrassments. In little Higgs theories, tree-level exchange of new particles tend to cause tension with precision electroweak constraints. Then the new states must be sufficiently heavy so that the hierarchy problem is reintroduced. By imposing "$T$-parity," new particles can only appear in loops for low-energy processes and the constraints can be easily avoided [36]. Then the lightest $T$-odd particle (LTP) becomes a candidate for dark matter. In technicolor models, the lightest technibaryon is stable (just like proton in QCD) and a dark matter candidate [37].

### 3.3. Extra dimensions

The source of the hierarchy problem is our thinking that there *is* physics at much shorter distances than $10^{-17}$ cm. What if there isn't? What if physics *ends* at $10^{-17}$ cm where quantum field theory stops applicable and is taken over by something more radical such as string theory? Normally we associate the ultimate limit of field theory with the Planck scale $d_{Pl} \approx 10^{-33}$ cm $= G_N^{1/2}$ where the gravity becomes as strong as other forces and its quantum effects can no longer be ignored. How then can the quantum gravity effects enter at a much larger distance scale such as $10^{-17}$ cm?

One way is to contemplate "large" extra dimensions of size $R$ [38]. Imagine there are $n$ extra dimensions in addition to our three-dimensional space. If you place two test masses at a distance $r$ much shorter than $R$, the field lines of gravity spread out into $3 + n$ dimensions and the force decreases as the surface area $r^{-(n+2)}$. However, if the distance is longer than $R$, there is a limit to which how much the field lines can spread because they are squeezed within the size $R$. Therefore, the force decreases only as $r^{-2}$ for $r \gg R$, reproducing the usual inverse square law of gravity. It turns out that the inverse square law is tested only down to 44 $\mu$m [39] (even though this is very impressive!) and extra dimensions smaller than that are allowed experimentally.

Matching two expressions for the gravitational force at $r = R$, we can related the Newton's constant in $n + 3$ dimensions to the Newton's constant $G_N$ in three

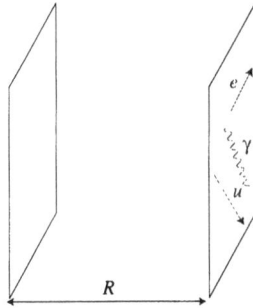

Fig. 6. Large extra dimensions. Even though the three-brane is drawn at the ends of extra dimensions, it does not have to be.

dimensions

$$G_{n+3}\frac{1}{R^{n+2}} = G_N\frac{1}{R^2},$$ (3.4)

and hence $G_{n+3} = G_N R^n$. In the natural unit $\hbar = c = 1$, the mass scale of gravity is related to the Planck scale by $M^{n+2}R^n = M_{Pl}^2$. Even if the true energy scale of quantum gravity is at $M \sim 1$ TeV, we may find an *apparent* scale of gravity to be $M_{Pl} \sim 10^{19}$ GeV. Then the required size of extra dimensions is

$$R = \begin{cases} 10^{15} \text{ cm} & (n = 1) \\ 10^{-1} \text{ cm} & (n = 2) \\ 10^{-6} \text{ cm} & (n = 3) \\ 10^{-12} \text{ cm} & (n = 6) \end{cases}$$ (3.5)

Obviously the $n = 1$ case is excluded because $R$ is even bigger than 1AU $\approx 10^{13}$ cm. The case $n = 2$ is just excluded by the small-scale gravity experiment, while $n \leq 3$ is completely allowed.

If we don't see the extra dimensions directly, what do they do to us? Let us look at the case of just one extra dimension with periodic boundary condition $y \rightarrow y + 2\pi R$. Then all particles have wave functions on the coordinate $y$ that satisfies $\psi(y + 2\pi R) = \psi(y)$. They can of course depend on the usual four-dimensional space time $x$, too. One can expand it in Fourier modes

$$\psi(x, y) = \sum_n \psi_n(x)e^{iny/R}.$$ (3.6)

The momentum along the $y$ direction is $p_y = -i\partial_y = n/R$, and the total energy of the particle is

$$E = \sqrt{\vec{p}^2 + p_y^2} = \sqrt{\vec{p}^2 + \left(\frac{n}{R}\right)^2}. \tag{3.7}$$

Namely that you find a tower of particles of mass $m_n = n/R$, called Kaluza–Klein states.

Of course, the standard model is tested down to $10^{-17}$ cm, and we have not found Kaluza–Klein excitation of electron, etc. This is not a problem if we are stuck on a three-dimensional sheet (three-*brane*) embedded inside the $n + 3$ dimensional space. The branes are important objects in string theory and it is easy to get particles with gauge interactions stuck on them. The brane may be freely floating inside extra dimensions or may be glued at singularities (*e.g.*, orbifold fixed points). The simplest way to use large extra dimensions is to assume that only gravity is spread out in extra dimensions, while the standard model particles are all on a three-brane.

Cosmology with large extra dimension is an iffy subject, however. The Kaluza–Klein excitation of gravitons can be produced in early universe and the cosmology would be different from the standard Friedmann univese (see, *e.g.*, [40]). I will not get into this discussion here.

Instead of models with large extra dimensions, models with small extra dimensions of size $R \approx 10^{-17}$ cm $\approx$ TeV$^{-1}$ are also interesting,[10] which allow for normal cosmology below TeV temperatures. This would also allow us standard model particles to live in extra dimensions, too, because our Kaluza–Klein excitations have been too heavy to be produced at accelerators so far. There are many versions of small extra dimensions.

One very popular version is warped extra dimension [41]. Instead of flat metric in the extra dimensions, it sets up an exponential behavior. It is something like Planck scale varies from $10^{19}$ GeV to TeV as you go across the 5th dimension. Therefore, physics does *end* at TeV if you on one end of the 5th dimension, while it keeps going to $10^{19}$ GeV on the other end. The hierarchy problem may be solved if Higgs resides on (or close to) the "TeV brane." This set up attracted a lot of attention because the bulk is actually a slice of anti-de Sitter space which has nice features of preserving supersymmetry, leading to AdS/CFT correspondence [42], etc. It is also possible to obtain quite naturally from string theory [43]. In a grand unified model from warped extra dimension, the proton longevity is

---

[10]Historically, unified theories and string theory assumed $R \approx d_{Pl} \approx 10^{-33}$ cm. TeV-sized extra dimensions are *much* larger than this, but I'm calling them "small" for the sake of distinction from the large extra dimensions.

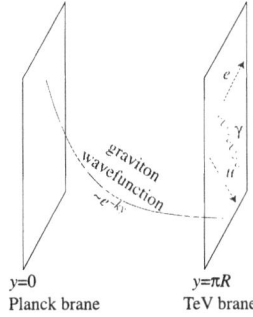

Fig. 7. Warped extra dimension. Even though the standard model particles are shown to be on the TeV brane, they may propagate in the bulk depending on the models.

an issue which is solved by a $\mathbb{Z}_3$ symmetry, and the lightest $\mathbb{Z}_3$-charged particle (LZP) is a candidate for dark matter [44].

It is also possible to have the "flat" extra dimension at the TeV scale and put all the standard model particles in the 5D bulk, called Universal Extra Dimension (UED) [45]. It is tricky to get chiral fermions in four dimensions if they are embedded in higher dimensional space. If you start out with five-dimensional Dirac equation

$$\left(i\gamma^\mu \partial_\mu + \gamma^5 \partial_y\right)\psi(x, y) = 0, \tag{3.8}$$

the Fourier-mode expansion for the mode $\psi_n(x)e^{-iny/R}$ gives

$$\left(i\gamma^\mu \partial_\mu - i\frac{n}{R}\gamma^5\right)\psi_n(x) = 0. \tag{3.9}$$

After a chiral rotation $\psi_n \to \psi_n e^{i\pi\gamma^5/2}\psi_n$, the second term turns into the usual mass term without $\gamma^5$. The problem is that here are always two eigenvalues $\gamma^5 = \pm 1$ and you find both left- and right-handed fermions with the same quantum numbers. Namely, you get Dirac fermion, not Weyl fermion. Then you don't get the standard model that distinguishes left from right. In terms of spectrum, what is on the left of Fig. 8 is the spectrum because the Fourier modes $n$ and $-n$ give the degenerate mass $n/R$ each of them with its own Dirac fermion.

The trick to get chiral fermions is to use an orbifold Fig. 9. Out of a circle $S^1$ ($y \in [-\pi R, \pi R]$), you identify points $y$ and $-y$ to get a half-circle $S^1/\mathbb{Z}_2$. There are two special points, $y = 0$ and $\pi R$, that are identified only with themselves called "fixed points." In addition, we take the boundary condition that $\psi(y) = -\gamma^5\psi(-y)$. For $n \neq 0$, we use $\cos ny/R$ for $\gamma^5 = -1$ and $\sin ny/R$ for $\gamma^5 = +1$, without the degeneracy between $n$ and $-n$. For $n = 0$, only $\gamma^5 = -1$

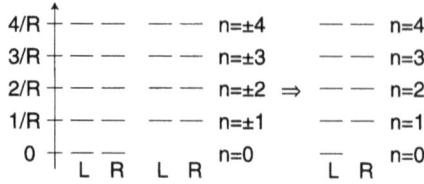

Fig. 8. The spectrum of fermions in the 5D bulk. After orbifold identification in Fig. 9, the spectrum is halved and one can obtain chiral fermions in 4D.

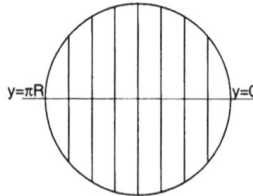

Fig. 9. The orbifold $S^1/\mathbb{Z}_2$. Points connected by the solid lines are identified.

survives with the wave function $\psi_0(y) = 1$. This way, we keep only a half of the states as shown in Fig. 8, and we can get chiral fermions. As a consequence, we find the system to have a $\mathbb{Z}_2$ symmetry under $y \to \pi R - y$, under which modes with even $n$ are even and odd $n$ odd. This $\mathbb{Z}_2$ symmetry is called KK parity and the lightest KK state (LKP) becomes stable. At the tree-level, all first Kaluza–Klein states are degenerate $m_1 = 1/R$.[11] Radiative corrections split their masses, and typically the first Kaluza–Klein excitation of the $U(1)_Y$ boson is the LKP [46]. Because the mass splittings are from the loop diagrams, they are small. Similarly to supersymmetry, there is a large number of new particles beyond the standard model, namely Kaluza–Klein excitations. Its collider phenomenology very much resembles that of supersymmetry and it is not trivial to tell them apart at the LHC (dubbed "bosonic supersymmetry" [47]).

## 4. Evidence for dark matter

Now we turn our attention to the problem of non-baryonic dark matter in the universe. Even though this is a sudden change in the topic, you will see soon that it is connected to the discussions we had on physics beyond the standard model. We first review basics of observational evidence for non-baryonic dark matter,

---

[11] Here I've ignored possible complications due to brane operators and electroweak symmetry breaking.

and then discuss how some of the interesting candidates are excluded. It leads to a paradigm that dark matter consists of unknown kind of elementary particles. By a simple dimensional analysis, we find that a weakly coupled particle at the TeV-scale naturally gives the correct abundance in the current universe. We will take a look at a simple example quite explicitly so that you can get a good feel on how it works. Then I will discuss more attractive dark matter candidates that arise from various models of physics beyond the standard model I discussed in the previous section.

The argument for the existence of "dark matter," namely mass density that is not luminous and cannot be seen in telescopes, is actually very old. Zwicky back in 1933 already reported the "missing mass" in Coma cluster of galaxies. By studying the motion of galaxies in the cluster and using the virial theorem (assuming of course that the galactic motion is virialized) he determined the mass distribution in the cluster and reported that a substantial fraction of mass is not seen. Since then, the case for dark matter has gotten stronger and stronger and most of us regard its existence established by now. I refer to a nice review for details [48] written back in 1997, but I add some important updates since the review.

Arguably the most important one is the determination of cosmological parameters by the power spectrum of CMB anisotropy. In the fit to the power-law flat $\Lambda$CDM model gives $\Omega_M h^2 = 0.127^{+0.007}_{-0.013}$ and $\Omega_B h^2 = 0.0223^{+0.0007}_{-0.0009}$ [49]. The point here is that these two numbers are *different*. Naively subtracting the baryon component, and adding the errors by quadrature, I find $(\Omega_M - \Omega_B)h^2 = 0.105^{+0.007}_{-0.013} \neq 0$ at a very high precision. This data alone says most of the matter component in the universe is not atoms, something else.

Another important way to determine the baryon density of the universe is based on Big-Bang Nucleosynthesis (BBN). The baryon density is consistent with what is obtained from the CMB power spectrum, $\Omega_B h^2 = 0.0216^{+0.0020}_{-0.0021}$ from five best measurements of deuterium abundance [50] using hydrogen gas at high redshift (and hence believed to be primordial) back-lit by quasars. This agrees very well with the CMB result, even though they refer to very different epochs: $T \sim 1$ MeV for BBN while $T \sim 0.1$ eV for CMB. This agreement gives us confidence that we know $\Omega_B$ very well.

A novel technique to determine $\Omega_M$ uses large-scale structure, namely the power spectrum in galaxy-galaxy correlation function. As a result of the acoustic oscillation in the baryon-photon fluid, the power spectrum also shows the "baryon oscillation" which was discovered only the last year [51]. Without relying on the CMB, they could determine $\Omega_M h^2 = 0.130 \pm 0.010$. Again this is consistent with the CMB data, confirming the need for non-baryonic dark matter.

I'd like to also mention a classic strong evidence for dark matter in galaxies. It comes from the study of rotation curves in spiral galaxies. The stars and gas rotate around the center of the galaxy. For example, our solar system rotates in our Milky Way galaxy at the speed of about 220 km/sec. By using Kepler's law, the total mass $M(r)$ within the radius $r$ and the rotation speed at this radius $v(r)$ are related by

$$v(r)^2 = G_N \frac{M(r)}{r}. \tag{4.1}$$

Once the galaxy runs out of stars beyond a certain $r$, the rotation speed is hence expected to decrease as $v(r) \propto r^{-1/2}$. This expectation is not supported by observation.

You can study spiral galaxies which happen to be "edge-on." At the outskirts of a galaxy, where you don't find any stars, there is cold neutral hydrogen gas. It turns out you can measure the rotation speed of this cold gas. A hydrogen atom has hyperfine splitting due to the coupling of electron and proton spins, which corresponds to the famous $\lambda = 21$ cm line emission. Even though the gas is cold, it is embedded in the thermal bath of cosmic microwave background whose temperature 2.7 K is *hot* compared to the hyperfine excitation $hc/k\lambda = 0.069$ K.[12] Therefore the hydrogen gas is populated in both hyperfine states and spontaneously emits photons of wavelength 21 cm by the M1 transition. This can be detected by radio telescopes. Because you are looking at the galaxy edge-on, the rotation is either away or towards us, causing Doppler shifts in the 21 cm line. By measuring the amount of Doppler shifts, you can determine the rotation speed. Surprisingly, it was found that the rotation speed stays constant well beyond the region where stars cease to exist.

I mentioned this classic evidence because it really shows galaxies are filled with dark matter. This is an important point as we look for signals of dark matter in our own galaxy. It is not easy to determine how much dark matter there is, however, because eventually the hydrogen gas runs out and we do not know how far the flat rotation curve extends. Nonetheless, it shows the galaxy to be made up of a nearly spherical "halo" of dark matter in which the disk is embedded.

## 5. What dark matter is not

We don't know what dark matter is, but we have learned quite a bit recently what it is *not*. I have already discussed that it is *not* ordinary atoms (baryons). I mention a few others of the excluded possibilities.

---

[12]If we had lived in a universe a hundred times larger, we would have lost this opportunity of studying dark matter content of the galaxies!

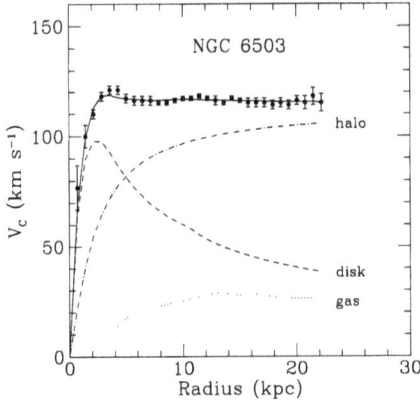

Fig. 10. Rotation curve of a spiral galaxy [52].

## 5.1. MACHOs

The first candidate for dark matter that comes to mind is some kind of astronomical objects, namely stars or planets, which are is too dark to be seen. People talked about "Jupiters," "brown dwarfs," etc. In some sense, that would be the most conservative hypothesis.[13] Because dark matter is not made of ordinary atoms, such astronomical objects cannot be ordinary stars either. But one can still contemplate the possibility that it is some kind of exotic objects, such as black holes. Generically, one refers to MACHOs which stand for MAssive Compact Halo Objects.

Black holes may be formed by some violent epochs in Big Bang (primordial black holes or PBHs) [53] (see also [54]). If the entire horizon collapses into a black hole, which is the biggest mass one can imagine consistent with causality, for example in the course of a strongly first order phase transition, the black hole mass would be

$$M_{\text{PBH}} \approx M_\odot \left( \frac{T}{100 \, \text{MeV}} \right)^{-2} \left( \frac{g_*}{10.75} \right)^{-1/2}. \tag{5.1}$$

Therefore, there is no causal mechanism to produce PBHs much larger than $10^3 M_\odot$ assuming that universe has been a normal radiation dominated universe for $T \lesssim 3 \, \text{MeV}$ to be compatible with Big-Bang Nucleosynthesis. Curiously, one finds $M_{\text{PBH}} \approx M_\odot$ if it formed at the QCD phase transition $T \approx 100 \, \text{MeV}$ [55]. On the other hand, PBHs cannot be too small because otherwise they emit Hawk-

---

[13] Somehow I can't call primordial black holes a "conservative" candidate without chuckling.

ing radiation of temperature $T = (8\pi G_N M_{PBH})^{-1}$ that would be visible. The limit from diffuse gamma ray background implies $M_{PBH} \gtrsim 10^{-16} M_\odot$.

How do we look for such invisible objects? Interestingly, it is not impossible using the gravitational microlensing effects [56]. The idea is simple. You keep monitoring millions of stars in nearby satellite galaxies such as Large Magellanic Cloud (LMC). Meanwhile MACHOs are zooming around in the halo of our galaxy at $v \approx 220$ km/s. By pure chance, one of them may pass very close along the line of sight towards one of the stars you are monitoring. Then the gravity would focus light around the MACHO, effectively making the MACHO a lens. You typically don't have a resolution to observe distortion of the image or multiple images, but the focusing of light makes the star appear temporarily brighter. This is called "microlensing." By looking for such microlensing events, you can infer the amount of MACHOs in our galactic halo.

I've shown calculations on the deflection angle by the gravitational lensing and the amplification in the brightness in the appendix. (Just for fun, I've also added some discussions on the strong lensing effects.) The bottom line is that you may expect the microlensing event at the rate of

$$\text{rate} \approx 5 \times 10^{-6} \frac{1}{\text{year}} \left( \frac{M_\odot}{M_{MACHO}} \right)^{1/2} \tag{5.2}$$

towards the LMC, with the duration of

$$\text{duration} \approx 6 \times 10^6 \text{sec} \left( \frac{M_{MACHO}}{M_\odot} \right)^{1/2} \left( \frac{\sqrt{d_1 d_2}}{25 \text{ kpc}} \right), \tag{5.3}$$

where $d_1$ ($d_2$) is the distance between the MACHO and us (the lensed star).

Two collaborations, the MACHO collaboration and the EROS collaboration, have looked for microlensing events. The basic conclusion is that MACHOs of mass $10^{-7}$–$30 \, M_\odot$ cannot make up 100% of our galactic halo (Fig. 11). See also [57, 58].

Even though the possibility of MACHO dark matter may not be completely closed, it now appears quite unlikely. The main paradigm for the dark matter of the universe has shifted from MACHOs to WIMPs.

## 5.2. Neutrinos

Having discovered neutrinos have finite mass, it is also natural to consider neutrinos to be dark matter candidate. As a matter of fact, neutrinos *are* a component of dark matter, contributing

$$\Omega_\nu h^2 = \frac{\sum_i m_{\nu_i}}{94 \text{ eV}}. \tag{5.4}$$

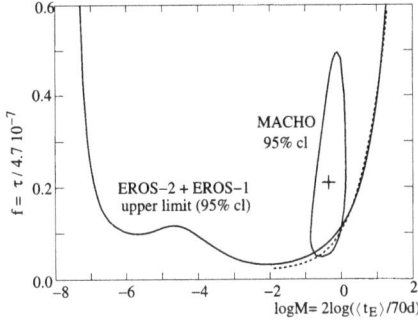

Fig. 11. Limit on the halo fraction $f$ of MACHOs from the EROS collaboration [57]. The spherical isothermal model of halo predicts the optical depth towards the LMC of $\tau = 4.7 \times 10^{-7}$. For more details, see the paper.

It is an attractive possibility if the particles which we already know to exist could serve as the required non-baryonic dark matter.

However, as Sergio Pastor discussed in his lectures, neutrinos are not good candidates for the bulk of dark matter for several reasons. First, there is an upper limit on neutrino mass from laboratory experiments (tritium beta decay) $m <$ 2 eV [59]. Combined with the smallness of mass-squared differences $\Delta m_\odot^2 = 8 \times 10^{-5}$ eV$^2$ and $\Delta m_\oplus^2 = 2.5 \times 10^{-3}$ eV$^2$, electron-volt scale neutrinos should be nearly degenerate. Then the maximum contribution to the matter density is $\Omega_\nu h^2 < (3 \times 2/94) < 0.064$. This is not enough.

Second, even if the laboratory upper limit on the neutrino mass turned out to be not correct, there is a famous Tremaine-Gunn argument [60]. For the neutrinos to dominate the halo of dwarf galaxies, you need to pack them so much that you would violate Pauli exclusion principle. To avoid this, you need to make neutrinos quite massive $\gtrsim 500$ eV so that you don't need so many of them [61]. This obviously contradicts the requirement that $\Omega_\nu < 1$.

Third, neutrinos are so light that they are still moving at speed of light (Hot Dark Matter) at the time when the structure started to form, and erase structure at small scales. Detailed study of large scale structure shows such a hot component of dark matter must be quite limited. The precise limit depends on the exact method of analyses. A relatively conservative limit says $\sum_i m_{\nu_i} < 0.62$ eV [62] while a more aggressive limit goes down to 0.17 eV [63]. Either way, neutrinos cannot saturate what is needed for non-baryonic dark matter.

In fact, what we want is Cold Dark Matter, which is already non-relativistic and slowly moving at the time of matter-radiation equality $T \sim 1$ eV. Naively a light (sub-electronvolt) particle would not fit the bill.

A less conservative hypothesis may be to postulate that there is a new heavy neutrino (4th generation). This is a prototype for WIMPs that will be discussed later. It turns out, however, that the direct detection experiments and the abundance do not have a compatible mass range. Namely the neutrinos are too strongly coupled to be the dark matter!

### 5.3. CHAMPs and SIMPs

Even though people do not talk about it any more, it is worth recalling that dark matter is unlikely be charged (CHAMP) [64] or strongly interacting (SIMP) [65]. I simply refer to papers that limit such possibilities, from a multitude of search methods that include search for anomalously heavy "water" molecule in the sea water, high-energy neutrinos from the center of the Earth from annihilated SIMPs accumulated there, collapsing neutron stars that accumulate CHAMPs, etc.[14]

## 6. WIMP dark matter

WIMP, or Weakly Interactive Massive Particle, is the main current paradigm for explaining dark matter of the universe. With MACHOs pretty much gone, it is indeed attractive to make a complete shift from astronomical objects as heavy as $M_\odot \approx 10^{57}$ GeV to "heavy" elementary particles of mass $\sim 10^2$ GeV. I will discuss why this mass scale is particularly interesting.

### 6.1. WIMP

The idea of WIMP is very simple. It is a relatively heavy elementary particle $\chi$ so that accelerator experiments so far did not have enough energy to create them, namely $m_\chi \gtrsim 10^2$ GeV. On the other hand, the Big Bang did once have enough energy to make them.

Let us follow the history from when $T \gtrsim m_\chi$. WIMPs were created as much as any other particles. Once the temperature dropped below $m_\chi$, even the universe stopped creating them. If they are stable, whatever amount that was produced was there, and the only way to get rid of them was to get them annihilating each other into more mundane particles (*e.g.*, quarks, leptons, gauge bosons). However, the universe expanded and there were fewer and fewer WIMPs in a given volume, and at some point WIMPs stopped finding each other. Then they

---

[14]I once got interested in the possibility that Jupiter is radiating heat more than it receives from the Sun because SIMPs are annihilating at its core [66]. It does not seem to explain heat from other Jovian planets, however, once empirical limits on SIMPs are taken into account.

could not annihilate any more and hence their numbers become fixed ("freeze-out"). This way, the universe could still be left with a certain abundance of WIMPs. This mechanism of getting dark matter is called "thermal relics."

Let us make a simple estimate of the WIMP abundance. In radiation dominated universe, the expansion rate is given by

$$H = \frac{\dot{a}}{a} = g_*^{1/2} \frac{T^2}{M_{Pl}} \left(\frac{\pi^2}{90}\right)^{1/2}, \tag{6.1}$$

where $M_{Pl} = 1/\sqrt{8\pi G_N} = 2.4 \times 10^{18}$ GeV is the reduced Planck scale. For simple estimates, we regard $(\pi^2/90)^{1/2} = 0.33 \approx 1$ and ignore many other factors of $O(1)$. Hence, $H \simeq g_*^{1/2} T^2/M_{Pl}$. The entropy density is correspondingly

$$s = g_* T^3 \left(\frac{2\pi^2}{45}\right)^{1/2} \simeq g_* T^3. \tag{6.2}$$

Given the thermally averaged annihilation cross section $\langle \sigma_{ann} v \rangle$, and the number density of WIMPs $n_\chi$, the annihilation rate of a WIMP is

$$\Gamma = \langle \sigma_{ann} v \rangle n_\chi. \tag{6.3}$$

The annihilation stops at the "freeze-out temperature" $T_f$ when $\Gamma \simeq H$, and hence

$$n_\chi(T_f) \simeq g_*^{1/2} \frac{T_f^2}{\langle \sigma_{ann} v \rangle M_{Pl}}. \tag{6.4}$$

The yield of WIMPs is defined by $Y_\chi = n_\chi/s$. This is a convenient quantity because it is conserved by the expansion of the universe as long as the expansion is adiabatic, *i.e.*, no new source of heat. This is due to the conservation of both the total entropy and total number of particles and their densities both scale as $1/a^3$. The estimate of the yield is

$$Y_\chi \sim g_*^{-1/2} \frac{1}{\langle \sigma_{ann} v \rangle T_f M_{Pl}} = g_*^{-1/2} \frac{x_f}{\langle \sigma_{ann} v \rangle m_\chi M_{Pl}}. \tag{6.5}$$

Here, we defined $T_f = m_\chi/x_f$. We will see later from more detailed calculations that $x_f \sim 20$. The abundance in the current universe is calculated using the yield and the current entropy density, divided by the current critical density,

$$\Omega_\chi = m_\chi \frac{n_\chi}{s} \frac{s_0}{\rho_c} \sim g_*^{-1/2} \frac{x_f}{\langle \sigma_{ann} v \rangle M_{Pl}} \frac{s_0}{\rho_c}. \tag{6.6}$$

We use $s_0 = 2890$ cm$^{-3}$ and $\rho_c = 1.05 \times 10^{-5} h^2$ GeV cm$^{-3}$, where the current Hubble constant is $H_0 = 100h$ km/sec/Mpc with $h \approx 0.65$. In order of obtain $\Omega_\chi h^2 \sim 0.12$, we need

$$\langle \sigma_{ann} v \rangle \sim g_*^{-1/2} x_f \frac{1.12 \times 10^{-10} \text{ GeV}^{-2}}{\Omega_\chi h^2} \sim 10^{-9} \text{ GeV}^{-2}. \tag{6.7}$$

Recall a typical annihilation cross section of a particle of mass $m_\chi$ by a relatively weak interaction of electromagnetic strength (*e.g.*, $e^+ e^- \to \gamma \gamma$) is

$$\sigma_{ann} v \sim \frac{\pi \alpha^2}{m_\chi^2}. \tag{6.8}$$

To obtain the correct abundance, what we need is

$$m_\chi \sim 300 \text{ GeV}. \tag{6.9}$$

This is a *very* interesting result. Namely, the correct abundance of thermal relics is obtained for a particle mass just beyond the past accelerator limits and where we expect new particles to exist because of the considerations of electroweak symmetry breaking and the hierarchy problem. In other words, it is exactly the right mass scale for a new particle!

In the next few sections, we will firm up this naive estimate by solving the Boltzmann equations numerically. We will also study a concrete model of a new particle for dark matter candidate and work out its annihilation cross section. In addition, we will see if we have a chance of "seeing" the dark matter particle in our galactic halo, or making it in future accelerator experiments. Then we will generalize the discussions to more theoretically attractive models of physics beyond the standard model.

### 6.2. Boltzmann equation

You have already seen Boltzmann equation in lectures by Sabino Matarrese and I don't repeat its derivations. We assume kinetic equilibrium, namely that each particle species has the Boltzmann distribution in the momentum space except for the overall normalization that is given by its number density. Considering the process of $\chi_1 \chi_2 \leftrightarrow \chi_3 \chi_4$, where $\chi_i$ refers to a certain elementary particle, the Boltzmann equation for the number density $n_1$ for the particle $\chi_1$ is

$$a^{-3} \frac{d(n_1 a^3)}{dt} = \langle \sigma v \rangle n_1^0 n_2^0 \left( \frac{n_3 n_4}{n_3^0 n_4^0} - \frac{n_1 n_2}{n_1^0 n_2^0} \right). \tag{6.10}$$

Here, $\sigma v$ is the cross section common for the process $\chi_1\chi_2 \to \chi_3\chi_4$ and its inverse process $\chi_3\chi_4 \to \chi_1\chi_2$ assuming the time reversal invariance. The number densities with superscript $^0$ refer to those in the thermal equilibrium.

In the case of our interest, $\chi_{3,4}$ are "mundane" light (relativistic) particles in the thermal bath, and hence $n_{3,4} = n_{3,4}^0$. In addition, we consider the annihilation $\chi\chi \leftrightarrow$ (mundane)$^2$, and hence $n_1 = n_2$. The Boltzmann equation simplifies drastically to

$$a^{-3}\frac{dn_\chi a^3}{dt} = \langle \sigma_{ann} v \rangle \left[ (n_\chi^0)^2 - (n_\chi)^2 \right]. \tag{6.11}$$

This time we pay careful attention to all numerical factors. We use

$$Y = \frac{n_\chi}{s}, \tag{6.12}$$

$$s = g_* T^3 \left( \frac{2\pi^2}{45} \right)^{1/2}, \tag{6.13}$$

$$H^2 = \frac{8\pi}{3} G_N g_* \frac{\pi^2}{30} T^4 = g_* \frac{\pi^2}{90} \frac{T^4}{M_{Pl}^2}, \tag{6.14}$$

$$x = \frac{m_\chi}{T}. \tag{6.15}$$

Even though we start out at temperatures $T > m_\chi$ when $\chi$ are relativistic, eventually the temperature drops below $m_\chi$ and we can use non-relativistic approximations. Then the equilibrium number density can be worked out easily as

$$n_\chi^0 = \int \frac{d^3 p}{(2\pi)^3} e^{-E/T} \qquad \left( E = m_\chi + \frac{\vec{p}^2}{2m_\chi} \right)$$

$$= e^{-m_\chi/T} \left( \frac{m_\chi T}{2\pi} \right)^{3/2} = e^{-x} \frac{m_\chi^3}{(2\pi x)^{3/2}}. \tag{6.16}$$

Therefore

$$Y_0 = \frac{n_\chi^0}{s} = \frac{1}{g_*} \frac{45}{2\pi^2} \left( \frac{x}{2\pi} \right)^{3/2} e^{-x} = 0.145 x^{3/2} e^{-x}. \tag{6.17}$$

Changing the variables from $n_\chi$ to $Y$ and $t$ to $x$, the Boltzmann equation becomes

$$\frac{dY}{dx} = -\frac{1}{x^2} \frac{s(m_\chi)}{H(m_\chi)} \langle \sigma_{ann} v \rangle (Y^2 - Y_0^2). \tag{6.18}$$

Here, we used $s(T) = s(m_\chi)/x^3$ and

$$dt = -\frac{1}{H(T)}\frac{dT}{T} = -\frac{m_\chi^2}{H(m_\chi)T^3}dT = \frac{1}{H(m_\chi)}x\,dx. \qquad (6.19)$$

It is useful to work out

$$\frac{s(m_\chi)}{H(m_\chi)} = \frac{2\pi^2}{45}\left(\frac{90}{\pi^2}\right)^{1/2}g_*^{1/2}m_\chi M_{Pl} = 1.32 g_*^{1/2}m_\chi M_{Pl}. \qquad (6.20)$$

Note that the annihilation cross section $\langle\sigma_{ann}v\rangle$ is insensitive to the temperature once the particle is non-relativistic $T \ll m_\chi$.[15] Therefore the whole combination $\frac{s(m_\chi)}{H(m_\chi)}\langle\sigma_{ann}v\rangle$ is just a dimensionless number. The only complication is that $Y_0$ has a strong dependence on $x$. We can further simplify the equation by introducing the quantity

$$y = \frac{s(m_\chi)}{H(m_\chi)}\langle\sigma_{ann}v\rangle Y. \qquad (6.21)$$

We obtain

$$\frac{dy}{dx} = -\frac{1}{x^2}(y^2 - y_0^2), \qquad (6.22)$$

with

$$y_0 = 0.192 g_*^{-1/2}M_{Pl}m_\chi\langle\sigma_{ann}v\rangle x^{3/2}e^{-x}. \qquad (6.23)$$

### 6.3. Analytic approximation

Here is a simple analytic approximation to solve Eq. (6.22). We assume $Y$ tracks $Y_0$ for $x < x_f$. On the other hand, we assume $y \gg y_0$ for $x > x_f$ because $y_0$ drops exponentially as $e^{-x}$. Of course this approximation has a discontinuity at $x = x_f$, but the transition between these two extreme assumptions is so quick that it turns out to be a reasonable approximation. Then we can analytically solve the equation for $x > x_f$ and we find

$$\frac{1}{y(\infty)} - \frac{1}{y(x_f)} = \frac{1}{x_f}. \qquad (6.24)$$

Since $y(\infty) \ll y(x_f)$, we obtain the simple estimate

$$y(\infty) = x_f. \qquad (6.25)$$

---

[15]This statement assumes that the annihilation is in the $S$-wave. If it is in the $l$-wave, $\langle\sigma_{ann}v\rangle \propto v^{2l} \propto x^{-l}$.

Given this result, we can estimate $x_f$ as the point where $y_0(x)$ drops down approximately to $x_f$,

$$0.192 g_*^{-1/2} M_{Pl} m_\chi \langle \sigma_{ann} v \rangle x_f^{3/2} e^{-x_f} \approx x_f, \qquad (6.26)$$

and hence

$$
\begin{aligned}
x_f &\approx \ln\left( \frac{0.192 m_\chi M_{Pl} \langle \sigma_{ann} v \rangle x_f^{1/2}}{g_*^{1/2}} \right) \\
&\approx 24 + \ln \frac{m_\chi}{100 \text{ GeV}} + \ln \frac{\langle \sigma_{ann} v \rangle}{10^{-9} \text{ GeV}^{-2}} - \frac{1}{2} \ln \frac{g_*}{100}.
\end{aligned} \qquad (6.27)
$$

## 6.4. Numerical integration

I've gone through numerical integration of the Boltzmann equation Eq. (6.22). Fig. 12 shows the $x$-evolution of $y$. You can see that it traces the equilibrium value very well early on, but after $x$ of about 20, it starts to deviate significantly and eventually asymptotes to a constant. This is exactly the behavior we expected in the analytic approximation studied in the previous section.

Fig. 13 shows the asymptotic values $y(\infty)$ which we call $x_f$. I understand this is a confusing notation, but we have to define the "freeze-out" in some way, and the analytic estimate in the previous section suggest that the asymptotic value $y(\infty)$ is nothing but the freeze-out value $x_f$. This is the result that enters the final estimate of the abundance and is hence the only number we need in the end anyway. It does not exactly agree with the estimate in the previous section, but does very well once I changed the offset in Eq. (6.27) from 24 to 20.43. Logarithmic dependence on $m_\chi$ is verified beautifully.

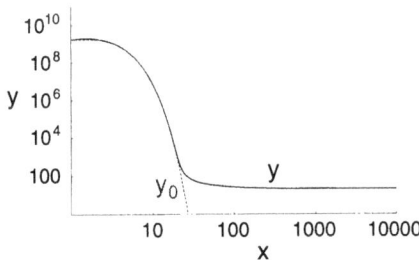

Fig. 12. Numerical solution to the Boltzmann equation Eq. (6.22) for $m = 100$ GeV, $g_* = 100$, $\langle \sigma_{ann} v \rangle = 10^{-9}$ GeV$^{-2}$. Superimposed is the equilibrium value $y_0$.

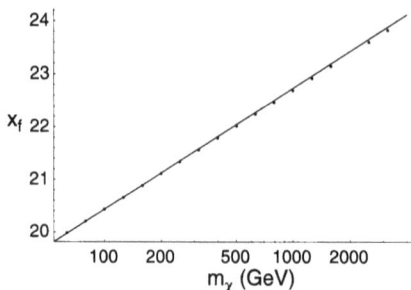

Fig. 13. $x_f$ values as a function of $m_\chi$, for $g_* = 100$, $\langle \sigma_{ann} v \rangle = 10^{-9}$ GeV$^{-2}$. The dots are the results of numerical integrations, while the solid line is just $\ln m_\chi$ with an offset so that $x_f = 20.43$ for $m_\chi = 100$ GeV.

Putting everything back together,

$$\rho_\chi = m_\chi n_\chi = m_\chi Y s = m_\chi \frac{H(m_\chi)}{s(m_\chi)} \frac{x_f}{\langle \sigma_{ann} v \rangle} s \tag{6.28}$$

As before, we use $s_0 = 2890$ cm$^{-3}$ and $\rho_c = 1.05 \times 10^{-5} h^2$ GeV cm$^{-3}$, where the current Hubble constant is $H_0 = 100h$ km/sec/Mpc with $h \approx 0.65$. To obtain $\Omega_M h^2 = 0.12$, we find $\langle \sigma_{ann} v \rangle = 1.6 \times 10^{-9}$ GeV$^{-2}$, confirming the simple estimate in Section 6.1.

### 6.5. The new minimal standard model

Now we would like to apply our calculations to a specific model, called the New Minimal Standard Model [2]. This is the model that can account for the empirical facts listed in Section 2.1 with the minimal particle content if you do not pay *any* attention to the theoretical issues mentioned in Section 2.2. It accomplishes this by adding only four new particles to the standard model;[16] very minimal indeed! The dark matter in this model is a real scalar field $S$ with an odd $\mathbb{Z}_2$ parity $S \to -S$, and its most general renormalizable Lagrangian that should be added to the Standard Model Eq. (2.1) is

$$\mathcal{L}_S = \frac{1}{2} \partial_\mu S \partial^\mu S - \frac{1}{2} m_S^2 S^2 - \frac{k}{2} |H|^2 S^2 - \frac{h}{4!} S^4. \tag{6.29}$$

The scalar field $S$ is the only field odd under $\mathbb{Z}_2$, and hence the $S$ boson is stable. Because of the analysis in the previous sections, we know that if $m_S$ is at the electroweak scale, it may be a viable dark matter candidate as a thermal relic. This is a model with only three parameters, $m_S$, $k$, and $h$, and actually the last

---

[16]The other three are the inflaton and two right-handed neutrinos.

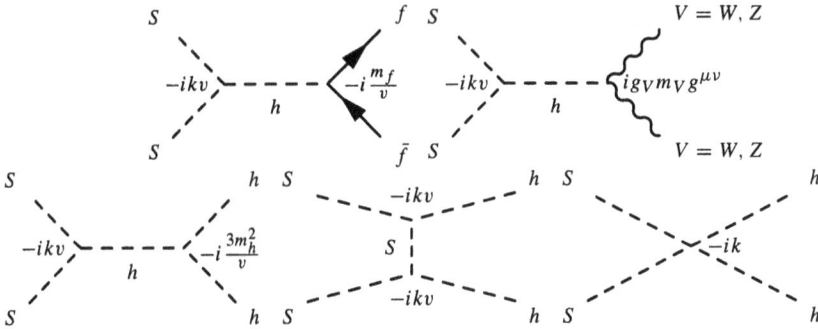

Fig. 14. Feynman diagrams for the annihilation of $S$ scalars. The final states in the first diagram can be any of the quark or lepton pairs $f\bar{f}$.

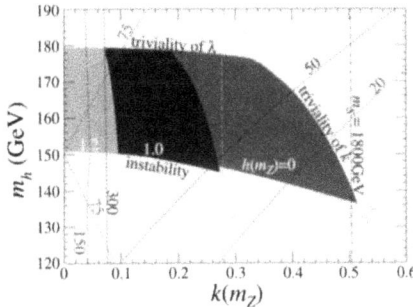

Fig. 15. The region of the NMSM parameter space $(k(m_Z), m_h)$ that satisfies the stability and triviality bounds, for $h(m_Z) = 0$, 1.0, and 1.2. Also the preferred values from the cosmic abundance $\Omega_S h^2 = 0.11$ are shown for various $m_S$. Taken from [2].

one is not relevant to the study of dark matter phenomenology. Therefore this is a very predictive model where one can work it out very explicitly and easily.

To calculate the dark matter abundance, what we need to know is the annihilation cross section of the scalar boson $S$. This was studied first in [68] and later in [69], but the third diagram was missing. In addition, there is a theoretical bounds on the size of couplings $k$ and $h$ so that they would stay perturbative up to high scales (*e.g.*, Planck scale). The cosmic abundance is determined by $m_S$ and $k$ in addition to $m_h$. Therefore on the $(k, m_h)$ plane, the correct cosmic abundance determines what $m_S$ should be. This is shown in Fig. 15. You can see that for a very wide range $m_S \simeq 5.5$ GeV–1.8 TeV, the correct cosmic abundance can be obtained within the theoretically allowed parameter space. For heavy $m_S \gg m_h$, the cross section goes like $k^2/m_S^2$ and is independent of $m_h$. This is why the $m_S$ contours are approximately straight vertically. For light $m_S \ll m_h$, the cross section goes like $k^2 m_S^2/m_h^4$. This is why the $m_S$ contours approximately have a fixed $k/m_h^2$ ratio. Note that when $m_S \simeq m_h/2$, the first two diagrams can hit the Higgs pole and the cross section can be very big even for small $k$. This resonance effect is seen in Fig. 15 where $m_S = 75$ GeV line reaches almost $k = 0$ for $m_h = 150$ GeV.

You may wonder why I am talking about $S$ as light as 5.5 GeV. Shouldn't we have seen it already in accelerator experiments? Actually, no. The only interaction the $S$ boson has is with the Higgs boson which we are yet to see. Therefore, we could not have produced the $S$ boson unless we had produced the Higgs boson. That is why even such a light $S$ boson does not contradict data. In other words, it wouldn't be easy to find this particle in accelerator experiments.

### 6.6. Direct detection experiments

How do we know if dark matter is indeed in the form of WIMP candidate you like? One thing we'd love to see is the direct detection of WIMPs. The idea is very simple. You place a very sensitive device in a quite location. WIMPs are supposed to be flying around in the halo of our galaxy with the typical speed of $\sim 220$ km/s $\sim 10^{-3}c$. Because they are only very weakly interacting, they can go through walls, rocks, even the entire Earth with little trouble, just like neutrinos. For a mass of $m_\chi \sim 100$ GeV, its typical kinetic energy is $E_{kin} = \frac{1}{2}m_\chi v^2 \sim 50$ keV. If the WIMP (ever) scatters off an atomic nucleus, the energy deposit is only (at most) of this order of magnitude. It is a *tiny* energy deposit that is very difficult to pick out against background from natural radioactivity (typically MeV energies). Therefore you have to make the device very clean, and also place it deep underground to be shielded from the cosmic-ray induced backgrounds, most importantly neutrons ejected from the rocks by cosmic-ray muons. One you've done all this, what you do is to wait to see this little "kick" in your detector.

Let us do an order of magnitude estimate. The local halo density is estimated to be about $\rho_\chi^{halo} \approx 0.3$ GeV/cm$^3$. The number density of WIMPs is $n_\chi = \rho_\chi^{halo}/m_\chi$. The flux of WIMPs is roughly $v n_\chi \approx 10^{-3}c n_\chi$. The elastic cross section of WIMP on neutron or proton may be spin-independent or spin-dependent. In the spin-independent case, the amplitude of the WIMP-nucleus cross section goes as $A$ (mass number) and hence the cross section on the nucleus $\sigma_A$ goes as $A^2$. Of course the detailed scaling is model-dependent, but in most phenomenological analyses (and also analyses of data) we assume $\sigma_A = A^2 \sigma_p$. Let us also assume $^{56}$Ge as the detector material so that $A = 56$. Then the expected event rate is

$$R = n_\chi^{halo} v \frac{m_{target}}{m_A} \sigma_A \approx \frac{10}{year} \frac{100 \text{ GeV}}{m_\chi} \frac{m_{target}}{100 \text{ kg}} \frac{A}{56} \frac{\sigma_p}{10^{-42} \text{ cm}^2}. \tag{6.30}$$

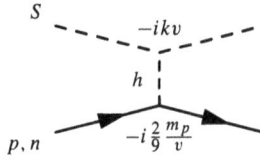

Fig. 16. Feynman diagrams for the scattering of the $S$-boson off a proton or neutron.

To prepare a very sensitive device as big as 100 kg and make it very clean is a big job. You can see that your wait may be *long*.

Now back to the New Minimal Standard Model. The scattering of the $S$ boson off a proton comes from the $t$-channel Higgs boson exchange. The coupling of the Higgs boson to the nucleon is estimated by the famous argument [70] using the conformal anomaly. The mass of the proton is proportional to the QCD scale $m_p \propto \Lambda_{QCD}$ ($m_u, m_d, m_s$ are ignored and hence this is the three-flavor scale). It is related to the Higgs expectation value through the one-loop renormalization group equation as (we do not consider higher loop effects here)

$$\Lambda_{QCD}^9 = m_c^{2/3} m_b^{2/3} m_t^{2/3} M^5 e^{-8\pi^2/g_s^2(M)} \tag{6.31}$$

where $M$ is some high scale and each quark mass is proportional to $v$. The coupling of the Higgs to the proton is given by expanding the vacuum expectation value as $v \to v + h$, and hence

$$y_{pph} = \frac{\partial m_p}{\partial v} = \frac{2}{9} \frac{m_p}{v}. \tag{6.32}$$

This allows us to compute the scattering cross section of the $S$ boson and the nucleon.

The result is shown in Fig. 16 as the red solid line. The point is that the elastic scattering cross section tends to be *very* small. Note that a hypothetical neutrino of the similar mass would have a cross section of $\sigma_p^\nu \sim G_F^2 m_\nu^2/\pi \sim 10^{-34}$ cm$^2$ which is much much bigger than this. This is the typical WIMP scattering cross section. Superimposed is the limit from the CDMS-II experiment [71] and hence the direct detection experiments are just about to reach the required sensitivity. In other words, this simple model is completely viable, and may be tested by future experiments. For the resonance region $m_S \simeq m_h/2$, the coupling $k$ is very small to keep enough abundance and hence the direct detection is very difficult.

The future of this field is not only to detect WIMPs but also understand its identity. For this purpose, you want to combine the accelerator data and the direct detection experiments. The direct detection experiments can measure the energy

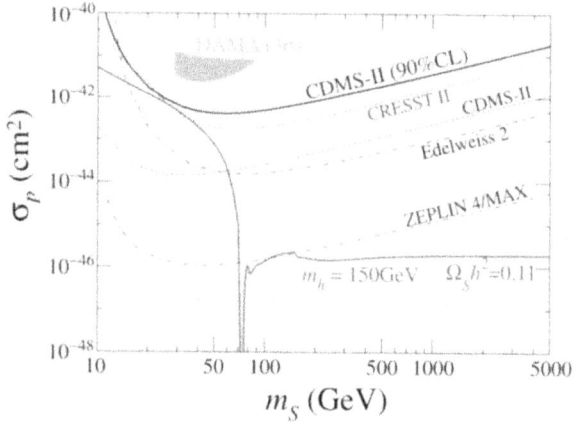

Fig. 17. The elastic scattering cross section of Dark Matter from nucleons in NMSM, as a function of the Dark Matter particle mass $m_S$ for $m_h = 150$ GeV. Note that the region $m_S > 1.8$ TeV is disallowed by the triviality bound on $k$. Also shown are the experimental bounds from CDMS-II [71] and DAMA [72], as well as improved sensitivities expected in the future [73]. Taken from [2].

deposit and hence the mass of the WIMP. It also measures the scattering cross section, even though it suffers from the astrophysical uncertainty in the estimate of the local halo density. On the other hand, assuming $m_S < m_h/2$, the Higgs boson decays *invisibly* $h \to SS$. Such an invisible decay of the light Higgs boson can be looked for at the LHC using the $W$-fusion process. Quarks from both sides radiate an off-shell $W$-boson that "fuse" in the middle to produce a Higgs boson. Because of the kick by the off-shell $W$-boson, the quarks acquire $p_T \sim m_W/2$ in the final state and can be tagged as "forward jets." Even though the Higgs boson would not be seen, you may "discover" it by the forward jets and missing $E_T$ [74]. The ILC can measure the mass of the Higgs precisely even when it decays dominantly invisibly (see, *e.g.*, [75]) and possibly its width. Combining it with the mass from the direct detection experiments, you can infer the coupling $k$ and calculate its cosmic abundance. It would be a very interesting test if it agrees with the cosmological data $\Omega_M \approx 0.23$. If it does, we can claim a victory; we finally understand what dark matter *is*!

### 6.7. Popular WIMPs

Superpartners of the photon and $Z$, and neutral Higgs bosons (there are two of them), mix among each other once $SU(2) \times U(1)$ symmetry breaks. Out of four such "neutralino" states, the lightest one is often the LSP[17] and is the most popular candidate for dark matter in the literature (see, *e.g.*, [77–79] for some of the recent papers). One serious problem with the supersymmetric dark matter is that there are many parameters in the model. Even the Minimal Supersymmetric Standard Model (MSSM) has 105 more parameters than the standard model. It is believed that the fundamental theory determines all these parameters or at least reduces the number drastically, and one typically ends up with five or so parameters in the study. Depending on what parameter set you pick, the phenomenology may be drastically different. For the popular parameter set called CMSSM (Constrained MSSM, also called minimal supergravity or mSUGRA) with four parameters and one sign, see a recent study in [76]. I do not go into detailed discussions about any of them. I rather mention a few generic points.

First, we *do* get viable dark matter candidates from sub-TeV supersymmetry as desired by the hierarchy problem. This is an important point that shouldn't be forgotten. Second, what exactly is the mass and composition of the neutralino depends on details of the parameter set. The supersymmetric standard model may not be minimal either; an extension with additional singlet called Next-to-Minimal Supersymmetric Standard Model (NMSSM) is also quite popular. Third, sub-TeV supersymmetry can be studied in great detail at LHC and (hopefully) ILC, so that we can measure their parameters very precisely (see [81] for the early work). One can hope to correlate the accelerator and underground data to fully test the nature of the dark matter [82]. Fourth, a large number of particles present in the model may lead to interesting effects we did not consider in the discussions above. I'll discuss one of such effects briefly below.

Universal Extra Dimension (UED) is also a popular model. I'm sure Geraldine Servant will discuss dark matter in this model in her lectures because she is one of the pioneers in this area [86]. Because the LKP is stable, it is a dark matter candidate. Typically the first KK excitation of the $U(1)_Y$ gauge boson is the LKP, and its abundance can be reasonable. Its prospect for direct and indirect detection experiments is also interesting. See a review article that came out after Les Houches [87].

The most striking effect of having many particle species is the *coannihilation* [83]. An important example in the case of supersymmetry is bino and stau. Bino is the superpartner of the $U(1)_Y$ gaugino (mixture of photino and zino), and its annihilation cross section tends to be rather small partly because it dominantly goes through the $P$-wave annihilation. If, however, the mass of the stau is not too far above bino, stau is present with the abundance suppressed only by $e^{-\Delta m/T}$ ($\Delta m = m_{\tilde{\tau}} - m_{\tilde{B}}$) assuming they are in chemical equilibrium $\tilde{B}\tau \leftrightarrow \gamma\tilde{\tau}$. There are models that suggest the mass splitting is indeed small. The cross sections

---

[17]Superpartner of neutrinos is not out of the question [80].

$\tilde{B}\tilde{\tau} \rightarrow \gamma\tau$ and $\tilde{\tau}\tilde{\tau}^* \rightarrow \gamma\gamma, f\bar{f}$ etc tend to be much larger, the former go-ing through the $S$-wave and the latter with many final states. Therefore, despite the Boltzmann suppression, these additional contributions may win over $\langle\sigma_{\tilde{B}\tilde{B}}v\rangle$. There are other cases where the mass splitting is expected to be small, such as higgsino-like neutralinos [84]. In the UED, the LKP is quite close in its mass to the low-lying KK states and again coannihilation is important.

### 6.8. Indirect detection experiments

On the experimental side, there are other possible ways of detecting signals of dark matter beyond the underground direct detection experiments and collider searches. They are *indirect detection* experiments, namely that they try to detect annihilation products of dark matter, not the dark matter itself. For annihilation to occur, you need some level of accumulation of dark matter. The possible sites are: galactic center, galactic halo, and center of the Sun. The annihilation products that can be searched for include gamma rays from the galactic center or halo, $e^+$ from the galactic halo, radio from the galactic center, anti-protons from the galactic halo, and neutrinos from the center of the Sun. Especially the neutrino signal complements the direct detection experiments because the sensitivity of the direct detection experiments goes down as $1/m_\chi$ because the number density goes down, while the sensitivity of the neutrino signal remain more or less flat for heavy WIMPs because the neutrino cross section rises as $E_\nu \propto m_\chi$. You can look at a recent review article [85] on indirect searches.

## 7. Dark horse candidates

### 7.1. Gravitino

Assuming $R$-parity conservation, superparticles decay all the way down to what-ever is the lightest with odd $R$-parity. We mentioned neutralino above, but an-other interesting possibility is that the LSP is the superpartner of the gravitino, namely gravitino. Since gravitino couples only gravitationally to other particles, its interaction is suppressed by $1/M_{Pl}$. It practically removes the hope of direct detection. On the other hand, it is a possibility we have to take seriously. This is especially so in models with gauge-mediated supersymmetry breaking [88].

The abundance of light gravitinos is given by

$$\Omega_{3/2}h^2 = \frac{m_{3/2}}{2\,\text{keV}} \tag{7.1}$$

if the gravitinos were thermalized. In most models, however, the gravitino is heavier and we cannot allow thermal abundance. The peculiar thing about a

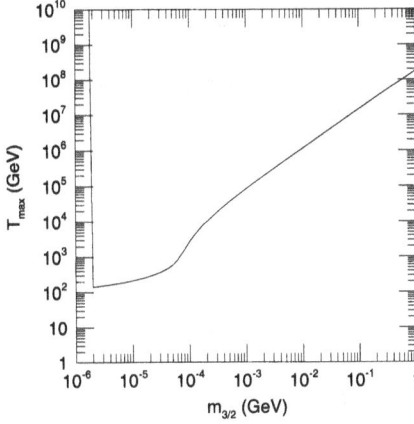

Fig. 18. The upper bound on $T_{max}$ as a function of the gravitino mass from the requirement that the relic stable gravitinos do not overclose the Universe. Taken from [90]. It assumes $h = 1$ and $\Omega_{3/2}h^2 < 1$ and hence the actual constraint is nearly an order of magnitude more stringent than this plot.

light gravitino is that the longitudinal (helicity $\pm 1/2$) components have a much stronger interaction $m_{SUSY}/(m_{3/2}M_{Pl})$ if $m_{3/2} \ll m_{SUSY}$. Therefore the production cross section scales as $\sigma \sim m_{SUSY}^2/(m_{3/2}M_{Pl})^2$, and the abundance scales as $Y_{3/2} \sim \sigma T^3/H(T) \sim m_{SUSY}^2 T/(m_{3/2}^2 M_{Pl})$. Therefore we obtain an upper limit on the reheating temperature after the inflation [89, 90]. If the reheating temperature is right at the limit, gravitino may be dark matter.

There is however another mechanism of gravitino production. The abundance of the "LSP"is determined the usual way as a WIMP, while it eventually decays into the gravitino. This decay can upset the Big-Bang Nucleosynthesis [89], but it may actually be helpful for a region of parameter space to ease some tension among various light element abundances [91]. Note that the "LSP" (or more correctly NLSP: Next-Lightest Supersymmetric Particle) may be even electrically charged or strongly coupled as it is not dark matter. When superparticles are produced at colliders, they decay to the "LSP" inside the detector, which escapes and most likely decays into the gravitino outside the detector. If the "LSP" is charged, it would leave a charged track with anomalously high $dE/dx$. It is in principle possible to collect the NLSP and watch them decay, and one may even be able to confirm the spin 3/2 nature of the gravitino [92] and its gravitational coupling to matter [93].

If the gravitino is heavier than the LSP, its lifetime is calcuated as

$$\tau(\tilde{G} \to \gamma\tilde{\gamma}) = 3.9 \times 10^5 \left(\frac{m_{3/2}}{\text{TeV}}\right)^{-3} \text{sec}. \tag{7.2}$$

It tends to decay after the BBN and upsets its success. Its production cross section scales as $\sigma \sim 1/M_{Pl}^2$ and hence $Y_{3/2} \sim T/M_{Pl}$. Depending on its mass and decay modes, one can again obtain upper limits on the reheating temperature. The case of hadronic decay for $m_{3/2} \sim 1$ TeV is the most limiting case that requires $T_{RH} \lesssim 10^6$ GeV [94], causing trouble to many baryogenesis models.

## 7.2. Axion

One of the puzzles about the standard model I discussed earlier is why $\theta \lesssim 10^{-10}$ in QCD. A very attractive solution to this problem is to promote $\theta$ to a dynamical field, so that when it settles to the minimum of the potential, it automatically makes $\theta$ effectively zero [95]. The dynamical field is called *axion*, and couples (after integrating out some heavy fields) as

$$\mathcal{L} = \left(\theta + \frac{a}{f_a}\right)\epsilon^{\mu\nu\rho\sigma}\mathrm{Tr}(G_{\mu\nu}G_{\rho\sigma}). \tag{7.3}$$

Here, $f_a$ is called axion decay constant which has a dimension of energy. The potential is given approximately as

$$V \sim m_\pi^2 f_\pi^2 \left[1 - \cos\left(\theta + \frac{a}{f_a}\right)\right], \tag{7.4}$$

and indeed the axion field settles to $a = -\theta f_a$ and the $G\tilde{G}$ term vanishes. The axion mass is therefore

$$m_a \approx 6\ \mu\mathrm{eV}\frac{10^{12}\ \mathrm{GeV}}{f_a}. \tag{7.5}$$

Various astrophysical limits basically require $f_a \gtrsim 10^{10}$ GeV and hence the axion is a very light boson (see, *e.g.*, [96]). Most of these limits come from the fact that the axion can carry away energy from stars and would cool them too quickly, such as white dwarfs, red giants, and SN1987A. Models of such high $f_a$ are called "invisible axion" models because then the axion coupling is very weak to other particles, avoids these limits, and hence is very difficult to observe. There are two popular versions, KSVZ [97, 98] and DFSZ [99, 100] models.

In the early universe $T \gg$ GeV, the axion potential looks so flat that it cannot tell where the minimum is.[18] Therefore we expect it starts out wherever it finds itself, mostly likely not at the minimum. The likely initial misplacement is of the order of $f_a$. Now we would like to know what happens afterwards.

---

[18]This is especially true for the axion because its mass originates from the QCD instaton effects which are suppressed by powers of the temperature in hot thermal bath [101].

Let us consider a scalar field in an expanding universe. Neglecting the spatial variation and considering the time dependence alone, the equation of motion is

$$\ddot{\phi} + 3H\dot{\phi} + V'(\phi) = 0. \tag{7.6}$$

For a quadratic potential (mass term) $V(\phi) = \frac{1}{2}m^2\phi^2$, the equation is particularly simple and homogeneous,

$$\ddot{\phi} + 3H\dot{\phi} + m^2\phi = 0. \tag{7.7}$$

It is useful to solve this equation for a constant (time-independent) $H$ first. For Fourier modes $\phi \sim e^{-i\omega t}$ (of course we take the real part later on), we need to solve

$$-\omega^2 - 3iH + m^2 = 0, \tag{7.8}$$

and we find

$$\omega_{\pm} = \frac{1}{2}\left[-3iH \pm \sqrt{-9H^2 + 4m^2}\right]$$

$$= \begin{cases} -3iH, & -i\frac{m^2}{3H} & (H \gg m) \\ m - i\frac{3}{2}H, & -m - i\frac{3}{2}H & (H \ll m) \end{cases} \tag{7.9}$$

Therefore for $H \gg m$ (early universe), one of the solutions damps quickly $\phi_+ \sim e^{-3Ht}$, while the other is nearly stationary $\phi_- \sim e^{-(m^2/3H)t}$. This is because it is "stuck" by the friction term $-3H\dot{\phi}$. On the other hand the field oscillates around the minimum as $\phi = \phi_0 e^{-imt}e^{-3Ht/2}$.

One can improve this analysis for time-dependent $H(t) = 1/2t$ and $a(t) \propto t^{1/2}$ in the radiation-dominated univese, by replacing $e^{-i\omega_{\pm}t}$ by $e^{-i\int^t \omega_{\pm}(t')dt'}$ (adiabatic approximation). When $H \gg m$ ($mt \ll 1$), we find

$$\phi_+ = \phi_0 e^{-3\int_{t_0}^t H(t')dt'} = \phi_0\left(\frac{t_0}{t}\right)^{-3/2} = \phi_0\left(\frac{a_0}{a}\right)^{-3},$$

$$\phi_- = \phi_0 e^{-m^2\int_{t_0}^t 2t'dt'/3} = \phi_0 e^{-m^2t^2/3} \approx \phi_0. \tag{7.10}$$

The second one is the solution that is stuck by the friction. On the other hand when $H \ll m$,

$$\phi_{\pm} = \phi_0 e^{\pm imt}e^{-\frac{3}{2}\int_{t_0}^t H(t')dt'} = \phi_0 e^{\pm imt}\left(\frac{t_0}{t}\right)^{3/4} = \phi_0 e^{\pm imt}\left(\frac{a_0}{a}\right)^{3/2}.$$

The field damps as $a^{-3/2}$, and its energy as $V = m^2\phi^2/2 \propto a^{-3}$, just like non-relativistic matter. In fact, a coherently oscillating homogeneous field can be regarded as a Bose–Einstein condensate of the boson at zero momentum state.

Therefore, the axion field can sit on the potential and does not roll down because of the large friction term $-3H\dot{\phi}$ when $H \gg m$. On the other hand for later universe $H \ll m$, it oscillates as a usual harmonic oscillator $e^{\pm imt}$ and dilutes as non-relativistic matter. This is why a very light scalar field can be a candidate for cold dark matter. Counterintuitive, but true. This way of producing cold dark matter is called "misalignment production" because it is due to the initial misalignment of the axion field relative to the potential minimum. Because the amount of misalignment is not known, we cannot predict the abundance of axion precisely. Assuming the misalignment of $O(f_a)$, $f_a \simeq 10^{12}$ GeV is the preferred range for axion dark matter.

There is a serious search going on for axion dark matter in the halo of our galaxy. In addition to the required coupling of the axion to gluons, most models predict its coupling to photons $a(\vec{E} \cdot \vec{B})$ of the similar order of magnitude. The ADMX experiment places a high-$Q$ cavity in a magnetic field. When an axion enters the cavity, this coupling would allow the axion to convert to a photon, which is captured resonantly by the cavity with a high sensitivity. By changing the resonant frequencies in steps, one can "scan" a range of axion mass. Their limit has just reached the KSVZ axion model [103, 104],[19] and an upgrade to reach the DFSZ axion model using SQUID is in the works. See [105] for more on axion microwave cavity searches.

## 7.3. Other candidates

I focused on the thermal reclic of WIMPs primarily because there is an attractive coincidence between the size of annihilation cross section we need for the correct abundance and the energy scale where we expect to see new partices from the points of view of electroweak symmetry breaking and hierarchy problem. Axion is not connected to any other known energy scale, yet it is well motivated from the strong CP problem. On the other hand, nature may not necessarily tell us a "motivation" for a particle she uses. Indeed, people have talked about many other possible candidates for dark matter. You may want to look up a couple of keywords: sterile neutrinos, axinos, warm dark matter, mixed dark matter, cold and fuzzy dark matter, $Q$-balls, WIMPZILLAs, etc. Overall, the candidates in this list range in their masses from $10^{-22}$ eV to $10^{22}$ eV, not to mention still possible MACHOs $\lesssim 10^{-7} M_\odot = 10^{59}$ eV. Clearly, we are making progress.

## 8. Cosmic coincidence

Whenever I think about what the univese is made of, including baryons, photons, neutrinos, dark matter, and dark energy, what bothers me (and many other people)

---

[19] They ignored theoretical uncertainty in the prediction of the axion-to-photon coupling, and the KSVZ model is not quite excluded yet [102].

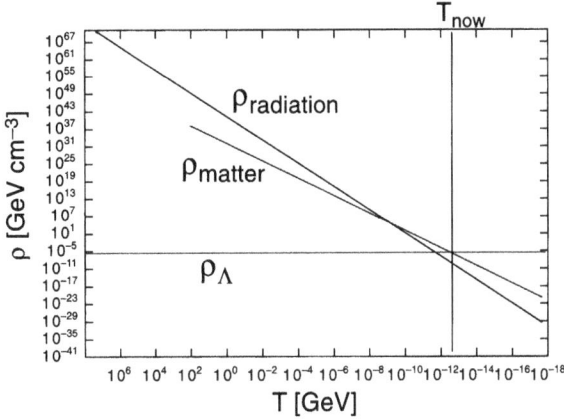

Fig. 19. The evolution of radiation, matter, and cosmological constant ($\Lambda$) components of the univese as the temperature drops over many orders of magnitude. "Now" is a very special moment when matter and $\Lambda$ are almost exactly the same, and the radiation is not that different either.

is this question: why do they have energy densities within only a few orders of magnitude? They could have been many orders of magnitude different, but they aren't. This question is related to the famous "Why now?" problem. The problem is clear in Fig. 19. As we think about evolution of various energy densities over many decades of temperatures, why do *we* live at this special moment when the dark matter and dark energy components become almost exactly the same? I feel like I'm back to Ptolemy from Copernicus. *We are special*, not in space any more, but in time. Is that really so?

In Fig. 19, the radiation component goes down as $T^{-4}$, while the matter $T^{-3}$. The cosmological constant is by definition constant $T^0$. Matter and $\Lambda$ meet *now*. When thinking about this problem, it is always tempting but dangerous to bring "us" into the discussion. Then we will be forced to talk about conditions for emergence of intelligent lifeforms, which we don't know very well about. Instead, it may be better to focus on physical quantities; namely the triple coincidence problem that three lines with different slopes seem to more or less meet at a point. In fact, dimensional analysis based on TeV-scale WIMP suggests

$$\rho_{\text{matter}} \sim \left( \frac{\text{TeV}^2}{M_{Pl}} \right)^3 T^3, \tag{8.1}$$

which agrees with $\rho_{\text{radiation}} \sim T^4$ at the temperature $T \sim \text{TeV}^2/M_{Pl} \approx \text{meV} = 10$ K; this is about now! In order for the cosmological constant to meet at the same time, we suspect there is a deep reason

$$\rho_\Lambda \sim \left(\frac{\mathrm{TeV}^2}{M_{Pl}}\right)^4 . \tag{8.2}$$

Indeed, $\rho_\Lambda^{1/4}$ is observationally about 2 meV, while $\mathrm{TeV}^2/M_{Pl} \approx 0.5$ meV. Maybe that is why we see a coincidence [106].

Actually, an exact coincidence does not leave a window for structure formation, which requires matter-dominated period. Fortunately, WIMP abundance is enhanced by weakness of the annihilation cross section, which goes like $1/\alpha^2$. This enhancement of matter relative to the triple coincidence gives us a window for matter domination and structure formation. May be that is why *we* seems to be in this triangle. But then, why is the baryon component also just a factor of five smaller than dark matter? Are they somehow related?

Oh well, we know so little.

## 9. Conclusions

In my lectures, I tried to emphasize that we are approaching an exciting time to cross new threshold of rich physics at the TeV energy scale in the next few years at the LHC. At the same time, the dark matter of the universe is now established to be not made of particles we know, requiring physics beyond the standard model. The main paradigm for dark matter now is WIMPs, TeV-scale particles produced by the Big Bang which naturally give the correct order of magnitude for its abundance. Even though nature may be tricking us by this coincidence, many of us (including I) think that there is indeed a new particle (or many of them) waiting to be discovered at the LHC (or ILC) that tells us something about the dark side of the universe. If this is so, I would feel lucky to be born to this age.

## Appendix A. Gravitational lensing

Gravitational lensing is an important tool in many studies in cosmology and astrophysics. In this appendix I introduce the deflection of light in a spherically symmetric gravitational field (Schwarzschild metric)

### Appendix A.1. Deflection angle

Using the Schwarzschild metric ($c = 1$)

$$ds^2 = \frac{r - r_S}{r} dt^2 - \frac{r}{r - r_S} dr^2 - r^2 d\theta^2 - r^2 \sin^2\theta \, d\phi^2 \tag{A.1}$$

where $r_S = 2G_N m$ is the Schwarzschild radius. The Hamilton–Jacobi equation[20] for light in this metric is

$$g^{\mu\nu} \frac{\partial S}{\partial x^\mu} \frac{\partial S}{\partial x^\nu}$$

$$= \frac{r}{r - r_S} \left( \frac{\partial S}{\partial t} \right)^2 - \frac{r - r_S}{r} \left( \frac{\partial S}{\partial r} \right)^2 - \frac{1}{r^2} \left( \frac{\partial S}{\partial \theta} \right)^2 - \frac{1}{r^2 \sin^2 \theta} \left( \frac{\partial S}{\partial \phi} \right)^2$$

$$= 0. \tag{A.2}$$

We separate the variables as

$$S(t, r, \theta, \phi) = S_1(t) + S_2(r) + S_3(\theta) + S_4(\phi) \tag{A.3}$$

where

$$\frac{r}{r - r_S} \left( \frac{dS_1}{dt} \right)^2 - \frac{r - r_S}{r} \left( \frac{dS_2}{dr} \right)^2 - \frac{1}{r^2} \left( \frac{dS_3}{d\theta} \right)^2 - \frac{1}{r^2 \sin^2 \theta} \left( \frac{dS_4}{d\phi} \right)^2 = 0.$$

Because the equation does not contain $t$ or $\phi$ explicitly, their functions must be constants,

$$\frac{dS_1}{dt} = -E, \tag{A.4}$$

$$\frac{dS_4}{d\phi} = L_z. \tag{A.5}$$

We can solve them immediately as

$$S_1(t) = -Et, \tag{A.6}$$

$$S_4(\phi) = L_z \phi. \tag{A.7}$$

Then Eq. (A.2) becomes

$$\frac{r}{r - r_S} E^2 - \frac{r - r_S}{r} \left( \frac{dS_2}{dr}(r) \right)^2 - \frac{1}{r^2} \left( \frac{dS_3}{d\theta}(\theta) \right)^2 - \frac{1}{r^2 \sin^2 \theta} L_z^2 = 0.$$

The $\theta$ dependence is only in the last two terms and hence

$$\left( \frac{dS_3}{d\theta}(\theta) \right)^2 + \frac{1}{\sin^2 \theta} L_z^2 = L^2 \tag{A.8}$$

is a constant which can be integrated explicitly if needed. Without a loss of generality, we can choose the coordinate system such that the orbit is on the $x$-$y$

---

[20]For an introduction to Hamilton–Jacobi equations, see http://hitoshi.berkeley.edu/221A/ classical2.pdf.

plane, and hence $L_z = 0$. In this case, $S_4(\phi) = 0$ and $S_3(\theta) = L\theta$. Finally, the equation reduces to

$$\frac{r}{r - r_S} E^2 - \frac{r - r_S}{r} \left( \frac{dS_2}{dr}(r) \right)^2 - \frac{L^2}{r^2} = 0. \tag{A.9}$$

Therefore,

$$S_2(r) = \int \sqrt{\frac{r^2}{(r - r_S)^2} E^2 - \frac{L^2}{r(r - r_S)}} \, dr. \tag{A.10}$$

Since $S(t, r, \theta, \phi) = S_2(r) - Et + L\theta$, $S_2$ can be regarded as Legendre transform $S_2(r, E, L)$ of the action, and hence the inverse Legendre transform gives

$$\frac{\partial S_2(r, E, L)}{\partial L} = -\theta. \tag{A.11}$$

Using the expression Eq. (A.10), we find

$$\theta(r) = \int_{r_c}^{r} \frac{L \, dr}{\sqrt{E^2 r^4 - L^2 r(r - r_S)}}. \tag{A.12}$$

The closest approach is where the argument of the square root vanishes,

$$E^2 r_c^4 - L^2 r_c(r_c - r_S) = 0. \tag{A.13}$$

It is useful to verify that the $m = 0$ ($r_S = 0$) limit makes sense. The closest approach is $E^2 r_c^4 - L^2 r_c^2 = 0$ and hence $r_c = L/E$, which is the impact parameter. The orbit Eq. (A.12) is

$$\theta(r) = \int_{r_c}^{r} \frac{L \, dr}{\sqrt{E^2 r^4 - L^2 r^2}} = \int_{r_c}^{r} \frac{r_c \, dr}{r\sqrt{r^2 - r_c^2}}. \tag{A.14}$$

Change the variable to $r = r_c \cosh \eta$, and we find

$$\theta(r) = \int_{0}^{\eta} \frac{r_c^2 \sin \eta d\eta}{r_c \cosh \eta r_c \sinh \eta} = \int_{0}^{\eta} \frac{d\eta}{\cosh \eta} = 2 \arctan \tanh \frac{\eta}{2}. \tag{A.15}$$

Hence $\tan \frac{\theta}{2} = \tanh \frac{\eta}{2}$, and

$$\cos \theta = \frac{1 - \tan^2 \theta/2}{1 + \tan^2 \theta/2} = \frac{1 - \tanh^2 \eta/2}{1 + \tanh^2 \eta/2} = \frac{1}{\cosh \eta} = \frac{r_c}{r}. \tag{A.16}$$

Therefore $r_c = r \cos \theta$ which is nothing but a straight line.

To find the deflection angle, we only need to calculate the asymptotic angle $\theta(r = \infty)$. Going back to Eq. (A.12), we need to calculate

$$\theta(\infty) = \int_{r_c}^{\infty} \frac{L dr}{\sqrt{E^2 r^4 - L^2 r(r - r_S)}}. \tag{A.17}$$

We would like to expand it up to the linear order in $r_S \ll r_c$. If you naively expand the integrand in $r_S$, the argument of the square root in the resulting expression can be negative for $r = r_c < L/E$. To avoid this problem, we change the variable to $r = r_c/x$:

$$\theta(\infty) = \int_0^1 \frac{L r_c dx}{\sqrt{E^2 r_c^4 - L^2 r_c(r_c - r_S x) x^2}}. \tag{A.18}$$

Using Eq. (A.13), we write $E^2 r_c^4$ and obtain

$$\theta(\infty) = \int_0^1 \frac{r_c dx}{\sqrt{r_c^2(1 - x^2) - r_c r_S(1 - x^3)}}. \tag{A.19}$$

Expanding it to the linear order in $r_S/r_c$, we find

$$\begin{aligned}
\theta(\infty) &= \int_0^1 \left( \frac{1}{\sqrt{1 - x^2}} + \frac{(1 + x + x^2) r_S}{2(1 + x)\sqrt{1 - x^2} \, r_c} + O(r_S)^2 \right) dx \\
&= \frac{\pi}{2} + \frac{r_S}{r_c}.
\end{aligned} \tag{A.20}$$

The deflection angle is $\Delta\theta = \pi - 2\theta(\infty) = 2\frac{r_S}{r_c} = 4G_N m/r_c$. It is easy to recover $c = 1$ by looking at the dimensions, and we find $\Delta\theta = 4G_N m/c^2 r_c$.

It is also useful to know the closest approach $r_c$ to the first order in $m$. We expand $r_c$ as $r_c = \frac{L}{E} + \Delta$. Then Eq. (A.13) gives

$$4\frac{L^3}{E}\Delta - 2\frac{L^3}{E}\Delta + \frac{E^3}{L} r_S + O(r_S)^2 = 0, \tag{A.21}$$

and hence

$$r_c = \frac{L}{E} - \frac{r_S}{2} + O(r_S)^2. \tag{A.22}$$

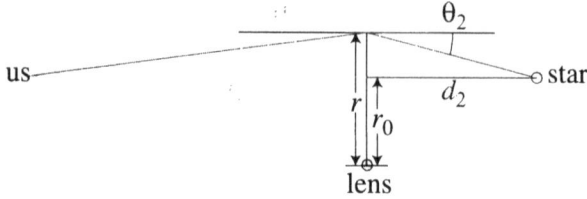

Fig. 20. The deflection of light due to a massive body close to the line of sight towards a star.

## Appendix A.2. Amplification in microlensing

Once the deflection angle is known, it is easy to work out the amplification using simple geometric optics. Throughout the discussion, we keep only the first order in very small angles. Just by looking at the geometry in Fig. 20, the deflection angle is

$$\Delta\theta = \theta_1 + \theta_2 = \frac{r - r_0}{d_1} + \frac{r - r_0}{d_2} = \frac{4G_N m}{r}. \tag{A.23}$$

Here, $r_0$ is the impact parameter. When $r_0 = 0$ (exactly along the line of sight), the solution is simple:

$$r(r_0 = 0) = R_0 \equiv \sqrt{4G_N m \frac{d_1 d_2}{d_1 + d_2}}. \tag{A.24}$$

This is what is called the Einstein radius, $R_0$ in Paczynski's notation [56]. For general $r_0$, Eq. (A.23) can be rewritten as

$$r(r - r_0) - R_0^2 = 0, \tag{A.25}$$

which is Eq. (1) in the Packzynski's paper. It has two solutions

$$r_\pm(r_0) = \frac{1}{2}\left(r_0 \pm \sqrt{r_0^2 + 4R_0^2}\right). \tag{A.26}$$

The solution with the positive sign is what is depicted in Fig. 20, while the solution with the negative sign makes the light ray go *below* the lens.

To figure out the amplification due to the gravitational lensing, we consider the finite aperture of the telescope (*i.e.*, the size of the mirror). We assume an infinitesimal circular aperture. From the point of view of the star, the finite aperture is an image on the deflection plane of size $\delta$, namely the plane perpendicular to the straight line from the star to the telescope where the lens is. The vertical aperture changes the impact parameter $r_0$ to a range $r_0 \pm \delta$ (size of the mirror is $\delta \times (d_1 + d_2)/d_2$). Correspondingly, the image of the telescope is at

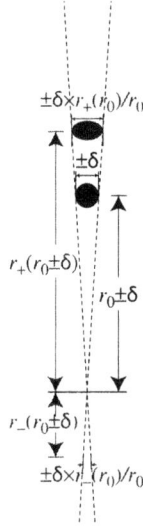

Fig. 21. The way the mirror of the telescope appears on the deflection plane from the point of view of the star. For the purpose of illustration, we took $R_0 = 2, r_0 = 3$.

$r_\pm(r_0 \pm \delta) = r_\pm(r_0) \pm \delta \frac{dr_\pm}{dr_0}$.[21] Using the solution Eq. (A.26), we find that the vertical aperture always appears squashed (see Fig. 21),

$$\delta \times \left| \frac{dr}{dr_0} \right| = \delta \times \left| \frac{1}{2} \left( 1 \pm \frac{r_0}{\sqrt{r_0^2 + 4R_0^2}} \right) \right| = \delta \times \frac{\sqrt{r_0^2 + 4R_0^2} \pm r_0}{2\sqrt{r_0^2 + 4R_0^2}} < \delta.$$

On the other hand, the horizontal aperture is scaled as

$$\delta \times \frac{r}{r_0}. \tag{A.27}$$

Because the amount of light that goes into the mirror is proportional to the elliptical aperture from the point of view of the star that emits light isotropically, the magnification is given by

$$A_\pm = \frac{r}{r_0} \left| \frac{dr}{dr_0} \right| = \frac{\left( \sqrt{r_0^2 + 4R_0^2} \pm r_0 \right)^2}{4r_0 \sqrt{r_0^2 + 4R_0^2}} = \frac{2r_0^2 + 4R_0^2 \pm 2r_0 \sqrt{r_0^2 + 4R_0^2}}{4r_0 \sqrt{r_0^2 + 4R_0^2}}.$$

---

[21] Note that this Taylor expansion is valid only when $\delta \ll r_0$. For $\delta \sim r_0$, we have to work it out more precisely; see next section.

The total magnificiation sums two images,

$$A = A_+ + A_- = \frac{r_0^2 + 2R_0^2}{r_0\sqrt{r_0^2 + 4R_0^2}} = \frac{u^2 + 2}{u\sqrt{u^2 + 4}} \tag{A.28}$$

with $u = r_0/R_0$.[22] Basically, there is a significant amplification of the brightness of the star when the lens passes through the line of sight within the Einstein radius.

### Appendix A.3. MACHO search

We estimate the frequency and duration of gravitational microlensing due to MA-CHOs in the galactic halo. The Large Magellanic Cloud is about 50kpc away from us, while we are about 8.5kpc away from the galactic center. The flat rotation curve for the Milky Way galaxy is about 220 km/sec (see Fig. 6 in [48]). The Einstein radius for a MACHO is calculated from Eq. (A.24),

$$R_0 = \sqrt{\frac{4G_N m}{c^2} \frac{d_1 d_2}{d_1 + d_2}} = 1.24 10^{12} \, m \left(\frac{m}{M_\odot}\right)^{1/2} \left(\frac{\sqrt{d_1 d_2}}{25\text{kpc}}\right). \tag{A.29}$$

To support the rotation speed of $v_\infty = 220$ km/sec in the isothermal model of halo, we need the velocity dispersion $\sigma = v_\infty/\sqrt{2}$. The average velocity transverse to the line of sight is

$$\langle v_x^2 + v_y^2 \rangle = 2\sigma^2 = v_\infty^2. \tag{A.30}$$

The time it takes a MACHO to traverse the Einstein radius is

$$\frac{R_0}{v_\infty} = 5.6 \times 10^6 \, \text{sec} \left(\frac{m}{M_\odot}\right)^{1/2} \left(\frac{\sqrt{d_1 d_2}}{25\text{kpc}}\right), \tag{A.31}$$

about two months for $m = M_\odot$ and $d_1 = d_2 = 25$ kpc. A microlensing event of duration shorter than a year can be in principle be seen.[23]

The remaining question is the frequency of such microlensing events. It is the probability of a randomly moving MACHO coming within the Einstein radius of a star in the LMC. We will make a crude estimate. The flat rotation curves requires $\frac{G_N M(r)}{r^2} = \frac{v_\infty^2}{r}$ and hence the halo density $\rho(r) = \frac{v_\infty^2}{4\pi G_N r^2}$. The number

---

[22]The singular behavior for $r_0 \to 0$ is due to the invalid Taylor expansion in $\delta$. This is practically not a concern because it is highly unlikely that a MACHO passes through with $r_0 \lesssim \delta_0$. Note that the true image is actually not quite elliptic but distorted in this case.

[23]MACHO collaboration did even more patient scanning to look for microlensing events longer than a year [67].

density of MACHOs, assuming they dominate the halo, is then $n(r) = \frac{v_\infty^2}{4\pi G_N m r^2}$. Instead of dealing with the Boltzmann (Gaussian) distribution in velocities, we simplify the problem by assuming that $\vec{v}_\perp^2 = v_x^2 + v_y^2 = \sigma^2$. From the transverse distance $r_\perp = \sqrt{x^2 + y^2}$, only the fraction $R_0/r_\perp$ heads the right direction for the distance $\sigma \Delta t$. Therefore the fraction of MACHOs that pass through the Einstein radius is

$$\int_0^{\sigma \Delta t} 2\pi r_\perp dr_\perp \frac{R_0}{r_\perp} = 2\pi R_0 \sigma \Delta t. \tag{A.32}$$

We then integrate it over the depth with the number density. The distance from the solar system to the LMA is not the same as the distance from the galactic center because of the relative angle $\alpha = 82°$. The solar system is away from the galactic center by $r_\odot = 8.5$ kpc. Along the line of sight to the LMA with depth $R$, the distance from the galactic center is given by $r^2 = R^2 + r_\odot^2 - 2Rr_\odot \cos \alpha$ with $\alpha = 82°$. Therefore the halo density along the line of sight is

$$n(r) = \frac{v_\infty^2}{4\pi G_N m (R^2 + r_\odot^2 - 2Rr_\odot \cos \alpha)} \tag{A.33}$$

The number of MACHOs passing through the line of sight towards a star in the LMA within the Einstein radius is

$$\int_0^{R_{LMC}} dR n(r) 2\pi R_0 \sigma$$

$$= \int_0^{R_{LMC}} dR \frac{v_\infty^2}{4\pi G_N m (R^2 + r_\odot^2 - 2Rr_\odot \cos \alpha)} 2\pi R_0 \sigma \Delta t \tag{A.34}$$

Pakzynski evaluates the optical depth, but I'd rather estimate a quantity that is directly relevant to the experiment, namely the rate of the microlensing events. Just by taking $\Delta t$ away,

$$\text{rate} = \int_0^{R_{LMC}} dR \frac{v_\infty^2}{4\pi G_N m (R^2 + r_\odot^2 - 2Rr_\odot \cos \alpha)} 2\pi R_0 \sigma$$

$$= \int_0^{R_{LMC}} dR \frac{v_\infty^2}{R^2 + r_\odot^2 - 2Rr_\odot \cos \alpha} \sqrt{\frac{R(R_{LMC} - R)}{G_N m R_{LMC}}} \frac{\sigma}{c}$$

$$= \frac{v_\infty^2 \sigma}{c\sqrt{G_N m R_{LMC}}} \int_0^{R_{LMC}} \frac{\sqrt{R(R_{LMC} - R)} \, dR}{R^2 + r_\odot^2 - 2Rr_\odot \cos \alpha}. \tag{A.35}$$

The integral can be evaluated numerically. For $R_{LMC} = 50$ kpc, $r_\odot = 8.5$ kpc, $\alpha = 82°$, Mathematica gives 3.05. Then with $\sigma = v_\infty/\sqrt{2}$, $v_\infty = 220$ km/sec, we find

$$\text{rate} = 1.69 \times 10^{-13} \sec^{-1} \left(\frac{M_\odot}{m}\right)^{1/2} = 5.34 \times 10^{-6} \text{year}^{-1} \left(\frac{M_\odot}{m}\right)^{1/2}.$$

Therefore, if we can monitor about a million stars, we may see 5 microlensing events for a solar mass MACHO per year, even more for ligher ones.

*Appendix A.4. Strong lensing*

Even though it is not a part of this lecture, it is fun to see what happens when $r_0 \lesssim \delta$. This can be studied easily with a slightly tilted coordinates in Fig. 22.

Using this coordinate system, we can draw a circle on the plane $(x, y) = (x_0, y_0) + \rho(\cos\phi, \sin\phi)$, and the corresponding image on the deflector plane is $(\tilde{x}, \tilde{y}) = (\tilde{x}_0, \tilde{y}_0) + \tilde{\rho}(\cos\phi, \sin\phi) = \frac{d_2}{d_1 + d_2}(x, y)$. The impact parameter is then $r_0 = \sqrt{\tilde{x}^2 + \tilde{y}^2}$ which allows us to calculate $r_\pm(r_0)$ using Eq. (A.26) for each $\phi$. Obviously $\phi$ is the same for the undistorted and distorted images. Fig. 23 shows

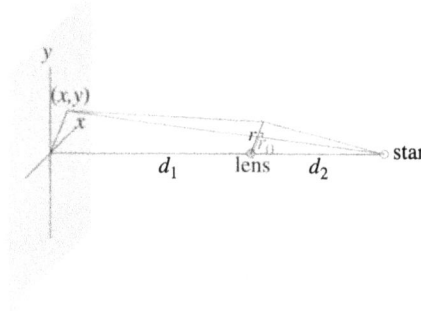

Fig. 22. A slightly different coordinate system to work out the distortion of images.

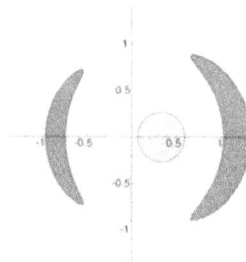

Fig. 23. A highly distorted image due to the gravitational lensing. Yellow circle is the undistorted image, while the two blue regions are the images distorted by the gravitational lensing.

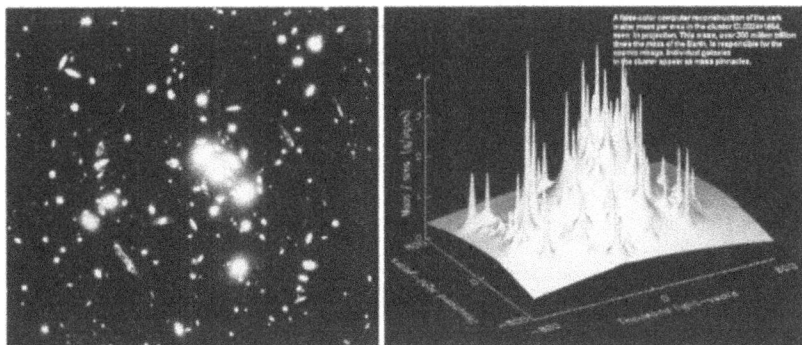

Fig. 24. A Hubble Space Telescope image of a gravitational lens formed by the warping of images of objects behind a massive concentration of dark matter. Warped images of the same blue background galaxy are seen in multiple places. The detailed analysis of lensing effects allows one to map out the mass distribution in the cluster that shows a smooth dark matter contribution not seen in the optical image. Taken from [107].

a spectacular example with $(x_0, y_0) = (1, 0)$, $\frac{d_2}{d_1+d_2} = \frac{1}{3}$, $\rho = 0.8$. Because $\rho \sim r_0$, the Taylor expansion does not work, and the image is far from an ellipse.

This kind of situation is not expected to occur for something as small as the mirror of a telescope, but may for something as big as a galaxy. When an image of a galaxy is distorted by a concentration of mass in the foreground, such as a cluster of galaxies, people have seen spectacular "strong lensing" effects, as shown in Fig. 24.

# References

[1] G. Steigman, Int. J. Mod. Phys. E **15**, 1 (2006) [arXiv:astro-ph/0511534].

[2] H. Davoudiasl, R. Kitano, T. Li and H. Murayama, Phys. Lett. B **609**, 117 (2005) [arXiv:hep-ph/0405097].

[3] H. Murayama, Int. J. Mod. Phys. A **19**, 1265 (2004) [arXiv:hep-ph/0312096].

[4] C.A. Baker *et al.*, Phys. Rev. Lett. **97**, 131801 (2006) [arXiv:hep-ex/0602020].

[5] M. Pospelov and A. Ritz, Annals Phys. **318**, 119 (2005) [arXiv:hep-ph/0504231].

[6] N. Arkani-Hamed and S. Dimopoulos, JHEP **0506**, 073 (2005) [arXiv:hep-th/0405159].

[7] V. Agrawal, S.M. Barr, J.F. Donoghue and D. Seckel, Phys. Rev. D **57**, 5480 (1998) [arXiv:hep-ph/9707380].

[8] R. Harnik, G.D. Kribs and G. Perez, Phys. Rev. D **74**, 035006 (2006) [arXiv:hep-ph/0604027].

[9] H. Murayama, Talk given at 22nd INS International Symposium on Physics with High Energy Colliders, Tokyo, Japan, 8-10 Mar 1994. Published in Proceedings, eds. S. Yamada and T. Ishii. World Scientific, 1995. 476p. [hep-ph/9410285].

[10] V.F. Weisskopf, Phys. Rev. **56**, 72 (1939).

[11]  M. Veltman, Acta Phys. Polon. **B12**, 437 (1981);
      S. Dimopoulos and S. Raby, Nucl. Phys. **B192**, 353 (1981);
      E. Witten, Nucl. Phys. **B188**, 513 (1981);
      M. Dine, W. Fischler and M. Srednicki, Nucl. Phys. **B189**, 575 (1981).

[12]  R. Barbieri and G.F. Giudice, Nucl. Phys. **B306**, 63 (1988);
      G.W. Anderson and D.J. Castano, Phys. Lett. **B347**, 300 (1995) [hep-ph/9409419]; Phys. Rev.
      **D 52**, 1693 (1995) [hep-ph/9412322].

[13]  H. Murayama, "Supersymmetry phenomenology," ICTP Summer School in Particle Physics,
      Trieste, Italy, 21 Jun – 9 Jul 1999. Published in PARTICLE PHYSICS 1999: Proceedings.
      Edited by G. Senjanovic and A.Yu. Smirnov. Singapore, World Scientific, 2000. 394p. (ICTP
      Series in Theoretical Physics, Vol. 16). arXiv:hep-ph/0002232.

[14]  S.P. Martin, arXiv:hep-ph/9709356.

[15]  L.E. Ibanez and G.G. Ross, Nucl. Phys. B **368**, 3 (1992).

[16]  H. Murayama and D.B. Kaplan, Phys. Lett. B **336**, 221 (1994) [arXiv:hep-ph/9406423].

[17]  C.D. Carone, L.J. Hall and H. Murayama, Phys. Rev. D **54**, 2328 (1996) [arXiv:hep-
      ph/9602364].

[18]  E. Farhi and L. Susskind, Phys. Rept. **74**, 277 (1981).

[19]  C.T. Hill and E.H. Simmons, Phys. Rept. **381**, 235 (2003) [Erratum-ibid. **390**, 553 (2004)]
      [arXiv:hep-ph/0203079].

[20]  C. Kolda and H. Murayama, JHEP **0007**, 035 (2000) [hep-ph/0003170].

[21]  C. Csaki, C. Grojean, H. Murayama, L. Pilo and J. Terning, Phys. Rev. D **69**, 055006 (2004)
      [arXiv:hep-ph/0305237].

[22]  C. Csaki, C. Grojean, L. Pilo and J. Terning, Phys. Rev. Lett. **92**, 101802 (2004) [arXiv:hep-
      ph/0308038].

[23]  H. Murayama, arXiv:hep-ph/0307293.

[24]  N. Arkani-Hamed, A.G. Cohen, E. Katz, A.E. Nelson, T. Gregoire and J.G. Wacker, JHEP
      **0208**, 021 (2002) [arXiv:hep-ph/0206020].

[25]  N. Arkani-Hamed, A.G. Cohen, E. Katz and A.E. Nelson, JHEP **0207**, 034 (2002) [arXiv:hep-
      ph/0206021].

[26]  D.B. Kaplan and H. Georgi, Phys. Lett. B **136**, 183 (1984).

[27]  H. Georgi and D.B. Kaplan, Phys. Lett. B **145**, 216 (1984).

[28]  N.S. Manton, Nucl. Phys. B **158**, 141 (1979).

[29]  L.J. Hall, Y. Nomura and D.R. Smith, Nucl. Phys. B **639**, 307 (2002) [arXiv:hep-ph/0107331].

[30]  G. von Gersdorff, N. Irges and M. Quiros, Nucl. Phys. B **635**, 127 (2002) [arXiv:hep-
      th/0204223].

[31]  C. Csaki, C. Grojean and H. Murayama, Phys. Rev. D **67**, 085012 (2003) [arXiv:hep-
      ph/0210133].

[32]  G. Burdman and Y. Nomura, Nucl. Phys. B **656**, 3 (2003) [arXiv:hep-ph/0210257].

[33]  R. Contino, Y. Nomura and A. Pomarol, Nucl. Phys. B **671**, 148 (2003) [arXiv:hep-
      ph/0306259].

[34]  K. Agashe, R. Contino and A. Pomarol, Nucl. Phys. B **719**, 165 (2005) [arXiv:hep-
      ph/0412089].

[35]  H.C. Cheng, J. Thaler and L.T. Wang, JHEP **0609**, 003 (2006) [arXiv:hep-ph/0607205].

[36]  H.C. Cheng and I. Low, JHEP **0408**, 061 (2004) [arXiv:hep-ph/0405243].

[37]  S. Nussinov, Phys. Lett. B **165**, 55 (1985).

[38] N. Arkani-Hamed, S. Dimopoulos and G. Dvali, Phys. Lett. **B429**, 263 (1998) [hep-ph/9803315];
I. Antoniadis, N. Arkani-Hamed, S. Dimopoulos and G. Dvali, Phys. Lett. **B436**, 257 (1998) [hep-ph/9804398].

[39] D.J. Kapner, T.S. Cook, E.G. Adelberger, J.H. Gundlach, B.R. Heckel, C.D. Hoyle and H.E. Swanson, Phys. Rev. Lett. **98**, 021101 (2007) [arXiv:hep-ph/0611184].

[40] L.J. Hall and D.R. Smith, Phys. Rev. D **60**, 085008 (1999) [arXiv:hep-ph/9904267].

[41] L. Randall and R. Sundrum, Phys. Rev. Lett. **83**, 3370 (1999) [arXiv:hep-ph/9905221].

[42] J.M. Maldacena, Adv. Theor. Math. Phys. **2**, 231 (1998) [Int. J. Theor. Phys. **38**, 1113 (1999)] [arXiv:hep-th/9711200].

[43] S.B. Giddings, S. Kachru and J. Polchinski, Phys. Rev. D **66**, 106006 (2002) [arXiv:hep-th/0105097].

[44] K. Agashe and G. Servant, Phys. Rev. Lett. **93**, 231805 (2004) [arXiv:hep-ph/0403143].

[45] T. Appelquist, H.C. Cheng and B.A. Dobrescu, Phys. Rev. D **64**, 035002 (2001) [arXiv:hep-ph/0012100].

[46] H.C. Cheng, K.T. Matchev and M. Schmaltz, Phys. Rev. D **66**, 036005 (2002) [arXiv:hep-ph/0204342].

[47] H.C. Cheng, K.T. Matchev and M. Schmaltz, Phys. Rev. D **66**, 056006 (2002) [arXiv:hep-ph/0205314].

[48] G.G. Raffelt, "Dark matter: Motivation, candidates and searches," 1997 European school of high energy physics, Proceedings edited by N. Ellis and M. Neubert. Geneva, Switzerland, CERN, 1998. 351p. (CERN-98-03) arXiv:hep-ph/9712538.

[49] D.N. Spergel *et al.* [WMAP Collaboration], arXiv:astro-ph/0603449.

[50] R.H. Cyburt, Phys. Rev. D **70**, 023505 (2004) [arXiv:astro-ph/0401091].

[51] D.J. Eisenstein *et al.* [SDSS Collaboration], Astrophys. J. **633**, 560 (2005) [arXiv:astro-ph/0501171].

[52] K.G. Begeman, A.H. Broeils and R.H. Sanders, Mon. Not. Roy. Astron. Soc. **249**, 523 (1991).

[53] B.J. Carr, Astrophys. J. **201**, 1 (1975).

[54] L.J. Hall and S. Hsu, Phys. Rev. Lett. **64**, 2848 (1990).

[55] K. Jedamzik, Phys. Rev. D **55**, 5871 (1997) [arXiv:astro-ph/9605152].

[56] B. Paczynski, Astrophys. J. **304**, 1 (1986).

[57] P. Tisserand *et al.* [EROS-2 Collaboration], arXiv:astro-ph/0607207.

[58] C. Alcock *et al.* [MACHO Collaboration], arXiv:astro-ph/9803082.

[59] W.M. Yao *et al.* [Particle Data Group], J. Phys. G **33**, 1 (2006).

[60] S. Tremaine and J.E. Gunn, Phys. Rev. Lett. **42**, 407 (1979).

[61] D.N. C. Lin and S.M. Faber, Astrophys. J. **266**, L21 (1983).

[62] S. Hannestad and G.G. Raffelt, JCAP **0611**, 016 (2006) [arXiv:astro-ph/0607101].

[63] U. Seljak, A. Slosar and P. McDonald, JCAP **0610**, 014 (2006) [arXiv:astro-ph/0604335].

[64] S. Dimopoulos, D. Eichler, R. Esmailzadeh and G.D. Starkman, Phys. Rev. D **41**, 2388 (1990).

[65] G.D. Starkman, A. Gould, R. Esmailzadeh and S. Dimopoulos, Phys. Rev. D **41**, 3594 (1990).

[66] M. Kawasaki, H. Murayama and T. Yanagida, Prog. Theor. Phys. **87**, 685 (1992).

[67] R.A. Allsman *et al.* [(Macho Collaboration) C Alcock], Astrop. J. Lett. **550**, L169 (2001), arXiv:astro-ph/0011506.

[68] J. McDonald, Phys. Rev. D **50**, 3637 (1994), actually considered a complex *S*.

[69] C.P. Burgess, M. Pospelov and T. ter Veldhuis, Nucl. Phys. B **619**, 709 (2001).

[70] M.A. Shifman, A.I. Vainshtein and V.I. Zakharov, Phys. Lett. B **78**, 443 (1978).

[71] [CDMS Collaboration], arXiv:astro-ph/0405033.

[72] R. Bernabei *et al.*, Riv. Nuovo Cim. **26N1**, 1 (2003).

[73] R. Gaitskell and V. Mandic, http://dmtools.berkeley.edu/limitplots/

[74] O.J. P. Eboli and D. Zeppenfeld, Phys. Lett. B **495**, 147 (2000) [arXiv:hep-ph/0009158].
L. Neukermans and B. Di Girolamo, Observing an invisibly decaying Higgs boson in ATLAS via vector boson fusion, ATLAS note, ATL-PHYS-2003-006.
V. Buscher and K. Jakobs, Int. J. Mod. Phys. A **20**, 2523 (2005) [arXiv:hep-ph/0504099].

[75] G. Weiglein *et al.* [LHC/LC Study Group], Phys. Rept. **426**, 47 (2006) [arXiv:hep-ph/0410364].

[76] J.R. Ellis, K.A. Olive, Y. Santoso and V.C. Spanos, Phys. Lett. B **565**, 176 (2003) [arXiv:hep-ph/0303043].

[77] J.R. Ellis, K.A. Olive, Y. Santoso and V.C. Spanos, JHEP **0605**, 063 (2006) [arXiv:hep-ph/0603136].

[78] H. Baer, E.K. Park, X. Tata and T.T. Wang, JHEP **0608**, 041 (2006) [arXiv:hep-ph/0604253].

[79] R. Trotta, R.R. de Austri and L. Roszkowski, New Astron. Rev. **51**, 316 (2007) [arXiv:astro-ph/0609126].

[80] N. Arkani-Hamed, L.J. Hall, H. Murayama, D.R. Smith and N. Weiner, Phys. Rev. D **64**, 115011 (2001) [arXiv:hep-ph/0006312].

[81] T. Tsukamoto, K. Fujii, H. Murayama, M. Yamaguchi and Y. Okada, Phys. Rev. D **51**, 3153 (1995).

[82] E.A. Baltz, M. Battaglia, M.E. Peskin and T. Wizansky, Phys. Rev. D **74**, 103521 (2006) [arXiv:hep-ph/0602187].

[83] K. Griest and D. Seckel, Phys. Rev. D **43**, 3191 (1991).

[84] S. Mizuta and M. Yamaguchi, Phys. Lett. B **298**, 120 (1993) [arXiv:hep-ph/9208251].

[85] J. Carr, G. Lamanna and J. Lavalle, Rept. Prog. Phys. **69**, 2475 (2006).

[86] G. Servant and T.M. P. Tait, Nucl. Phys. B **650**, 391 (2003) [arXiv:hep-ph/0206071].

[87] D. Hooper and S. Profumo, arXiv:hep-ph/0701197.

[88] M. Dine, A.E. Nelson and Y. Shirman, Phys. Rev. D **51**, 1362 (1995) [arXiv:hep-ph/9408384];
M. Dine, A.E. Nelson, Y. Nir and Y. Shirman, Phys. Rev. D **53**, 2658 (1996) [arXiv:hep-ph/9507378].

[89] T. Moroi, H. Murayama and M. Yamaguchi, Phys. Lett. B **303**, 289 (1993).

[90] A. de Gouvea, T. Moroi and H. Murayama, Phys. Rev. D **56**, 1281 (1997) [arXiv:hep-ph/9701244].

[91] J.L. Feng, A. Rajaraman and F. Takayama, Phys. Rev. D **68**, 063504 (2003) [arXiv:hep-ph/0306024].

[92] W. Buchmuller, K. Hamaguchi, M. Ratz and T. Yanagida, Phys. Lett. B **588**, 90 (2004) [arXiv:hep-ph/0402179].

[93] J.L. Feng, A. Rajaraman and F. Takayama, Int. J. Mod. Phys. D **13**, 2355 (2004) [arXiv:hep-th/0405248].

[94] K. Kohri, T. Moroi and A. Yotsuyanagi, Phys. Rev. D **73**, 123511 (2006) [arXiv:hep-ph/0507245].

[95] R.D. Peccei and H.R. Quinn, Phys. Rev. Lett. **38**, 1440 (1977).

[96] H. Murayama, G. Raffelt, C. Hagmann, K. van Bibber and L.J. Rosenberg, "Axions and other very light bosons," in [59].

[97] J.E. Kim, Phys. Rev. Lett. **43**, 103 (1979).

[98]  M.A. Shifman, A.I. Vainshtein and V.I. Zakharov, Nucl. Phys. B **166**, 493 (1980).

[99]  A.R. Zhitnitsky, Sov. J. Nucl. Phys. **31**, 260 (1980) [Yad. Fiz. **31**, 497 (1980)].

[100]  M. Dine, W. Fischler and M. Srednicki, Phys. Lett. B **104**, 199 (1981).

[101]  D.J. Gross, R.D. Pisarski and L.G. Yaffe, Rev. Mod. Phys. **53**, 43 (1981).

[102]  M. Buckley and H. Murayama, in preparation.

[103]  S.J. Asztalos *et al.*, Phys. Rev. D **69**, 011101 (2004) [arXiv:astro-ph/0310042].

[104]  L.D. Duffy *et al.*, Phys. Rev. D **74**, 012006 (2006) [arXiv:astro-ph/0603108].

[105]  R. Bradley *et al.*, Rev. Mod. Phys. **75**, 777 (2003).

[106]  N. Arkani-Hamed, L.J. Hall, C.F. Kolda and H. Murayama, Phys. Rev. Lett. **85**, 4434 (2000) [arXiv:astro-ph/0005111].

[107]  http://www.bell-labs.com/org/physicalsciences/projects/darkmatter/darkmatter1.html.

# SHORT TOPICAL LECTURES

## Effective Field Theories and Gravitational Radiation

*W. Goldberger*

## Holographic Cosmology

*T. Hertog*

## Neutrino Physics and Cosmology

*S. Pastor*

## Cosmic Microwave Background: Observational Status

*S. Prunet*

## Structure Formation with Numerical Simulations

*R. Teyssier*

## Giving Mass to the Graviton

*P. Tinyakov*

Course 7

# EFFECTIVE FIELD THEORIES
# AND GRAVITATIONAL RADIATION

## Walter D. Goldberger

*Department of Physics, Yale University, New Haven, CT 06520, USA*

*F. Bernardeau, C. Grojean and J. Dalibard, eds.*
*Les Houches, Session LXXXVI, 2006*
*Particle Physcis and Cosmology: The Fabric of Spacetime*
© *2007 Published by Elsevier B.V.*

# Contents

## Abstract

These lectures give an overview of the uses of effective field theories in describing gravitational radiation sources for LIGO or LISA. The first lecture reviews some of the standard ideas of effective field theory (decoupling, matching, power counting) mostly in the context of a simple toy model. The second lecture sets up the problem of calculating gravitational wave emission from non-relativistic binary stars by constructing a tower of effective theories that separately describe each scale in the problem: the internal size of each binary constituent, the orbital separation, and the wavelength of radiated gravitons.

## 1. Lecture I

### 1.1. Introduction and motivation

These lectures describe the uses of effective field theory (EFT) methods to solve problems in gravitational wave physics. Many of the signals relevant to gravitational wave experiments such as LIGO [1], VIRGO [2] and the planned LISA [3] correspond to astrophysical sources whose evolution involve a number of distinctly separated length scales. In order to compute signal templates that capture the physics accurately, it is important to systematically account for effects arising at all these different scales. It is exactly this sort of problem that is best treated by EFT methods, analogous to the EFTs constructed to unravel multiple scale problems in high energy physics or condensed matter.

For the sake of concreteness, I will focus in these lectures on the EFT formulation of the slow "inspiral" phase of compact binary stars (that is, with neutron star (NS) or black hole (BH) constituents). A more comprehensive review of gravitational wave sources and phenomenology can be found in the lectures by A. Buonnano at this school (also see, e.g., ref. [4]). The inspiral phase plays an important role in gravitational wave physics, and corresponds to the period in the evolution of the binary in which the system is non-relativistic, the bound orbit slowly decaying due to the emission of gravitational radiation.

An appealing feature of the binary inspiral phase is that, theoretically, it is a very tractable problem. In principle, the dynamics is calculable as a perturbative expansion of the Einstein equations in the parameter $v \ll 1$, a typical three-velocity.[1] To get a handle on the scales involved, it is worthwhile to calculate some of the features of the binary inspiral at the order of magnitude level. Let's consider inspiral events seen by LIGO, which for illustration we will assume operates in the frequency band between 10 Hz and 1 kHz. At such frequencies, the relevant inspiral sources correspond to neutron star binaries, with $m_{NS} \sim 1.4 m_\odot$, ($m_\odot$ is the solar mass) or perhaps black holes with typical mass $m_{BH} \sim 10 m_\odot$. Since the time evolution is non-relativistic, it is to a good approximation treated in terms of Newtonian gravity. In particular, the orbital parameters are related by

$$v^2 \sim \frac{G_N m}{r} \equiv \frac{r_s}{2r}, \tag{1.1}$$

where $r$ is the orbital radius, and we have introduced the Schwarzschild radius $r_s = 2 G_N m$. The frequency of the radiation emitted during this phase is of order the orbital frequency $2\pi v \sim v/r$, so for a binary inspiral that scans the LIGO band, the typical orbital radius is measured in kilometers

$$r(10 \text{ Hz}) \sim 300 \text{ km} \left( \frac{m}{m_\odot} \right)^{1/3} \quad \rightarrow \quad r(1 \text{ kHz}) \sim 14 \text{ km} \left( \frac{m}{m_\odot} \right)^{1/3}, \tag{1.2}$$

which is therefore well separated from the size of the compact objects, $r_s \sim 1 \text{ km } m/m_\odot$. The expansion parameter $v$ also evolves as the binary sweeps the detector frequency band:

$$v(10 \text{ Hz}) \sim 0.06 \left( \frac{m}{m_\odot} \right)^{1/3} \quad \rightarrow \quad v(1 \text{ kHz}) \sim 0.3 \left( \frac{m}{m_\odot} \right)^{1/3}, \tag{1.3}$$

indicating that the velocity is a good expansion parameter up to the last few cycles of orbit seen by the detector. As $v \rightarrow 1$, the inspiral phase ends, and the evolution must be treated by numerical simulations.

During the inspiral phase, the bound orbit is unstable to the emission of gravitational radiation. The power emitted in gravitational waves is well approximated by the quadrupole radiation formula. Taking the orbit to be circular, the power radiated is

$$\frac{dE}{dt} = \frac{32}{5} G_N^{-1} v^{10}. \tag{1.4}$$

---

[1] Even though we will be doing classical physics for most of these lectures, I use particle physics units $c = \hbar = 1$ throughout.

The mechanical energy of the binary is just $E = -\frac{1}{2}mv^2$ to leading order in $v$, so by conservation of energy

$$\frac{d}{dt}\left(-\frac{1}{2}mv^2\right) = -\frac{32}{5}G_N^{-1}v^{10}, \tag{1.5}$$

we obtain an estimate for the duration of the inspiral event seen by LIGO

$$\Delta t = \frac{5}{512}r_s\left[\frac{1}{v_i^8} - \frac{1}{v_f^8}\right] \sim 5\,\text{min.}\left(\frac{m}{m_\odot}\right)^{-8/3}, \tag{1.6}$$

and the number of orbital cycles as the signal sweeps the detector band

$$N \sim \int_{t_i}^{t_f} \omega(t)dt = \frac{1}{32}\left[\frac{1}{v_i^5} - \frac{1}{v_f^5}\right] \sim 4 \times 10^4\left(\frac{m}{m_\odot}\right)^{-5/3}\text{radians}, \tag{1.7}$$

with $\omega$ the orbital angular frequency.

Eqs. (1.2), (1.3), (1.6) give an estimate for the typical length, velocity, and duration of an inspiral event that sweeps the LIGO frequency band. Inspiral events seen by LISA follow the same dynamics. However, because LISA will operate in a frequency range complementary to LIGO, $10^{-5}$ Hz $< \nu < 1$ Hz, the sources correspond to objects of much large mass (e.g., BH/BH binaries with $m_{BH} \sim 10^{5-8}m_\odot$ or BH/NS systems with $m_{BH} \sim 10^{5-7}$).

For either LIGO or LISA, the number of orbital cycles spent in the detector frequency band is quite large, see e.g. Eq. (1.7). Thus even a slight deviation between theoretical calculations of the gravitational wave phase (the waveform "templates") and the data will become amplified over the large number of cycles of evolution. Consequently, inspiral wave signals carry detailed information about gravitational dynamics. In fact, it has been determined that LIGO will be sensitive to corrections that are order $v^6$ in the velocity expansion beyond the leading order quadrupole radiation predictions [5].

The procedure for computing corrections to the motion of the binary system in the non-relativistic limit by an iterated expansion of the Einstein equations is called the post-Newtonian expansion of general relativity [6]. What makes the calculations difficult (and interesting) is that there is a proliferation of physically relevant effects occurring at *different length scales*,

$r_s$ = size of compact objects,

$r$ = orbital radius,

$\lambda$ = wavelength of emitted radiation,

all controlled by the same expansion parameter $v$:

$$\frac{r_s}{r} \sim v^2 \quad \frac{r}{\lambda} \sim v, \tag{1.8}$$

where the first estimate follows from Kepler's law and the second from the multipole expansion of the radiation field coupled to non-relativistic sources. Thus to do systematic calculations at high orders in $v$, it is necessary to deal with physics at many different scales. A convenient way of doing this is to construct a chain of EFTs that capture the relevant physics at each scale separately.

The goal of these lectures is to recast the problem of computing gravitational wave observables for non-relativistic sources in terms of EFTs. In the next section we briefly review the EFT logic, using a standard particle physics example, the low energy dynamics of Goldstone bosons in a theory with spontaneously broken global symmetry, to illustrate the main ideas. In sec. 2, we turn to the EFT formulation of the binary inspiral problem. The presentation follows [7], which is based on similar EFTs for non-relativistic bound states in QED and QCD [8]. However, the focus here is not on detailed calculations, but in describing how to "integrate out" physics at each of the scales $r_s$, and $r$ to obtain a theory with well defined rules for calculating gravitational wave observables (at the scale $\lambda$) as an expansion in $v$. Possible extensions of the ideas presented here to other problems in gravitational wave physics are discussed in the conclusions, sec. 3.

### 1.2. Effective field theories: a review

EFTs are indispensable for treating problems that simultaneously involve two or more widely separated scales. As a typical application, suppose one is interested in calculating the effects of some sort of short distance physics, characterized by a scale $\Lambda$, on the dynamics at a low energy scale $\omega \ll \Lambda$. In the EFT description of such a problem, the effects of $\Lambda$ on the low energy physics become simple, making it possible to construct a systematic expansion in the ratio $\omega/\Lambda \ll 1$. Here we briefly review the basic ideas that go into the construction and applications of EFTs. More detailed reviews can be found in refs. [9–12].

The key insight that makes EFTs possible is the following: consider for instance a field theory with light degrees of freedom collectively denoted by $\phi$ (e.g, a set of massless fields) and some heavy fields $\Phi$ with mass near the UV scale $\Lambda$. The interactions of these fields are described by some action functional $S[\phi, \Phi]$. Suppose also that we are only interested in describing experimental observables involving only the light modes $\phi$. Then it makes sense to integrate out the modes $\Phi$ to obtain an effective action that describes the interactions of the light fields among each other

$$e^{i S_{eff}[\phi]} = \int D\Phi(x)\, e^{i S[\phi, \Phi]}. \tag{1.9}$$

It turns out that in general, $S_{eff}[\phi]$ can be expressed as a *local* functional of the fields $\phi(x)$,

$$S_{eff}[\phi] = \sum_i c_i \int d^4x \, \mathcal{O}_i(x), \tag{1.10}$$

for (in general an infinite number) local operators $\mathcal{O}_i(x)$. The coefficients $c_i$ are usually referred to as "Wilson coefficients". If $\mathcal{O}_i(x)$ has mass dimension $\Delta_i$, then the Wilson coefficients, evaluated at a renormalization point $\mu$ of order $\Lambda$, scale as powers of $\Lambda$

$$c_i(\mu = \Lambda) = \frac{\alpha_i}{\Lambda^{\Delta_i - 4}}, \tag{1.11}$$

with $\alpha_i \sim \mathcal{O}(1)$. From this observation we conclude that the short distance physics can have two types of effects on the dynamics at energies $\omega \ll \Lambda$:
• Renormalization of the coefficients of operators with mass dimension $\Delta \leq 4$.
• Generation of an infinite tower of irrelevant (i.e. $\Delta > 4$) operators with coefficients scaling as in Eq. (1.11).

This result regarding the structure of the low energy dynamics encoded in $S_{eff}[\phi]$ is usually referred to as *decoupling*. This concept has its roots in the work of K. Wilson on the renormalization group [13]. Decoupling as used by practicing field theorists was first made rigorous in [14]. The idea is useful because it states that the dependence of the low energy physics on the scale $\Lambda$ is extremely simple. All UV dependence appears directly in the coefficients of the effective Lagrangian, and determining the $\Lambda$ dependence of an observable follows from *power counting* (essentially a generalized form of dimensional analysis). Effective Lagrangians are typically used in one of two ways:

1. The "full theory" $S[\phi, \Phi]$ is known: In this case, integrating out the heavy modes gives a simple way of systematically analyzing the effects of the heavy physics on low energy observables. Because only the low energy scale appears explicitly in the Feynman diagrams of the EFT, amplitudes are easier to calculate and to power count than in the full theory.
   **Examples:** Integrating out the $W$, $Z$ bosons from the $SU(2)_L \times U(1)_Y$ Electroweak Lagrangian at energies $E \ll m_{W,Z}$ results in the Fermi theory of weak decays plus corrections suppressed by powers of $E^2/m_{W,Z}^2$. It is easier to analyze electromagnetic or QCD corrections to weak decays in the four-Fermi theory than in the full Electroweak Lagrangian, as the graphs of Fig. 1 indicate. See, e.g., ref. [15]. Another example is the use of EFTs to calculate heavy particle threshold corrections to low energy gauge couplings [16]. This has applications, for instance, in Grand Unified Theories and in QCD.

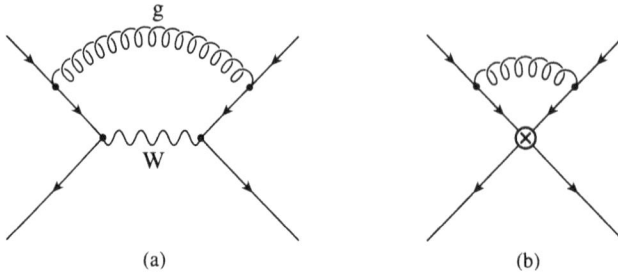

Fig. 1. QCD correction to a typical four-Fermi process in the Standard Model, calculated in (a) the full theory with propagating $W$ boson s, and (b) the effective Fermi theory of weak interactions. Graph (b) reproduces graph (a) up to corrections suppressed by powers of $E^2/m_W^2 \ll 1$.

2. The full theory is unknown (or known but strongly coupled): Whatever the physics at the scale $\Lambda$ is, by decoupling it must manifest itself at low energies as an effective Lagrangian of the form Eq. (1.10). If the symmetries (eg. Poincare, gauge, global) that survive at low energies are known, then the operators $\mathcal{O}_i(x)$ appearing in $S_{eff}[\phi]$ must respect those symmetries. Thus by writing down an effective Lagrangian containing the most general set of operators consistent with the symmetries, we are necessarily accounting for the UV physics in a completely *model independent way*.

**Examples**: The QCD chiral Lagrangian below the scale $\Lambda_{\chi SB}$ of $SU(3)_L \times SU(3)_R \to SU(3)_V$ chiral symmetry breaking [17, 18]. Here the full theory, QCD, is known, but because $\Lambda_{\chi SB}$ is of order the scale $\Lambda_{QCD} \sim 1$ GeV where the QCD coupling is strong, it is impossible to perform the functional integral in Eq. (2.16) analytically. Another example is general relativity below the scale $m_{Pl} \sim 10^{19}$ GeV. This theory can be used to calculate, e.g., graviton-graviton scattering at energies $E \ll m_{Pl}$. Above those energies, however, scattering amplitudes calculated in general relativity start violating unitarity bounds, and the effective field theory necessarily breaks down. Thus general relativity is an effective Lagrangian for quantum gravity below the strong coupling scale $m_{Pl}$. The EFT interpretation of general relativity is reviewed in more detail in refs. [19, 20].

Finally, it is believed that the Standard Model itself is an effective field theory below scales of order $\Lambda = 1$ TeV or so (see lectures by H. Murayama at this school). This scale manifests itself indirectly, in the form of $SU(2)_L \times U(1)_Y$ gauge invariant operators of dimension $\Delta > 4$ constructed from Standard Model fields [21], certain linear combinations of which have been constrained experimentally using collider data from the LEP experiments at CERN and from SLD at SLAC (see [22] for a recent analysis of precision electroweak constraints using effective Lagrangians). If there is indeed new physics at the

TeV scale, it will be seen directly, at the CERN LHC which is due to come on line in the next few years.

In either of these two classes of examples, integrating out the heavy physics as in Eq. (2.16) results in a an effective Lagrangian that contains, in general, an *infinite* number of operators $\mathcal{O}_i(x)$. However, because

1. An operator $\mathcal{O}(x)$ with $[\mathcal{O}] = \Delta_{\mathcal{O}} > 4$ contributes to an observable at relative order

$$\left(\frac{\omega}{\Lambda}\right)^{\Delta_{\mathcal{O}}-4} \ll 1, \tag{1.12}$$

2. A given observable can only be determined up to a finite experimental resolution $\epsilon \ll 1$,

one may typically truncate the series in Eq. (1.10) after a finite number of operators, those with mass dimension $\Delta \leq N + 4$, where

$$\epsilon = \text{expt. error} = \left(\frac{\omega}{\Lambda}\right)^{N}. \tag{1.13}$$

Therefore $S_{eff}[\phi]$ is predictive as long as there are more observables than operators with mass dimension $\Delta \leq N + 4$.

To make these general remarks more concrete, it is worth studying a toy example in some detail. We'll consider the quantum field theory of a single complex scalar field $\phi(x)$ with Lagrangian (the "full theory")

$$\mathcal{L} = |\partial_\mu \phi|^2 - V(\phi), \tag{1.14}$$

where

$$V(\phi) = \frac{\lambda}{2}\left(|\phi|^2 - v^2/2\right)^2. \tag{1.15}$$

This theory is invariant under the global $U(1)$ symmetry

$$U(1) : \phi(x) \rightarrow e^{i\alpha}\phi(x), \tag{1.16}$$

for some constant phase $\alpha$, as well as a discrete charge conjugation symmetry

$$C : \phi(x) \rightarrow \phi^*(x). \tag{1.17}$$

Classically, the ground state of this theory is determined by the minimum of $V(\phi)$. Because of the $U(1)$ symmetry the ground state is degenerate, and the vacuum manifold is the circle $|\phi|^2 = v^2/2$ in the complex $\phi$ plane. To study fluctuations about the vacuum, we expand the fields about any of these (equivalent) vacuua, for instance the point

$$\langle\phi\rangle = \frac{v}{\sqrt{2}}. \tag{1.18}$$

Expanding about this point spontaneously breaks the $U(1)$ symmetry, resulting in one Goldstone boson. It is convenient to write the original field $\phi(x)$ as

$$\phi(x) = \frac{1}{\sqrt{2}}\left(v + \rho(x)\right)e^{i\pi(x)/v}. \tag{1.19}$$

Under the original symmetries, the new fields transform as

$$U(1) : \begin{cases} \rho(x) \to \rho(x), \\ \pi(x)/v \to \pi(x)/v + \alpha, \end{cases} \tag{1.20}$$

and

$$C : \begin{cases} \rho(x) \to \rho(x), \\ \pi(x) \to -\pi(x). \end{cases} \tag{1.21}$$

The Lagrangian in terms of the new fields is given by

$$\mathcal{L} = \frac{1}{2}(\partial_\mu\rho)^2 + \frac{1}{2}\left(1 + \frac{\rho}{v}\right)^2 (\partial_\mu\pi)^2 - \frac{\lambda}{2}(v\rho + \rho^2/2)^2, \tag{1.22}$$

so the spectrum of excitations about the vacuum consists of a "modulus" field $\rho(x)$ with tree level mass given by $m_\rho^2 = \lambda v^2$ and the Goldstone boson $\pi(x)$ with $m_\pi = 0$. (We take $\lambda \ll 1$ so that a perturbative treatment is valid).

Suppose that we are interested in working out the predictions of this theory at energies $\omega \ll m_\rho$. At such scales, only the dynamics of the massless field $\pi(x)$ is non-trivial, and following our general discussion it is convenient to write an effective Lagrangian. The general structure of this effective Lagrangian is dictated by the symmetries of the original theory. In particular the $U(1)$, realized non-linearly as the shift symmetry $\pi(x)/v \to \pi(x)/v + \alpha$, restricts the effective Lagrangian to be a function of $\partial_\mu\pi$ only. Therefore the symmetries of the low energy theory alone explain a number of consequences of the low energy dynamics, for instance

• The field $\pi(x)$ must be massless, since a mass term would break the shift symmetry. This is just the statement of Goldstone's theorem in the context of the low energy EFT.

• The field $\pi(x)$ is derivatively self-coupled. This implies in particular that scattering amplitudes are "soft", vanishing as powers of the typical energy $\omega$ in the limit $\omega \to 0$.

In addition to the constraints from the non-linearly realized $U(1)$, there is also a constraint from the charge conjugation symmetry $\pi \to -\pi$ which says that the EFT must be even in $\pi$. Thus the effective Lagrangian must be of the general form

$$\mathcal{L}_{EFT} = \frac{1}{2}(\partial_\mu \pi)^2 + \frac{c_8}{4\Lambda^4}(\partial_\mu \pi \, \partial^\mu \pi)^2 + \cdots, \qquad (1.23)$$

where only the leading operators with two and four $\pi$'s have been displayed. In this equation $\Lambda \sim m_\rho$ and $c_8$ is some dimensionless constant of order $\lambda$.

### 1.2.1. Matching

Since we know the full theory, Eq. (1.14), it is possible to explicitly perform the functional integral of Eq. (1.9) in order to obtain the EFT parameters like $\Lambda$ and $c_8$ in terms of the couplings $\lambda$, $v$ appearing in the full theory Lagrangian. Rather than calculating this integral, the low energy parameters are fixed in practice through a procedure referred to as *matching*.

To perform a matching calculation, one simply calculates some observable, for instance a scattering amplitude among the light particles, in two ways. First one calculates the amplitude in the full theory, expanding the result in powers of $\omega/\Lambda$. One then calculates the same quantity in the effective field theory, adjusting the EFT parameters in order to reproduce the full theory result.

As an example, consider $\pi\pi \to \pi\pi$ scattering in our toy model. In the full theory, the amplitude to leading order in $\lambda$ is given by the diagrams of Fig. 2. Reading off the Feynman rules from the full theory Lagrangian, Eq. (1.14), one finds

$$i\mathcal{A}_{full} = \left[\frac{2i}{v}k_1 \cdot k_2\right]\frac{i}{s - m_\rho^2}\left[\frac{2i}{v}k_3 \cdot k_4\right] + \text{crossings}, \qquad (1.24)$$

where $k_1, k_2$ are the initial (incoming) momenta, and $k_3, k_4$ are the final (outgoing) momenta. Introducing the usual Mandelstam variables $s = (k_1 + k_2)^2$,

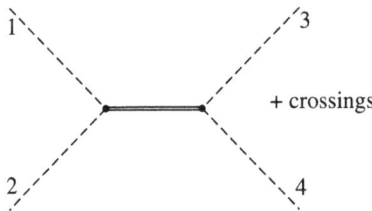

Fig. 2. Leading order contribution to $\pi\pi \to \pi\pi$ scattering in the full theory. The intermediate double line corresponds to the $\rho$ field propagator.

$t = (k_1 - k_3)^2$, $u = (k_1 - k_4)^2$, and expanding in $s, t, u \ll m_\rho^2$, this becomes

$$i \mathcal{A}_{full} \simeq \frac{4i}{v^2 m_\rho^2}\left[\left(\frac{s}{2}\right)^2 + \left(\frac{t}{2}\right)^2 + \left(\frac{u}{2}\right)^2\right] + \mathcal{O}(s^3/m_\rho^6) \qquad (1.25)$$

In the effective field theory, the amplitude arises from the leading dimension eight interaction in Eq. (1.23),

$$i \mathcal{A}_{EFT} = \frac{ic_8}{\Lambda^4}\left[\left(\frac{s}{2}\right)^2 + \left(\frac{t}{2}\right)^2 + \left(\frac{u}{2}\right)^2\right]. \qquad (1.26)$$

Thus taking $\Lambda = m_\rho$, the matching condition for the coefficient $c_8$ is given by

$$c_8(\mu = \Lambda) = 4\lambda + \mathcal{O}(\lambda^2). \qquad (1.27)$$

### 1.2.2. Corrections to the matching coefficients

In general, the coefficients in the effective Lagrangian are calculable as a series expansion in the parameters of the full theory. In our example, the coefficients receive corrections at all orders in the parameter $\lambda \ll 1$, of which the tree level matching calculation in the above example gives only the first term. It is instructive to see how one goes about computing corrections to the Wilson coefficients, as this illustrates certain general properties of effective field theories.

In the example above, consider for instance the coefficient of the kinetic term for the $\pi$ field in the low energy EFT,

$$\mathcal{L}_{EFT} = \frac{1}{2}Z(\partial_\mu \pi)^2 + \cdots. \qquad (1.28)$$

To leading order $Z = 1$. Corrections of order $\lambda$ are obtained by comparing the one-loop $\pi$ field propagator in the full and the effective theories, adjusting $Z$ so that the two calculations agree. Rather than calculate the full propagator, it is enough to consider the "one light-particle irreducible" two-point function. In the full theory this is the sum of graphs that cannot be made disconnected by cutting one $\pi$ field propagator. In the EFT, it is just the usual 1PI graphs. The relevant diagrams are shown in the full and effective theories in Fig. 3, and Fig. 4 respectively. In the full theory we find at one-loop

$$\text{Fig. 3(a)} = \frac{i\lambda}{16\pi^2}k^2\left[\frac{1}{\epsilon} - \gamma + \frac{3}{2} - \ln\left(\frac{m_\rho^2}{4\pi\mu^2}\right)\right] + \mathcal{O}(k^4/m_\rho^2), \qquad (1.29)$$

$$\text{Fig. 3(b)} = -\frac{i\lambda}{16\pi^2}k^2\left[\frac{1}{\epsilon} - \gamma + 1 - \ln\left(\frac{m_\rho^2}{4\pi\mu^2}\right)\right], \qquad (1.30)$$

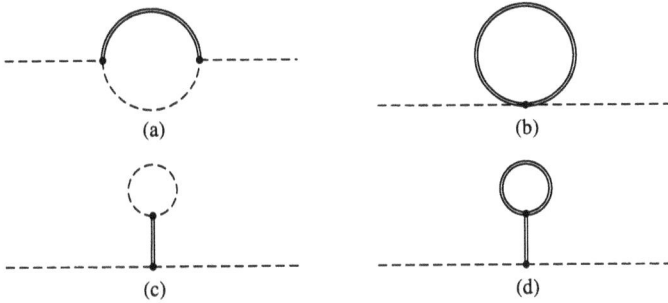

(a)　　　　　　　　(b)

(c)　　　　　　　　(d)

Fig. 3. Feynman graphs contributing to the $\pi$ field self-energy in the full theory with a propagating $\rho$ field.

where the graphs have been regulated by dimensional regularization in $d = 4-2\epsilon$ dimensions. The third diagram is

$$\text{Fig. 3(c)} = \frac{2i\lambda k^2}{m_\rho^4} \int \frac{d^d q}{(2\pi)^d} = 0, \tag{1.31}$$

where we have made use of the fact that in dimensional regularization

$$\lim_{n \to 0} \int \frac{d^d q}{(2\pi)^d} \frac{1}{(q^2)^n} = \lim_{n \to 0} \frac{i}{(4\pi)^{d/2}} \frac{\Gamma(n - d/2)}{\Gamma(n)} (0)^{d/2-n} \to 0. \tag{1.32}$$

The last diagram is given by

$$\text{Fig. 3(d)} = \frac{3i\lambda}{16\pi^2} k^2 \left[ \frac{1}{\epsilon} - \gamma + 1 - \ln\left( \frac{m_\rho^2}{4\pi\mu^2} \right) \right]. \tag{1.33}$$

Note that the $1/\epsilon$ divergences in Fig. 3(a) and Fig. 3(b) cancel each other. This is equivalent to the statement that the full theory does not have wavefunction renormalization at one-loop order (it is just $\phi^4$ after all). On the other hand, the tadpole graph in Fig. 3(d), which corresponds to a radiative correction to the VEV of the original field $\phi$ requires a tree graph (not shown in the figure) with a counterterm insertion to regulate the logarithmic divergence. After subtracting this $1/\epsilon$ pole using the counterterm contribution, one finds the following result for the renormalized $\pi$ field self-energy in the full theory at one-loop (working in the $\overline{MS}$ scheme):

$$-i\Pi_\pi(k^2) = \frac{i\lambda}{16\pi^2} k^2 \left[ \frac{7}{2} - 3\ln\left( \frac{m_\rho^2}{\mu^2} \right) \right] + \mathcal{O}(k^4/m_\rho^2). \tag{1.34}$$

Fig. 4. Feynman graphs contributing to the $\pi$ field self-energy in the low energy EFT. Graph (b), including one insertion of the coupling $c_8$ vanishes in dimensional regularization.

In the EFT, the self-energy of the Goldstone boson is given by the graphs of Fig. 4. These are:

$$\text{Fig. 4(a)} = i(Z - 1)k^2, \tag{1.35}$$

and

$$\text{Fig. 4(b)} = 0, \tag{1.36}$$

by Eq. (1.32). Setting the full and EFT results equal, we find

$$Z = 1 + \frac{\lambda}{16\pi^2}\left[\frac{7}{2} - 3\ln\left(\frac{m_\rho^2}{\mu^2}\right)\right] + \mathcal{O}(\lambda^2). \tag{1.37}$$

The one-loop matching calculation presented here serves to illustrate two general features of matching calculations:
• Integrating out heavy modes not only generates new irrelevant (dimension $\Delta > 4$) operators in the effective theory. It also renormalizes the coefficients of the operators with $\Delta \leq 4$.
• Loop graphs in matching calculations typically contain logarithms of $m_\rho/\mu$. In order to avoid possible large logarithms that could render perturbation theory invalid, one must choose a matching scale $\mu$ that is of the same order as the masses of the fields that are being integrated out.
This last point in particular implies that it is the Wilson coefficients $c_i(\mu \sim \Lambda)$ that exhibit simple scaling in powers of the UV scale $\Lambda$. Given the RG equations for the full theory coupling constants, one may also use any other $\mu > \Lambda$ as the matching scale. Possible large logarithms that may arise from this choice are then resummed by RG running. For example, using

$$\mu\frac{dZ(\mu)}{d\mu} = \frac{3\lambda(\mu)}{8\pi^2} + \mathcal{O}(\lambda^2), \tag{1.38}$$

and the one-loop RG equation for the full theory coupling constant $\lambda$,

$$\mu\frac{d\lambda(\mu)}{d\mu} = \frac{5\lambda^2(\mu)}{8\pi^2}, \tag{1.39}$$

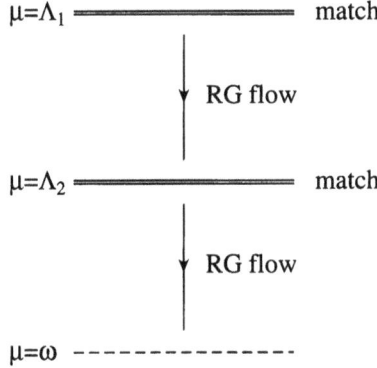

Fig. 5. Construction of a low energy EFT for a theory with scales $\Lambda_1 \gg \Lambda_2 \gg \omega$.

one finds the relation

$$Z(\mu) = Z(\Lambda) + \frac{3}{5} \ln \frac{\lambda(\mu)}{\lambda(\Lambda)}. \tag{1.40}$$

Thus in this theory, using the "wrong" matching scale has a relatively mild effect, although this is not generally true in other theories, e.g. QCD. In any case, it is simplest in practice to just match directly at $\mu \sim \Lambda$.

Likewise, to calculate observables at low energy scale $\omega \ll \Lambda$, it is better to evaluate loop graphs in the EFT at a renormalization point $\mu \simeq \omega$, indicating that in the EFT one should use the couplings $c_i(\mu \sim \omega)$. These can be obtained in terms of the coefficients $c_i(\mu \sim \Lambda)$ obtained through matching by renormalization group (RG) evolution between the scales $\Lambda$ and $\omega$ within the EFT.

For a theory with multiple scales, the procedure is similar. A typical example is shown in Fig. 5, which depicts the construction of an EFT at a low scale $\omega \ll \Lambda_2 \ll \Lambda_1$ starting from a theory of light fields $\phi$ coupled to heavy fields $\Phi_1$, $\Phi_2$ (masses of order $\Lambda_1$, $\Lambda_2$). One first constructs an $EFT_1$ for $\phi$, $\Phi_2$ (regarded as approximately massless) by integrating out the fields $\Phi_1$. This generates a theory defined by its coupling constants at the renormalization scale $\mu \sim \Lambda_1$. This $EFT_1$ is then used to RG evolve the couplings down to the threshold $\mu \sim \Lambda_2$, at which point $\Phi_2$ is treated as heavy and removed from the theory. This finally generates an $EFT_2$ for the light fields $\phi$ which can be used to calculate at the scale $\omega$. In $EFT_2$, logarithms of $\omega/\Lambda_2$ can be resummed by RG running.[2]

---

[2] There are exceptions to the construction described in this paragraph, for example NRQCD, the EFT that describes the non-relativistic limit of the strong interactions. Although this theory contains several widely separated scales, these are correlated implying that RG running must be performed in one stage, rather than in the two-stage picture presented here [23]. See [12] for a pedagogical review. This will not be an issue in our discussion of the non-relativistic gravity.

**Exercise**: Calculate the one-loop correction to the Wilson coefficient $c_8(\mu)$.

*1.2.3. Power counting*

In every EFT, there is a rule for determining which operators are needed at a given order in the expansion parameter $\omega/\Lambda \ll 1$. This rule is called *power counting*. It is slightly unusual, in that a typical power counting scheme assigns the same counting to tree and loop contributions to a given observable. This is unlike the situation in renormalizable field theories where the perturbative expansion is equivalent to an expansion in powers of $\hbar$ (the loop expansion).

To see how this works, we develop the power counting rules of our toy EFT. In this case, power counting is simply a matter of keeping track of operator mass dimensions.[3] We simply make the obvious assignment

$$k^\mu \sim \omega \Rightarrow \partial_\mu \sim \omega \Rightarrow x^\mu \sim \omega^{-1}. \tag{1.41}$$

Since the kinetic operator is leading in the expansion, it should scale as

$$\int d^4x (\partial_\mu \pi)^2 \sim \left(\frac{\omega}{\Lambda}\right)^0, \tag{1.42}$$

indicating that $\pi(x) \sim \omega$, and therefore the scaling of an operator is simply its mass dimension.

Given these power counting rules, we now know how to assign powers of $\omega/\Lambda$ to terms in the expansion of any low energy observable. To see how this works let us consider again $\pi\pi \rightarrow \pi\pi$ scattering. By the LSZ reduction formula, the $S$-matrix element for this process is of the form

$$S \sim \left[\int d^4x_i e^{-ik_i \cdot x_i} \partial_i^2\right] \langle 0|T\pi(x_1)\cdots\pi(x_4)|0\rangle. \tag{1.43}$$

Thus to decide the order in which a given operator contributes to the scattering amplitude, we need to know how its contribution to the four-point function $\langle T\pi(x_1)\cdots\pi(x_4)\rangle$ sales in $\omega/\Lambda$.

For example, the leading non-trivial contribution to scattering, from a single insertion of the dimension eight operator $\mathcal{O}_8 = (\partial_\mu\pi\partial^\mu\pi)^2$ gives a term whose magnitude is

$$\text{Fig. 6(a)} = \left\langle T\pi(x_1)\cdots\pi(x_4)\left[\frac{ic_8}{\Lambda^4}\int d^4x\mathcal{O}_8(x)\right]\right\rangle_0$$

$$\sim \omega^4\left[\frac{\omega^{-4}\times\omega^8}{\Lambda^4}\right] = \omega^4\left(\frac{\omega}{\Lambda}\right)^4. \tag{1.44}$$

---

[3] This need not always be the case, as we will see in the next lecture.

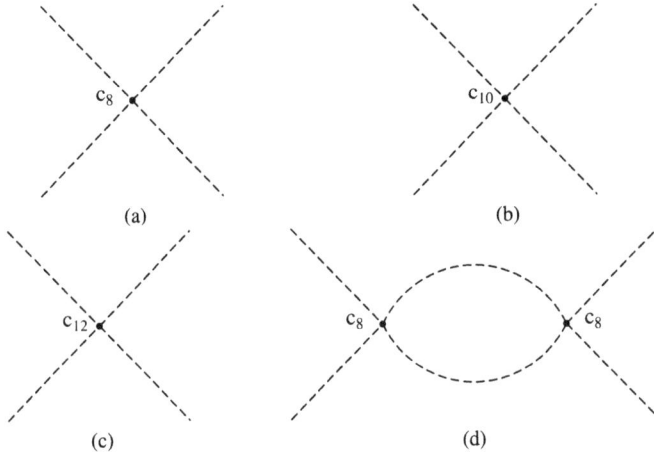

Fig. 6. Feynman graphs contributing to the $\pi\pi \to \pi\pi$ in the effective theory. Graph (a) corresponds to one insertion of the dimension eight operator $(\partial_\mu \pi \partial^\mu \pi)^2$. Graphs (b), (c) correspond to insertions of a generic dimension ten and twelve operator, respectively. Note that the $\omega/\Lambda$ power counting scheme implies that the one-loop diagram (d) comes in at the same order as the tree graph (c).

(The subscript 0 means we are calculating the correlation functions on the RHS in the free field theory, using Wick's theorem). At higher orders in the expansion, the four-point function receives corrections from the diagrams in Fig. 6(b), (c), (d). In particular, a typical contribution from a dimension ten operator $\mathcal{O}_{10}$ goes like

$$\text{Fig. 6(b)} = \left\langle T\pi(x_1)\cdots\pi(x_4)\left[\frac{ic_{10}}{\Lambda^6}\int d^4x \mathcal{O}_{10}(x)\right]\right\rangle_0 \sim \omega^4\left(\frac{\omega}{\Lambda}\right)^6, \quad (1.45)$$

and likewise a single insertion of a dimension twelve operator $\mathcal{O}_{12}$ (Fig. 6(c)) comes in at relative order $\omega^8/\Lambda^8$. Note however that at this order there is an additional contribution from one-loop diagrams with two insertions of the dimension eight operator:

$$\text{Fig. 6(d)} = \left\langle T\pi(x_1)\cdots\pi(x_4)\left[\frac{ic_8}{\Lambda^4}\int d^4x \mathcal{O}_8(x)\right]^2\right\rangle_0 \sim \omega^4\left(\frac{\omega}{\Lambda}\right)^8, \quad (1.46)$$

thus in practice loop graphs can contribute at the same order as tree level insertions.

$$\mu = 1/r_s \frac{\text{BH,NS +GR}}{\text{pt. particle +GR}} \quad \text{match}$$

RG flow

$$\mu = 1/r \frac{\text{bd. state}}{\text{mult. + rad.}} \quad \text{match}$$

RG flow

$$\mu = v/r \text{ -- -- -- -- -- -- -- --}$$

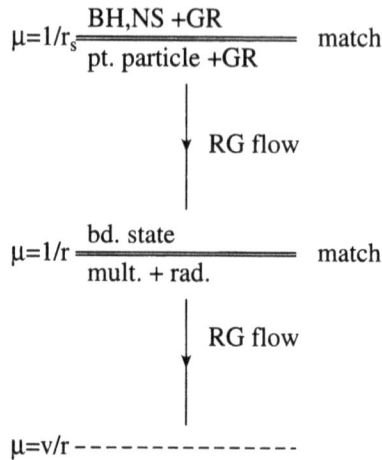

Fig. 7. Construction of EFT for binary stars in two stages.

## 2. Lecture II

### 2.1. The binary inspiral as an EFT calculation

As discussed in the previous lecture, there are three scales involved in the binary inspiral problem: the internal structure scale $r_s$, the orbital distance $r$ and the radiation wavelength $r/v$. The goal is to calculate gravitational wave observables, e.g. the radiated power, arising from physics at the scale $r/v$. Following our previous discussion, a convenient way of doing this is to formulate the problem in terms of EFTs. Since the problem has two intermediate scales, the construction of the EFT that describes the radiation modes must proceed in stages, as shown in Fig. 7.

The starting point is the theory of an isolated black hole or neutron star coupled to gravity. Thus in this theory, the relevant degrees of freedom are field perturbations (gravitational, electromagnetic) propagating in the BH/NS background geometry. Of these degrees of freedom, the field modes of interest in the non-relativistic binary dynamics have wavelengths ($\lambda \sim r/v$) much larger than the scale that characterizes the internal structure of the compact object (e.g., the Schwarzschild radius $r_s$). Given our previous discussion, it therefore makes sense to integrate out the internal structure of the object by matching onto a new theory that captures the relevant degrees of freedom. This theory is simply the theory of a point particle coupled to the gravitational fields (plus whatever other massless fields there are in the problem). We describe how to construct this EFT (identifying which modes to keep and constructing the Lagrangian) in sec. 2.2.

This point particle EFT is the correct theory for length scales all the way to the orbital radius $r$. At the orbital scale, the gravitational field can be viewed as a superposition of "potential" modes that act over short distances, mediating the forces that form the bound system, and "radiation modes" which correspond to the gravitational waves that propagate out to the detector. It is convenient once again to integrate out the potential modes by matching onto an EFT of radiation gravitons coupled to a composite object of size $r \ll \lambda$. This new EFT consists of a point particle, together with a set of multipole mass moments generated by the mechanical plus gravitational energy of the two-particle bound state. These multipoles are coupled to gravity in a way consistent with gauge invariance under long wavelength coordinate transformations. We discuss the matching onto, and power counting within this theory in sec. 2.4.

Note that in Fig. 7, we have also indicated that the parameters in each of the EFTs relevant to the binary exhibit RG flows as a function of a renormalization scale. This RG flow, which is purely classical, unfortunately does not arise until order $v^6$ in the binary dynamics, and is beyond the scope of the discussion in these lectures.

## 2.2. The EFT for isolated compact objects

An isolated compact object probed by long wavelength fields can be approximated as a point particle. Even if we do not know the internal structure of the compact object, it is possible to write down an EFT that describes its interactions with external fields which accounts for finite size effects in a *systematic* fashion.

Recall from our general discussion that to build an EFT, one needs to do two things:

• Identify the relevant degrees of freedom at the scale of interest.
• Construct the most general Lagrangian for these degrees of freedom that is consistent with the symmetries.

A black hole[4] interacting with long wavelength gravitational fields can be viewed to a first approximation as a point particle probe of the background geometry. The degrees of freedom necessary to describe such a system are

1. The gravitational field $g_{\mu\nu}(x)$.

2. The black hole's worldline coordinate $x^{\mu}(\lambda)$, which is a function of an arbitrary affine parameter $\lambda$.

3. An orthonormal frame $e_a^{\mu}(\lambda)$ localized on the particle worldline. This describes the orientation of the object relative to local inertial frames. It encodes how the particle is spinning relative to the gravitational field.

---

[4]Neutron stars may contain additional low frequency modes that must be kept in the point particle EFT. See [24] for a review of NS/BH spectroscopy.

For the sake of simplicity, I will ignore the effects of spin in the subsequent discussion. The construction of effective Lagrangians involving spin degrees of freedom can be found in [25].

These degrees of freedom must couple in all ways allowed by the symmetries of the problem. These are

1. General coordinate invariance, $x^\mu \to x^{\bar\mu}(x)$.

2. Worldline reparametrization invariance (RPI), $\lambda \to \bar\lambda(\lambda)$.

I will make one additional symmetry assumption, again just to keep the discussion as simple as possible:

3. $SO(3)$ invariance. This guarantees that the compact object is perfectly spherical. In particular it has no permanent moments relative to its own rest frame.

This last assumption, together with the omission of spin degrees of freedom, means that the EFT that we are constructing is the appropriate one for describing Schwarzschild black holes interacting with external gravitational fields.

It is straightforward to write down effective Lagrangians that are invariant under these symmetries. To take care of coordinate invariance, we just write down Lagrangians that transform as coordinate scalars constructed from $g_{\mu\nu}(x)$ and $dx^\mu/d\lambda$. A simple way of ensuring RPI is to use the proper time variable

$$d\tau^2 = g_{\mu\nu}\big(x(\lambda)\big)dx^\mu dx^\nu, \tag{2.1}$$

as the worldline parameter. Since proper time is physical (i.e., measurable) it must be invariant under worldline reparametrizations.

The effective action consistent with these criteria is then

$$S_{eff}\big[x^\mu, g_{\mu\nu}\big] = S_{EH}[g] + S_{pp}[x, g], \tag{2.2}$$

where we take the usual action for the gravitational field[5]

$$S_{EH} = -2m_{Pl}^2 \int d^4x \sqrt{g}R(x), \tag{2.3}$$

with $m_{Pl}^{-2} = 32\pi G_N$, and $R(x)$ is the Ricci scalar. $S_{pp}$ is given by

$$S_{pp} = -m \int d\tau + \cdots, \tag{2.4}$$

---

[5]In addition to the Einstein-Hilbert term, the action for gravity may contain additional powers of the curvature suppressed by powers of the scale $m_{Pl}$ where new physics is expected to come in. These play no role in our discussion.

where $m$ is the particle mass, and we have suppressed temporarily any possible curvature dependent terms in the point particle action. Extremizing $S_{pp}$ gives rise to the usual geodesic motion of a test particle in a gravitational field

$$\delta \left[ -m \int d\tau \right] = 0 \Longrightarrow \ddot{x}^{\mu} + \Gamma^{\mu}_{\ \alpha\beta} \dot{x}^{\alpha} \dot{x}^{\beta} \equiv \dot{x}^{\alpha} D_{\alpha} \dot{x}^{\mu} = 0, \qquad (2.5)$$

($\dot{x}^{\mu} \equiv dx^{\mu}/d\tau$). Including only one power of the curvature, $S_{pp}$ has two additional terms

$$S_{pp} = -m \int d\tau + c_R \int d\tau R + c_V \int d\tau R_{\mu\nu} \dot{x}^{\mu} \dot{x}^{\nu} \cdots . \qquad (2.6)$$

Note however that the leading equations of motion for the gravitational field, which follow varying $S_{EH}$ and neglecting sources imply that $R_{\mu\nu}(x) = 0$. Thus operators constructed from the Ricci curvature are "redundant" operators, and can be omitted from $S_{pp}$ without affecting the physical consequences of the theory. Technically, this is because it is possible to perform a field redefinition of $g_{\mu\nu}(x)$ which sets the coefficients of these operators to zero. See the appendix for details.

Since terms with $R_{\mu\nu}$ can be omitted, all that is left are operators constructed from the Riemann tensor (specifically its traceless part, the Weyl tensor). The simplest such operators involve two powers of the curvature and can be written as

$$S_{pp} = -m \int d\tau + c_E \int d\tau E_{\mu\nu} E^{\mu\nu} + c_B \int d\tau B_{\mu\nu} B^{\mu\nu} + \cdots . \qquad (2.7)$$

The tensors $E_{\mu\nu}$, $B_{\mu\nu}$ denote the decomposition of the Riemann tensor $R_{\mu\nu\alpha\beta}$ into components of electric and magnetic type parity respectively. They are the gravitational analog of the decomposition of the electromagnetic field strength $F_{\mu\nu}$ into electric and magnetic fields. Explicitly, they are given by

$$E_{\mu\nu} = R_{\mu\alpha\nu\beta} \dot{x}^{\alpha} \dot{x}^{\beta}, \qquad (2.8)$$

$$B_{\mu\nu} = \epsilon_{\mu\alpha\beta\rho} \dot{x}^{\rho} R^{\alpha\beta}_{\ \ \rho\nu} \dot{x}^{\rho}. \qquad (2.9)$$

These tensors are purely spatial in the particle rest frame and for a background with $R_{\mu\nu} = 0$ also satisfy $E^{\mu}_{\ \mu} = B^{\mu}_{\ \mu} = 0$.

The operators $\int d\tau E_{\mu\nu} E^{\mu\nu}$ and $\int d\tau B_{\mu\nu} B^{\mu\nu}$ are the first in an infinite series of terms that systematically encode the internal structure of the black hole. One way to see that these terms describe finite size effects is to calculate their effect on the motion of a particle moving in a background field $g_{\mu\nu}$. The variation of $S_{pp}$ including the terms quadratic in the curvature gives

$$\delta\left[-m\int d\tau\right] = -\delta\left[c_E\int d\tau\, E_{\mu\nu}E^{\mu\nu} + c_B\int d\tau\, B_{\mu\nu}B^{\mu\nu} + \cdots\right]$$
$$\implies \dot{x}^\alpha D_\alpha \dot{x}^\mu \neq 0. \tag{2.10}$$

In other words, due to the curvature couplings, the particle no longer moves on a geodesic. However, geodesic deviation implies stretching by tidal forces, which occurs when one considers the motion of extended objects in a gravitational field. In fact, an explicit matching calculation, described in Sec. 2.6, predicts that the coefficients $c_{E,B} \sim m_{Pl}^2 r_s^5$, vanishing rapidly as the size of the black hole goes to zero.

The EFT in Eq. (2.2) describes the dynamics of one extended object in a gravitational field. To describe the motion of several objects, we simply include a separate point particle action for each

$$S_{eff}[x_a, g] = S_{EH}[g] + \sum_a S_{pp}^a[x_a, g], \tag{2.11}$$

where the index $a$ runs over all the particles moving in the field $g_{\mu\nu}$.

### 2.3. Calculating observables

In principle, Eq. (2.11) correctly captures the physics of an arbitrary system of extended objects for scales $\mu < 1/r_s$. It can therefore be used to calculate all observables measured by gravitational wave detectors.

If we decompose the typical gravitational wave signal as

$$h(t) = A(t)\cos\phi(t), \tag{2.12}$$

then interferometric detectors, such as LIGO/VIRGO and LISA, are particularly sensitive to the phase $\phi(t) \sim 2\int^t d\tau\,\omega(\tau)$ of the gravitational wave. This is usually calculated using the "adiabatic approximation": Consider a non-relativistic binary inspiral, and suppose we have calculated to some order in the $v \ll 1$ expansion the quantities

$$E(v) = \text{mechanical energy of binary,}$$
$$P(v) = \text{power emitted in gravitational waves,}$$

as functions of the orbital parameters, such as the velocity $v(t)$, or equivalently the orbital frequency $\omega(t)$. Energy conservation

$$\frac{dE}{dt} = -P, \tag{2.13}$$

then gives a differential equation that can be used to solve for $\omega(t)$ which in turn gives the frequency of the GW wave signal $\omega_{GW}(t) \simeq 2\omega(t)$ and consequently the phase $\phi(t)$ (the factor of two arises from the helicity-two nature of the graviton). We implicitly did this calculation to leading order in $v$ in Sec. 1.1. There we found

$$\phi(t) = 2\int_{t_i}^{t} d\tau\, \omega(\tau) = \frac{1}{16}\left[\frac{1}{v(t)^5} - \frac{1}{v(t_i)^5}\right], \tag{2.14}$$

for equal mass stars in a circular orbit.

Although the observables we are interested in calculating are purely classical, let us pretend for the moment that we are doing quantum field theory. Write

$$g_{\mu\nu} = \eta_{\mu\nu} + \frac{h_{\mu\nu}}{m_{Pl}}, \tag{2.15}$$

and calculate the functional $S_{eff}(x_a)$ defined by the path integral

$$\exp\left[i\,S_{eff}(x_a)\right] = \int Dh_{\mu\nu}(x)\exp\left[i\,S_{EH}(h) + i\,S_{pp}(h, x_a)\right], \tag{2.16}$$

with the particle worldlines $x_a^{\mu}(\tau)$ that source $h_{\mu\nu}$ held fixed. It turns out that $S_{eff}(x_a)$ is a generating function for the quantities of interest. In particular, the classical limit of the variation

$$\delta\left[\mathrm{Re}\,S_{eff}(x_a)\right] = 0 \tag{2.17}$$

gives rise to the equations of motion for the worldlines $x_a^{\mu}(\tau)$. From this one can derive an energy function $E(v)$ in the usual way. Furthermore

$$\frac{1}{T}\mathrm{Im}\,S_{eff}(x_a) = \frac{1}{2}\int dE\,d\Omega\,\frac{d^2\Gamma}{d\Omega\,dE}, \tag{2.18}$$

measures the total *number* of gravitons emitted as worldlines $x_a^{\mu}(\tau)$ evolve over a time $T \to \infty$. Although graviton number is not a well defined observable classically, the power emitted can be obtained by integrating the differential rate $d\Gamma$ over the energy of the emitted graviton

$$P = \text{tot. power} = \int dE\,d\Omega\, E\,\frac{d^2\Gamma}{d\Omega\,dE}. \tag{2.19}$$

To see how the calculation of observables via Eq. (2.16) works, we will consider a toy gravity model consisting of a scalar graviton field $\phi(x)$ interacting with several point particles. The scalar $\phi$ couples to the point particles with strength proportional to mass

$$S = \frac{1}{2}\int d^4x\, \partial_\mu\phi\partial^\mu\phi - \sum_a m_a \int d\tau_a \left[ 1 + \frac{\phi}{2\sqrt{2}m_{Pl}} \right], \qquad (2.20)$$

or equivalently

$$S = -\sum_a m_a \int d\tau_a + \int d^4x \left[ \frac{1}{2}\partial_\mu\phi\partial^\mu\phi + J(x)\phi(x) \right], \qquad (2.21)$$

with

$$J(x) \equiv -\sum_a \frac{m_a}{2\sqrt{2}m_{Pl}} \int d\tau_a \delta^4(x - x_a). \qquad (2.22)$$

Then the functional that generates the observables in this model is given by

$$S_{eff}(x_a) = -\sum_a m_a \int d\tau_a - i \ln Z[J], \qquad (2.23)$$

where

$$Z[J] = \int D\phi(x) \exp\left[ i \int d^4x \frac{1}{2}\partial_\mu\phi\partial^\mu\phi + J(x)\phi(x) \right]. \qquad (2.24)$$

This is a Gaussian integral so it can be easily calculated explicitly. Up to an irrelevant constant, the result is

$$\ln Z[J] = -\frac{1}{2}\int d^4x d^4y\, J(x) D_F(x - y) J(y). \qquad (2.25)$$

In the case of real gravity, with non-linear self interactions, the analog of $Z[J]$ will not have such a simple expression. Nevertheless, the perturbative expansion of the generating function in real gravity has a simple diagrammatic interpretation. As a warmup to constructing the diagrammatic rules in real gravity, it is useful to recall how Eq. (2.25) is recovered by summing up Feynman diagrams. Treat the coupling $\int d^4x\, J(x)\phi(x)$ perturbatively, and introduce the Feynman rule

$$\phantom{xxxxxx} = i \int d^4x\, J(x)e^{-ik\cdot x} = -i\sum_a \frac{m_a}{2\sqrt{2}m_{Pl}} \int d\tau_a e^{-ik\cdot x_a}.$$

Then by expanding the interaction term in Eq. (2.24), we see that $Z[J]$ has a diagrammatic expansion in terms of "ladder diagrams"

$$Z[J] = \underrightarrow{\phantom{xxx}} + \phantom{x} + \phantom{x} + \cdots$$

$$= \exp\left(\phantom{xxx}\right). \qquad (2.26)$$

Note that the intermediate particle lines have no propagators associated with them (they just depict the time evolution of the particle worldlines), so diagrams with multiple "rungs" are simply products of the diagram with a single scalar exchange. Specifically, the diagram with $n$ intermediate graviton propagators is given by

$$\phantom{xxx} \cdots \phantom{xxx} = \frac{1}{n!}\left[\frac{1}{2}\int d^4x\, d^4y (i\,J(x)) D_F(x-y)(i\,J(y))\right]^n, \qquad (2.27)$$

where the factor of $1/2^n n!$ appearing here is the symmetry factor associated with the diagram. Summing up contributions from diagrams with any number $n$ of scalar lines, we see that the series in Eq. (2.26) reproduces Eq. (2.25).

Notice that the exponentiation of the diagrams contributing to $Z[J]$ implies that $S_{eff}(x_a)$ receives a contribution only from diagrams that remain connected after the particle worldlines are stripped off, which in this theory is just the diagram with a single graviton exchange. The fact that $S_{eff}(x_a)$ is given by diagrams that do not break up after removing all worldlines remains true in real gravity, where the graphs can have graviton self-interaction vertices.

The effective action in the toy gravity model is thus given by

$$S_{eff}(x_a) = -\sum_a m_a \int d\tau_a + \frac{i}{2}\sum_{a,b}\frac{m_a m_b}{8m_{Pl}^2}\int d\tau_a d\tau_b D_F(x_a - x_b). \qquad (2.28)$$

Eq. (2.28) contains all the information about the classical particle dynamics. For illustration, let's evaluate it for the case where the motion is non-relativistic. In this case the integral

$$\int d\tau_a d\tau_b D_F(x_a - x_b) = \int d\tau_a d\tau_b \int \frac{d^4k}{(2\pi)^4}\frac{i}{k^2 + i\epsilon}e^{-ik\cdot(x_a - x_b)} \qquad (2.29)$$

can be split up into contributions from two regions of momenta:

• **Potential**: This is the region corresponding to gravitons with spacelike momenta of the form

$$k^\mu \rightarrow \left(k^0 \sim \frac{v}{r}, \mathbf{k} \sim \frac{1}{r}\right), \qquad (2.30)$$

where $r$ is the typical separation between the particles and $v \ll 1$ is the typical three-velocity. Gravitons with this momentum configuration mediate nearly instantaneous exchanges between the point particles. Note that potential gravitons can never be on-shell, $k^2 \neq 0$. Therefore they never contribute to the imaginary part of Eq. (2.28), which happens when the $i\epsilon$ term in the propagator becomes important, i.e., when $k^2 = 0$.

• **Radiation**: Radiation gravitons correspond to on-shell modes with

$$k^\mu \rightarrow \left( k^0 \sim \frac{v}{r}, \mathbf{k} \sim \frac{v}{r} \right). \tag{2.31}$$

Because these modes satisfy $k^2 \simeq 0$, they are responsible for giving rise to the imaginary part of $S_{eff}(x_a)$ and therefore the radiation that propagates out to the detector.

Since only potential modes contribute to the real part of $S_{eff}(x_a)$, we may calculate in the NR limit by expanding the propagator as

$$\frac{1}{k_0^2 - \mathbf{k}^2} = -\frac{1}{\mathbf{k}^2}\left[ 1 + \frac{k_0^2}{\mathbf{k}^2} + \cdots \right] = -\frac{1}{\mathbf{k}^2}\left[ 1 + \mathcal{O}(v^2) \right]. \tag{2.32}$$

Using

$$\int \frac{d^4 k}{(2\pi)^4} e^{-ik\cdot x} \frac{1}{\mathbf{k}^2} = \frac{1}{4\pi |\mathbf{x}|}\delta(x^0), \tag{2.33}$$

and expanding $d\tau_a = dx^0\sqrt{1 - \mathbf{v}_a^2} \simeq dx^0[1 - \frac{1}{2}\mathbf{v}_a^2]$, we find to leading order in $v$

$$\mathrm{Re}\, S_{eff}[x_a] = \frac{1}{2}\sum_a \int dx^0 m_a \mathbf{v}_a^2 - \frac{1}{2}\sum_{a,b} \int dx^0 \frac{G_N m_a m_b}{|\mathbf{x}_a - \mathbf{x}_b|} + \cdots, \tag{2.34}$$

which is just the Lagrangian for classical NR particles interacting through a Newtonian potential. Note that the second term contains divergent self-energy contributions whenever $a = b$ in the sum. These divergences can be absorbed by renormalization into the particle masses. Formally, they can just be set to zero, by evaluating the momentum integral in dimensional regularization using the formula

$$\int \frac{d^d \mathbf{k}}{(2\pi)^d} e^{-i\mathbf{k}\cdot\mathbf{x}} \frac{1}{(\mathbf{k}^2)^\alpha} = \frac{1}{(4\pi)^{d/2}} \frac{\Gamma(d/2 - \alpha)}{\Gamma(\alpha)} \left( \frac{\mathbf{x}^2}{4} \right)^{\alpha - d/2}, \tag{2.35}$$

and taking the limit $\mathbf{x} \rightarrow 0$ before setting $d = 3$.

To calculate the imaginary part of $S_{eff}(x_a)$ we use

$$\text{Im}\frac{1}{k^2 + i\epsilon} = -i\pi\delta(k^2),$$ (2.36)

which physically has the effect of ensuring that only on-shell particles contribute to the radiated power. Then

$$\text{Im}S_{eff}(x_a) = \frac{1}{16m_{Pl}^2}\int\frac{d^3k}{(2\pi)^3}\frac{1}{2|\mathbf{k}|}\left|\sum_a m_a\int d\tau_a e^{-ik\cdot x_a}\right|^2_{k^0=|\mathbf{k}|},$$ (2.37)

and the differential power is

$$\frac{dP}{d\Omega d|\mathbf{k}|} = \frac{1}{T}\frac{G_N}{4\pi^2}k^2\left|\sum_a m_a\int d\tau_a e^{-ik\cdot x_a}\right|^2_{k^0=|\mathbf{k}|}.$$ (2.38)

This is exactly what one would find classically, by solving the wave equation for $\phi$ with source term given by Eq. (2.22) and then calculating its energy-momentum tensor at an asymptotically large distance from the source. It can also be calculated from the tree-level amplitude for single graviton emission from the source particles

$$i\mathcal{A}(\text{vac.}\to\phi) = -i\sum\frac{m_a}{2m_{Pl}\sqrt{2}}\int d\tau_a e^{ik\cdot x_a}.$$ (2.39)

The differential *probability* for the emission of one graviton is then

$$d\text{Prob}(\text{vac.}\to\phi) = |\mathcal{A}(\text{vac.}\to\phi)|^2\frac{d^3k}{(2\pi)^3}\frac{1}{2|\mathbf{k}|}.$$ (2.40)

Dividing this result by the observation time $T\to\infty$ to get a rate, and multiplying by the graviton energy $|\mathbf{k}|$ to convert to power reproduces Eq. (2.38). So indeed the functional integral formula for $S_{eff}(x_a)$ knows about the classical observables for the particle ensemble.

**Exercise**: In the case of electrodynamics coupled to point particles,

$$S = -\frac{1}{4}\int d^4x F_{\mu\nu}F^{\mu\nu} - \sum_a m_a\int d\tau_a + \sum_a eQ_a\int dx_a^\mu A_\mu,$$ (2.41)

show by calculating $\text{Im}S_{eff}(x_a)$ that the radiated power in photons is given by

$$\frac{dP}{d\Omega d|\mathbf{k}|} = \frac{1}{T}\frac{\alpha}{4\pi^2}k^2\left|\sum_a Q_a\int dx_a^\mu e^{-ik\cdot x_a}\right|^2_{k^0=|\mathbf{k}|},$$ (2.42)

where $\alpha = e^2/(4\pi)$. Verify that, upon time averaging, this reproduces the usual electric dipole radiation formula in the non-relativistic limit.

Gravitational radiation in general relativity can be calculated by a straightforward generalization of the methods used for the scalar gravity model. Real gravity is of course more complex, but the complications are mainly computational. If we expand $g_{\mu\nu}$ as in Eq. (2.15) and plug into the gravitational action we find an infinite series of terms that are schematically of the form

$$-2m_{Pl}^2 \int d^4x \sqrt{g} R(x) \ \rightarrow \ \int d^4x \left[ (\partial h)^2 + \frac{h(\partial h)^2}{m_{Pl}} + \frac{h^2(\partial h)^2}{m_{Pl}^2} + \cdots \right],$$

$$= (\sim\!\sim)^{-1} + \text{(diagram)} + \text{(diagram)} + \cdots,$$

(2.43)

leading to graviton self-interactions with Feynman vertices containing any number of graviton lines. We will not need the detailed form of these terms in what follows. A derivation of the Feynman rules for gravity can be found in the lectures by Veltman [26]. See also [19]. In addition, the gravitational field has non-linear interactions with the point particle, e.g.,

$$-m \int d\bar{\tau} \sqrt{1 + \frac{h_{\mu\nu}\dot{x}^\mu \dot{x}^\nu}{m_{Pl}}} = -m \int d\bar{\tau} - \frac{m}{2m_{Pl}} \int d\bar{\tau} h_{\mu\nu}\dot{x}^\mu \dot{x}^\nu$$

$$-\frac{m}{8m_{Pl}^2} \int d\bar{\tau} \left( h_{\mu\nu}\dot{x}^\mu \dot{x}^\nu \right)^2 + \cdots,$$

(2.44)

where $d\bar{\tau}^2 = \eta_{\mu\nu}dx^\mu dx^\nu$. This gives rise to vertices

$$\text{(diagram)} + \text{(diagram)} + \text{(diagram)} + \cdots.$$

(2.45)

As a result of this, the diagrammatic expansion of $S_{eff}(x_a)$ in the gravitational case is much richer than in the simple model considered above. The first few terms in the expansion are shown in Fig. 8. The problem with this (covariant) form of the perturbative series is that it is not optimal for the calculation of observables in the $v \ll 1$ limit. For example, consider the diagram in Fig. 8(c). At what order in $v$ does it contribute? In the language of the previous lecture, we do not have a velocity *power counting* scheme for the diagrams appearing in Fig. 8. Since in the end we are interested in computing gravitational wave signals to a

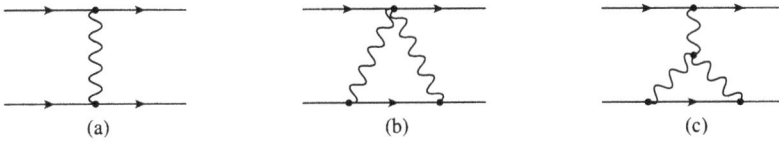

Fig. 8. The first few diagrams contributing to $S_{eff}(x_a)$ in Lorentz covariant perturbation theory.

fixed order in $v$, it is important to develop a set of rules that assigns a unique power of $v$ to each diagram in the theory.

## 2.4. Integrating out the orbital scale

The reason why the diagrams do not scale as definite powers of $v$ is that there are multiple scales in the theory. Although we have successfully integrated out the internal structure scale $r_s$, the momentum integrals in the diagrams of Fig. 8 have contributions both from potential ($v/r$, $1/r$) and radiation ($v/r$, $v/r$) gravitons. But potential modes are never on-shell (they have $k^2 \sim 1/r$) and do not belong in the effective theory at long distance scales.

To remedy this problem, we will explicitly treat the potential and radiation modes separately, by splitting up the gravitational field as

$$h_{\mu\nu}(x) = \bar{h}_{\mu\nu}(x) + H_{\mu\nu}(x). \tag{2.46}$$

Here, the field $\bar{h}_{\mu\nu}$ represents the long-wavelength radiation modes. Schematically, it satisfies

$$\partial_\alpha \bar{h}_{\mu\nu} \sim \frac{v}{r} \bar{h}_{\mu\nu}, \tag{2.47}$$

meaning that the field $\bar{h}_{\mu\nu}$ varies slowly over spacetime, with a typical length scale $r/v$. It can therefore be regarded as a slowly varying background field in which the potential modes $H_{\mu\nu}$, with

$$\partial_0 H_{\mu\nu} \sim \frac{v}{r} H_{\mu\nu}, \quad \partial_i H_{\mu\nu} \sim \frac{1}{r} H_{\mu\nu}, \tag{2.48}$$

propagate. Actually it is useful to re-write $H_{\mu\nu}$ in terms of its Fourier transform

$$H_{\mu\nu}(x) = \int \frac{d^3\mathbf{k}}{(2\pi)^3} e^{i\mathbf{k}\cdot\mathbf{x}} H_{\mathbf{k}\mu\nu}(x_0). \tag{2.49}$$

This allows us to explicitly disentangle hard momenta

$$\mathbf{k} \sim 1/r, \tag{2.50}$$

from the long wavelength scale

$$x^\mu \sim \frac{r}{v}, \tag{2.51}$$

and to treat all derivatives acting on fields on the same footing, $\partial_\mu \sim v/r$.

To derive an EFT that has manifest velocity power counting rules, we need to integrate out the potential modes. The basic idea is the following: we calculate $S_{eff}(x_a)$ (and hence all observables), in two stages. In the first stage, we perform the integral over $H_{\mathbf{k}\mu\nu}$,

$$e^{i S_{NR}(\bar{h}, x_a)} = \int D H_{\mathbf{k},\mu\nu}(x^0) e^{i S_{EH}(\bar{h}+H) + S_{pp}(\bar{h}+H, x_a)} \tag{2.52}$$

where the field $\bar{h}_{\mu\nu}$ is treated as a background field. This gives the quantity $S_{NR}(\bar{h}, x_a)$, which formally contains the two-body forces between the point particles, written as an explicit expansion in powers of $v$, and the couplings of the particles to radiation. In this theory, the short distance scale $r$ appears explicitly, in the coefficients of operators. In other words, Eq. (2.52) is simply a formal way of doing the *matching* to the long-distance EFT containing $\bar{h}_{\mu\nu}$ and the particle worldlines. Given $S_{NR}(\bar{h}, x)$, the second stage in the calculation is to perform the functional integral over $\bar{h}_{\mu\nu}$. Feynman diagrams are easier to calculate in this theory than in the full theory, as momentum integrals only contain the scale $r/v$.

Note that in Eq. (2.52) we have dropped gauge fixing terms necessary to make sense of the path integral. We will not need the explicit form of such terms in the subsequent discussion, but it is perhaps worth noting that it is very convenient to do the gauge fixing in a way that preserves the invariance under diffeomorphisms that transform the background metric $\bar{h}_{\mu\nu}$ (this is called the background field method [27]). If such a gauge fixing scheme is chosen, the action for $\bar{h}_{\mu\nu}$ is guaranteed to be gauge invariant. This places strong restrictions on the form of the EFT that describes radiation.

Diagrammatically, $S_{NR}(\bar{h}, x_a)$ is given by a sum over diagrams that have the following topological properties:

1. Diagrams must remain connected if the particle worldlines are stripped off.

2. Diagrams may *only* contain internal lines corresponding to propagators for the potential modes $H_{\mathbf{k}\mu\nu}(x_0)$. Diagrams *cannot* contain external potential graviton lines.

3. Diagrams may *only* contain external $\bar{h}_{\mu\nu}$ lines. Since the functional integral in Eq. (2.52) views the field $\bar{h}_{\mu\nu}$ as a background, diagrams *cannot* contain propagators corresponding to internal radiation graviton lines.

The point of splitting the original graviton $h_{\mu\nu}$ into the new modes $\bar{h}_{\mu\nu}$, $H_{\mathbf{k}\mu\nu}$ is that the diagrams in terms of these new variables can be assigned definite powers of the expansion parameter $v$. The power counting rules for determining how many powers of $v$ to assign to a given diagram follow simply from the fact that the three momentum of a potential graviton scales as $\mathbf{k} \sim 1/r$, since this is the range of the force it mediates, and that the spacetime variation of a radiation graviton is $x^{\mu} \sim r/v$. These two observations are sufficient to consistently assign powers of $v$ to any term in the action, and by extension to the Feynman rules.

First, let's determine the power counting for diagrams that only contain potential modes. We can fix the scaling of $H_{\mathbf{k}\mu\nu}$ in terms of kinematic variables by looking at its propagator. To obtain the propagator, we need the quadratic terms in $H_{\mathbf{k}\mu\nu}$ from the gravitational Lagrangian. Including an appropriate gauge fixing term, whose form is not necessary for our purposes here (see [7] for details) the relevant terms are

$$S_{H^2} = -\frac{1}{2} \int dx^0 \frac{d^3\mathbf{k}}{(2\pi)^3} \left[ \mathbf{k}^2 H_{\mathbf{k}\mu\nu} H_{-\mathbf{k}}^{\mu\nu} - \frac{\mathbf{k}^2}{2} H_{\mathbf{k}} H_{-\mathbf{k}} \right], \tag{2.53}$$

where $H_{\mathbf{k}} = H^{\alpha}{}_{\alpha\mathbf{k}}$. Terms with time derivatives have been dropped, as $\partial_0$ is down relative to $\mathbf{k}$ by a power of $v$. Because of this, the potential propagator is instantaneous,

$$\langle T H_{\mathbf{k}\mu\nu}(x^0) H_{\mathbf{q}\alpha\beta}(0) \rangle = -\frac{i}{\mathbf{k}^2} (2\pi)^3 \delta^3(\mathbf{k} + \mathbf{q}) \delta(x^0) P_{\mu\nu;\alpha\beta}, \tag{2.54}$$

where the tensor structure is given by

$$P_{\mu\nu;\alpha\beta} = \frac{1}{2} \left[ \eta_{\alpha\mu}\eta_{\beta\nu} + \eta_{\alpha\nu}\eta_{\beta\mu} - \eta_{\mu\nu}\eta_{\alpha\beta} \right]. \tag{2.55}$$

Eq. (2.54) is essentially the Fourier transform of the Newton potential, as one would expect. Since $P_{\mu\nu;\alpha\beta}$ is $\mathcal{O}(1)$, the scaling of $H_{\mathbf{k}\mu\nu}$ is simply

$$H_{\mathbf{k}\mu\nu}^2 \sim \left(\frac{1}{r}\right)^{-2} \times \left(\frac{1}{r}\right)^{-3} \times \left(\frac{r}{v}\right)^{-1} = r^4 v, \tag{2.56}$$

so that

$$H_{\mathbf{k}\mu\nu} \sim r^2 \sqrt{v}. \tag{2.57}$$

Given the scaling rule for $H_{\mathbf{k}\mu\nu}$, it is now possible to assign powers of velocity to diagrams with no external radiation gravitons. For example, the coupling of

$H_{\mathbf{k}00}$ to a NR particle is

$$-\frac{m}{2m_{Pl}}\int dx^0 \frac{d^3\mathbf{k}}{(2\pi)^3} e^{i\mathbf{k}\cdot\mathbf{x}(x^0)} H_{\mathbf{k}00} \sim \frac{m}{m_{Pl}}\left(\frac{r}{v}\right)\left(\frac{1}{r}\right)^3 (r^2 v^{1/2})$$

$$= v^{-1/2}\frac{m}{m_{Pl}}. \tag{2.58}$$

The scaling of $m/m_{Pl}$ is fixed by the virial theorem,

$$v^2 \sim \frac{G_N m}{r} \Rightarrow \frac{m^2}{m_{Pl}^2} \sim mv^2 r = Lv, \tag{2.59}$$

where we have used $G_N \sim 1/m_{Pl}^2$, and have introduced the orbital angular momentum $L = mvr$. The interaction in Eq. (2.58) then scales as

$$\hspace{1cm} = -\frac{m}{2m_{Pl}}\int dx^0 \frac{d^3\mathbf{k}}{(2\pi)^3} e^{i\mathbf{k}\cdot\mathbf{x}(x^0)} H_{\mathbf{k}00} \sim L^{1/2}, \tag{2.60}$$

so the exchange diagram with two insertions of this term, which leads to the Newton potential between the point particles, scales as a power of the angular momentum $L$,

$$\hspace{1cm} \sim L. \tag{2.61}$$

It can be shown that any diagram without external radiation lines scales like $L^n v^m$, with $n \leq 1$ and $m \geq 0$. The bound on $n$ is saturated by diagrams without graviton loops. Thus $L$ is the loop counting parameter of this theory, and since we are interested in the classical limit $\hbar \to 0$, it is a good approximation to drop diagrams with graviton loops (formally, such diagrams are down by powers of $\hbar/L \ll 1$).

Power counting rules for potential graviton self-interactions can be derived along similar lines. Consider the potential three-graviton vertex. Expanding the gravitational Lagrangian to cubic order in $H_{\mathbf{k}\mu\nu}$, one finds terms with the structure

$$S_{H^3} \sim \frac{1}{m_{Pl}}\int dx^0 (2\pi)^3 \delta^3\left(\sum_i \mathbf{k}_i\right)\mathbf{k}^2 \prod_{i=1}^{3}\frac{d^3\mathbf{k}_i}{(2\pi)^3} H_{\mathbf{k}_i}, \tag{2.62}$$

where we have inserted the expansion Eq. (2.49) into the Einstein-Hilbert action and performed the $d^3\mathbf{x}$ integral. The factor of $\mathbf{k}^2$ comes from the fact that every term in the action has two derivatives on the metric. From this we find

$$\sim \frac{1}{m_{Pl}} \times \left(\frac{r}{v}\right) \times \left(\frac{1}{r}\right)^{-3} \times \left(\frac{1}{r}\right)^2 \times \left[\left(\frac{1}{r^3}\right) \times (r^2\sqrt{v})\right]^3$$

$$= \frac{v^2}{\sqrt{L}}, \tag{2.63}$$

and therefore the diagram

$$\sim (\sqrt{L})^2 \times \frac{v^2}{\sqrt{L}} \times \sqrt{L} = Lv^2, \tag{2.64}$$

gives rise to a term in the two-body potential that is suppressed by $v^2$ relative to the leading order Newtonian exchange diagram. The full set of potentials at order $v^2$ arise from the diagrams in Fig. 9. In addition to the three graviton term just discussed, there are additional velocity suppressed vertices from the expansion of the particle proper time Lagrangian in powers of $v$,

$$S_{pp} \Rightarrow \frac{m}{m_{Pl}} \int dx^0 \left[-\frac{1}{2}h_{00} - h_{0i}v_i - \frac{1}{4}h_{00}\mathbf{v}^2 - \frac{1}{2}h_{ij}v^i v^j\right]. \tag{2.65}$$

The sum of these diagrams gives the corrections to the two-body NR Lagrangian at order $v^2$

$$L_{v^2} = \frac{1}{8}\sum_a m_a v_a^4 + \frac{G_N m_1 m_2}{|\mathbf{x}_1 - \mathbf{x}_2|}\left[3(\mathbf{v}_1^2 + \mathbf{v}_2^2) - 7\mathbf{v}_1 \cdot \mathbf{v}_2\right.$$

$$\left. - \frac{(\mathbf{v}_1 \cdot \mathbf{x}_{12})(\mathbf{v}_2 \cdot \mathbf{x}_{12})}{|\mathbf{x}_1 - \mathbf{x}_2|}\right] - \frac{G_N^2 m_1 m_2 (m_1 + m_2)}{2|\mathbf{x}_1 - \mathbf{x}_2|^2}, \tag{2.66}$$

where $\mathbf{x}_{12} = \mathbf{x}_1 - \mathbf{x}_2$. This Lagrangian was first obtained by Einstein, Infeld and Hoffman [28] in 1938. In this equation, the first term is just the first relativistic correction to the particle kinetic energies, the second term arises from diagrams with a single graviton exchanged by velocity-dependent vertices, and the last term is from the diagrams of Fig. 9(c), (d). Diagrams at higher orders in $v$ can be calculated and power counted by the same methods outlined here.

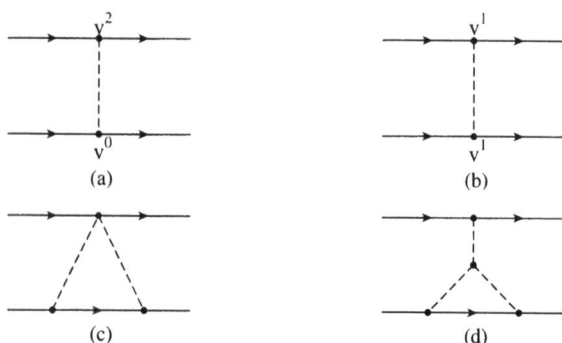

Fig. 9. Diagrams contributing to the order $v^2$ corrections to the two-body potential.

## 2.5. Radiation

Integrating out the non-dynamical potential modes generates the gravitational forces between the non-relativistic particles. These arise from diagrams with no external factors of the radiation field $\bar{h}_{\mu\nu}$. The functional integral in Eq. (2.52) also generates, from the diagrams with one or more external radiation graviton, the couplings of matter to radiation.

Incorporating radiation is fairly straightforward. First note that the propagator for the radiation field, in a suitable gauge, is given by

$$\langle T\bar{h}_{\mu\nu}(x)\bar{h}_{\alpha\beta}(0)\rangle = \int \frac{d^4k}{(2\pi)^4} \frac{i}{k^2+i\epsilon} e^{-ik\cdot x} P_{\mu\nu;\alpha\beta}. \qquad (2.67)$$

Since $k^\mu \sim v/r$, Eq. (2.67) implies that radiation modes should scale as

$$\bar{h}_{\mu\nu} \sim \frac{v}{r}. \qquad (2.68)$$

This rule allows one to power count terms in the action containing the radiation field. However, in order to obtain an EFT for radiation that has manifest velocity power counting, the decomposition of the graviton into potential and radiation modes is not sufficient. It is necessary also to multipole expand the couplings of the radiation field to either potentials or to the point particles,

$$\begin{aligned}
\bar{h}_{\mu\nu}(\mathbf{x}, x^0) &= \bar{h}_{\mu\nu}(\mathbf{X}, x^0) + \delta\mathbf{x}^i \partial_i \bar{h}_{\mu\nu}(\mathbf{X}, x^0) \\
&+ \frac{1}{2}\delta\mathbf{x}^i \delta\mathbf{x}^j \bar{h}_{\mu\nu}(\mathbf{X}, x^0) + \cdots,
\end{aligned} \qquad (2.69)$$

where $\mathbf{X}$ is an arbitrary point, for example the center of mass of the multi-particle system, $\mathbf{X}_{cm} = \sum_a m_a \mathbf{x}_a / \sum_a m_a$. The reason for this is familiar in the case

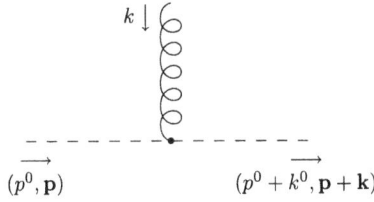

$k \downarrow$

$(p^0, \mathbf{p})$           $(p^0 + k^0, \mathbf{p} + \mathbf{k})$

Fig. 10. The interaction of a potential and a radiation mode.

of couplings to matter sources: consider the amplitude for graviton emission by an ensemble of point particles ($v_a^\mu$ is the four-velocity, $\epsilon_{\mu\nu}(k)$ is the graviton polarization)

$$i\mathcal{A} = \sum_a \underbrace{\phantom{xxx}} = -i \sum_a \frac{m_a}{2m_{Pl}} \int d\tau_a e^{ik \cdot x_a} v_a^\mu v_a^\nu \epsilon_{\mu\nu}^*(k). \qquad (2.70)$$

The final state graviton is on-shell, so $k^0 = |\mathbf{k}|$. On the other hand, if the particles are non-relativistic, $x^0 \gg |\mathbf{x}_a|$ (measuring positions relative to the center of mass). Then

$$e^{-ik \cdot x_a} = e^{-ik_0 x^0} \left[ 1 + i\mathbf{k} \cdot \mathbf{x}_a - \frac{1}{2}(\mathbf{k} \cdot \mathbf{x}_a)^2 + \cdots \right]. \qquad (2.71)$$

Note that if $|\mathbf{k}| \sim v/r$ and $|\mathbf{x}_a| \sim r$, $\mathbf{k} \cdot \mathbf{x}_a \sim v$. So the exponential contains an infinite number of powers of $v$, and the amplitude does not scale homogeneously in velocity. This is avoided by multipole expanding $\bar{h}_{\mu\nu}$ at the level of the Lagrangian. Likewise, if we consider a graph containing potential as well as radiation modes

$$\text{Fig. 10} \sim \frac{1}{(\mathbf{p} + \mathbf{k})^2} = \frac{1}{\mathbf{p}^2}[1 - 2\mathbf{p} \cdot \mathbf{k} + \cdots], \qquad (2.72)$$

we find that because $|\mathbf{p}| \sim 1/r$ and $|\mathbf{k}| \sim v/r$, the propagator for the potential mode with three-momentum $\mathbf{p} + \mathbf{k}$ contains an infinite number of powers of $v$. To ensure that this does not happen, it is necessary to arrange that radiation gravitons do not impart momentum to the potential modes. This is exactly what occurs if we plug in the expansion Eq. (2.69) into the couplings generated by the gravitational Lagrangian. The necessity to multipole expand potential-radiation couplings was first pointed out in the context of non-relativistic gauge theories in ref. [29].

Given these ingredients, we are finally in a position to calculate the matching coefficients for the radiation graviton EFT at the scale $\mu \simeq 1/r$. Matching boils down to calculating Eq. (2.52), or equivalently comparing diagrams with any number of external radiation modes in the full theory and the EFT. For example, consider the graphs with one external $\bar{h}_{\mu\nu}$. In the theory with both potentials and radiation, the leading one is

$$\sum_a \; \begin{array}{c}\text{[diagram]}\end{array}_{v^0} \Rightarrow \begin{array}{c}\text{[diagram]}\end{array}_{\mu^0} \quad -\frac{1}{2m_{pl}} \int dx^0 \sum_a m_a \bar{h}_{00}(0, x^0). \qquad (2.73)$$

In this equation, the double lines on the graph on the right indicates that this is a vertex in the EFT below $\mu \simeq 1/r$, where the separation $r$ between the point particles (a short distance scale) cannot be resolved. It is the vertex obtained by matching at the potential scale, and it corresponds to the term in the EFT shown on the right hand side of the equation. At this order, the matching is just the zeroth order multipole expansion (performed about the center of mass), i.e. which generates the coupling of $\bar{h}_{00}$ to the mass monopole $m = \sum_a m_a$. At the next order, the effective theory vertex contains terms both from the multipole expansion and from the explicit velocity dependence of the point particle gravitational couplings. The result is

$$\sum_a \; \begin{array}{c}\text{[diagram]}\end{array}_{v^1} \Rightarrow \begin{array}{c}\text{[diagram]}\end{array}_{\mu^1} \quad -\frac{1}{2m_{pl}} \int dx^0 \left[ \mathbf{X}_{cm}^i \partial_i \bar{h}_{00} + 2\mathbf{P}^i \bar{h}_{0i} \right], \qquad (2.74)$$

where $\mathbf{P} = \sum_a m_a \dot{\mathbf{x}}_a$ is the total linear momentum of the system. In the CM frame $\mathbf{X}_{cm} = \mathbf{P}_{cm} = 0$, so this coupling vanishes, i.e., there is no dipole radiation in general relativity.

Things are more interesting in the one-graviton sector at order $v^2$. The graphs at this order are

$$\sum_a \; \begin{array}{c}\text{[diagram]}\end{array}_{v^2} + \begin{array}{c}\text{[diagram]}\end{array} + \begin{array}{c}\text{[diagram]}\end{array} \Rightarrow \begin{array}{c}\text{[diagram]}\end{array}_{v^2} \qquad (2.75)$$

where the first graph depicts the terms from the multipole expansion and from the point-particle couplings containing two powers of the velocity, the second graph (an almost identical mirror image graph not shown) comes from the two-graviton couplings of the point particle, and the vertex in the third term is from the $\bar{h}H^2$ terms in the Einstein-Hilbert Lagrangian. The sum of these graphs gives the second order terms in the EFT, $S_{v^2}[\bar{h}] = \int dx^0 L_{v^2}[\bar{h}]$, where

$$L_{v^2}[\bar{h}] = -\frac{1}{2m_{Pl}}\bar{h}_{00}\left[\frac{1}{2}\sum_a m_a v_a^2 - \frac{G_N m_1 m_2}{|\mathbf{x}_1 - \mathbf{x}_2|}\right] - \frac{1}{2m_{Pl}}\epsilon_{ijk}\mathbf{L}_k \partial_j \bar{h}_{0i}$$

$$+ \frac{1}{2m_{Pl}}\sum_a m_a x_a^i x_a^j R_{0i0j}. \tag{2.76}$$

Here, the first term is just the coupling of $\bar{h}_{00}$ to the Newtonian energy of the two-particle system. This term can be regarded as a kinetic plus gravitational correction to the mass monopole of the source

$$\sum_a m_a \rightarrow \sum_a m_a\left(1 + \frac{1}{2}v_a^2\right) - \frac{G_N m_1 m_2}{|\mathbf{x}_1 - \mathbf{x}_2|}. \tag{2.77}$$

The second term is a coupling of the graviton to the total mechanical angular momentum of the system, $\mathbf{L} = \sum_a \mathbf{x}_a \times m_a \mathbf{v}_a$. Both the mass monopole and the angular momentum are conserved at this order in the velocity expansion, and therefore cannot source on-shell radiation. Rather they source the long range static gravitational field of the two-body system. The last term, on the other hand, is a coupling of the moment $\sum_a m_a x_a^i x_a^j$ to the (linearized) Riemann tensor of the radiation field. Note that for on-shell radiation $R_{00} = R_{0i0i} = 0$, so only the traceless moment

$$Q^{ij} = \sum_a m_a\left(x_a^i x_a^j - \frac{1}{3}x_a^2 \delta_{ij}\right) \tag{2.78}$$

is a source of gravitational waves. It is interesting to note that in order to obtain Eq. (2.76), we needed to include not only the multipole expansion of the couplings of $\bar{h}_{\mu\nu}$ to the point particles, but also diagrams with graviton self-interactions. This is forced upon us by the power counting rules of the theory. It would be simply inconsistent to drop the second two graphs in Eq. (2.76) and keep only the first. Thus even the leading order quadrupole coupling to radiation is sensitive to the non-linearities present in relativistic gravity. Physically, this is perhaps not too surprising, as the graviton couples to all sources of energy-momentum, including the energy-momentum stored in the gravitational field of the particles themselves.

**Exercise**: Use the power counting rules to check that

$\sim v^{5/2}$, $\sim \dfrac{v^{5/2}}{\sqrt{L}}$, $\tag{2.79}$

so that the second two graphs in Eq. (2.75) each scale as $\sqrt{L}v^{5/2}$. This is suppressed by $v^2$ relative to the graph in Eq. (2.73).

Working out higher order terms in the EFT is similar, and there is no conceptual problem in carrying out the expansion to any desired order in $v$. Once we have calculated the terms in $S_{NR}[\bar{h}]$, it is straightforward to compute physical observables such as the gravitational radiation power output. As discussed in Sec. 2.3, these can be obtained from

$$\exp[i S_{eff}(x_a)] = \int D\bar{h}_{\mu\nu}(x) e^{i S_{NR}[\bar{h}]}, \tag{2.80}$$

which is simply the sum over diagrams, calculated using the Feynman rules from $S_{NR}[\bar{h}]$, that have no external graviton lines. For example, the leading order quadrupole radiation formula follows from

$$\mathrm{Im} \quad \rightarrow P = \frac{G_N}{5} \langle \dddot{Q}_{ij} \dddot{Q}_{ij} \rangle, \tag{2.81}$$

where the brackets denote a time average. In order to compute the power, one needs to know the time evolution of the moment $Q_{ij}(t)$. This can be obtained by solving the equations of motion for the $x_a(t)$, which follow from the terms in $S_{NR}[\bar{h}]$ with no powers of $\bar{h}_{\mu\nu}$.

## 2.6. Finite size effects

At some order in the velocity expansion, the internal structure of the binary star constituents play a role in the dynamics. In the case of black hole binaries, it is easy to determine the order in $v$ for which this happens. As discussed in Sec. 2.2, finite size effects are encoded in non-minimal couplings to the curvature. Such terms are built out of invariants constructed from the Riemann tensor, the simplest ones being those of Eq. (2.7),

$$S = \cdots + c_E \int d\tau \, E_{\mu\nu} E^{\mu\nu} + c_B \int d\tau \, B_{\mu\nu} B^{\mu\nu} + \cdots. \tag{2.82}$$

The coefficients $c_{E,B}$ can be obtained by a matching calculation. One simply compares some observable in the point particle theory to the analogous quantity in the full theory, adjusting the coefficients to ensure that the two calculations agree.

For the purposes of calculating the coefficients, $c_{E,B}$, a convenient observable is the $S$-matrix element for a graviton to scatter off the BH background. In the

EFT, this amplitude is schematically of the form

$$i\mathcal{A} = \cdots + \quad + \cdots \sim \cdots + i\frac{c_{E,B}}{m_{Pl}^2}\omega^4 + \cdots. \tag{2.83}$$

Here, $\omega$ is the energy of the incoming graviton. For the point particle EFT to be valid, it must satisfy $r_s\omega \ll 1$. The factor of $1/m_{Pl}^2$ is due to the fact that we are looking at a two graviton process. We have not shown, for example, the leading order term in the amplitude, due to graviton scattering off the field created by the mass term $-m\int d\tau$ in the point particle action. Although the leading effect of $c_{E,B}$ is from interference with this term, the above equation predicts that the total scattering cross section contains a term

$$\sigma(\omega)_{EFT} \sim \cdots + \frac{c_{E,B}^2}{m_{Pl}^4}\omega^8 + \cdots. \tag{2.84}$$

The only scale in the full theory is the BH radius $r_s$, so the cross section must be of the form

$$\sigma(\omega)_{BH} = r_s^2 f(r_s\omega), \tag{2.85}$$

where $f(r_s\omega)$ is a function that for $r_s\omega \ll 1$ can be expanded in powers of $r_s\omega$ (possibly times logs). Thus we expect to find a term

$$\sigma(\omega)_{BH} \sim \cdots + r_s^{10}\omega^8 + \cdots, \tag{2.86}$$

and therefore $c_{E,B} \sim m_{Pl}^2 r_s^5$. Using this scaling, the power counting rules in the non-relativistic limit indicate that the finite size operators first contribute to the dynamics through their effect on the two-body forces, through the diagram

$$c_{E,B} \quad \sim Lv^{10}, \tag{2.87}$$

so is down by ten powers of $v$ relative to the Newton potential. One concludes that finite size effects are completely irrelevant for binary inspirals. Actually, this is not completely true, as realistic black holes usually have non-zero spin, and the inclusion of spin tends to enhance finite size effects [25]. Also, absorption by the black hole horizon, which necessitates the introduction of additional worldline modes in the EFT [30] arises at order $v^8$ for non-spinning black holes [31] and at order $v^5$ [32] when the black hole has spin.

## 3. Conclusions

These lectures presented an introduction to the basic ideas of EFTs and their use in understanding the evolution of coalescing binary stars in the non-relativistic regime. In treating such systems, one encounters a wide range of physically important length scales, from the internal size of the binary constituents to the gravitational radiation wavelength. In order to make sense of physics at all these scales, it is extremely convenient to formulate the problem one scale at a time, by constructing a tower of EFTs as outlined in these lectures.

One topic that was not discussed in these lectures is the issue of ultraviolet divergences in the non-relativistic expansion. Such divergences arise even classically, in the computation of the Feynman diagram that contribute to a given observable. Physically, the presence of these divergences can be attributed to the singular nature of the point particle limit. In the EFT approach, these can be handled by the usual regularization and renormalization procedure found in textbooks: divergences get cutoff using any convenient regulator (e.g. dimensional regularization) and renormalized into the coefficients of operators in the point particle Lagrangian. Since by construction the point particle EFT has all possible operators consistent with symmetries, all short distance singularities can be removed from the theory. At high enough orders in the expansion, some of the EFT coefficients are logarithmically renormalized, leading to RG flows that can be exploited to compute terms in the velocity expansion that are logarithmically enhanced. See [7] for more details.

We expect that the technology of EFTs should be applicable to other astrophysical sources of relevance to gravitational wave physics. For example, it is expected that LISA will detect gravitational waves from the motion of a small object, i.e. a neutron star, around a super-massive black hole. In this case, there is a small expansion parameter, namely the ratio of the neutron star size to the curvature length of the black hole background. It seems natural to construct the expansion in this parameter using EFT methods.

The point-particle EFTs discussed here may also potentially play a role in the problem of tracking the evolution of colliding black holes in numerical relativity. In particular, the EFT approach to parameterizing black hole internal structure could be useful in addressing the question of how to systematically handle the curvature singularities present in numerical simulations. Work on the ideas outlined here is underway.

### Acknowledgments

I would like to thank the students at Session 86 of the Les Houches summer school for their excellent questions, and the organizers, particularly Christophe

Grojean, for their hospitality. I also thank Ira Rothstein and Witek Skiba for comments on the manuscript, and JiJi Fan for carefully reading parts of the draft. This work is supported in part by DOE grant DE-FG-02-92ER40704.

## Appendix A.  Redundant operators

In general, an effective Lagrangian contains all operators constructed from the light degrees of freedom that are invariant under the symmetries of the low energy theory. In practice, some operators are redundant, and may be dropped without altering the physical predictions of the EFT. In fact, operators that vanish by the leading order equations of motion may be omitted from the list of operators appearing in the effective Lagrangian. Such operators are sometimes called redundant operators.

The basic reason for this is simple. If an operator vanishes by the leading order equations of motion (i.e., it vanishes "on-shell"), this means that one can redefine the EFT fields in such a way that the Lagrangian written in terms of the new fields does not contain the redundant operator. But field redefinitions have no effect on physical observables ($S$-matrix elements) so the Lagrangian without the redundant operator is equivalent to the original Lagrangian.

To see how this works in practice, consider in the EFT for the massless $\pi$ field discussed in Lecture I, the dimension-ten operator

$$\mathcal{O}_{10}(x) = (\partial^2 \pi) F[\pi, \partial_\mu \pi], \tag{A.1}$$

with

$$F = (\partial_\alpha \partial_\beta \pi) \partial^\alpha \pi \partial^\beta \pi. \tag{A.2}$$

$\mathcal{O}_{10}$ potentially contributes to to the amplitude for $\pi\pi \to \pi\pi$ scattering at order $(\omega/\Lambda)^6$. Because to leading order in the $\omega/\Lambda$ expansion the equation of motion for $\pi$ is $\partial^2 \pi(x) = 0$, this operator actually gives a vanishing contribution to scattering. This is obvious at tree level, but in fact $\mathcal{O}_{10}(x)$ may also be dropped from loop diagram contributions to scattering amplitudes as well.

Starting from the Lagrangian including $\mathcal{O}_{10}(x)$

$$S = \frac{1}{2} \int d^4x (\partial_\mu \pi)^2 + \cdots + \frac{c_{10}}{\Lambda^6} \int d^4x \mathcal{O}(x) + \cdots, \tag{A.3}$$

define a new field $\bar{\pi}(x)$ by

$$\pi(x) = \bar{\pi}(x) + \delta\pi(x). \tag{A.4}$$

Then

$$\int d^4x(\partial_\mu\pi)^2 = \int d^4x(\partial_\mu\bar\pi)^2 - 2\int d^4x\partial^2\bar\pi\,\delta\pi(x) + \mathcal{O}(\delta\pi^2). \tag{A.5}$$

So if we choose

$$\delta\pi(x) = \frac{c_{10}}{\Lambda^6}F[\bar\pi,\partial_\mu\bar\pi], \tag{A.6}$$

we see that the shift in the leading order (dimension four) term in the Lagrangian precisely cancels the term with the $\mathcal{O}_{10}(x)$ in the original Lagrangian. Note however that the effects of this operator are not completely gone. For example, inserting the field redefinition into the dimension eight operator $\mathcal{O}_8(x) = (\partial_\mu\pi\partial^\mu\pi)^2$ generates operators of dimension fourteen and higher

$$\left(\partial_\mu\pi\partial^\mu\pi\right)^2 \to \left(\partial_\mu\pi\partial^\mu\pi\right)^2 + \frac{2c_{10}}{\Lambda^6}\left(\partial_\mu\pi\partial^\mu\pi\right)\partial_\nu\pi\partial^\nu F + \cdots \tag{A.7}$$

However, when constructing the original EFT, one must include all possible operators consistent with the symmetries at every order in the $\omega/\Lambda$ expansion. Thus, the shift in $\mathcal{O}_8(x)$ due to the field redefinition can be absorbed into the coefficients of a operators that are already present in the theory. The same is true for the substitution of the field redefinition into any other operator, and we conclude that in terms of the new variables, the effects of the redundant operator $\mathcal{O}_{10}(x)$ are completely spurious.

The fact that operators that vanish on-shell may be omitted is very general and very useful. A systematic discussion is given by Politzer in [33]. See also [34]. We used this result in the EFT that describes an extended object coupled to gravity. The effective theory for the point particle contains the terms

$$c_R\int d\tau\,R + c_V\int d\tau\,R_{\mu\nu}\dot x^\mu\dot x^\nu, \tag{A.8}$$

which are redundant due to the fact that at leading order (ignoring coupling to sources) the Einstein equations imply $R_{\mu\nu}(x) = 0$. Suppose we re-define the metric appearing in the original Einstein-Hilbert Lagrangian, $g_{\mu\nu} \to g_{\mu\nu}+\delta g_{\mu\nu}$, with

$$\delta g_{\mu\nu}(x) = \frac{1}{2m_{Pl}^2}\int d\tau\frac{\delta^4(x-x(\tau))}{\sqrt{g}}\left[\xi_V\dot x^\mu\dot x^\nu - (\xi_R - \frac{1}{2}\xi_V)g_{\mu\nu}\right]. \tag{A.9}$$

When plugged into the gravitational action $S_{EH}$ this redefinition induces the shifts

$$c_{R,V} \to c_{R,V} + \xi_{R,V}. \tag{A.10}$$

Thus by adjusting $\xi_{R,V}$ we can make the coefficients $c_{R,V}$ whatever we like, including zero, without affecting the physical predictions of the theory.

# References

[1] A. Abramovici *et al.*, "Ligo: The Laser Interferometer Gravitational Wave Observatory," Science **256**, 325 (1992).

[2] A. Giazotto, "The VIRGO project: a wide band antenna for gravitational wave detection," Nucl. Instrum. Meth. A **289**, 518 (1990).

[3] K. Danzmann and A. Rudiger, "Lisa Technology – Concept, Status, Prospects," Class. Quant. Grav. **20**, S1 (2003).

[4] C. Cutler and K.S. Thorne, "An overview of gravitational-wave sources," arXiv:gr-qc/0204090.

[5] C. Cutler *et al.*, "The Last three minutes: issues in gravitational wave measurements of coalescing compact binaries," Phys. Rev. Lett. **70**, 2984 (1993) [arXiv:astro-ph/9208005].

[6] For a review and further references, see L. Blanchet, "Gravitational radiation from post-Newtonian sources and inspiralling compact binaries," Living Rev. Rel. **5**, 3 (2002) [arXiv:gr-qc/0202016].

[7] W.D. Goldberger and I.Z. Rothstein, "An effective field theory of gravity for extended objects," Phys. Rev. D **73**, 104029 (2006) [arXiv:hep-th/0409156]. See also W. D. Goldberger and I. Z. Rothstein, "Towers of gravitational theories," Gen. Rel. Grav. **38**, 1537 (2006) [arXiv:hep-th/0605238].

[8] W.E. Caswell and G.P. Lepage, "Effective Lagrangians For Bound State Problems In QED, QCD, And Other Field Theories," Phys. Lett. B **167**, 437 (1986).

[9] J. Polchinski, "Effective Field Theory And The Fermi Surface," arXiv:hep-th/9210046.

[10] D.B. Kaplan, "Effective field theories," arXiv:nucl-th/9506035.

[11] A.V. Manohar, "Effective field theories," arXiv:hep-ph/9606222.

[12] I.Z. Rothstein, "TASI lectures on effective field theories," arXiv:hep-ph/0308266.

[13] For a review, see K.G. Wilson and J.B. Kogut, "The Renormalization group and the epsilon expansion," Phys. Rept. **12**, 75 (1974).

[14] T. Appelquist and J. Carazzone, "Infrared Singularities And Massive Fields," Phys. Rev. D **11**, 2856 (1975).

[15] An extensive review of EFT methods in weak processes is A.J. Buras, "Weak Hamiltonian, CP violation and rare decays," arXiv:hep-ph/9806471.

[16] S. Weinberg, "Effective Gauge Theories," Phys. Lett. B **91**, 51 (1980); L.J. Hall, "Grand Unification Of Effective Gauge Theories," Nucl. Phys. B **178**, 75 (1981).

[17] S. Weinberg, "Phenomenological Lagrangians," PhysicaA **96**, 327 (1979).

[18] J. Gasser and H. Leutwyler, "Chiral Perturbation Theory To One Loop," Annals Phys. **158**, 142 (1984); J. Gasser and H. Leutwyler, "Chiral Perturbation Theory: Expansions In The Mass Of The Strange Quark," Nucl. Phys. B **250**, 465 (1985).

[19] J.F. Donoghue, "Introduction to the Effective Field Theory Description of Gravity," arXiv:gr-qc/9512024;

[20] C.P. Burgess, "Quantum gravity in everyday life: General relativity as an effective field theory," arXiv:gr-qc/0311082; C.P. Burgess, "Introduction to Effective Field Theory," arXiv:hep-th/0701053.

[21] W. Buchmuller and D. Wyler, "Effective Lagrangian Analysis Of New Interactions And Flavor Conservation," Nucl. Phys. B **268**, 621 (1986); B. Grinstein and M.B. Wise, "Operator analysis for precision electroweak physics," Phys. Lett. B **265**, 326 (1991).

[22] Z. Han and W. Skiba, "Effective theory analysis of precision electroweak data," Phys. Rev. D **71**, 075009 (2005) [arXiv:hep-ph/0412166].

[23] M.E. Luke, A.V. Manohar and I.Z. Rothstein, "Renormalization group scaling in nonrelativistic QCD," Phys. Rev. D **61**, 074025 (2000) [arXiv:hep-ph/9910209].

[24] K.D. Kokkotas and B.G. Schmidt, "Quasi-normal modes of stars and black holes," Living Rev. Rel. **2**, 2 (1999) [arXiv:gr-qc/9909058]; H.P. Nollert, Class. Quant. Grav. **16**, R159 (1999).

[25] R.A. Porto, "Post-Newtonian corrections to the motion of spinning bodies in NRGR," Phys. Rev. D **73**, 104031 (2006) [arXiv:gr-qc/0511061]; R.A. Porto and I.Z. Rothstein, "The hyperfine Einstein-Infeld-Hoffmann potential," Phys. Rev. Lett. **97**, 021101 (2006) [arXiv:gr-qc/0604099].

[26] M. Veltman, in *Methods in Field Theory, Proceedings of the Les Houches Summer School, 1975,* eds. R. Balian and J. Zinn-Justin, North Holland, 1976.

[27] B.S. Dewitt, "Quantum Theory Of Gravity. Ii. The Manifestly Covariant Theory," Phys. Rev. **162**, 1195 (1967); L.F. Abbott, "The Background Field Method Beyond One Loop," Nucl. Phys. B **185**, 189 (1981).

[28] A. Einstein, L. Infeld and B. Hoffmann, "The Gravitational Equations And The Problem Of Motion," Annals Math. **39**, 65 (1938).

[29] B. Grinstein and I.Z. Rothstein, Phys. Rev. D **57**, 78 (1998) [arXiv:hep-ph/9703298].

[30] W.D. Goldberger and I.Z. Rothstein, "Dissipative effects in the worldline approach to black hole dynamics," Phys. Rev. D **73**, 104030 (2006) [arXiv:hep-th/0511133].

[31] E. Poisson and M. Sasaki, "Gravitational radiation from a particle in circular orbit around a black hole. 5: Black hole absorption and tail corrections," Phys. Rev. D **51**, 5753 (1995) [arXiv:gr-qc/9412027].

[32] H. Tagoshi, S. Mano and E. Takasugi, "Post-Newtonian expansion of gravitational waves from a particle in circular orbits around a rotating black hole: Effects of black hole absorption," Prog. Theor. Phys. **98**, 829 (1997) [arXiv:gr-qc/9711072].

[33] H.D. Politzer, "Power Corrections At Short Distances," Nucl. Phys. B **172**, 349 (1980).

[34] H. Georgi, "On-Shell Effective Field Theory," Nucl. Phys. B **361**, 339 (1991).

Course 8

# HOLOGRAPHIC COSMOLOGY

## Thomas Hertog

*Theory Division, CERN, CH-1211 Geneva 23, Switzerland*
*and*
*APC, 11 Place Marcelin Berthelot, 75005 Paris, France*

*F. Bernardeau, C. Grojean and J. Dalibard, eds.*
*Les Houches, Session LXXXVI, 2006*
*Particle Physcis and Cosmology: The Fabric of Spacetime*
© *2007 Published by Elsevier B.V.*

# Contents

## 1. Introduction

One of the main goals of quantum gravity is to provide a better understanding of the big bang singularity in cosmology. This is essential for cosmology to be a truly predictive science, that explains how the distinctive features of the universe emerged from the early quantum gravitational phase and why they are what they are. A long-standing issue is whether cosmological singularities represent a true beginning or end of evolution. More generally, one would like to understand how semiclassical spacetime and our usual notion of time arise from the past singularity. Various suggestions have been made for how this quantum gravity transition happens. These include the no boundary wave function [1] that describes creation ex nihilo, and the chaotic initial conditions proposed in [2]. Alternatively, it is possible that evolution essentially continues through the singularity, with an immediate transition from a big crunch to a big bang [3].

Since our usual notions of space and time are likely to break down near cosmological singularities, a more promising approach to study the problem of initial conditions in cosmology is to find a dual description in terms of more fundamental variables. In string theory we do not yet have a dual description of real cosmologies, but we do have the celebrated AdS/CFT correspondence [4] which provides a non-perturbative definition of string theory on asymptotically anti-de Sitter (AdS) spacetimes in terms of a conformal field theory (CFT). In this lecture I will describe solutions of $\mathcal{N} = 8$, $D = 4$ supergravity where smooth, asymptotically AdS initial data emerge from a big bang in the past and evolve to a big crunch singularity in the future [5]. I will then discuss to what extent the AdS/CFT duality provides a precise framework in which the quantum nature of the cosmological singularities can be understood.

## 2. Framework

We consider the low energy limit of M theory with $AdS_4 \times S^7$ boundary conditions. The massless sector of the compactification of $D = 11$ supergravity on $S^7$ is $\mathcal{N} = 8$ gauged supergravity in four dimensions. The bosonic part of this theory involves the graviton, 28 gauge bosons in the adjoint of $SO(8)$, and 70 real scalars, and admits $AdS_4$ as a vacuum solution. It is possible to consistently truncate this theory to include only gravity and a single scalar with action

$$S = \int d^4x \sqrt{-g} \left[ \frac{1}{2}R - \frac{1}{2}(\nabla\phi)^2 + 2 + \cosh(\sqrt{2}\phi) \right] \tag{2.1}$$

where we have set $8\pi G = 1$ and chosen the gauge coupling so that the AdS radius is one. The potential has a maximum at $\phi = 0$ corresponding to an $AdS_4$ solution with unit radius. It is unbounded from below, but small fluctuations have $m^2 = -2$, which is above the Breitenlohner-Freedman bound $m_{BF}^2 = -9/4$ [6], so with the usual boundary conditions $AdS_4$ is stable.

We will work in global coordinates in which the $AdS_4$ metric takes the form

$$ds_0^2 = \bar{g}_{\mu\nu}dx^\mu dx^\nu = -(1+r^2)dt^2 + \frac{dr^2}{1+r^2} + r^2 d\Omega_2 \tag{2.2}$$

In all asymptotically AdS solutions, the scalar $\phi$ decays at large radius as

$$\phi(r) = \frac{\alpha}{r} + \frac{\beta}{r^2} \tag{2.3}$$

where $\alpha$ and $\beta$ can depend on the other coordinates. The standard boundary conditions correspond to either $\alpha = 0$ or $\beta = 0$ [6], but the choice $\beta_k = -k\alpha^2$ (with $k$ an arbitrary constant) is another possible boundary condition that preserves all the asymptotic AdS symmetries [7]. One can consider even more general boundary conditions $\beta = \beta(\alpha)$. Although these will generically break some of the asymptotic AdS symmetries, they are invariant under global time translations. Hence there is still a conserved total energy.

The usual definition of energy in AdS diverges whenever $\alpha \neq 0$. This is because the backreaction of the scalar field causes certain metric components to fall off slower than usual. In particular, one has [7]

$$g_{rr} = \frac{1}{r^2} - \frac{(1+\alpha^2/2)}{r^4} + O(1/r^5) \quad g_{tt} = -r^2 - 1 + O(1/r)$$

$$g_{tr} = O(1/r^2) \qquad\qquad\qquad g_{ab} = \bar{g}_{ab} + O(1/r)$$

$$g_{ra} = O(1/r^2) \qquad\qquad\qquad g_{ta} = O(1/r) \tag{2.4}$$

The expression for the conserved mass, however, depends on the asymptotic behavior of the fields and is defined as follows. Let $\xi^\mu$ be a timelike vector which asymptotically approaches a (global) time translation in AdS. The Hamiltonian takes the form

$$H = \int_\Sigma \xi^\mu C_\mu + \text{surface terms} \tag{2.5}$$

where $\Sigma$ is a spacelike surface, $C_\mu$ are the usual constraints, and the surface terms should be chosen so that the variation of the Hamiltonian is well defined. It turns out that when $\alpha \neq 0$, there is an additional scalar surface contribution to the variation of the Hamiltonian, which exactly cancels the divergence of the usual gravitational surface term. The total charge can therefore be integrated, yielding

$$Q[\xi] = Q_G[\xi] + r \oint \frac{\alpha^2}{2} d\Omega + \oint [\alpha\beta + W(\alpha)] d\Omega \tag{2.6}$$

where we have defined

$$W(\alpha) = \int_0^\alpha \beta(\tilde{\alpha}) d\tilde{\alpha} \tag{2.7}$$

and where $Q_G[\xi]$ is the usual gravitational surface term. For spherically symmetric solutions $\alpha$ and $\beta$ are independent of angles, and the total mass becomes [7]

$$M = 4\pi (M_0 + \alpha\beta + W) \tag{2.8}$$

where $M_0$ is just the coefficient of the $1/r^5$ term in $g_{rr}$.


## 3. Anti-de Sitter cosmologies

We now construct a set of big bang/big crunch AdS cosmologies that are solutions of (2.1) with boundary conditions

$$\beta_k = -k\alpha^2 \tag{3.1}$$

on the scalar field. One first finds an $O(4)$-invariant Euclidean instanton solution of the form

$$ds^2 = \frac{d\rho^2}{b^2(\rho)} + \rho^2 d\Omega_3 \tag{3.2}$$

and $\phi = \phi(\rho)$. The field equations determine $b$ in terms of $\phi$

$$b^2(\rho) = \frac{2V\rho^2 - 6}{\rho^2\phi'^2 - 6} \tag{3.3}$$

and the scalar field $\phi$ itself obeys

$$b^2\phi'' + \left(\frac{3b^2}{\rho} + bb'\right)\phi' - V_{,\phi} = 0 \tag{3.4}$$

where prime denotes $\partial_\rho$.

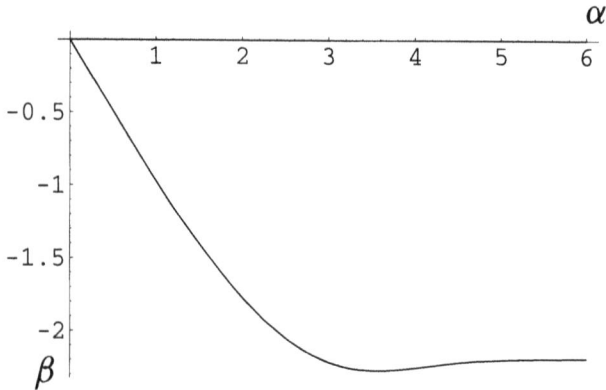

Fig. 1. The function $\beta_i$ obtained from the instantons.

Regularity at the origin requires $\phi'(0) = 0$. Thus the instanton solutions can be labeled by the value of $\phi$ at the origin. For each $\phi(0)$, one can integrate (3.4) and get an instanton. Asymptotically one finds $\phi(\rho) = \alpha/\rho + \beta/\rho^2$, where $\alpha$ and $\beta$ are now constants. Hence for each $\phi(0)$ one obtains a point in the $(\alpha, \beta)$ plane. Repeating for all $\phi(0)$ yields a curve $\beta_i(\alpha)$ where the subscript indicates this is associated with instantons. This curve is plotted in Fig. 1.

The slice through the instanton obtained by restricting to the equator of the $S^3$ defines time symmetric initial data for a Lorentzian solution. The Euclidean radial distance $\rho$ simply becomes the radial distance $r$ in this initial data. So given a choice of boundary condition $\beta(\alpha)$, one can obtain suitable initial data by first selecting the instanton corresponding to a point where the curve $\beta_i(\alpha)$ intersects $\beta(\alpha)$, and then taking a slice through this instanton.

Clearly all AdS-invariant boundary conditions (3.1) admit precisely one instanton solution. Furthermore, the mass (2.8) of the resulting initial data is given by

$$M = 4\pi \left( M_0 - \frac{4}{3}k\alpha^3 \right). \tag{3.5}$$

From the asymptotic form of (3.3) it follows that $M_0 = 4k\alpha^3/3$. Thus this initial data corresponds to a zero mass solution, consistent with its interpretation as the solution $AdS_4$ decays into.

With $\beta_k = -k\alpha^2$ boundary conditions, the evolution of the initial data defined by the instanton geometry is simply obtained by analytic continuation [8]. The origin of the Euclidean instanton becomes the lightcone of the Lorentzian solution. Outside the lightcone, the solution is given by (3.2) with $d\Omega_3$ replaced by three dimensional de Sitter space. The scalar field $\phi$ remains bounded in this region. On the light cone we have $\phi = \phi(0)$ and $\partial_t \phi = 0$ (since $\phi_{,\rho} = 0$ at the origin in the instanton). Inside the lightcone, the $SO(3, 1)$ symmetry ensures that the solution evolves like an open FRW universe,

$$ds^2 = -dt^2 + a^2(t)d\sigma_3 \tag{3.6}$$

where $d\sigma_3$ is the metric on the three dimensional unit hyperboloid. Under evolution $\phi$ rolls down the negative potential. This causes the scale factor $a(t)$ to vanish in finite time, producing a big crunch singularity.

To verify that this analytic continuation indeed satisfies our boundary condition, we must do a coordinate transformation in the asymptotic region outside the light cone. The relation between the usual static coordinates (2.2) for $AdS_4$ and the $SO(3, 1)$ invariant coordinates,

$$ds^2 = \frac{d\rho^2}{1 + \rho^2} + \rho^2(-d\tau^2 + \cosh^2 \tau d\Omega_2) \tag{3.7}$$

is

$$\rho^2 = r^2 \cos^2 t - \sin^2 t \tag{3.8}$$

Hence the asymptotic behavior of $\phi$ in global coordinates is given by

$$\phi(r) = \frac{\tilde{\alpha}}{r} - \frac{k\tilde{\alpha}^2}{r^2} + O(r^{-3}) \tag{3.9}$$

where $\tilde{\alpha} = \alpha/\cos t$. This clearly satisfies the boundary condition (3.1), but $\tilde{\alpha}$ is now time dependent and blows up as $t \to \pi/2$, when the singularity hits the boundary.

For the purpose of understanding cosmological singularities in M theory, one can forget the origin of this solution as the analytic continuation of an instanton. We have simply found an explicit example of asymptotically AdS initial data which evolves to a big crunch. Since the initial data is time symmetric, there is also a big bang in the past (see Fig. 2).

## 4. Dual field theory evolution

We now turn to the dual field theory interpretation. M theory on spacetimes which asymptotically approach $AdS_4 \times S^7$ is dual to the 2+1 conformal field theory (CFT) describing the low energy excitations of a stack of $N$ M2-branes. In this correspondence, scalar modes with $\beta = 0$ boundary conditions correspond to

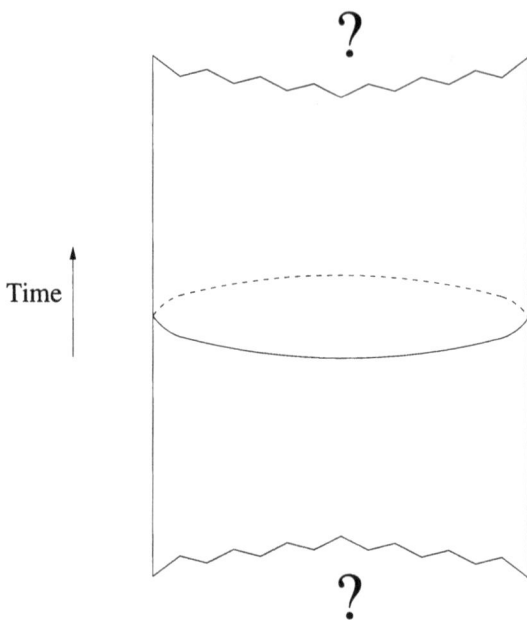

Fig. 2. Anti-de Sitter big bang/big crunch cosmologies.

physical states, and adding nonzero $\beta$ corresponds to modifying the CFT action. Our bulk scalar $\phi$ is dual to a dimension one operator $\mathcal{O}$. One way of obtaining this CFT is by starting with the field theory on a stack of D2-branes and taking the infrared limit. In that description,

$$\mathcal{O} = \frac{1}{N} Tr T_{ij} \varphi^i \varphi^j \tag{4.1}$$

where $T_{ij}$ is symmetric and traceless and $\varphi^i$ are the adjoint scalars.

The field theory dual to the AdS invariant boundary conditions $\beta_k = -k\alpha^2$ is obtained by adding a triple trace term to the action [9]

$$S = S_0 - \frac{k}{3} \int \mathcal{O}^3 \tag{4.2}$$

The extra term has dimension three, and hence is marginal and preserves conformal invariance, at least to leading order in $1/N$. In general, imposing nontrivial boundary conditions $\beta(\alpha)$ in the bulk corresponds to adding a multi-trace interaction $\int W(\mathcal{O})$ to the CFT action, such that after formally replacing $\mathcal{O}$ by its expectation value $\alpha$ one has [9]

$$\beta = \frac{\delta W}{\delta \alpha} \tag{4.3}$$

We have seen that the coefficient $\tilde{\alpha}$ of the bulk solution (3.9) diverges as $t \to \pi/2$, when the big crunch singularity hits the boundary. Since the coefficient of $1/r$ is interpreted as the expectation value of $\mathcal{O}$ in the dual CFT, this shows that to leading order in $1/N$, $\langle \mathcal{O} \rangle$ diverges in finite time.

A qualitative explanation for this field theory behavior is the following. The term we have added to the action is not positive definite. Since the energy associated with the asymptotic time translation in the bulk can be negative [7], the dual field theory should also admit negative energy states and have a spectrum unbounded from below. This shows that the usual vacuum must be unstable, and that there are (nongravitational) instantons which describe its decay. After the tunneling, the field rolls down the potential and becomes infinite in finite time.

A semiclassical analysis supports this reasoning. If we neglect for a moment the nonabelian structure and identify $\mathcal{O}$ with $\varphi^2$, we are led to consider a single scalar field theory with standard kinetic term and potential,

$$V = \frac{1}{8}\varphi^2 - \frac{k}{3}\varphi^6 \tag{4.4}$$

where the quadratic term corresponds to the conformal coupling of $\varphi$ to the curvature of $S^2$,[1] and the second term represents the second term in (4.2). Although this is clearly a huge simplification of the field theory, at the classical level it captures the bulk behavior in a surprisingly quantitative way. In particular, it admits the following exact homogeneous classical solution,

$$\varphi(t) = \frac{C}{\cos^{1/2} t} \tag{4.5}$$

where $C = (3/8k)^{1/4}$. This solution has zero energy since the field starts at rest where the potential vanishes. Hence it is analogous to the solution obtained by analytically continuing the instanton. Since $\varphi^2$ is identified with $\tilde{\alpha}$ on the bulk side, the time dependence of this solution agrees with that predicted from supergravity, including the fact that the field diverges at $t \to \pi/2$.

In this model field theory, the usual vacuum at $\varphi = 0$ is perturbatively stable but nonperturbatively unstable. There are (nongravitational) instantons which describe the semiclassical decay of the usual vacuum. For small $k$, the potential barrier is large, and the instanton action is large. So tunneling is suppressed. For

---

[1] Since we have set the AdS radius equal to one, the dual field theory lives on $S^2 \times R$ where the sphere also has unit radius.

large $k$, the barrier is small and tunneling is not suppressed. (This agrees with the action of the gravitational instantons in the bulk.) After the tunneling, the field rolls down the potential and becomes infinite in finite time. So a semiclassical analysis suggests that the CFT does not have well defined evolution for all time.

This conclusion changes dramatically, however, if one considers the quantum mechanics of the potential (4.4). That is, we again concentrate on the homogeneous mode $\varphi(t) = x(t)$ only, but now treat it quantum mechanically. The quantum mechanics of unbounded potentials of this type is well understood and discussed in detail in [10]. Like the classical trajectories, a wave packet with an energy distribution peaked at some value $E$ that moves in a potential of the form $-kx^6$ (with $k > 0$) will reach infinity in finite time. Thus a packet can 'disappear' and probability is apparently lost. But in quantum mechanics this problem can be dealt with by constructing a self-adjoint extension of the Hamiltonian $H = -(1/2)(d^2/dx^2) + V(x)$, which is done by carefully specifying its domain. Once a domain is chosen, the Hamiltonian is self-adjoint and unitary time evolution is guaranteed.

Yet the center of a wave packet follows essentially the classical trajectory and still reaches infinity in finite time. What happens [10] is that a right-moving wave packet bounces off infinity and reappears as a left-moving wave packet. Hence the quantum mechanics of the potential (4.4) in the dual field description of our AdS cosmologies indicates that the big crunch is not an endpoint of evolution. Furthermore, it shows that for exactly homogeneous initial data there is a bounce through the singularity, as envisioned in the pre-big bang and cyclic universe models [3].

It is natural to ask if one should expect this conclusion to hold also in the full field theory. It is not known if self-adjoint extensions can be constructed in field theories with potentials of this form. However, even if one can define a unitary evolution, the full field theory evolution may be rather different, because our discussion of the quantum mechanics obviously neglects the possibility of particle creation.

It is well known that in theories where the scalar field rolls down from the top of its effective potential towards the true minimum, particles are produced in great numbers while the field is rolling down. This phenomenon is called tachyonic preheating [11]. It happens essentially because the effective negative mass term in the potential causes long wavelength quantum fluctuations to grow exponentially. Tachyonic preheating is so efficient that in many theories most of the initial potential energy density is converted into the energy of scalar particles well before the field reaches the true minimum. Thus a prolonged stage of oscillations of the homogeneous component of the scalar field around the true minimum of the potential does not exist in spontaneous symmetry breaking.

Hence if $1/N$ corrections regularize the dual field theory by creating a stable ground state at large $\phi$, then tachyonic preheating occurs also in the dual field theory description of our AdS cosmologies, where the supergravity initial data correspond to a homogeneous field theory configuration high up the potential. The field theory evolution will then not be dual to a bounce through the singularity. Indeed this would require the miraculous conversion of all the energy back into the homogeneous mode! Instead, one expects the system to thermalize around the true vacuum. On the bulk side, this means that quantum gravity effects turn the cosmological singularity into a black hole[2] with scalar hair [5].

However, if $1/N$ corrections do not regularize the field theory potential the situation remains unclear. The phenomenon of tachyonic preheating means inhomogeneities grow while the field rolls down, but it is not clear if the energy is actually converted into the energy of scalar particles.

## 5. Discussion

We have seen that the dual description of an AdS cosmology involves a field theory with a potential that is unbounded from below. We have used various approaches to study its dynamics. In a semiclassical analysis, the evolution ends in finite time. In a full quantization of the homogeneous mode, evolution continues for all time and suggests a bounce. However this may be an artefact of throwing away all the inhomogeneous degrees of freedom. If one regulates the potential so that it has a global minimum, one sees all the inhomogeneous modes become excited and the system evolves into a thermal state about the true vacuum. However, the corresponding bulk evolution now produces a large black hole with scalar hair rather than a cosmological singularity. The cosmological singularity arises then as the limit of a specific class of hairy black holes as the regulator $\epsilon$ is taken to zero. This suggests that the approach to a big crunch is naturally viewed in the dual theory as a similar evolution to an equilibrium state, but one which does not describe a semiclassical spacetime. If this is indeed the case then the big bang and the emergence of semiclassical spacetime would be an exceedingly rare fluctuation from a generic equilibrium state in quantum gravity.

There seems to be interesting ground to explore, however, between the cyclic evolution of the quantum mechanics of the homogeneous (minisuperspcae) system on the one hand, and the evolution toward equilibrium when the field theory is regularized on the other hand. In particular, we do not yet have a good understanding of the full field theory evolution if it turns out $1/N$ corrections do not regularize the potential.

---

[2] This is possible because the regularization of the field theory corresponds (via 4.3) to a modification of the bulk boundary conditions, which alters the evolution of the instanton initial data.

# References

[1]  J.B. Hartle, S.W. Hawking, "The Wave Function of the Universe," Phys. Rev. **D28** (1983) 2960.

[2]  T. Damour, M. Henneaux, "Chaos in Superstring Cosmology," Phys. Rev. Lett. **85** (2000) 920, hep-th/0003139; T. Damour, M. Henneaux and H. Nicolai, "Cosmological billiards," Class. Quant. Grav. **20** (2003) R145, hep-th/0212256.

[3]  M. Gasperini, G. Veneziano, "Pre-Big Bang in String Cosmology," Astropart. Phys. **1** (1993) 317, hep-th/9211021; J. Khoury, B.A. Ovrut, N. Seiberg, P.J. Steinhardt, N. Turok, "From Big Crunch to Big Bang," Phys. Rev. **D65** (2002) 086007, hep-th/0108187; P.J. Steinhardt, N. Turok, "Cosmic Evolution in a Cyclic Universe," Phys. Rev. **D65** (2002) 126003, hep-th/0111098.

[4]  J.M. Maldacena, "The large N limit of superconformal field theories and supergravity," Adv. Theor. Math. Phys. **2** (1998) 231, hep-th/9711200.

[5]  T. Hertog, G.T. Horowitz, "Holographic Description of AdS Cosmologies," JHEP **0504** (2005) 005, hep-th/0503071; T. Hertog, G.T. Horowitz, "Towards a Big Crunch Dual," JHEP **0407** (2004) 073, hep-th/0406134.

[6]  P. Breitenlohner and D.Z. Freedman, "Stability In Gauged Extended Supergravity," Annals Phys. **144** (1982) 249.

[7]  T. Hertog, K. Maeda, "Black Holes with Scalar Hair and Asymptotics in $N = 8$ Supergravity," JHEP **0407** (2004) 051, hep-th/0404261.

[8]  S.R. Coleman and F. De Luccia, "Gravitational Effects On And Of Vacuum Decay," Phys. Rev. D **21** (1980) 3305.

[9]  E. Witten, "Multi-Trace Operators, Boundary Conditions, and AdS/CFT Correspondence," hep-th/0112258.

[10] M. Carreau, E. Fahri, S. Gutmann, P. F. Mende, "The Functional Integral for Quantum Systems with Hamiltonians Unbounded from Below," Ann. Phys. **204** (1990) 186.

[11] G. Felder, J. Garcia-Bellido, P.B. Greene, L. Kofman, A. Linde, I. Tkachev, "Dynamics of Symmetry Breaking and Tachyonic Preheating," Phys. Rev. Lett. **87** (2001) 011601, hep-ph/0012142.

Course 9

# NEUTRINO PHYSICS AND COSMOLOGY

## Sergio Pastor

*Instituto de Física Corpuscular (CSIC-Universitat de València)*
*Ed. Institutos de Investigación, Apdo. 22085, E-46071 Valencia, Spain*

*F. Bernardeau, C. Grojean and J. Dalibard, eds.*
*Les Houches, Session LXXXVI, 2006*
*Particle Physcis and Cosmology: The Fabric of Spacetime*
© *2007 Published by Elsevier B.V.*

# Contents

# 1. Introduction

In this contribution I summarize the topics discussed in my two lectures at the School on Neutrino Cosmology, one of the best examples of the very close ties that have developed between nuclear physics, particle physics, astrophysics and cosmology. I tried to present the most interesting aspects, but many others that were left out can be found in the reviews [1–3]. A more general review on the connection between particle physics and cosmology can be found in [4].

We begin with a description of the properties and evolution of relic neutrinos and their influence on Primordial Nucleosynthesis. We devote the main part of this contribution to the impact of massive neutrinos on cosmological observables, that is used to extract bounds on neutrino masses from present data and eventually a measurement thanks to the improved sensitivities of future experiments.

# 2. The cosmic neutrino background

The existence of the cosmic neutrino background (CNB) is a generic feature of the standard hot big-bang model, in number only slightly below that of relic photons that form the cosmic microwave background (CMB). The CNB has not been detected yet, but its presence is indirectly established through its influence on cosmological observables. Here we review its evolution and main properties.

## 2.1. Relic neutrino production and decoupling

At large temperatures frequent weak interactions kept cosmic neutrinos of any flavour ($\nu_e$, $\nu_\mu$, $\nu_\tau$) in equilibrium until these processes became ineffective in the course of the expansion. While coupled to the rest of the primeval plasma (relativistic $e^\pm$ and photons), neutrinos had the momentum spectrum

$$f_{\mathrm{eq}}(p, T) = \left[\exp((p - \mu_\nu)/T) + 1\right]^{-1}, \tag{2.1}$$

i.e. a Fermi-Dirac spectrum with temperature $T$, which is just one example of the general case of particles in equilibrium as shown e.g. in [5]. We have included a neutrino chemical potential $\mu_\nu$ that would exist in the presence of a neutrino asymmetry, but it can be shown [6] that its contribution can be safely ignored.

As the Universe cools, the weak interaction rate falls below the expansion rate and one says that neutrinos decouple from the rest of the plasma. An estimate of the decoupling temperature $T_{\text{dec}}$ can be found by equating the thermally averaged value of the weak interaction rate $\Gamma_\nu = \langle \sigma_\nu n_\nu \rangle$ (here $\sigma_\nu \propto G_F^2$ is the cross section of the electron-neutrino processes with $G_F$ the Fermi constant and $n_\nu$ is the neutrino number density) with the expansion rate given by the Hubble parameter $H^2 = 8\pi\rho/3M_P^2$. Here $\rho \propto T^4$ is the total energy density, dominated by relativistic particles, and $M_P = 1/G^{1/2}$ is the Planck mass. If we approximate the numerical factors to unity, with $\Gamma_\nu \approx G_F^2 T^5$ and $H \approx T^2/M_P$, we obtain the rough estimate $T_{\text{dec}} \approx 1$ MeV. More accurate calculations give slightly higher values of $T_{\text{dec}}$ which are flavour dependent since electron neutrinos and antineutrinos are in closer contact with electrons and positrons, see e.g. [1].

Although neutrino decoupling is not described by a unique $T_{\text{dec}}$, it can be approximated as an instantaneous process, a very simple and reasonably accurate description (see e.g. [5] or [7]). In this approximation, the spectrum in Eq. (2.1) is preserved after decoupling, since both neutrino momenta and temperature redshift identically with the expansion of the Universe. In other words, the number density of non-interacting neutrinos remains constant in a comoving volume since decoupling. As we see later, active neutrinos cannot possess masses much larger than 1 eV, so they were ultra-relativistic at decoupling. This is why the momentum distribution in Eq. (2.1) does not depend on the neutrino masses, even after decoupling, i.e. there is no neutrino energy in the exponential of $f_{\text{eq}}(p)$.

When calculating quantities related to relic neutrinos, one must consider the various possible degrees of freedom per flavour. If neutrinos are massless there are two degrees of freedom per flavour, one for neutrinos and one for antineutrinos (or negative and positive helicity states for Majorana neutrinos). Instead, for Dirac neutrinos there are in principle twice more degrees of freedom, corresponding to the two helicity states. However, the extra degrees of freedom have a vanishingly small density with respect to the usual left-handed neutrinos (unless neutrinos have masses above the keV range, as explained in Sec. 6.4 of [1], but such large masses are excluded for active neutrinos). Thus the relic density of active neutrinos does not depend on their nature, either Dirac or Majorana.

Shortly after neutrino decoupling the temperature drops below the electron mass, favouring $e^\pm$ annihilations that heat the photons. If one assumes that this entropy transfer did not affect the neutrinos because they were already completely decoupled, it is easy to calculate the change in the photon temperature before any $e^\pm$ annihilation and after the electron-positron pairs disappear by assuming entropy conservation of the electromagnetic plasma. The result is $T_\gamma(\text{after})/T_\gamma(\text{before}) = (11/4)^{1/3} \simeq 1.40102$, which is also the ratio between the temperatures of relic photons and neutrinos. The evolution of this ratio during

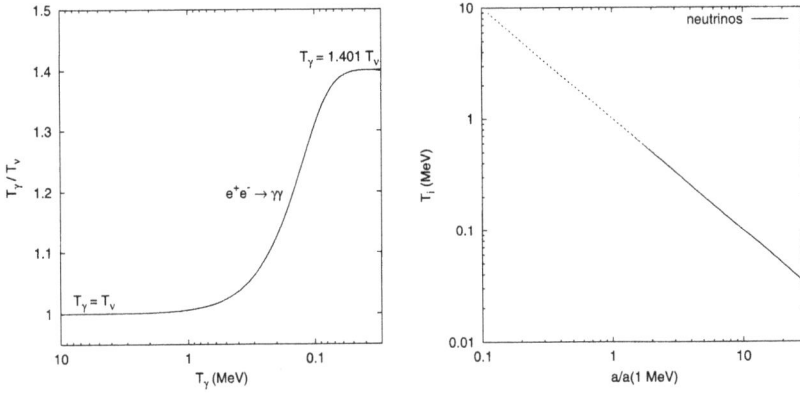

Fig. 1. Photon and neutrino temperatures during the process of $e^{\pm}$ annihilations: evolution of their ratio (left) and their decrease with the expansion of the Universe (right).

the process of $e^{\pm}$ annihilations is shown in the left panel of Fig. 1, while one can see in the right plot how in this epoch $T_\gamma$ decreases with the expansion less than the inverse of the scale factor $a$ (instead $T_\nu$ always falls as $1/a$).

It turns out that the standard picture of neutrino decoupling is slightly modified: the processes of $\nu$ decoupling and $e^{\pm}$ annihilations are sufficiently close in time so that some relic interactions exist, which lead to non-thermal distortions in the neutrino spectra at the per cent level and a slightly smaller increase of the comoving photon temperature, as noted in a series of works (see the full list given in [1]). The most recent analysis [8] includes the effect of flavour neutrino oscillations on the neutrino decoupling process, and finds slightly larger energy densities with respect to the instantaneous decoupling approximation (0.73% or $\nu_e$'s and 0.52% for $\nu_{\mu,\tau}$'s) and a value of the comoving photon temperature after $e^{\pm}$ annihilations which is a factor 1.3978 larger, instead of 1.40102. These changes modify the contribution of relativistic relic neutrinos to the total energy density which is taken into account using $N_{\mathrm{eff}} \simeq 3.046$, as defined later in Eq. (4.1). In practice, these distortions have small consequences on the evolution of cosmological perturbations, and for many purposes can be safely neglected.

Any quantity related to relic neutrinos can be calculated after decoupling with the spectrum in Eq. (2.1) and $T_\nu$. For instance, the number density per flavour $n_\nu = (3/11)\, n_\gamma = (6\zeta(3)/11\pi^2)\, T_\nu^3$ is fixed by the temperature, leading to a present value of 113 neutrinos and antineutrinos of each flavour per cm$^3$. Instead, the energy density for massive neutrinos should in principle be calculated numerically, with two well-defined analytical limits: $\rho_\nu = (7\pi^2/120)(4/11)^{4/3}\, T_\gamma^4$ for massless neutrinos and $\rho_\nu = m_\nu n_\nu$ when $m_\nu \gg T_\nu$.

## 2.2. Background evolution

Let us discuss the evolution of the CNB after decoupling in the expanding Universe, which is described by the Friedmann-Robertson-Walker metric [7]

$$ds^2 = dt^2 - a(t)^2 \delta_{ij} dx^i dx^j, \tag{2.2}$$

where we assumed negligible spatial curvature. Here $a(t)$ is the scale factor, related to the redshift $z$ as $a = 1/(1+z)$. General relativity tells us the relation between the metric and the matter and energy in the Universe via the Einstein equations, whose time-time component is the Friedmann equation

$$\left(\frac{\dot{a}}{a}\right)^2 = H^2 = \frac{8\pi G}{3}\rho = H_0^2 \frac{\rho}{\rho_c^0}, \tag{2.3}$$

that gives the Hubble rate in terms of the total energy density $\rho$. At any time, the critical density $\rho_c$ is defined as $\rho_c = 3H^2/8\pi G$, and the current value $H_0$ of the Hubble parameter gives the critical density today $\rho_c^0 = 1.8788 \times 10^{-29} h^2$ g cm$^{-3}$, where $h \equiv H_0/(100 \text{ km s}^{-1} \text{ Mpc}^{-1})$. The different contributions to $\rho$ evolve following the energy conservation law in an expanding Universe $\dot{\rho} = -3H(\rho + p)$, where $p$ is the pressure. Thus the homogeneous density of photons $\rho_\gamma$ scales like $a^{-4}$, that of non-relativistic matter ($\rho_{cdm}$ for cold dark matter and $\rho_b$ for baryons) like $a^{-3}$, and the cosmological constant density $\rho_\Lambda$ is of course time-independent. Instead, neutrinos contribute to the radiation density at early times but behave as matter after the non-relativistic transition.

The evolution of all densities is shown on the left plot of Figure 2, starting at MeV temperatures until now. We also display the characteristic times for the end of Primordial Nucleosynthesis and for photon decoupling or recombination. The evolution of the density fractions $\Omega_i \equiv \rho_i/\rho_c$ is shown on the right panel, where it is easier to see which of the Universe components is dominant, fixing its expansion rate: first radiation in the form of photons and neutrinos (Radiation Domination or RD), then matter which can be CDM, baryons and massive neutrinos at late times (Matter Domination or MD) and finally the cosmological constant density takes over at low redshift (typically $z < 0.5$). Massive neutrinos are the only particles that present the transition from radiation to matter, when their density is clearly enhanced (upper solid lines in Fig. 2). Obviously the contribution of massive neutrinos to the energy density in the non-relativistic limit is a function of the mass (or the sum of all masses for which $m_i \gg T_\nu$), and the present value $\Omega_\nu$ could be of order unity for eV masses (see Sec. 5).

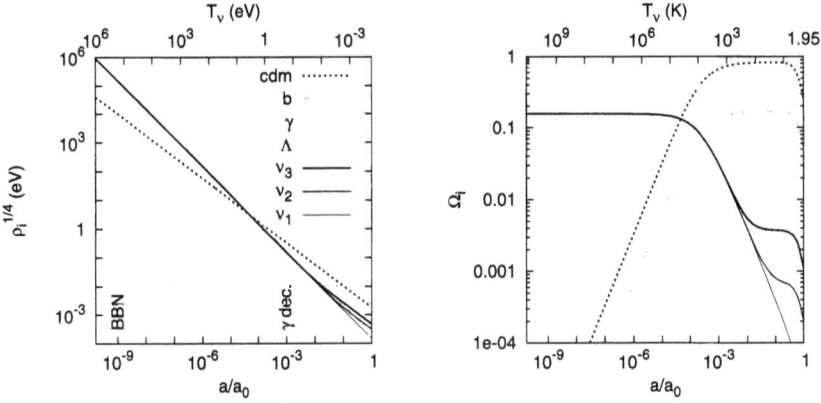

Fig. 2. Evolution of the background energy densities (left) and density fractions $\Omega_i$ (right) from the time when $T_\nu = 1$ MeV until now, for each component of a flat $\Lambda$MDM model with $h = 0.7$ and current density fractions $\Omega_\Lambda = 0.70$, $\Omega_b = 0.05$, $\Omega_\nu = 0.0013$ and $\Omega_{cdm} = 1 - \Omega_\Lambda - \Omega_b - \Omega_\nu$. The three neutrino masses are $m_1 = 0$, $m_2 = 0.009$ eV and $m_3 = 0.05$ eV.

## 3. Neutrinos and Primordial Nucleosynthesis

In the course of its expansion, when the early Universe was only less than a second old, the conditions of temperature and density of its nucleon component were such that light nuclei could be created via nuclear reactions (for a recent review, see [9]). During this epoch, known as Primordial or Big-Bang Nucleosynthesis (BBN), the primordial abundances of light elements were produced: mostly $^4$He but also smaller quantities of less stable nuclei such as D, $^3$He and $^7$Li. Heavier elements could not be produced because of the rapid evolution of the Universe and its small nucleon content, related to the value of the baryon asymmetry which normalized to the photon density, $\eta_b \equiv (n_b - n_{\bar{b}})/n_\gamma$, is about a few times $10^{-10}$. Measuring these primordial abundances today is a very difficult task, because stellar process may have altered the chemical compositions, but data on the $^4$He, D and $^7$Li abundances exist and can be compared with the theoretical predictions to learn about the conditions of the Universe at such an early period, testing any non-standard physics or cosmology [10].

The physics of BBN is well understood, since in principle only involves the Standard Model of particle physics and the time evolution of the expansion rate as given by the Friedmann equation. In the first phase of BBN the weak processes that keep the neutrons and protons in equilibrium ($n + \nu_e \leftrightarrow p + e^-$ and $n + e^+ \leftrightarrow p + \bar{\nu}_e$) freeze and the neutron-to-proton ratio becomes a constant, later diminished due to neutron decays. This ratio largely fixes the produced $^4$He abundance. Later the primordial abundances of light elements are produced and their values, that can be quite precisely found with BBN numerical codes (see e.g. [11]), depend on the competition between the nuclear reaction rates and the expansion rate. At present there exist a nice agreement with the observed D abundance for a value of the baryon asymmetry $\eta_b = 6.1 \pm 0.6$ [9], which also agrees with the region determined by CMB and large-scale structure data (LSS). Instead, the predicted primordial abundance of $^4$He tends to be a bit larger than the observed value, but it is difficult to consider this as a serious discrepancy, because the accuracy of the observations of $^4$He is limited by systematic uncertainties.

There are two main effects of relic neutrinos at BBN. The first one is that they contribute to the relativistic energy density of the universe (if $m_\nu \ll T_\nu$), thus fixing the expansion rate. This is why BBN gave the first allowed range of the number of neutrino species before accelerators (see Sec. 4). On the other hand, BBN is the last period of the Universe sensitive to neutrino flavour, since electron neutrinos and antineutrinos play a direct role in the $n \leftrightarrow p$ processes.

## 4. Extra radiation and the effective number of neutrinos

Together with photons, in the standard case neutrinos fix the expansion rate during the cosmological era when the Universe is dominated by radiation. Their contribution to the total radiation content can be parametrized in terms of the effective number of neutrinos $N_{\text{eff}}$, through the relation

$$\rho_r = \rho_\gamma + \rho_\nu = \left[ 1 + \frac{7}{8} \left( \frac{4}{11} \right)^{4/3} N_{\text{eff}} \right] \rho_\gamma, \qquad (4.1)$$

where we normalized to the photon energy density because its value today is known from the measurement of the CMB temperature. This equation is valid for complete neutrino decoupling and holds as long as all neutrinos are relativistic.

We know that the number of light neutrinos sensitive to weak interactions (flavour or active neutrinos) equals three from the analysis of the invisible $Z$-boson width at LEP, $N_\nu = 2.984 \pm 0.008$ [12], and we saw in a previous section from the analysis of neutrino decoupling that they contribute as $N_{\text{eff}} \simeq 3.046$. Any departure of $N_{\text{eff}}$ from this last value would be due to non-standard neutrino features or to the contribution of other relativistic relics. For instance, the energy density of a hypothetical scalar particle $\phi$ in equilibrium with the same temperature as neutrinos would be $\rho_\phi = (\pi/30) T_\nu^4$, leading to a departure of $N_{\text{eff}}$ from the standard value of 4/7. A detailed discussion of cosmological scenarios where $N_{\text{eff}}$ is not fixed to three can be found in [1] or [10].

We saw before that the expansion rate during BBN fixes the produced abundances of light elements, in particular that of $^4$He. Thus the value of $N_{\text{eff}}$ is constrained at the BBN epoch from the comparison of theoretical predictions and experimental data on the primordial abundances. In addition, a non-standard value of $N_{\text{eff}}$ would modify the transition epoch from RD to MD, which has some consequences on cosmological observables such as the power spectrum of CMB anisotropies, leading to independent bounds on the radiation content. These are two complementary ways of constraining $N_{\text{eff}}$ at very different epochs.

Here, as an example, we describe the results of a very recent analysis [13] (see the references therein for a list of recent works), that considers both BBN and CMB/LSS data to calculate the allowed regions on the plane defined by $N_{\text{eff}}$ and the baryon contribution to the present energy density. For the BBN analysis they include the $^4$He and D abundances, leading to the allowed range $N_{\text{eff}} = 3.1^{+1.4}_{-1.2}$ (95% C.L.). They also consider the most recent cosmological data on CMB temperature anisotropies and polarization, Large Scale galaxy clustering from SDSS and 2dF and luminosity distances of type Ia Supernovae (we will describe these measurements later in Sec. 7). The allowed range in this case is $N_{\text{eff}} = 5.2^{+2.7}_{-2.2}$, while in a less conservative analysis adding data on the Lyman-$\alpha$ absorption clouds and the Baryonic Acoustic Oscillations (BAO) from SDSS, is $N_{\text{eff}} = 4.6^{+1.6}_{-1.5}$. These ranges are in reasonable agreement with the standard prediction of $N_{\text{eff}} \simeq 3.046$, and show that there exists an allowed region of $N_{\text{eff}}$ values that is common at early (BBN) and more recent epochs, although some tension remains particularly when adding Lyman-$\alpha$ and BAO data. However, the reader should be cautious in the interpretation of this tension as an indication for relativistic degrees of freedom beyond the contribution of flavour neutrinos.

## 5. Massive neutrinos

### 5.1. Flavour oscillations and neutrino masses

Nowadays there exist compelling evidences for flavour neutrino oscillations from a variety of experimental data on solar, atmospheric, reactor and accelerator neutrinos. These are very important results, because the existence of flavour change implies that neutrinos mix and have non-zero masses, which in turn requires particle physics beyond the Standard Model. Oscillation experiments measure two differences of squared neutrino masses (these are $3\sigma$ ranges from [14])

$$\Delta m^2_{21} = \left(7.9^{+1.0}_{-0.8}\right) \times 10^{-5} \text{ eV}^2 \quad |\Delta m^2_{31}| = (2.6 \pm 0.6) \times 10^{-3} \text{ eV}^2 \quad (5.1)$$

where $\Delta m^2_{ij} = m^2_i - m^2_j$. Unfortunately oscillation experiments are insensitive to the absolute scale of neutrino masses, since the knowledge of $\Delta m^2_{21} > 0$ and $|\Delta m^2_{31}|$ leads to the two possible schemes shown in Fig. 1 of [3], known as

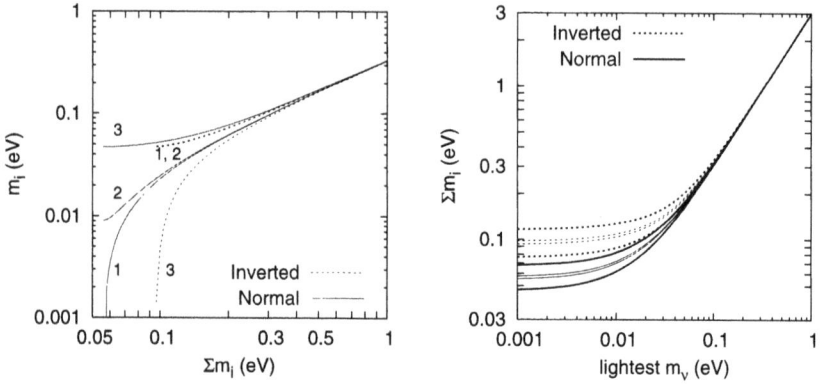

Fig. 3. Expected values of neutrino masses according to the values in Eq. (5.1). Left: individual neutrino masses as a function of the total mass for the best-fit values of the $\Delta m^2$. Right: ranges of total neutrino mass as a function of the lightest state within the $3\sigma$ regions (thick lines) and for a future determination at the 5% level (thin lines).

normal (NH) and inverted (IH) hierarchies, characterized by the sign of $\Delta m^2_{31}$, but leaves one neutrino mass unconstrained. For small values of the lightest neutrino mass $m_0$, i.e. $m_1$ ($m_3$) for NH (IH), the mass states follow a hierarchical scenario, while for masses much larger than the differences all neutrinos share in practice the same mass and then we say that they are degenerate. In general, the relation between the individual masses and the total neutrino mass can be found numerically, as shown in Fig. 3.

There are two types of laboratory experiments searching for the absolute scale of neutrino masses. The neutrinoless double beta decay $(Z, A) \rightarrow (Z + 2, A) + 2e^-$ is a rare nuclear processes where lepton number is violated . If the process is mediated by a light neutrino, the results from neutrinoless double beta decay experiments are converted into an upper bound or a measurement of $m_{\beta\beta} = |\sum_i U^2_{ei} m_i|$ which depends on the two Majorana phases that appear in lepton-number-violating processes (see [15] for a review). Instead, $\beta$ decay experiments, which involve only the kinematics of electrons, are in principle the best strategy for measuring directly the neutrino mass [16]. The current limits from tritium beta decay apply only to the range of degenerate neutrino masses, so that $m_\beta \simeq m_0$, where $m_\beta = (\sum_i |U_{ei}|^2 m_i^2)^{1/2}$ is the relevant parameter for beta decay experiments. The bound at 95% CL is $m_0 < 2.05$–$2.3$ eV from the Troitsk and Mainz experiments, respectively. This value is expected to be improved by the KATRIN project to reach a discovery potential for 0.3–0.35 eV masses (or a sensitivity of 0.2 eV at 90% CL). Taking into account this upper bound and the minimal values of the total neutrino mass in the NH (IH), the sum of neutrino masses is restricted to the approximate range $0.06(0.1)$ eV $\lesssim \sum_i m_i \lesssim 6$ eV. As we discuss in the next sections, cosmology is at first order sensitive to the total neutrino mass $\sum_i m_i$ if all states have the same number density, providing information on $m_0$ but blind to neutrino mixing angles or possible CP violating phases. Thus cosmological results are complementary to terrestrial experiments. The interested reader can find the allowed regions in the parameter space defined by any pair of parameters $(\sum_i m_i, m_{\beta\beta}, m_\beta)$ in [17].

Could neutrino oscillations have an efect on any cosmological epoch? We saw that in the standard cosmological picture all flavour neutrinos were produced with the same energy spectrum, so we do not expect any effect from oscillations among these three states (up to the small spectral distortions calculated in [8]). But there are two cases where neutrino oscillations could have cosmological consequences: flavour oscillations with non-zero relic neutrino asymmetries and active-sterile neutrino oscillations. For more details, see Sec. 5 in [18].

## 5.2. Neutrinos as dark matter

Nowadays the existence of Dark Matter (DM), the dominant non-baryonic component of the matter density in the Universe, is well established. A priori, massive neutrinos are excellent DM candidates, because we are certain that they exist, with a present number density of 339 relic neutrinos and antineutrinos per cm$^3$, and their energy density in units of the critical value of the energy density is

$$\Omega_\nu = \frac{\rho_\nu}{\rho_c^0} = \frac{\sum_i m_i}{93.14\, h^2 \text{ eV}}. \tag{5.2}$$

Here $\sum_i m_i$ includes all masses of the neutrino states which are non-relativistic today. It is also useful to define the neutrino density fraction $f_\nu$ with respect to the total matter density, $f_\nu \equiv \rho_\nu/(\rho_{\text{cdm}} + \rho_b + \rho_\nu) = \Omega_\nu/\Omega_m$.

Let us find the possible present values of $\Omega_\nu$ in agreement with the neutrino masses shown in Fig. 3 and the approximate bounds from beta decay experiments. Note that at least two of the neutrino states are non-relativistic today, since both $(\Delta m_{31}^2)^{1/2} \simeq 0.05$ eV and $(\Delta m_{21}^2)^{1/2} \simeq 0.009$ eV are larger than the temperature $T_\nu \simeq 1.96$ K $\simeq 1.7 \times 10^{-4}$ eV. If the third neutrino state is very light and still relativistic, its relative contribution to $\Omega_\nu$ is negligible. One finds that $\Omega_\nu$ is restricted to the approximate range $0.0013(0.0022) \lesssim \Omega_\nu \lesssim 0.13$, where we included $h \approx 0.7$. This applies only to the standard case of three light active neutrinos, while in general a cosmological upper bound on $\Omega_\nu$ has been used since the 1970s to constrain the possible values of neutrino masses. For instance, if we demand that neutrinos should not be heavy enough to overclose the Universe ($\Omega_\nu < 1$), we obtain an upper bound $\sum_i m_i \lesssim 45$ eV. Moreover, since from present analyses of cosmological data we know that $\Omega_m \simeq 0.3$, the neutrino masses should obey the stronger bound $\sum_i m_i \lesssim 15$ eV, a bound which is roughly only a factor 2 worse than the bound from tritium beta decay, but of course with the caveats that apply to any cosmological limit.

Dark matter particles with a large velocity dispersion such as that of neutrinos are called hot dark matter (HDM). The role of neutrinos as HDM particles has been widely discussed since the 1970s, and the reader can find a historical review in [19]. It was realized in the mid-1980s that HDM affects the evolution of cosmological perturbations in a particular way: it erases the density contrasts on wavelengths smaller than a mass-dependent free-streaming scale. In a universe dominated by HDM, this suppression contradicts various observations. For instance, large objects such as superclusters of galaxies form first, while smaller structures like clusters and galaxies form via a fragmentation process. This top-down scenario is at odds with the fact that galaxies seem older than clusters. Given the failure of HDM-dominated scenarios, the attention then turned to cold dark matter (CDM) candidates, i.e. particles which were fully non-relativistic at the epoch when the universe became matter-dominated, which provided a better agreement with observations. Still in the mid-1990s it appeared that a small mixture of HDM in a universe dominated by CDM fitted better the observational data on density fluctuations at small scales than a pure CDM model. However, within the presently favoured $\Lambda$CDM model dominated at late times by a cosmological constant (or some form of dark energy) there is no need for a significant contribution of HDM. Instead, one can use the available cosmological data to find how large the neutrino contribution can be, as we will see later.

To conclude this section, we would like to mention the case of a sterile neutrino with a mass in the keV range and a very small mixing with the flavour neutrinos. Such *heavy* neutrinos could be produced by active-sterile oscillations but not fully thermalized, so that they could play the role of dark matter and replace the usual CDM component. But due to their large thermal velocity (slightly smaller than that of active neutrinos), they would behave as Warm Dark Matter and erase small-scale cosmological structures. Their mass can be bounded from below using Lyman-$\alpha$ forest data, and from above using X-ray observations. The viability of this scenario is currently under careful examination (see e.g. [20] for a recent discussion and a list of references).

## 6. Effects of neutrino masses on cosmology

Here we briefly describe the main cosmological observables and the effects caused by neutrino masses. For a more detailed discussion of the effects of massive neutrinos on the evolution of cosmological perturbations, see Secs. 4.5 and 4.6 of [3].

## 6.1. Brief description of cosmological observables

Although there exist many different types of cosmological measurements, here we will restrict the discussion to those that are more important for obtaining an upper bound or eventually a measurement of neutrino masses.

First of all, we have the CMB temperature anisotropy power spectrum, defined as the angular two-point correlation function of CMB maps $\delta T/\bar{T}(\hat{n})$ ($\hat{n}$ is a direction in the sky) and usually expanded in Legendre multipoles

$$\left\langle \frac{\delta T}{\bar{T}}(\hat{n}) \frac{\delta T}{\bar{T}}(\hat{n}') \right\rangle = \sum_{l=0}^{\infty} \frac{(2l+1)}{4\pi} C_l \, P_l(\hat{n} \cdot \hat{n}'), \tag{6.1}$$

For Gaussian fluctuations, all the information is encoded in the multipoles $C_l$ which probe correlations on angular scales $\theta = \pi/l$. Since each neutrino family can only have a mass of the order of 1 eV, the transition of relic neutrinos to the non-relativistic regime is expected to take place after the time of recombination between electrons and nucleons, i.e. after photon decoupling. The shape of the CMB spectrum is related mainly to the physical evolution *before* recombination, so it will be only marginally affected by neutrino masses, except for an indirect effect through the modified background evolution. The CMB polarization adds interesting complementary information to the temperature power spectrum, and currently we have some less precise data on the temperature × E-polarization (TE) correlation function and the E-polarization self-correlation spectrum (EE).

The current Large Scale Structure (LSS) of the Universe is probed by the matter power spectrum, observed with various techniques described in the next section (directly or indirectly, today or in the near past at redshift $z$). It is defined as the two-point correlation function of non-relativistic matter fluctuations in Fourier space $P(k, z) = \langle |\delta_m(k, z)|^2 \rangle$, where $\delta_m = \delta\rho_m/\bar{\rho}_m$. Usually $P(k)$ refers to the matter power spectrum evaluated today (at $z = 0$). In the case of several fluids (e.g. CDM, baryons and non-relativistic neutrinos), the total matter perturbation can be expanded as $\delta_m = \sum_i \bar{\rho}_i \, \delta_i / \sum_i \bar{\rho}_i$, and represents indifferently the energy or mass power spectrum of non-relativistic matter. The shape of the matter power spectrum is affected by the free-streaming caused by small neutrino masses of $\mathcal{O}(eV)$ and thus it is the key observable for constraining $m_\nu$ with cosmological methods. We will show later in Fig. 4 the typical shape of both the CMB temperature anisotropy spectrum $C_l$ and the matter power spectrum $P(k)$.

## 6.2. Neutrino free-streaming

After thermal decoupling, relic neutrinos constitute a collisionless fluid, where the individual particles free-stream with a characteristic velocity that, in average, is the thermal velocity $v_{th}$. It is possible to define a horizon as the typical distance on which particles travel between time $t_i$ and $t$. When the Universe was dominated by radiation or matter and $t \gg t_i$, this horizon is asymptotically equal to $v_{th}/H$, up to a numerical factor of order one. Similar to the definition of the Jeans length (see Sec. 4.4 in [3]), we can define the neutrino free-streaming wavenumber and length as

$$k_{FS}(t) = \left( \frac{4\pi G \bar{\rho}(t) a^2(t)}{v_{th}^2(t)} \right)^{1/2}, \quad \lambda_{FS}(t) = 2\pi \frac{a(t)}{k_{FS}(t)}. \tag{6.2}$$

As long as neutrinos are relativistic, they travel at the speed of light and their free-streaming length is simply equal to the Hubble radius. When they become non-relativistic, their thermal velocity decays like $v_{th} \equiv \langle p \rangle / m \simeq 3T_\nu/m \simeq 150(1+z)(1 \text{ eV}/m) \text{ km s}^{-1}$, where we used for the present neutrino temperature $T_\nu^0 \simeq (4/11)^{1/3} T_\gamma^0$ and $T_\gamma^0 \simeq 2.726$ K. This gives for the free-streaming wavelength and wavenumber during matter or $\Lambda$ domination

$$\lambda_{FS}(t) = 7.7 \frac{1+z}{\sqrt{\Omega_\Lambda + \Omega_m (1+z)^3}} \left( \frac{1 \text{ eV}}{m} \right) h^{-1} \text{Mpc}, \tag{6.3}$$

$$k_{FS}(t) = 0.82 \frac{\sqrt{\Omega_\Lambda + \Omega_m (1+z)^3}}{(1+z)^2} \left( \frac{m}{1 \text{ eV}} \right) h \text{ Mpc}^{-1}, \tag{6.4}$$

where $\Omega_\Lambda$ and $\Omega_m$ are the cosmological constant and matter density fractions, respectively, evaluated today. So, after the non-relativistic transition and during matter domination, the free-streaming length continues to increase, but only like $(aH)^{-1} \propto t^{1/3}$, i.e. more slowly than the scale factor $a \propto t^{2/3}$. Therefore, the comoving free-streaming length $\lambda_{FS}/a$ actually decreases like $(a^2 H)^{-1} \propto t^{-1/3}$. As a consequence, for neutrinos becoming non-relativistic during matter domination, the comoving free-streaming wavenumber passes through a minimum $k_{nr}$ at the time of the transition, i.e. when $m = \langle p \rangle = 3T_\nu$ and $a_0/a = (1+z) = 2.0 \times 10^3 (m/1 \text{ eV})$. This minimum value is found to be

$$k_{nr} \simeq 0.018 \, \Omega_m^{1/2} \left( \frac{m}{1 \text{ eV}} \right)^{1/2} h \text{ Mpc}^{-1}. \tag{6.5}$$

The physical effect of free-streaming is to damp small-scale neutrino density fluctuations: neutrinos cannot be confined into (or kept outside of) regions smaller than the free-streaming length, for obvious kinematic reasons. There exists a gravitational back-reaction effect that also damps the metric perturbations on those scales. Instead, on scales much larger than the free-streaming scale the neutrino velocity is effectively vanishing and after the non-relativistic transition the neutrino perturbations behave like CDM perturbations: modes with $k < k_{nr}$ are never affected by free-streaming and evolve like in a pure $\Lambda$CDM model.

## 6.3. Impact of massive neutrinos on the matter power spectrum

The small initial cosmological perturbations in the early Universe evolve, under the linear regime at any scale at early times and on the largest scales more recently, and produce the structures we see today. We can not review here all the details (see [3] and references therein), but we will emphasize the main effects caused by massive neutrinos in the framework of the standard cosmological scenario: a $\Lambda$ Mixed Dark Matter ($\Lambda$MDM) model, where Mixed refers to the inclusion of some HDM component.

First let us describe the changes in the background evolution of the Universe. Since massless neutrinos are always radiation, in this case the present value of the matter contribution $\Omega_m^0$ is equal to the contribution of CDM and baryons. Instead, massive neutrinos contribute to radiation at early times but to matter after becoming non-relativistic, so now to $\Omega_m^0$, reducing the values of $\Omega_{CDM}^0$ and $\Omega_b^0$ with respect to the massless case. As a result, if these massive neutrinos have not yet become non-relativistic at the time of radiation/matter equality, then this transition is delayed. The consequence of a late equality for the LSS matter power spectrum is the following: since on sub-Hubble scales the matter density contrast $\delta_m$ grows more efficiently during MD than during RD, the matter power spectrum is suppressed on small scales relatively to large scales.

At the perturbation level, we also saw that free-streaming damps small-scale neutrino density fluctuations. This produces a direct effect on $P(k)$ (see Sec. 4.5 of [3]) that depends on the value $k$ with respect to $k_{nr}$ in Eq. (6.5),

$$P(k) = \left\langle \left( \frac{\delta\rho_{cdm} + \delta\rho_b + \delta\rho_\nu}{\rho_{cdm} + \rho_b + \rho_\nu} \right)^2 \right\rangle = \begin{cases} \langle \delta_{cdm}^2 \rangle & k < k_{nr} \\ [1 - \Omega_\nu/\Omega_m]^2 \langle \delta_{cdm}^2 \rangle & k \gg k_{nr} \end{cases}$$

Thus the role of the neutrino masses would be simply to cut the power spectrum by a factor $[1 - \Omega_\nu/\Omega_m]^2$ for $k \gg k_{nr}$. However, it turns out that the presence of neutrinos actually modifies the evolution of the CDM and baryon density contrasts and the suppression factor is enhanced, more or less by a factor four.

In conclusion, the combined effect of the shift in the time of equality and of the reduced CDM fluctuation growth during matter domination produces an attenuation of small-scale perturbations for $k > k_{nr}$. It can be shown that for small values of $f_\nu$ this effect can be approximated in the large $k$ limit by the well-known linear expression $P(k)^{f_\nu}/P(k)^{f_\nu=0} \simeq -8 f_\nu$ [21]. For comparing with the data, one could use instead some better analytical approximations to the full MDM or $\Lambda$MDM matter power spectrum, valid for arbitrary scales and redshifts, as listed in [3]. However, nowadays the analyses are performed using the matter power spectra calculated by Boltzmann codes such as CMBFAST [22] or CAMB [23], that solve numerically the evolution of the cosmological perturbations. An example of $P(k)$ with and without massive neutrinos is shown in Fig. 4, where the effect

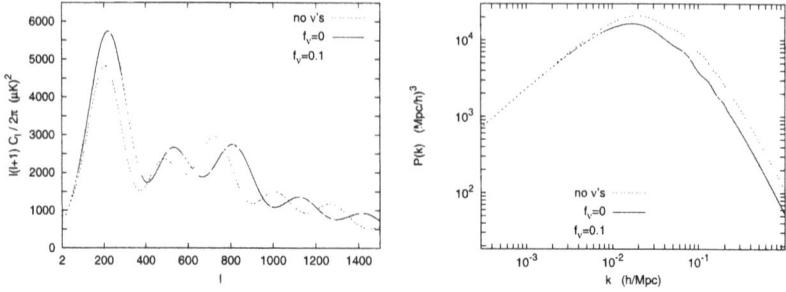

Fig. 4. CMB temperature anisotropy spectrum $C_l^T$ and matter power spectrum $P(k)$ for three models: a neutrinoless $\Lambda$CDM model, a more realistic $\Lambda$CDM model with three massless neutrinos, and finally a $\Lambda$MDM model with three massive degenerate neutrinos and a total density fraction $f_\nu = 0.1$. In all models, the values of other cosmological parameters have been kept fixed.

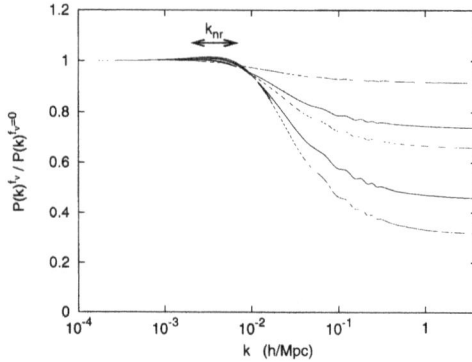

Fig. 5. Ratio of the matter power spectrum including three degenerate massive neutrinos with density fraction $f_\nu$ to that with three massless neutrinos. The parameters $(\omega_m, \Omega_\Lambda) = (0.147, 0.70)$ are kept fixed, and from top to bottom the curves correspond to $f_\nu = 0.01, 0.02, 0.03, \ldots, 0.10$. The individual masses $m_\nu$ range from 0.046 to 0.46 eV, and the scale $k_{nr}$ from $2.1 \times 10^{-3} h\,\mathrm{Mpc}^{-1}$ to $6.7 \times 10^{-3} h\,\mathrm{Mpc}^{-1}$ as shown on the top of the figure.

of $m_\nu$ at large $k$'s can be clearly visible. Such suppression is probably better seen in Fig. 5, where we plot the ratio of the matter power spectrum for $\Lambda$MDM over that of $\Lambda$CDM, for different values of $f_\nu$ and three degenerate massive neutrinos, but for fixed parameters $(\omega_m, \Omega_\Lambda)$.

Is it possible to mimic the effect of massive neutrinos on the matter power spectrum with some combination of other cosmological parameters? If so, one would say that a parameter degeneracy exists, reducing the sensitivity to neutrino masses. This possibility depends on the interval $[k_{min}, k_{max}]$ in which the $P(k)$ can be accurately measured. Ideally, if we could have $k_{min} \leq 10^{-2} h \, \text{Mpc}^{-1}$ and $k_{max} \geq 1 \, h \, \text{Mpc}^{-1}$, the effect of the neutrino mass would be non-degenerate, because of its very characteristic step-like effect. In contrast, other cosmological parameters like the scalar tilt or the tilt running change the spectrum slope on all scales. The problem is that usually the matter power spectrum can only be accurately measured in the intermediate region where the mass effect is neither null nor maximal: in other words, many experiments only have access to the transition region in the step-like transfer function. In this region, the neutrino mass affects the slope of the matter power spectrum in a way which can be easily confused with the effect of other cosmological parameters. Because of these parameter degeneracies, the LSS data alone cannot provide significant constraints on the neutrino mass, and it is necessary to combine them with other cosmological data, in particular the CMB anisotropy spectrum, which could lift most of the degeneracies. Still, for exotic models with e.g. extra relativistic degrees of freedom, a constant equation-of-state parameter of the dark energy different from $-1$ or a non-power-law primordial spectrum, the neutrino mass bound can become significantly weaker.

### 6.4. Impact of massive neutrinos on the CMB anisotropy spectrum

For neutrino masses of the order of 1 eV (about $f_\nu \leq 0.1$) the three neutrino species are still relativistic at the time of photon decoupling, and the direct effect of free-streaming neutrinos on the evolution of the baryon-photon acoustic oscillations is the same in the $\Lambda$CDM and $\Lambda$MDM cases. Therefore, the effect of the mass is indirect, appearing only at the level of the background evolution: the fact that the neutrinos account today for a fraction $\Omega_\nu$ of the critical density implies some change either in the present value of the spatial curvature, or in the relative density of other species. If neutrinos were heavier than a few eV, they would already be non-relativistic at decoupling. This case would have more complicated consequences for the CMB, as described in [24]. However, we will see later that this situation is disfavoured by current upper bounds on the neutrino mass.

Let us describe one example: we choose to maintain a flat Universe with fixed $(\omega_b = \Omega_b h^2, \omega_m = \Omega_m h^2, \Omega_\Lambda)$. Thus, while $\Omega_b$ and $\Omega_\Lambda$ are constant, $\Omega_{cdm}$ is constrained to decrease as $\Omega_\nu$ increases. The main effect on the CMB anisotropy spectrum results from a change in the time of equality. Since neutrinos are still relativistic at decoupling, they should be counted as radiation instead of matter around the time of equality (when $\rho_b + \rho_{cdm} = \rho_\gamma + \rho_\nu$). This gives $a_{eq} = \Omega_r/(\Omega_b + \Omega_{cdm})$, where $\Omega_r$ stands for the radiation density extrapolated until today *assuming that all neutrinos would remain massless*, given by Eq. (4.1) with $N_{eff} \simeq 3.04$. So, when $f_\nu$ increases, $a_{eq}$ increases proportionally to $[1 - f_\nu]^{-1}$: equality is postponed. This produces an enhancement of small-scale perturbations, especially near the first acoustic peak. Also, postponing the time of equality increases slightly the size of the sound horizon at recombination. These two features explain why in Fig. 4 the acoustic peaks are slightly enhanced and shifted to the left in the $\Lambda$MDM case.

Since the effect of the neutrino mass on CMB fluctuations is indirect and appears only at the background level, one could think that by changing the value of other cosmological parameters it would be possible to cancel exactly this effect. It can be actually shown that in the simplest $\Lambda$MDM model, with only seven cosmological parameters, one cannot vary the neutrino mass while keeping fixed $a_{\rm eq}$ and all other quantities governing the CMB spectrum. Therefore, it is possible to constrain the neutrino mass using CMB experiments alone [3, 25], although neutrinos are still relativistic at decoupling. This conclusion can be altered in more complicated models with extra cosmological parameters. For instance, allowing for an open Universe or varying the number of relativistic degrees of freedom one needs to add LSS data to remove the parameter degeneracies.

## 7. Current bounds on neutrino masses

In this section we review how the available cosmological data is used to get information on the absolute scale of neutrino masses, complementary to laboratory experiments. Note that the upper bounds on the sum of neutrino masses in the next subsections are all based on the Bayesian inference method, and are given at 95% C.L. after marginalization over all free cosmological parameters. We refer the reader to Sec. 5.1 of [3] for a detailed discussion on this statistical method, as well as for most of the references for the experimental data or parameter analysis.

### 7.1. CMB anisotropies

The experimental situation of the measurement of the CMB anisotropies is dominated by the third-year release of WMAP data (WMAP3), which improved the already precise TT and TE angular power spectra of the first-year release (WMAP1), and adds a detection of the E-polarization self-correlation spectrum (EE). On similar or smaller angular scales than WMAP, we have results from experiments that are either ground-based (ACBAR, VSA, CBI, DASI, ...) or balloon-borne (ARCHEOPS, BOOMERANG, MAXIMA, ...). We saw in the previous section that the signature on the CMB spectrum of a neutrino mass smaller than 0.5 eV is small but does not vanish due to a background effect, changes some characteristic times and scales in the evolution of the Universe, and affects mainly the amplitude of the first acoustic peak as well as the location of all the peaks. Therefore, it is possible to constrain neutrino masses using CMB experiments only, down to the level at which this background effect is masked by instrumental noise or cosmic variance, or by parameter degeneracies in the case of some cosmological models beyond the minimal $\Lambda$MDM framework.

We assume that the total neutrino mass $M_\nu \equiv \sum_i m_i$ is the only additional parameter with respect to a flat $\Lambda$CDM cosmological model characterized by 6 parameters (this applies to all bounds reviewed in this section, unless specified otherwise). In this framework, many analyses support the conclusion that a sensible bound on $M_\nu$ exists using CMB data only, of order of 2–3 eV. This is an important result, since it does not depend on the uncertainties from LSS data.

## 7.2. Galaxy redshift surveys

We have seen that free-streaming of massive neutrinos produces a direct effect on the formation of cosmological structures. As shown in Fig. 5, the presence of neutrino masses leads to an attenuation of the linear matter power spectrum on small scales. It was shown in the seminal paper [21] that an efficient way to probe eV neutrino masses was to use data from large galaxy redshift surveys (GRS), which measure the distance to a large number of objects, giving us a 3D picture of the universe. At present, we have data from two large projects: the 2 degree Field (2dF) GRS, whose final results were obtained from more than 220,000 galaxy redshifts, and the Sloan Digital Sky Survey (SDSS), which will be completed soon with data from one million galaxies.

One of the main goals of a GRS is to reconstruct the power spectrum of matter fluctuations on very large scales, whose cosmological evolution is described entirely by linear perturbation theory. However, the linear power spectrum must be reconstructed from individual galaxies which underwent a strongly non-linear evolution. A simple analytic model of structure formation suggests that on large scales, the galaxy-galaxy correlation function should be proportional to the linear matter density power spectrum, up to a constant light-to-mass bias ($b$). This parameter can be obtained from independent methods, which tend to confirm that the linear biasing assumption is correct, at least in first approximation. A conservative way to use the data from a GRS in a cosmological analysis is to take the bias as a free parameter, i.e. to consider only the shape of the matter power spectrum at the corresponding scales and not its amplitude (denoted as galaxy clustering data). An upper limit on $M_\nu$ between 0.8 and 1.7 eV is found from the analysis of galaxy clustering data (SDSS and/or 2dF, with the bias as a free parameter) added to CMB data. In general, one obtains weaker bounds on neutrino masses using preliminary SDSS results instead of 2dF data, but this conclusion could change with future SDSS data. The bounds on neutrino masses are more stringent when the amplitude of the matter power spectrum is fixed with a measurement of the bias: the upper limits on $M_\nu$ are reduced to values of order 0.5–0.9 eV although some analyses also add Lyman-$\alpha$ data.

Finally, a GRS performed in a large volume can also be sensitive to the imprint created by the baryon acoustic oscillations (BAO) at large scales on the power spectrum of non-relativistic matter. Since baryons are only a subdominant component of matter, the BAO feature is manifested as a small single peak in the galaxy correlation function in real space that was recently detected from the analysis of the SDSS luminous red galaxy (LRG) sample. The observed position of this baryon oscillation peak provides a way to measure the angular diameter distance out to the typical LRG redshift of $z = 0.35$, which in turn can be used to constrain the parameters of the underlying cosmological model. Ref. [26] included the BAO-SDSS measurement and obtained a bound of 0.44 eV on $M_\nu$.

### 7.3. Lyman-$\alpha$ forest

The matter power spectrum on small scales can also be inferred from data on the Lyman-$\alpha$ forest: the absorption of photons traveling from distant quasars ($z \sim 2$–3) by the neutral H in the intergalactic medium. As the Universe expands, photons are continuously red-shifted along the line of sight, and can be absorbed when they reach a wavelength of 1216 Å in the rest-frame of the intervening medium. Therefore, the quasar spectrum contains a series of absorption lines, whose amplitude as a function of wavelength traces back the density and temperature fluctuations of neutral H along the line of sight. It is then possible to infer the matter density fluctuations in the linear or quasi-linear regime, but one needs to recover the matter power spectrum from the spectrum of the transmitted flux, a task that requires the use of hydro-dynamical simulations for the corresponding cosmological model. This is a difficult procedure, and given the various systematics involved in the analysis the robustness of Lyman-$\alpha$ forest data is still a subject of intense discussion between experts. In any case, the recovered $P(k)$ is again sensitive to the suppression of growth of mass fluctuations caused by massive neutrinos, and in many cosmological analyses the Lyman-$\alpha$ data is added to CMB and other LSS data. For a free bias, one finds that Lyman-$\alpha$ data help to reduce the upper bounds on $M_\nu$ down to the 0.5–0.7 eV level. But those analyses that include Lyman-$\alpha$ data and fix the bias do not always lead to a lower limit, ranging from 0.4 to 0.7 eV. Finally, [26] found the upper bound $M_\nu < 0.30$ eV adding simultaneously Ly$\alpha$ and BAO data, both from SDSS.

### 7.4. Summary and discussion of current bounds

The upper bounds on $M_\nu$ from the previous subsections are representative of an important fact: a single cosmological bound on neutrino masses does not exist. Fig. 6 is a graphical summary, where the bands are the cosmological bounds

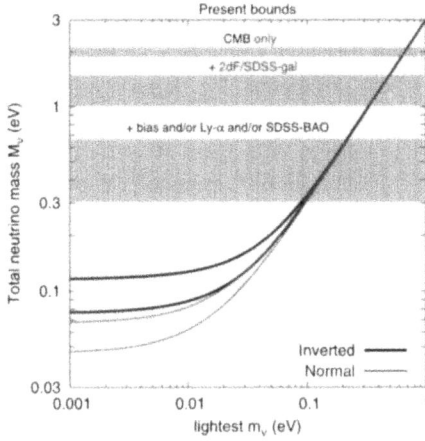

Fig. 6. Current upper bounds (95%CL) from cosmological data on the sum of neutrino masses, compared to the values in agreement at a $3\sigma$ level with neutrino oscillation data in Eq. (5.1).

grouped according to the included set of data and whose thickness roughly describe the spread of values obtained from similar cosmological data: 2–3 eV (CMB only), 0.9–1.7 eV (CMB & 2dF/SDSS-gal) or 0.3–0.9 eV with the inclusion of a measurement of the bias and/or Lyman-$\alpha$ forest data and/or SDSS-BAO. Other cosmological data sets are usually added to CMB/LSS data, such as the measurement of the present value of the Hubble parameter by the Key Project of the HST, giving $h = 0.72 \pm 0.08 \, (1\sigma)$, which excludes low values of $h$ and leads to a stronger upper bound on the total neutrino mass. One can also include the constraints on the current density of the dark energy component deduced from the redshift dependence of type Ia supernovae luminosity, which measures the late evolution of the expansion rate of the Universe. For a flat Universe with a cosmological constant, these constraints can be translated into bounds on $\Omega_m$.

One can see from Fig. 6 that current cosmological data probe the region of three degenerate neutrino masses. A conservative upper bound is approximately 1 eV from CMB results combined only with 2dF/SDSS galaxy clustering data (i.e. the shape of the matter power spectrum for the relevant scales). The addition of further data leads to an improvement of the bounds, which reach the lowest values when data from Lyman-$\alpha$ and/or BAO-SDSS are included or the bias is fixed. In such cases the contribution of a total neutrino mass of the order 0.3–0.6 eV seems already disfavoured. In [17] these cosmological bounds are compared with those coming from tritium $\beta$ decay and neutrinoless double beta.

We remind the reader that the impressive cosmological bounds on $M_\nu$ shown in Fig. 6 may change if additional cosmological parameters are allowed beyond the $\Lambda$CDM model. This could be the case whenever a new parameter degeneracy with the neutrino masses arises. For instance, in the presence of extra radiation the bound on $M_\nu$ gets less stringent [27]. These results are interesting for the 4-neutrino mass schemes that also incorporate the LSND results [28], which would require a fourth sterile neutrino with mass of $\mathcal{O}(\text{eV})$ (see e.g. [29] for a recent cosmological analysis). Other possible parameter degeneracy exists between $M_\nu$ and the parameter $w$, that characterizes the equation of state of the dark energy component $X$ ($p_X = w\rho_X$), but it can be broken adding BAO data [26]. Finally, the cosmological implications of neutrino masses could be very different if the spectrum or evolution of the cosmic neutrino background was non-standard (see the discussion in Sec. 5.7 of [3]).

## 8. Future sensitivities on neutrino masses from cosmology

In the near future we will have more precise data on cosmological observables from various experimental techniques: CMB anisotropies measured with ground-based experiments or satellites such as PLANCK, galaxy redshift surveys, galaxy cluster surveys, … In particular, it has been recently emphasized the potentiality for measuring small neutrino masses of weak lensing experiments, which will look for the lensing effect caused by the large scale structure of the neighboring universe, either on the CMB signal [30] or on the apparent shape of galaxies (measured by cosmic shear surveys, see e.g. [31]). If the characteristics of these future experiments are known with some precision, it is possible to assume a "fiducial model", i.e. a cosmological model that would yield the best fit to future data, and to estimate the error bar on a particular parameter that will be obtained after marginalizing the hypothetical likelihood distribution over all the other free parameters (see Sec. 6 of [3] for further details). Technically, the simplest way to forecast this error is to compute a Fisher matrix, a technique has been widely used in the literature, for many different models and hypothetical datasets, now complemented by Monte Carlo methods.

Fig. 7 is a graphical summary of the forecast sensitivities to neutrino masses of different cosmological data compared to the allowed values of neutrino masses in the two possible 3-neutrino schemes. One can see that there are very good prospects for testing neutrino masses in the degenerate and quasi-degenerate mass regions above 0.2 eV or so. A detection at a significant level of the minimal value of the total neutrino mass in the inverted hierarchy scheme will demand the combination of CMB lensing and cosmic shear surveys, whose more ambitious projects will provide a $2\sigma$ sensitivity to the minimal value in the case of normal hierarchy (of order 0.05 eV). The combination of CMB observations with future

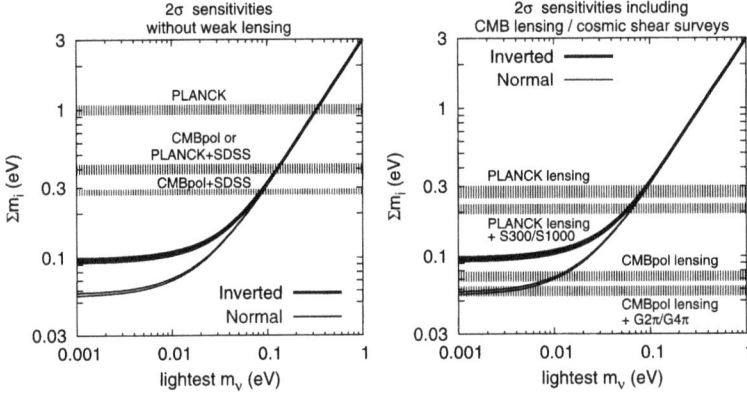

Fig. 7. Forecast $2\sigma$ sensitivities to the total neutrino mass from future cosmological experiments compared to the values in agreement with present neutrino oscillation data in Eq. (5.1) (assuming a future determination at the 5% level). Left: sensitivities expected for future CMB experiments (without lensing extraction), alone and combined with the completed SDSS galaxy redshift survey. Right: sensitivities expected for future CMB experiments including lensing information, alone and combined with future cosmic shear surveys. Here CMBpol refers to a hypothetical CMB experiment roughly corresponding to the INFLATION PROBE mission.

galaxy cluster surveys [32], derived from the same weak lensing observations, as well as X-ray and Sunyaev–Zel'dovich surveys, should yield a similar sensitivity.

## 9. Conclusions

Neutrinos, despite the weakness of their interactions and their small masses, can play an important role in Cosmology that we have reviewed in this contribution. In addition, cosmological data can provide information on the properties of these elusive particles, such as the effective number of neutrinos and the total neutrino mass. We have seen that the analysis of current cosmological data constrains the sum of neutrino masses, complementary to terrestrial experiments such as tritium beta decay and neutrinoless double beta decay experiments. In the next future, thanks to the data from new cosmological experiments we could even hope to test the minimal values of neutrino masses guaranteed by the present evidences for flavour neutrino oscillations. For this and many other reasons, we expect that neutrino cosmology will remain an active research field in the next years.

## Acknowledgments

I thank the organizers of the School for their invitation and hospitality. Many of the topics discussed here were developed in an enjoyable collaboration with

Julien Lesgourgues. This work was supported by the European Network of Theoretical Astroparticle Physics ILIAS/N6 under contract number RII3-CT-2004-506222 and the Spanish grants FPA2005-01269 and GV/05/017 of Generalitat Valenciana, as well as by a Ramón y Cajal contract of MEC.

## References

[1]  A.D. Dolgov, Phys. Rep. **370** (2002) 333.

[2]  S. Hannestad, Ann. Rev. Nucl. Part. Sci. **56** (2006) 137.

[3]  J. Lesgourgues and S. Pastor, Phys. Rep. **429** (2006) 307.

[4]  M. Kamionkowski and A. Kosowsky, Ann. Rev. Nucl. Part. Sci. **49** (1999) 77.

[5]  E.W. Kolb and M.S. Turner, *The Early Universe* (Addison-Wesley, 1990).

[6]  A.D. Dolgov et al., Nucl. Phys. B **632** (2002) 363.

[7]  S. Dodelson, *Modern Cosmology* (Academic Press, 2003).

[8]  G. Mangano et al., Nucl. Phys. B **729** (2005) 221.

[9]  G. Steigman, Int. J. Mod. Phys. E **15** (2006) 1.

[10]  S. Sarkar, Rep. Prog. Phys. **59** (1996) 1493.

[11]  A. Cuoco et al., Int. J. Mod. Phys. A **19** (2004) 4431.

[12]  W.M. Yao et al. [Particle Data Group], J. Physics G **33** (2006) 1.

[13]  G. Mangano et al., astro-ph/0612150.

[14]  M. Maltoni et al., New J. Phys. **6** (2004) 122.

[15]  S.R. Elliott and P. Vogel, Ann. Rev. Nucl. Part. Sci. **52** (2002) 115.

[16]  K. Eitel, Nucl. Phys. Proc. Suppl. **143** (2005) 197.

[17]  G.L. Fogli et al., Prog. Part. Nucl. Phys. **57** (2006) 742.

[18]  S. Pastor, lectures at 61st SUSSP (Taylor & Francis Academic Pub., 2007), in press.

[19]  J.R. Primack, astro-ph/0112336.

[20]  M. Viel et al., Phys. Rev. Lett. **97** (2006) 071301.

[21]  W. Hu et al., Phys. Rev. Lett. **80** (1998) 5255.

[22]  U. Seljak and M. Zaldarriaga, Astrophys. J. **469** (1996) 437, see also http://cmbfast.org.

[23]  A. Lewis et al., Astrophys. J. **538** (2000) 473, see also http://camb.info.

[24]  S. Dodelson et al., Astrophys. J. **467** (1996) 10.

[25]  K. Ichikawa et al., Phys. Rev. D **71** (2005) 043001.

[26]  A. Goobar et al., JCAP **0606** (2006) 019.

[27]  P. Crotty et al., Phys. Rev. D **69** (2004) 123007.

[28]  A. Aguilar et al. [LSND Collaboration], Phys. Rev. D **64** (2001) 112007.

[29]  S. Dodelson et al., Phys. Rev. Lett. **97** (2006) 041301.

[30]  J. Lesgourgues et al., Phys. Rev. D **73** (2006) 045021.

[31]  S. Hannestad et al., JCAP **0606** (2006) 025.

[32]  S. Wang et al., Phys. Rev. Lett. **95** (2005) 011302.

Course 10

# COSMIC MICROWAVE BACKGROUND: OBSERVATIONAL STATUS

S. Prunet

*Institut d'Astrophysique de Paris*
*98 bis Boulevard Arago*
*75014 Paris, France*

*F. Bernardeau, C. Grojean and J. Dalibard, eds.*
*Les Houches, Session LXXXVI, 2006*
*Particle Physcis and Cosmology: The Fabric of Spacetime*
© *2007 Published by Elsevier B.V.*

# Contents

## 1. Introduction

During the last fifteen years, since their discovery by COBE-DMR [46], the measurement of the cosmic microwave background (CMB) anisotropies has dramatically improved, thanks to complementary techniques (ground-based and balloon-borne, as well as spatial experiments) and detector technologies (phase-sensitive radiometers and bolometers).

A cosmic-variance limited measurement of the first accoustic peak of the CMB anisotropies power spectrum has recently been achieved by the WMAP satellite first year release [22]; this data set, together with several measurements by small-scale experiments (CBI, VSA, ACBAR)[1] represents the best measurements of the temperature anisotropies as of today.

In the mean time, the (fainter) polarization of the CMB has been discovered by DASI [27] at intermediate scales, and soon confirmed by measurements of the Boomerang, VSA, CBI and CAPMAP experiments. In its three-years release, the WMAP team has published the first measurement of the polarization power spectrum at large scales. This difficult measurement is, as we will see, of prime importance to put constraints on the reionization history of the Universe, and as a side effect, has a big impact on the constraints of inflationary parameters such as the primordial density fluctuations amplitude $A_s$ and spectral index $n_s$.

In the next section, I will discuss the early stages of the CMB temperature power spectrum measurements, with its main cosmological implications, such as the constraints on the Universe mean curvature; I will also show how a progressive improvement of the data sets in terms of size and signal-to-noise ratio (SNR) has triggered the development of new data processing techniques, progressively shifting the emphasis from SNR to systematic errors control.

In the second section, I will more specifically discuss WMAP first year release of data, and its cosmological implications. In a third section, after a bried reminder of the physical origin of the CMB polarization, I will discuss its early measurements by ground-based and balloon-borne experiments at intermediate scales.

The fourth section will be devoted to the WMAP three years release (WMAP3 hereafter), with an emphasis on polarization. I will try to illustrate the impor-

---

[1] Designed as the "WMAP-ext" data set in [22].

tance of polarization measurements to help break parameters degeneracies on one hand; on the other hand, emphasis will be put on the role of systematic effects on polarization measurements, in particular the role of foreground emissions from the Galaxy, and proper noise modelling. Finally, the near-term future of polarization measurements will be discussed.

## 2. CMB temperature anisotropies: the "early" days

The discovery of anisotropies in the CMB at large scales by COBE-DMR [46], and its implications on the physics of the early Universe, has triggered an unprecedented observational effort to map with increasing precision the CMB anisotropies. Indeed, the COBE measurements, concentrated on the super-Hubble scales at last scattering where the Sachs-Wolfe effect dominates, give direct information on the power spectrum of the primordial fluctuations of the metric.

Another important prediction of the CMB theory is the existence of acoustic peaks in the CMB power spectrum, induced by coherent standing waves in the plasma. Detailed measurement of the phase, shapes and relatives amplitudes of these peaks would give invaluable information on the nature of the primordial perturbations, the matter content of the Universe, its mean curvature (see S. Matarrese's lecture for a detailed description of CMB physics). The COBE-DMR measurements therefore opened a race to the discovery of the first acoustic peak.

Given the faint level of the CMB anisotropies, the main problem of ground-based observations is the atmospheric emission; the experimental effort was thus divided into two concurrent methods: differential measurements (interferometers) that eliminate the sky noise part that is common to both beams of an antenna pair, and total power, balloon-borne measurements made in the upper stratosphere.

Among the main interferometric experiments, one can cite TOCO [33], DASI [18], CBI [38, 40], VSA [12, 43], and BIMA [7] at the smallest scales. The most successful balloon-borne, total power measurements benefited from very sensitive bolometric detectors, cryogenically cooled to ∼300mK. Among them, the MAXIMA [19, 30], BOOMERanG [9, 35] and ARCHEOPS [3, 50] experiments played a leading role in mapping the first peaks of the power spectrum. Bolometric measurements from the ground at the South Pole were also successfully made at small angular scales [28], concurrencing the interferometric measurements.

The observational situation at the end of 2002 is summarized in figure 1. One can see that already at that time, the first peak structure was clearly defined. This

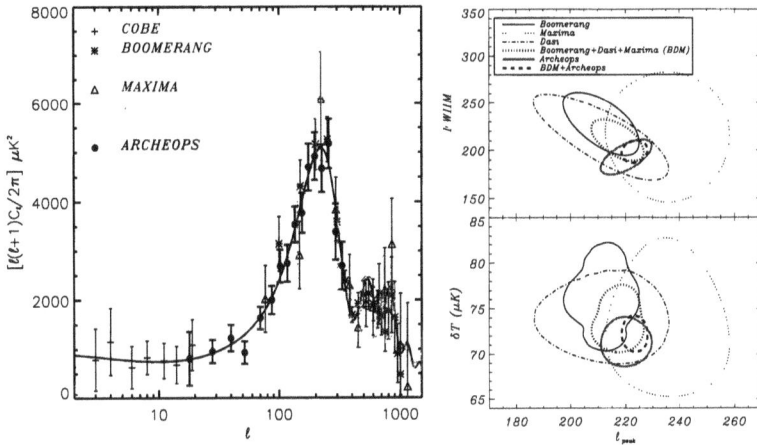

Fig. 1. CMB anisotropies power spectrum measurements, as of end 2002. The first acoustic peak is clearly visible, however the limited size of the surveys leaves space for improvement. Figure taken from [4].

measurement allowed to put strong constraints on the mean curvature of the universe, which is related, by the Friedman equation, to its mean energy density [29]. Subsequent measurements of the second peak allowed, by comparing its amplitude to that of the first peak, to put constrain on the mean baryon content [8]. Figure 1, right panel, shows the measured position of the first peak in multipole space and its corresponding error, for the different experiments.

## 3. WMAP first release

During year 2003, the WMAP team released their first power spectrum estimate [22], providing a coherent picture of the Sachs-Wolfe plateau and the first acoustic peak, with unprecedented accuracy. WMAP is a NASA satellite mission that is dedicated to the observation of cosmic microwave background anisotropies. Launched in 2001, it operates at the L2 sun-earth Lagrange point in a very clean and stable environment, allowing the first full-sky survey of CMB anisotropies, cosmic-variance limited down to sub-degree scales.

WMAP is a radiometer-based, differential experiment, with two back-to-back, off-axis telescopes. It operates with a fast precessing scanning strategy centered on the anti-solar direction that allows rapid coverage of the sky and hence high redundancy, which is of prime importance when assessing the quality of data in terms of systematic errors levels. Full-sky maps in five bands (23, 33, 41, 61, and

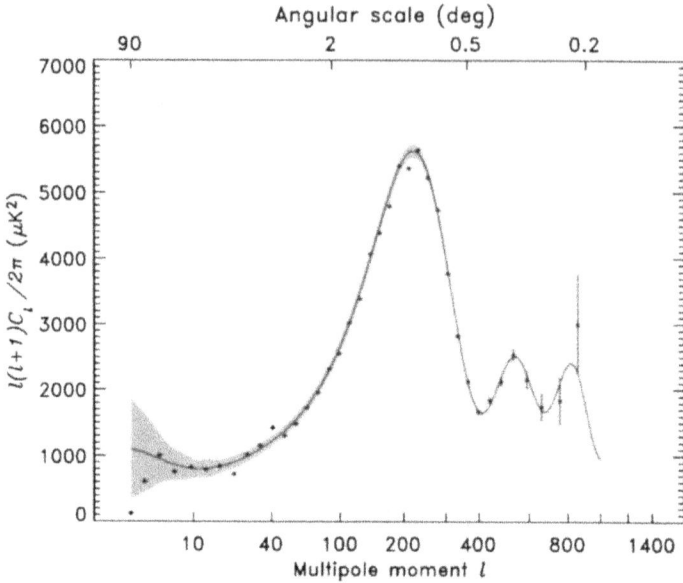

Fig. 2. Temperature power spectrum of WMAP first year release. The grey area corresponds to the 1$\sigma$ error region, the solid line is WMAP best-fit model. Figure from [22]

94 GHz) were produced, thus allowing for a multi-channel foreground emission cleaning of the data prior to power spectrum estimation [2].

The main results for the temperature power spectrum are shown in figure 2. Compared to figure 1, one clearly sees the improvement on scales down to $\ell \simeq$ 600. The methodology of data processing to obtain maps and power spectra are explained in details in [20, 22]. The large size of the dataset lead the WMAP team to choose fast data processing methods; for instance the maps were made using an iterative procedure first described in [51], and the power spectrum was computed using the MASTER method described in [23], rather than the optimized maximum-likelihood method developed by [36]. This is in fact a general trend in CMB data processing as data set grow larger and as the error budget becomes increasingly dominated by systematic errors that are most easily dealt with Monte-Carlo methods.

This dramatic improvement on the power spectrum measurement had of course an important impact on the cosmological parameters constraints, as described in detail in [48]. Interestingly though, while the Bayesian confidence regions shrunk in size, the recovered parameters were overall quite compatible with those inferred from previous experiments; however, the large scale anisotropies

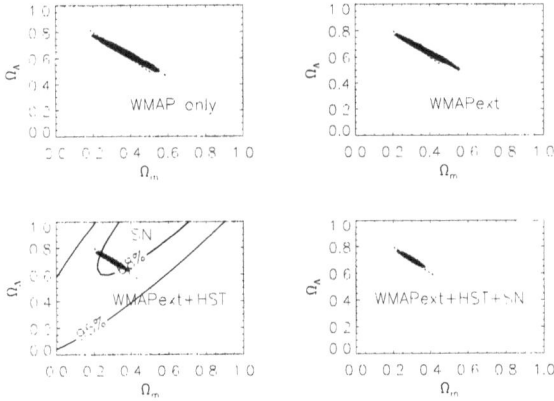

Fig. 3. 68%and 96%Bayesian confidence regions in the $\Omega_\Lambda - \Omega_m$ plane, respectively for WMAP data only, *WMAPext* data (WMAP, CBI, ACBAR), *WMAPext* and the HST Hubble prior, and *WMAPext* and HST and SNIa priors. Figure from [48].

as measured by COBE, being impressively recovered, show the same intriguing features, namely a low quadrupole amplitude [22, 49]—see however the WMAP3 section and the maximum-likelihood measurements of the spectrum, e.g. [13, 45]—and a correlation of the directions on the sky of the quadrupole and the octupole [49], see however [5].

Figure 3 shows the inferred Bayesian confidence regions in the $\Omega_\Lambda - \Omega_m$ plane, with different CMB data sets, as well as external prior information coming from HST Hubble constant measurements [17], or SN1a constraints (e.g. [42]). As can be seen in the figure, the total energy density, and hence the curvature, is very well constrained by WMAP alone, with little added information from the small scale experiments, as this constraint on the curvature comes primarily from the position of the first acoustic peak. However, the geometric degeneracy [14] between $\Omega_\Lambda$ and $\Omega_m$ is clear from the figure upper panels, it is broken by adding prior information on the Hubble constant as can be seen from the lower left panel.

## 4. Polarization of the CMB – discovery and first measurements

In year 2002, the DASI team [27] made a five sigma detection of a polarized signal that was compatible with the E-mode polarization power spectrum predicted from the best-fit cosmological model of that time. This very important discovery was later confirmed by the same team in a three-year data analysis [31]. In the mean time, other measurements of the E-mode power spectra, as well as tem-

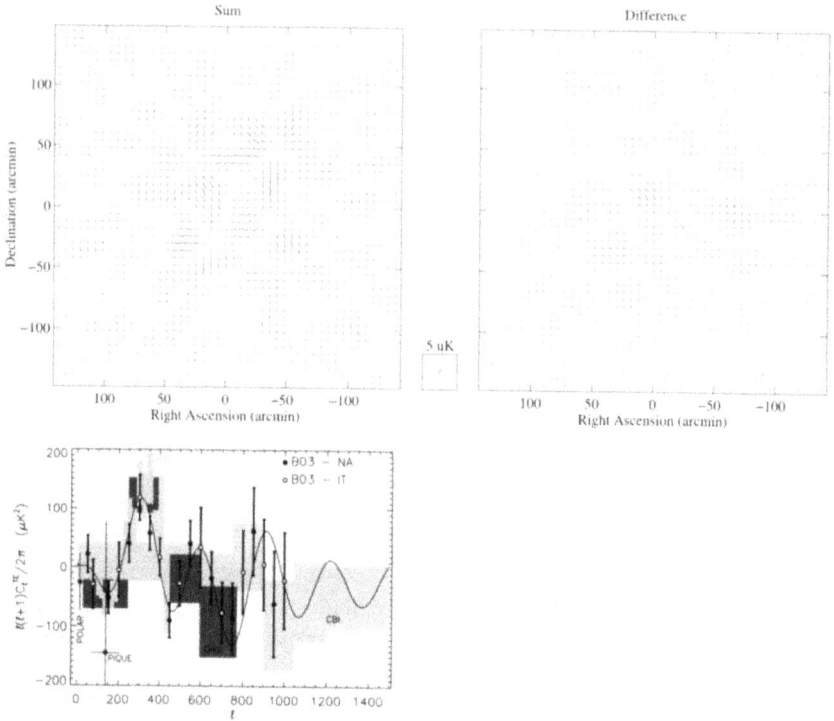

Fig. 4. *Top panel*: Sum and difference of the first-year DASI polarization fields. *Lower-left panel*: TE measurements summary, as of 2005. *Lower-right panel*: EE measurements summary, as of 2005. Figures taken respectively from [ and  27, 39] and [34].

perature E-mode (TE) cross-correlation were made by different, interferometric and balloon-borne experiments [1, 10, 11, 34, 39, 41], as well as WMAP in their first-year release [26]. The first measurement of the EE power spectrum at large scales was done by the WMAP team in their three-years release, we will come back in details on this later.

Polarization of the CMB bears new, important and complementary information to the temperature anisotropies. Generated at recombination by the scattering of quadrupolar radiation anisotropies off free electrons, it does not have contributions from the gravitational fields on the way to the observer (i.e. no integrated Sachs-Wolfe effect), and thus has no power on angular scales larger than the sound horizon at recombination, except a tail coming from projection effects. Moreover, the anisotropies in the polarization field have a single source (the quadrupolar anisotropy of the radiation in the electron's rest frame), when the temperature anisotropies result from the superposition of acoustic and Doppler contributions (see e.g. [24]). This results in sharper peaks in the polarization power spectrum, and the relative phase and amplitudes of these peaks compared to the temperature power spectrum peaks give information on the nature (adiabatic, isocurvature, see [6]) of the primordial scalar fluctuations.

In addition, the tensorial nature of the polarization field allow to decompose it into scalar (E) and pseudo-scalar (B) modes, that have different behaviour under parity transformations. It has been shown [25, 44] that polarization B modes cannot be generated by primordial density fluctuations. Since vector modes decay with time, the only possible, non-negligible, source of B mode polarization is believed to be the gravitational waves of primordial origin. Interestingly, this result is a consequence of geometry only (and the chirality of B mode polarization), and thus makes no assumption about a particular mechanism for the generation of the primordial metric fluctuations (with the exception of active sources of vector modes like topological defects, but these have been ruled out as a viable scenario for structure formation). The measurement of the CMB polarization has therefore a great potential to constrain early universe physics and the mechanisms to generate primordial metric fluctuations (e.g. inflationary models); it has therefore recently become the next observational target of the discipline, and ushered the development of new generations of detectors.

Finally, the absence of power at large scale in E-mode polarization make it a sensitive probe of reionization history, where rescattering of photons off free electrons at low redshift creates additional anisotropies at large scales in the polarization field. This makes the TE and EE power spectra sensitive namely to the Thomson optical depth to recombination ($\tau$), and helps to break the partial degeneracies that exist between $\tau$, the amplitude of scalar perturbations $A_s$, and the scalar spectral index $n_s$ (see e.g. [52]).

## 5. WMAP three year release

In year 2006, the WMAP team has produced a new release of data based on three years of cumulated observations. This release consists of an update of the CMB temperature anisotropies and their power spectrum [21], as well as a first measurement of the EE polarization power spectrum [37] at large angular scales. As mentioned previously, this measurement is directly relevant to constraining the value of the optical depth to recombination $\tau$, which in turn has implications on the constraints on inflationary parameters [47].

Let us begin with the update on the temperature power spectrum. As described in [21], the three year update of the temperature power spectrum is broadly compatible with the first year results, with however a noticeable correction at low

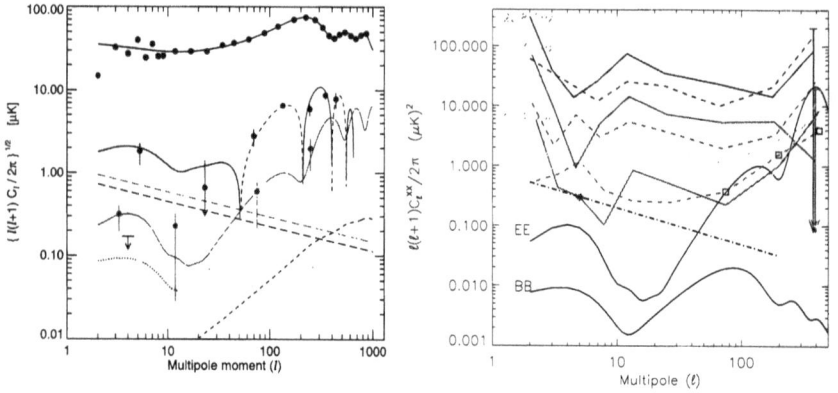

Fig. 5. WMAP3 polarization results. *Left panel*: final power spectra, after foreground cleaning. Black is temperature, red is TE correlation, green is EE power spectrum, and the blue upper limit is for BB. The dashed lines give an estimate of the foreground power at 65 GHz. *Right panel*: EE (solid lines) and BB (dashed lines) raw power spectra (including foregrounds and noise) from K-band (magenta) down to Q-band (green). This figure illustrate the importance of foreground contamination of large-scale CMB polarization. Figures from [37].

multipoles, due to a change in the data processing technique, an updated estimate of the instrumental beams, as well as a change in the radiometers gain model. Following [13], the WMAP team applied a hybrid power spectrum estimator to the data, where the low multipole power is computed using a maximum likelihood approach, and the high multipole power using a fast edge-corrected quadratic estimator with a new weighting scheme. The difference is most important at low multipole, and the higher value of the quadrupole using this estimator somewhat alleviates the debate on the discrepency with the best-fit LCDM. The new WMAP code in fact computes directly the large scales contribution to the likelihood directly from the maps, where its structure is well-known.

The new polarization results are described in details in [37]. For the first time, the EE polarization power spectrum is measured at large angular scales and the TE power spectrum is updated, taking into account a much more detailed analysis of the instrumental noise properties. The results are summarized in figure 5, where the left panel shows the final power spectra for temperature and polarization, and the right panel shows the importance of foreground power in polarization. The procedure followed by the WMAP team to clean the polarized foregrounds is a simple template substraction of the polarized synchrotron and dust emission. The K-band map is taken as the synchrotron template, and a synthetic dust template is created using the [15] intensity model 8, together with a model for the polarization degree and orientation from starlight data [16]. This simple procedure allows in principle to propagate the foreground residual power

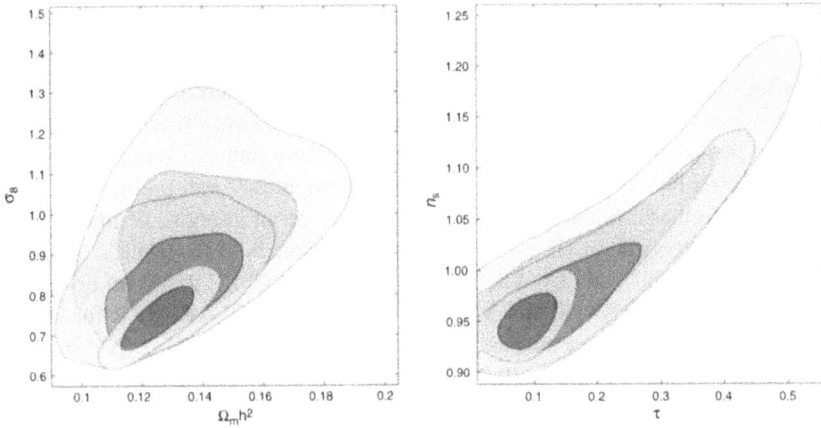

Fig. 6. Parameters constraints improvements using WMAP three year data (including polarization), as compared to WMAP first year data. The left panel shows the marginalized posterior distribution in the $\sigma_8 - \Omega_m h^2$ plane. The right panel shows the marginalized distribution in the $n_S - \tau$ plane. Grey contours indicate the confidence regions for WMAP first year release. Orange regions are for the *WMAPext* data set (see [48]). Blue regions are for the WMAP three-year release, including the polarization measurements. As can be seen from both panels, the determination of $\tau$ from polarization helps constraining other cosmological parameters. Figure from [47].

in the likelihood estimation (see for instance [22] for the temperature case), however, it is not clear from [37] that this procedure was followed when analysing polarized data where foreground contamination, as suggested by Figure 5 (left panel), is a much bigger issue than in the temperature case.

The assessment of the residual contamination level of CMB polarization at large angular scales has very important implications for the cosmological interpretation of the WMAP3 data, as the measurement of $\tau$ (more precisely, breaking the degeneracy between $\tau$, $n_S$ and $A_S$) is almost entirely driven by the amplitude of the first "bump" in the EE power spectrum at $\ell < 10$.

In their cosmological interpretation of their data set, the WMAP team use the measurement of $\tau$, together with the other cosmological parameters, to put important constraints on cosmological and inflationary parameters [47], see figure 6. It is interesting to note that most of these new constraints are indeed related to the measurement of the polarized power spectra which are almost equivalent to a prior distribution on $\tau$ [32]. Given the importance for cosmology of large-scale CMB polarization measurements, and the inherent difficulty of these measurements due to instrumental systematic errors and Galactic foreground contamination, it is fair to say that a large effort from the CMB community is still needed to put these results on a firmer ground.

## 6. Conclusion

The past years have seen considerable progress in the measurements of the CMB anisotropies, both in temperature and polarization. However, if temperature anisotropy measurements have already reached a mature state, with important cosmological implications, the polarization measurements are still preliminary, and need to be further refined. Fortunately, a large set of on-going or planned CMB polarization experiments should remedy this situation, both from space (Planck, http://planck.esa.int), and from the ground (see e.g. BICEP, QUAD for on-going experiments, and EBEX, SPIDER, CLOVER, QUIET, POLARBEAR for future experiments). The Planck mission will be the first spatial CMB experiment to include bolometric detectors, allowing in principle a combined gain in sensitivity and angular resolution of $\sim$40 compared to WMAP; these specifications should be sufficient for a precise mapping of E-mode polarization, but are marginal for B-mode detection, given the current upper-limits on the primordial gravitational wave background.

As the quantum photon noise is already contributing a large fraction of the current detectors noise budget, the only solution to reach the increased sensitivities needed to achieve the observational goals (precise mapping of scalar E-mode polarization, and possible detection of tensor B-mode polarization) lies in the multiplication of the number of detectors by orders of magnitude. This will be possible in the future with the construction of closely packed focal plane arrays of detectors; these new instrumental technologies will soon be tested on a number of ground-based and balloon-borne experiments.

## References

[1]  D. Barkats, C. Bischoff, P. Farese, *et al.*, "First Measurements of the Polarization of the Cosmic Microwave Background Radiation at Small Angular Scales from CAPMAP," Astrophys. J. Letter **619**, L127 (2005).
[2]  C.L. Bennett, R.S. Hill, G. Hinshaw, *et al.*, "First-Year Wilkinson Microwave Anisotropy Probe (WMAP) Observations: Foreground Emission," Astrophys. J. Suppl. Ser. **148**, 97 (2003).
[3]  A. Benoit, P. Ade, A. Amblard, *et al.*, "The cosmic microwave background anisotropy power spectrum measured by Archeops," Astr. & Astrophys. *in press* **399**, L19 (2003).
[4]  A. Benoit, P. Ade, A. Amblard, *et al.*, "Cosmological constraints from Archeops," Astr. & Astrophys. *in press* **399**, L25 (2003).
[5]  P. Bielewicz, K.M. Gorski, and A.J. Banday, "Low-order multipole maps of cosmic microwave background anisotropy derived from WMAP," Mon. Not. R. Astr. Soc. **355**, 1283 (2004).
[6]  M. Bucher, K. Moodley, and N. Turok, "Constraining Isocurvature Perturbations with Cosmic Microwave Background Polarization," *Phys. Rev. Lett.* **87**, 191301 (2001).
[7]  K.S. Dawson, W.L. Holzapfel, J.E. Carlstrom, *et al.*, "Measurement of Arcminute-Scale Cosmic Microwave Background Anisotropy with the Berkeley-Illinois-Maryland Association Array," Astrophys. J. **581**, 86 (2002).

[8] P. de Bernardis, P.A.R. Ade, J.J. Bock, *et al.*, "Multiple Peaks in the Angular Power Spectrum of the Cosmic Microwave Background: Significance and Consequences for Cosmology," Astrophys. J. **564**, 559 (2002).

[9] P. de Bernardis, P.A.R. Ade, J.J. Bock, *et al.*, "A flat Universe from high-resolution maps of the cosmic microwave background radiation," Nature **404**, 955 (2000).

[10] A. de Oliveira-Costa, M. Tegmark, C. O'dell, *et al.*, "Large-scale polarization of the microwave background and foreground," Phys. Rev. D **68**, 083003 (2003).

[11] A. de Oliveira-Costa, M. Tegmark, M. Zaldarriaga, *et al.*, "First attempt at measuring the CMB cross-polarization," Phys. Rev. D **67**, 023003 (2003).

[12] C. Dickinson, R.A. Battye, P. Carreira, *et al.*, "High-sensitivity measurements of the cosmic microwave background power spectrum with the extended Very Small Array," Mon. Not. R. Astr. Soc. **353**, 732 (2004).

[13] G. Efstathiou, "A maximum likelihood analysis of the low cosmic microwave background multipoles from the Wilkinson Microwave Anisotropy Probe," Mon. Not. R. Astr. Soc. **348**, 885 (2004).

[14] G. Efstathiou and J.R. Bond, "Cosmic confusion: degeneracies among cosmological parameters derived from measurements of microwave background anisotropies," Mon. Not. R. Astr. Soc. **304**, 75 (1999).

[15] D.P. Finkbeiner, M. Davis, and D.J. Schlegel, Astrophys. J. **524**, 867 (1999).

[16] P. Fosalba, A. Lazarian, S. Prunet, and J.A. Tauber, "Statistical Properties of Galactic Starlight Polarization," Astrophys. J. **564** 762 (2002).

[17] W.L. Freedman, B.F. Madore, B.K. Gibson, *et al.*, "Final Results from the Hubble Space Telescope Key Project to Measure the Hubble Constant," Astrophys. J. **553** 47 (2001).

[18] N.W. Halverson, E.M. Leitch, C. Pryke, *et al.*, "Degree Angular Scale Interferometer First Results: A Measurement of the Cosmic Microwave Background Angular Power Spectrum," Astrophys. J. **568**, 38 (2002).

[19] S. Hanany, P. Ade, A. Balbi, *et al.*, "MAXIMA-1: A Measurement of the Cosmic Microwave Background Anisotropy on Angular Scales of $10' - 5$deg," Astrophys. J. Letter **545**, L5 (2000).

[20] G. Hinshaw, C. Barnes, C.L. Bennett, *et al.*, "First-Year Wilkinson Microwave Anisotropy Probe (WMAP) Observations: Data Processing Methods and Systematic Error Limits," Astrophys. J. Suppl. Ser. **148**, 63 (2003).

[21] G. Hinshaw, M.R. Nolta, C.L. Bennett, *et al.*, astro-ph/0603451.

[22] G. Hinshaw, D.N. Spergel, L. Verde, *et al.*, Astrophys. J. Suppl. Ser. **148**, 135 (2003).

[23] E. Hivon, K.M. Gorski, C.B. Netterfield, *et al.*, "MASTER of the Cosmic Microwave Background Anisotropy Power Spectrum: A Fast Method for Statistical Analysis of Large and Complex Cosmic Microwave Background Data Sets," Astrophys. J. **567**, 2 (2002).

[24] W. Hu and M. White, "CMB anisotropies: Total angular momentum method," Phys. Rev. D **56**, 596 (1997).

[25] M. Kamionkowski, A. Kosowsky, and A. Stebbins, "A Probe of Primordial Gravity Waves and Vorticity," *Phys. Rev. Lett.* **78**, 2058 (1997).

[26] A. Kogut, D.N. Spergel, C. Barnes, *et al.*, "First-Year Wilkinson Microwave Anisotropy Probe (WMAP) Observations: Temperature-Polarization Correlation," Astrophys. J. Suppl. Ser. **148**, 161 (2003).

[27] J.M. Kovac, E.M. Leitch, C. Pryke, *et al.*, "Detection of polarization in the cosmic microwave background using DASI," Nature **420**, 772 (2002).

[28] C.L. Kuo, P.A.R. Ade, J.J. Bock, *et al.*, "High-Resolution Observations of the Cosmic Microwave Background Power Spectrum with ACBAR," Astrophys. J. **600**, 32 (2004).

[29] A.E. Lange, P.A. Ade, J.J. Bock, *et al.*, "Cosmological parameters from the first results of Boomerang," Phys. Rev. D **63** 042001 (2001).

[30] A.T. Lee, P. Ade, A. Balbi, *et al.*, "A High Spatial Resolution Analysis of the MAXIMA-1 Cosmic Microwave Background Anisotropy Data," Astrophys. J. Letter **561**, L1 (2001).

[31] E.M. Leitch, J.M. Kovac, N.W. Halverson, *et al.*, "Degree Angular Scale Interferometer 3 Year Cosmic Microwave Background Polarization Results," Astrophys. J. **624**, 10 (2005).

[32] A. Lewis, "Observational constraints and cosmological parameters," astro-ph/0603753.

[33] A.D. Miller, R. Caldwell, M.J. Devlin, *et al.*, "A Measurement of the Angular Power Spectrum of the Cosmic Microwave Background from L = 100 to 400," Astrophys. J. Letter **524**, L1 (1999).

[34] T.E. Montroy, P.A.R. Ade, J.J. Bock, *et al.*, "A Measurement of the CMB EE Spectrum from the 2003 Flight of BOOMERANG," Astrophys. J. , **647**, 813 (2006).

[35] C.B. Netterfield, P.A.R. Ade, J.J. Bock, *et al.*, "A Measurement by BOOMERANG of Multiple Peaks in the Angular Power Spectrum of the Cosmic Microwave Background," Astrophys. J. **571**, 604 (2002).

[36] S.P. Oh, D.N. Spergel, and G. Hinshaw, "An Efficient Technique to Determine the Power Spectrum from Cosmic Microwave Background Sky Maps," Astrophys. J. **510**, 551 (1999).

[37] L. Page, G. Hinshaw, E. Komatsu, *et al.*, astro-ph/0603450.

[38] T.J. Pearson, B.S. Mason, A.C.S. Readhead, *et al.*, "The Anisotropy of the Microwave Background to l = 3500: Mosaic Observations with the Cosmic Background Imager," Astrophys. J. **591**, 556 (2003).

[39] F. Piacentini, P.A.R. Ade, J.J. Bock, *et al.*, "A Measurement of the Polarization-Temperature Angular Cross-Power Spectrum of the Cosmic Microwave Background from the 2003 Flight of BOOMERANG," Astrophys. J. **647**, 833 (2006).

[40] A.C.S. Readhead, B.S. Mason, C.R. Contaldi, *et al.*, "Extended Mosaic Observations with the Cosmic Background Imager," Astrophys. J. **609**, 498 (2004).

[41] A.C.S. Readhead, S.T. Myers, T.J. Pearson, *et al.*, "Polarization Observations with the Cosmic Background Imager," *Science* **306**, 836 (2004).

[42] A.G. Riess, P.E. Nugent, R.L. Gilliland, *et al.*, "The Farthest Known Supernova: Support for an Accelerating Universe and a Glimpse of the Epoch of Deceleration," Astrophys. J. **560**, 49 (2001).

[43] P.F. Scott, P. Carreira, K. Cleary, *et al.*, "First results from the Very Small Array – III. The cosmic microwave background power spectrum," Mon. Not. R. Astr. Soc. **341**,1076 (2003).

[44] U. Seljak and M. Zaldarriaga, "Signature of Gravity Waves in the Polarization of the Microwave Background," *Phys. Rev. Lett.* **78**, 2054 (1997).

[45] A. Slosar, U. Seljak, and A. Makarov, "Exact likelihood evaluations and foreground marginalization in low resolution WMAP data," Phys. Rev. D **69**, 123003 (2004).

[46] G.F. Smoot, C.L. Bennett, A. Kogut, *et al.*, "Structure in the COBE differential microwave radiometer first-year maps," Astrophys. J. Letter **396**, L1 (1992).

[47] D.N. Spergel, R. Bean, O. Dore', *et al.*, astro-ph/0603449.

[48] D.N. Spergel, L. Verde, H.V. Peiris, *et al.*, Astrophys. J. Suppl. Ser. **148**, 175 (2003).

[49] M. Tegmark, A. de Oliveira-Costa, and A.J. Hamilton, "High resolution foreground cleaned CMB map from WMAP," Phys. Rev. D **68**, 123523 (2003).

[50] M. Tristram, G. Patanchon, J.F. Macías-Pérez, *et al.*, "The CMB temperature power spectrum from an improved analysis of the Archeops data," Astr. & Astrophys. *in press* **436**, 785 (2005).

[51] E.L. Wright, "Scanning and Mapping Strategies for CMB Experiments," astro-ph/9612006.

[52] M. Zaldarriaga, D.N. Spergel, and U. Seljak, "Microwave Background Constraints on Cosmological Parameters," Astrophys. J. **488**, 1 (1997).

Course 11

# STRUCTURE FORMATION
# WITH NUMERICAL SIMULATIONS

## R. Teyssier

*DAPNIA, Service d'Astrophysique, CEA Saclay Batiment 709,*
*91191 Gif–sur–Yvette, France*

*F. Bernardeau, C. Grojean and J. Dalibard, eds.*
*Les Houches, Session LXXXVI, 2006*
*Particle Physcis and Cosmology: The Fabric of Spacetime*
© *2007 Published by Elsevier B.V.*

# Contents

## 1. Introduction

Numerical simulations have been a key player in our recent progresses in physical cosmology. Due to the non–linear nature of gravitational dynamics and gas physics, high–resolution, multi–scale schemes are now routinely used on massively parallel computers, to perform simulations of ever increasing size and complexity. In this lecture, I will present briefly the so–called hierarchical scenario of structure formation, and how its well-posed nature sets an ideal framework for numerical simulations. I will then present the current computational technics used to simulate both the collisionless component, known as dark matter, and the dissipative one, namely the baryons. I will finally present some "computational discoveries" that were made in the last decades on the internal structure of dark matter haloes, focusing on the gas components and how well we understand its internal structure. I will conclude by presenting new results obtained by a very large simulation of galaxy formation on the Mare Nostrum supercomputer in Barcelona, a typical example of "Grand Challenge Project" in computational cosmology.

## 2. The hierarchical scenario

The current cosmological picture, the so–called "Big Bang" theory, is now widely accepted as our best model to explain the increasing number of observational data coming from our distant universe. Within the framework of an isotropic and homogeneous expanding universe, the hierarchical scenario of structure formation can be summarized as the followings: primordial density fluctuations that arise spontaneously from quantum foam are expanded up to cosmic scales through a mecanism yet to be fully understood, called "inflation". These density perturbations start to grow through the effect of the gravitational instability in the expanding universe. Initially, the amplitude of these growing fluctuations is very small, around 1 for $10^5$ when the universe reaches the epoch of recombination, the first observational window for astronomers, seen as a cosmic microwave background. These primordial fluctuations are actually seen on the 3 Kelvin radiation on the sky as temperature fluctuations. Current experiment, such as the WMAP satellite or the future Planck mission, are able to measure with great precision the amplitude and the power spectrum of these fluctuations. This picture of the universe at its infancy, when it was only 380 000 years old, can be considered as our initial conditions for the cosmic evolution. This situation is rather unique in astrophysics: we have a precise mathematical description of our initial conditions, we have a simple set of equation to solve (self-gravitating fluid dynamics), and we can observe the outcome of the cosmic evolution using astronomical observations of the local univere. Our only problem is therefore to compute accurately the predictions of the gravitational instability scenario, and confront our theoretical view to the real universe. The hierarchical scenario is based on the gravitational dynamics of a collisionless fluid, governed by the Vlasov–Poisson equations, and on the fluid dynamics of a gaseous component, governed by the Euler equations. The cosmological plasma is also subject to various inelastic collisions, for which radiative atomic transitions occur. This turns into a "cooling term" in the energy equation, a fundamental ingredient for the collapse of galactic disks. Structure formation can therefore be recasted into a self–gravitating, radiative hydrodynamics, initial value problem. This is however not the case at very small scales, inside the centrifugally supported galactic discs, where star formation occurs. New processes have to be accounted for, with various recipies that will be presented in the section on galaxy formation physics.

## 3. Computational cosmology

Numerical simulations have played a major role in understanding some important aspects of structure formation in the universe. Gravitational dynamics and fluid dynamics are both highly non–linear processes, for which analytical tools are quite difficult to design. The present computing power allows us to solve the model equations on discrete mesh points or particles, with larger and larger configurations. The number of physical processes that are solved by current state–of–the–art cosmological computational codes is also growing at the same rate, with magneto–hydrodynamics (MHD) and radiative transfer (RT) being the most recently introduced players. The collisionless component (dark matter and stars) is modelled using "N body codes": the phase–space is discretized by macro-particles for which individual trajectories are computed in a Lagrangian way. There is basically 2 types of N body schemes, TREE codes and PM codes. TREE codes are computing the $1/r^2$ Newtonian force directly, using an "oct tree" to group together distant particles into a larger one of equivalent mass and position [1, 2]. This turns into a $N \ln N$ algorithm. A force softening parameter $\epsilon$ is however needed, in order to ensure that the system remains non collisonal. The value of $\epsilon$ is usually chosen constant in space and between $1/10$ and $1/100$ of the mean interparticular distance. PM codes stand for the "Particle–Mesh" method [3], for which the density field is computed on a grid, regular or adaptative, using tri–linear interpolation of particle clouds at each mesh point. The Poisson equation is then solved using Fast Fourier Transform or various relaxation methods to determine the gravitational potential, which is finite differenced and interpolated back to get the particle acceleration. Modern PM schemes are usually based on adaptive grids, such as the one obtained with the Adaptive Mesh Refinement scheme [4–6]. In this case, the force softening length is naturally given by the mesh size, and is therefore adapted to the local particle number. In regions with large number of particles, the force softening is small, say smaller than $1/100$ of the mean interparticular distance, while in low density regions where the local particle density is low, the force softening can be as high as the mean particular distance it self. We see that these two numerical schemes, TREE codes and PM-AMR have different approach of the same problem. These differences turn into different systematic effects in the numerical results: PM-AMR tend to have a poor spatial resolution in low mass haloes (less than 100 particles, say), while TREE codes tend to have a poorer resolution in the core of massive haloes. These two types of codes are also coupled to different types of hydrodynamics solvers. TREE codes hydro solver are based on the Smooth Particles Hydrodynamics technics, while PM-AMR preferently use shock–capturing, Godunov–type solver. Here again, each approach has its own advantages and drawbacks. SPH is striclty Galilean invariant, so that results are independant on the center–of–mass velocity of a galaxy, while Godunov schemes tend to have more numerical diffusion in fast moving objects. On the other hand, Godunov schemes can deal with flow discontinuities such as shocks or contact surfaces, representing them as sharp boundaries, while SPH tend to smear them out, producing artificial mixing or surface tension [7]. The rule is nowadays to compare precisely the results obtained by each solver for the same set of simulations, in order to assess which conclusions are robusts and which code is the best for a given application. Most groups in computational cosmology spend a lot of effort in mastering both types of codes and compare them. The new frontier in computational cosmology is the inclusion of new physical processes, such as MHD and RT. Several groups are designing such algorithms, both for TREE-SPH and PM-AMR solvers, in order to improve the degree of realism in the predictions of the hierarchical scenario. RT, featuring an inhomogeneous radiation background, is likely to play an important role in the cooling and heating of gas into proto–galaxies, during and after reionization, while magnetic fields in galaxy, growing through the effect of $\alpha$–type dynamos, is likely to control the efficiency of star formation in the interstellar medium or influence the fate of AGN–driven bubbles in galaxy clusters. Such processes are currently being included in most cosmological codes.

## 4. Internal structure of dark matter haloes

One of the most famous discoveries made in recent years in numerical simulations is the internal structure of dark matter haloes. While the dark matter density
profile is now reasonabily understood, using the Navarro, Frenk & White analytical fit [8], the gas distribution has been matter of a longer debate. Let us give
first the analytical form of the NFW profile,

$$\rho(r) = \frac{\rho_s}{r(1 + r/r_s)^2} \tag{4.1}$$

where $r_s$ is the scaling radius and $\rho_s$ is the scaling density. These 2 parameters
are equivalent to $M_{200} \propto \rho_s r_s^3$, the Virial mass of the halo and $c = r_{200}/r_s$,
the concentration parameter. Although the exact shape of the very inner part of
the density profile is still discussed, the overall agreement with simulation results is usually better than 20%. Determining the gas density profile is a much
more difficult task. First of all, the gas physics is more complex, with cooling
and star formation taking place in the very central part of the dark matter halo,
altering drastically the gas distribution. A first important step would be therefore to restrict ourselves to pure adiabatic gas dynamics, with $pdV$ work and
shock heating. This regime is also likely to be valid for large X-ray clusters (at
least those without a cooling flow in their center). Even in this restrictive case,
our understanding of the gas distribution is still at its infancy. Numerical simulations of a single large halo have been performed with AMR and SPH in the
mid-90's within the so–called "Santa Barbara Cluster Comparison Project" [9].
Large discrepancy of the inner gas density profiles were found. It was later on understood that some older SPH schemes had energy conservation issues. Modern
SPH scheme were designed in order to overcome this problem. This results into
a better agreement among the codes [10]. We still need however some kind of
analytical tool to explain and validate these slowly converging numerical results.
We must rely on physical approximations to render the problem tractable. The
idea is to use the hydrostatic equilibrium equation, assuming that the gas is almost at rest within the dark matter gravitational potential given by Equation 4.1.
This approach was pionneered by Suto and collaborators [11], then extended by
Komatsu and Seljak [12] et finally validated by Ascasibar and collaborators [10]
using numerical simulations. In order to solve the hydrostatic equation, one needs
an additional constraint. The idea is to assume that the gas follows a polytropic
equation of state, namely that $P \propto \rho^\Gamma$ where $\Gamma$ is the polytropic index. Note
that this index is different from the adiabatic index $\gamma$ of the underlying perfect
gas, usually chosen to be equal to $5/3$. A polytrope is a fluid for which no energy
equation is required. The pressure is imposed by some kind of external heating

and cooling phenomena. This assumption translates into

$$T(r) = T_0 \left( \frac{\rho(r)}{\rho_0} \right)^{\Gamma - 1} \tag{4.2}$$

Posing $\tilde{\rho} = \rho/\rho_s$, the hydrostatic equation becomes

$$\frac{\partial \tilde{\rho}^{\Gamma-1}}{\partial x} = -\alpha \frac{\left( \ln(1+x) - \frac{x}{1+x} \right)}{x^2}, \quad \text{where} \quad \alpha = \frac{4\pi G \rho_s r_s^2 (\Gamma - 1) \mu m_p}{\Gamma k_B T_0} \tag{4.3}$$

We get the temperature profile using the constraint $\rho \to 0$ when $r \to \infty$, so that

$$T(r) = T_0 \frac{\ln(1+x)}{x}, \quad \text{with} \quad \frac{k_B T_0}{\mu m_p} = 4\pi G \rho_s r_s^2 \frac{\Gamma - 1}{\Gamma} \tag{4.4}$$

It is worth mentionning that the shape of the temperature profile is independant of $\Gamma$, while the shape density profile is not. We still need to determine the value of $\rho_0$ and $\Gamma$. The first approach, proposed by Ascasibar *et al.* [10], consists in considering $\Gamma$ as a mere "fitting parameter", as it is already the case for the concentration parameter $c$. For each halo, the density–temperature relation has to be determine and fitted to a power law. Finally, one obtains $\rho_0$ by imposing the mass of baryons to be equal to a fixed fraction of the total mass $M_g = \Omega_b/\Omega_m M_{200}$.

Another approach, more predictive, was proposed by Komatsu and Seljak [12]: the idea is to compute the value of $\Gamma$ that corresponds to a gas density profile in the outer regions similar to the NFW dark matter density profile. The physical explanation is that in the outskirts of relaxed haloes, gas and dark matter dynamics have to be similar, since only gravity matters. The procedure followed by Komatsu and Seljak [12] is slightly complicated, but at the end of the day, one gets a predicted value for $\Gamma$ as a function of the halo mass. This predicted value is in good agreement with the value fitted on simulations by Ascasibar *et al.* [10]. I propose here a third approach, similar to the one proposed by Komatsu and Seljak but easier to implement in practice. The idea is to define a typical radius, noted $r_{eq}$, beyond which the hydrostatic equilibrium is not valid anymore. It is well known from numerical simulations that this assumption is valid only in regions of high enough density contrast, say $\Delta > \Delta_{eq} = 1000\bar{\rho}$. Outside this central region, the density profile is following the NFW analytical formula, while inside it, the polytropic hydrostatic solution is valid. The value of $\Gamma$ is determined by imposing that these two regimes are smoothly connected at the critial radius. More precisely we impose that the derivative of the ratio $\rho_{gaz}/\rho_{cdm}$

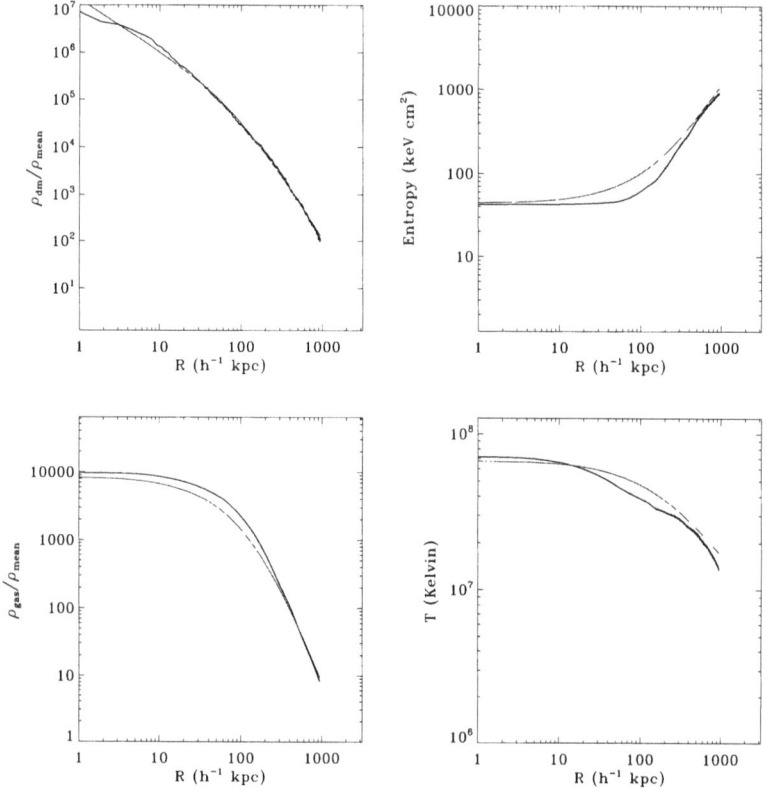

Fig. 1. Comparison of various profiles measured in numerical simulations (black) and computed according to analytical models (red) for the dark matter density (upper-left), gas density (lower-left) gas entropy (uper-right) and gas temperature (lower-right). See text for details

vanishes exactly at $x = x_{eq}$. We obtain after some tedious calculations

$$\Gamma = 1 + \frac{(1 + x_{eq}) \ln(1 + x_{eq}) - x_{eq}}{(1 + 3x_{eq}) \ln(1 + x_{eq})}, \quad \text{where} \quad x_{eq} \simeq \sqrt{5}\, c \tag{4.5}$$

This model is very predictive, as the one proposed by Komatsu and Seljak, since no additional parameter was introduced to reproduce the gas distribution. As shown in Figure 1, the agreement with adiabatic gas and N body simulation of individual clusters is very good. This precise knowledge of the internal structure of both components, gas and dark matter, will allow us to apply this analytical approach to compute observational predictions, such as Sunyaev–Zeldovich effect on the cosmic microwave background (see Refregier and Teyssier [13]).

## 5. The halo model

Numerical simulations have been used to determine a fundamental statistical tool in cosmology, the mass function (MF) of dark matter haloes. The MF gives the number density of haloes in a given mass range and at a given redshift. This function was determined for the first time in the late 80's using N body simulations by Efstathiou et al. [14]. It was shown at that time that the Press & Schechter [15] toy model, derived in the mid-70's, give a suprisingly good fit to the numerical data. More advanced models of the mass function were developped later on, with emphasis on both the statistical aspects [16, 17] and the dynamical ones [18, 19]. These new models provide now a very good fit (better that 20%) to the halo mass function measured in N body simulation [20]. The number of halos with mass M per unit comoving volume et per unit mass at redshift $z$ is given by

$$\frac{dn}{dM} = \frac{\overline{\rho}}{M}\frac{d\nu}{dM}f(\nu), \quad \text{where} \quad \nu(M,z) \equiv \frac{\delta_c}{\sigma(M,z)} \tag{5.1}$$

and $\overline{\rho}$ is the average matter density in the universe and $\sigma(M,z)$ is the variance of the perturbed density field smoothed on scale $R \propto M^{1/3}$ by a suitable low-pass filter. The dynamical parameter $\delta_c$ equal to 1.68 with a weak cosmological dependance we neglect here. The original Press & Schechter mass function, though less accurate than the one used in the recent literature, is obtained using

$$f(\nu) = \sqrt{\frac{2}{\pi}}e^{-\nu^2/2}, \quad \text{which gives} \quad \int_0^\infty d\nu f(\nu) = 1. \tag{5.2}$$

The last integral is important, since it imposes that all the matter in the universe is contained in dark matter halos. Each halo, labelled by its total mass $M_{200}$, is then assigned a density profile following the NFW model (see Section 4). The concentration parameter $c \equiv r_{200}/r_s$ vary strongly between haloes, even within a given mass range. We take here for each halo of mass $M$ the *average concentration parameter*, as measured in N body simulation (see [21]),

$$c(M,z) \simeq g(z)\left[\frac{M}{M_*(z)}\right]^{-1/4} \quad \text{with} \quad g(z) \simeq 10.3(1+z)^{-1/3} \tag{5.3}$$

$M_*(z)$ is the *non linear mass scale* and corresponds to $\sigma(M_*,z) = 1.686$. Based on the mass function and on the average density profiles of dark matter haloes, it is possible to compute the 2-point statistics of the cosmological density fields, using the so-called *halo model*. Following the ideas originally exposed in [22, 23], the dark matter density field is represented as the sum of individual NFW density profiles. In this framework, the power spectrum can be written as the sum of two terms

$$P(k, z) = P_1(k, z) + P_2(k, z), \tag{5.4}$$

The first contribution, called "1-halo term" or "Poisson term" corresponds to the density profile auto-correlation (inside each halo boundaries), writes

$$P_1(k) = \int \frac{dn}{dM} \left( \frac{\tilde{\rho}}{\overline{\rho}} \right)^2 dM, \quad \text{and} \quad \tilde{\rho}(k, M) = 4\pi \int r^2 \rho(r, M) \frac{\sin kr}{kr} dr \tag{5.5}$$

and the second contribution, called "2-halo term" or "clustering term", comes from the correlation of the halo density profile with the large scale density field, and writes,

$$P_2(k) = \left[ \int_0^\infty dM \frac{dn}{dM} b(M) \frac{\tilde{\rho}(k, M)}{\overline{\rho}} \right]^2 P_{\text{lin}}(k) \tag{5.6}$$

and the "bias function" $b(M, z) = 1 + (\nu^2 - 1)/\delta_c$ is given the Mo & White formalism [24]. This analytical approach provides a very good fit to N body simulations (see [13] for a quantitative comparison). The exact same methodology can be applied for the gas distribution, using the density, temperature or pressure profiles derived in the previous section. The pressure is of primary importance in cosmology, since it is responsible for the the distorsion of the Cosmic Microwave Background by foreground hot ionized gas concentrations, e.g. the Sunayev–Zeldovich effect. We will not perform this calculation here, but all the necessary ingredient have been discussed. Here again, we can see in Figure 2 that the agreement between simulation results and the halo model predictions is quite reasonnable.

The halo model turns out to be one of the most succesful approach to describe the gravitational clustering in the non–linear regime. It relies heavily on numerical simulations, since most of the model parameters are fitted to them, but gives nevertheless a good understanding of the main physical processes at the origin of the hierarchical model.

## 6. Galaxy formation physics

One of the most exciting field of research in cosmology is to understand galaxy formation within the hierarchical scenario of structure formation. We have already stressed the importance of numerical simulations to describe the self-gravitating dynamics of both dark matter and gas. The main difference between the

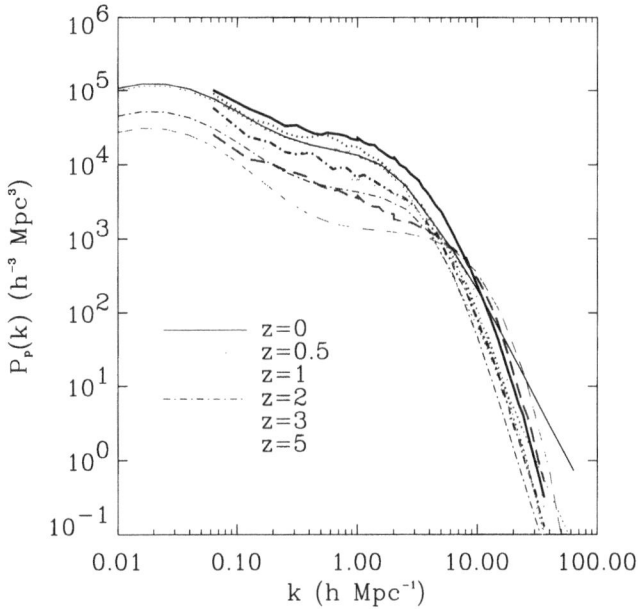

Fig. 2. Power spectrum of the gas pressure measured at different redshift (from 0 to 5) in a N body and hydrodynamical simulation (thick lines) and compared to the halo model analytical predictions (thin lines).

2 components comes from the dissipative processes at work during microscopic collisions in the cosmological plasma. These collisions are inelastic, in the sense that energy and momentum are lost, in the form of photons escaping the optically thin region from which they are emitted. These radiative processes are often referred to as "gas cooling". This is a key ingredient in galaxy formation: in the last section, we have described a dark matter halo, with gas in thermal equilibrium within the potential well. Because of radiative processes, this equilibrium is in fact no longer valid, as soon as the cooling time is shorter than the dynamical time, or the sound crossing time in the halo. Since radiative processes, such as excitation cooling or recombination cooling, are dominant for a plasma temperature between $10^4$ to $10^6$ K, corresponding to typical atomic levels energies, the corresponding halo mass, according to Equation 4.4, will sit around $10^9$ to $10^{12}$ $M_\odot$, precisely the mass of galactic haloes. Since pressure support is lost, the gas will sink into the center of their host halo and reaches antoher type of dynamical equilibrium: the centrifugal equilibrium. Large discs will form, similar to the observed spiral galaxies. Although we believe that this overall picture is probably correct, the details are very difficult to sort out. Numerical simulations are here even more crucial, in order to understand various complex issues during

disc formation: angular momentum trasnsport, disc stability during or in between mergers, etc.

Additional complexity arises when one wishes to understand the nature of the dense and cold gas populating galactic discs. Various phenomena, like thermal instabilities, turbulence and magnetic dynamos give rise to the formation of a very complex medium with a very large range of scales involved. This medium is well-studied since astrophysics exists and is known as "the interstellar medium". The precise numerical description of the ISM is completely out–of–reach from current cosmological simulations. We therefore rely on a simplified sub–cell approach, to mimic the large scale outcome of the most advanced ISM theories and observations. These models are the followings:

1. the star formation rate is proportional to the local dynamical time, with an unknown efficiency calibrated on local observations (Kennicutt law [25])

2. the ISM follows a polytropic equation of state, incorporating a multiphase model [26, 27], in which thermal energy is injected by randomly exploding supernovae

3. the ISM needs a permanent turbulent energy injection, in which kinetic energy is injected by randomly exploding supernovae

The last ingredient is often referred to as "wind model", and is responsible for the apparition of galactic winds or galactic fountains [28]. State–of–the–art galaxy formation physics can be summarized in these 3 ingredients (with a fourth one based on a simplified treatment of metal enrichment and diffusion by supernovae and galactic winds). Figure 3 shows the results of a RAMSES simulation performed by Yohan Dubois in [28] for a low mass, isolated halo, in which gas cools into a rotating discs from which a strong supernovae–driven wind escape. This interesting simulations proves that it is very difficult for such a wind to escape from the galaxy, mainly beacause of the ram pressure of infalling cosmological gas that confines the wind energy in the vicinity of the disc. Note that these simulations are highly idealized, since the proper, three-dimensional and filamentary environments of galactic halos is not described.

## 7. The Mare Nostrum simulation

Due to the ever increasing size of present-day supercomputers, it has now be-come realistic to design large–scale, 3D simulations with the galaxy formation physics we have just described implemented. These ambitious projects requires the mobilisation of astrophysicicts, but also computer scientists of various exper-tises. They are often referred to as "Grand Challenge Projects". Such a project

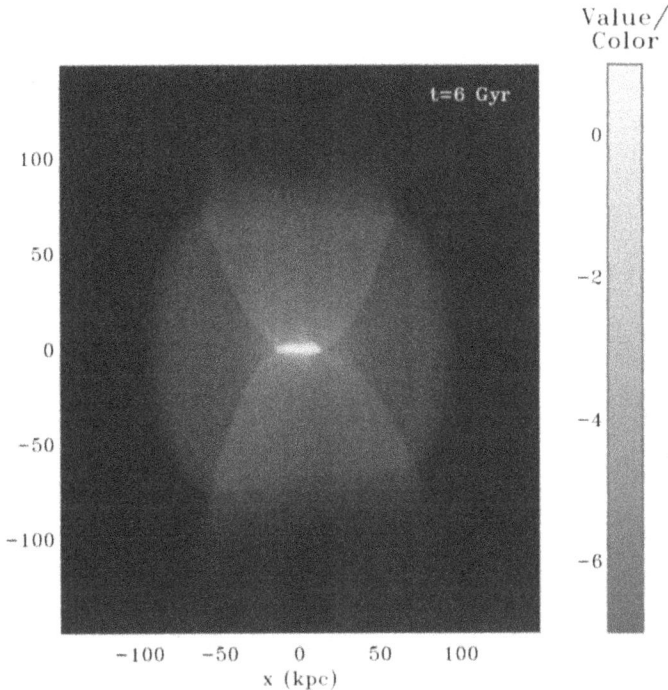

Fig. 3. Density map showing a cut through a simulated galaxy, whose dense rotating and star–forming discs blows a well–defined, supernovae–driven, metal–rich galactic winds [28].

has been prepared and selected at the Barcelona Supercomputing Center, and is one of the most exciting and important project in computational cosmology in the last decade. The goal is to perform the largest cosmological simulations so far, with 2 different, state-of-the-art numerical methods (SPH and AMR) and a wealth of physical ingredients to study galaxy formation with unprecedented accuracy. This project is a collaboration between the Horizon team, a French Science Foundation funded project, and a European team led by Gustavo Yepes from the University of Madrid. Each team has access to a different code (RAM-SES versus GADGET) based on different numerical techniques (AMR versus SPH). As we have discussed in the first section, each code has its own advantage and drawbacks. Since both team are using the same set of initial conditions, provided by our global coordinator Gustavo Yepes, we are in the unique position of being able to directly compare the outcome of both simulations.

After the first run, we were able to visualize the data, discovering an impressive level of details. This has reminded us of the first very large N body simu-

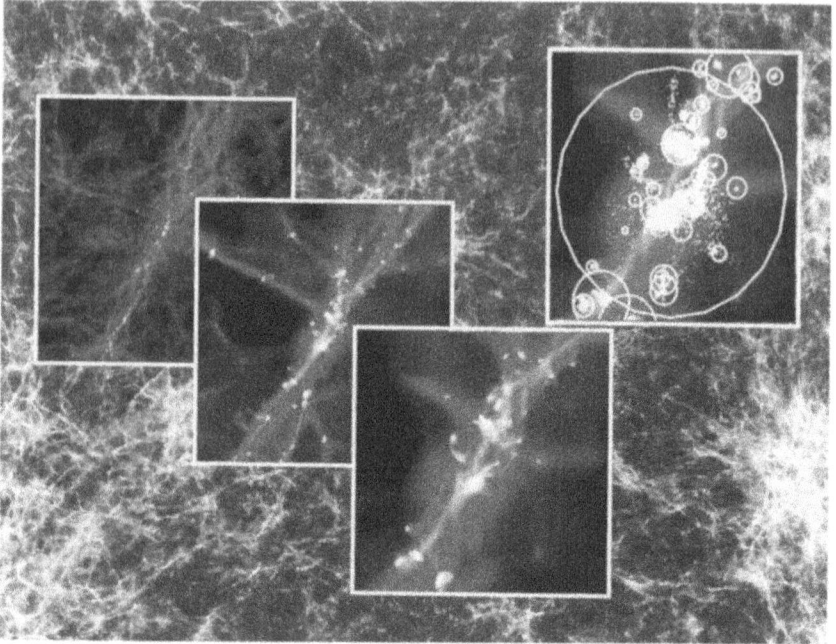

Fig. 4. Density map of the Mare Nostrum simulation performed with the RAMSES code. A sequence of zoom into a filament shows the level of details one can obtained inside a single dark matter halo, identified as the large yellow circle in the upper right insert. The other circles corresponds to different galaxies hosted by the same parent halo.

lations, such as the Hubble Volume Simulation performed in 2000 by the Virgo consortium, except that we have now a precise description of gas dynamics, star formation and other complex physical processes. The large-scale density distribution (shown in Figure 4), is typical of self-gravitating Gaussian random fields, with a filamentary structure often referred to as the "Cosmic Web". It is worth mentioning that these filaments are in quasi pressure equilibrium, so that, in this respect, the Mare Nostrum simulation can be considered as "converged". These filaments constitute the "Lyman alpha" forest that appears in the distant quasars' spectral energy distribution. The current epoch of the simulation (redshift 4) is however slightly too early to be able to compare directly with observed quasars' spectra, so we will need to wait for further snapshots to do so.

At smaller scales, it is possible to discover in our simulation one very important aspect in galaxy formation, namely the accretion flows around forming galaxies. Figure 4 shows a zoom on one massive halo where several galaxies are visible as dense disc-like features in the upper image. Clearly visible around the halo are elongated filaments, often referred to as "cold stream", which are feeding the central galaxies with fresh gas. This gas will be processed into stars by the sub-cell model for star formation occurring in the galactic discs. In between these filaments exist hot shocks, typical of large haloes such as the one shown in the image. What we are witnessing here is the transition from the "cold mode" to the "hot mode" of accretion, bimodality believed to be at the origin of the "blue" versus "red color" dichotomy in the present-day galaxy population [29]. We have been able to observe such a phenomenon in the Mare Nostrum simulation, because of the large box size, allowing rare events to occur in the simulated volume and the good shock-capturing properties of our AMR scheme (Godunov method) that allows for a accurate treatment of shock heating.

## 8. Conclusion

We have described in details the impact of computational cosmology in our understanding of large scale and galaxy formation. We have stressed the importance of having both high-resolution simulations with as realistic physical ingredient as possible, but we have also presented a powerful analytical approach, namely the halo model. This reasonabily accurate tool allows for a nice description of the main aspects of structure formation. Both numerical and analytical approaches will shed light on the still poorly understood processes leading to the formation of galaxies in the universe, and ultimately to our own galaxy, the Milky Way.

### Acknowledgments

The Horizon Project is supported by the "Agence National pour la Recherche" under project number NT05-4-41478. The galactic wind simulation was performed at CEA supercomputing Center CCRT. The Mare Nostrum simulation was run at Barcelona Supercomputing Centre under project id AECT-2006-3-0011 and AECT-2006-4-0009.

### References

[1] J. Barnes and P. Hut, 1996, *A hierarchical O(N logN) force-calculation algorithm.* Nature, 324, 446.

[2] F.R. Bouchet and L. Hernquist, 1988, *Cosmological simulations using the hierarchical tree method. ApJS*, 68, 521.

[3] R.W. Hockney and J.W. Eastwood, 1981, *Computer Simulation Using Particles*, New York: McGraw-Hill.

[4] G.L. Bryan and M.L. Norman. *Simulating x-ray clusters with adaptive mesh refinement*, 1997, In *ASP Conf. Ser. 123: Computational Astrophysics; 12th Kingston Meeting on Theoretical Astrophysics*, 363.

[5] A.V. Kravtsov, A.A. Klypin, and A.M. Khokhlov, 1997, *Adaptive refinement tree: A new high-resolution N-body code for cosmological simulations. ApJS*, 111, 73.

[6] R. Teyssier, 2002, *Cosmological hydrodynamics with adaptative mesh refinement*, A&A, 385, 337.

[7] O. Agertz et al., 2007, *Fundamental differences between SPH and grid methods*, MNRAS, submitted, astro-ph/0610051.

[8] J.J. Navarro, C.S. Frenk, & S.D.M. White, 1996, *The Structure of Cold Dark Matter Halos.*, ApJ, 462, 563.

[9] C.S. Frenk, et al., 1999, *The Santa Barbara Cluster Comparison Project: A Comparison of Cosmological Hydrodynamics Solutions*, ApJ, 525, 554.

[10] Y. Ascasibar, G. Yepes, V. Müller, S. & Gottlöber, 2003, *The radial structure of galaxy groups and clusters*, MNRAS, 346, 731.

[11] Y. Suto, S. Sasaki, N. & Makino, 1998, *Gas Density and X-Ray Surface Brightness Profiles of Clusters of Galaxies from Dark Matter Halo Potentials: Beyond the Isothermal beta-Model*, ApJ, 509, 544.

[12] E. Komatsu, U. & Seljak, 2001, *Universal gas density and temperature profile*, MNRAS, 327, 1353.

[13] A. Refregier and R. Teyssier, 2002 *Numerical and analytical predictions for the large-scale Sunyaev-Zel'dovich effect. Phys. Rev. D*, 66(4), 043002.

[14] Efstathiou, G., Frenk, C.S., White, S.D.M. & Davis, M., 1988, MNRAS, 235, 715.

[15] Press, W.H. & Schechter, P., 1974, ApJ, 188, 425.

[16] J.R. Bond, S. Cole, G. Efstathiou, and N. Kaiser, 1991 *Excursion set mass functions for hierarchical Gaussian fluctuations*, ApJ, 379, 440.

[17] Lacey, C. and Cole, S. 1994, MNRAS, 271, 676.

[18] E. Audit, R. Teyssier, and J.-M. Alimi, 1997, *Non-linear dynamics and mass function of cosmic structures. I. Analytical results.*, A&A, 325, 439.

[19] R.K. Sheth & G. Tormen, 1991, /mnras 308, 119.

[20] M. White, 2002, *The Mass Function*, ApJ, 143, 241.

[21] J.S. Bullock T.S. Kolatt, Y. Sigad, R.S. Somerville, A.V. Kravtsov, A.A. Klypin, J.R. Primack and A. Dekel. 2001, *Profiles of dark haloes: evolution, scatter and environment.* MNRAS, 321, 559.

[22] U. Seljak, Mon. Not. R. Astron. Soc. 318, 203 (2000).

[23] C.P. Ma and J.N. Fry, 2000, ApJ, 543, 503.

[24] H.J. Mo & S.D.M. White, 1996, Mon. Not. R. Astron. Soc. 282, 347.

[25] R.C. Kennicutt, 1998 *The Global Schmidt Law in Star-forming Galaxies.* ApJ, 498, 541.

[26] G. Yepes, R. Kates, A. Khokhlov, and A. Klypin, 1997 *Hydrodynamical simulations of galaxy formation: effects of supernova feedback.* MNRAS, 284, 235.

[27] V. Springel and L. Hernquist, 2003, *Cosmological smoothed particle hydrodynamics simulations: a hybrid multiphase model for star formation.* MNRAS, 339, 289.

[28] Y. Dubois and R. Teyssier, 2006, *Metal enrichment in galactic winds.* MNRAS, CRAL-Conference Series I "Chemodynamics: from first stars to local galaxies", Lyon 10–14 July 2006, France, Eds. Emsellem, Wozniak, Massacrier, Gonzalez, Devriendt, Champavert, EAS Publications Series.

[29] A. Dekel and Y. Birnboim, 2006, *Galaxy bimodality due to cold flows and shock heating.* MNRAS, 368, 2.

Course 12

# GIVING MASS TO THE GRAVITON

## P. Tinyakov

*Service de Physique Théorique, Université Libre de Bruxelles, CP225,
blv. du Triomphe, B-1050 Bruxelles, Belgium*

*F. Bernardeau, C. Grojean and J. Dalibard, eds.*
*Les Houches, Session LXXXVI, 2006*
*Particle Physcis and Cosmology: The Fabric of Spacetime*
© *2007 Published by Elsevier B.V.*

# Contents

# 1. Introduction

The question of whether the gravitational interaction is described by the Einstein theory of relativity at all scales is of both theoretical and practical interest. On theory side, the attempts to construct an alternative model, successful or not, serve to better understanding of the fundamental principles lying behind the theory of gravity. The requirement of general covariance fixes the form of the gravitational Lagrangian almost uniquely. There exist a very few modifications of gravity which do not involve higher-derivative terms, the most known being scalar-tensor models of the Brans-Dicke type [1]. A natural question then is whether the general covariance can be broken, say, spontaneously, in a manner similar to the Higgs mechanism in gauge theories. If that were the case one would expect, by analogy with the gauge theory, that the graviton gets a non-zero mass. More generally, the question is whether at all one can construct a consistent theory of gravity where the graviton has a non-zero mass. Whatever is the answer, it will certainly contribute to better understanding of gravity.

On phenomenological side, the conventional theory of gravity is completely successful at scales of order and below the solar system size up to scales of order a fraction of a millimeter. At larger scales there is a hint of a problem: one needs to introduce the (otherwise undetected) dark matter in order to explain the rotation curves of the galaxies and galaxy clusters. At cosmological scales yet another form of matter — the one behaving like the cosmological constant — is also needed [2]. With these two additions the Einstein's theory apparently works quite well at all scales. However, it is disturbing that the new components are only needed to correct the gravitational interaction at very large scales. Moreover, at those scales the new components must play a dominant role in order to fit the observations.

Before accepting the existence of the new forms of matter it is natural to wonder whether the gravitational interaction itself can be modified at large distances so as to explain the existing observations without the need of the dark matter and the dark energy. Whether likely or not, this is a logical possibility.

Perhaps one of the first alternatives to the dark matter was the model known as MOND (modified Newtonian dynamics) [3,4]. In this model one postulates the existence of a critical acceleration at which the $1/r^2$ fall off of the Newtonian force changes to a slower dependence. There is an ongoing discussion in the literature whether this model is viable phenomenologically (see, e.g., Ref. [5] and references therein) and whether it can be generalized to a fully relativistic theory [6].

The idea to modify the gravitational interaction at large distances lies behind several recent attempts to find alternative models of gravity. One of the first such attempts was performed in Ref. [7] in the context of extra dimensions. The model developed there involved branes with a negative tension and was later shown to possess ghosts [8, 9]. Another attempt employing extra dimensions is the DGP model [10]. This model possesses interesting cosmological solutions [11, 12]. However it is still debated whether it is consistent theoretically [13–15]. Yet another approach is based on the actions which are singular in the low curvature limit, the so-called $f(R)$ gravities (see, e.g., [16]). These models are widely discussed now in the cosmological context [17–19], however their ability to pass the solar system tests is questioned [20, 21] (see, however, Ref. [22]). One should mention also the bi-gravity models which involve two metric tensors [23–25]. None of these models possesses massive gravitons.

In these lectures I will concentrate on the question of whether the graviton can be given a mass. This question has a long history dating back to late 30th when Fierz and Pauli [26] have constructed a Lorentz-invariant mass term for the spin 2 field and shown that the resulting theory has no ghosts. It was noted much later [27, 28] that the model constructed by Fierz and Pauli has a problem which makes it phenomenologically unacceptable: it predicts the wrong value of the light bending by massive bodies, and moreover, this (wrong) prediction persists in the limit of vanishing graviton mass. This phenomenon of the absence of a smooth zero-mass limit is known as the van Dam-Veltman-Zakharov (vDVZ) discontinuity. The bending of light has been experimentally measured and shown to be in agreement with the Einstein's general relativity, thus ruling out the Fiertz-Pauli model. The conclusion was that the graviton mass has to be strictly zero.

The status of the graviton mass seemed clear until it was noticed [29] that the discontinuity argument leading to the contradiction with the experiment has a loophole. The light bending in the massive case was calculated within the linear approximation, while the model is actually non-linear at the relevant scales. It was shown that the onset of the non-linear regime at non-zero graviton mass happens at much larger distances from the source than one would naively expect. What precisely is the corresponding distance scale may depend on the non-linear terms in the action [30], but it cannot be made short enough. Another manifestation of the same phenomenon is the strong coupling between longitudinal polarizations of massive gravitons which sets in at unacceptably low energies [31].

A new approach to the modification of gravity has been developed recently which involves the spontaneous breaking of the Lorentz symmetry. The first model of this type is the so-called ghost condensate model [32]. Although the graviton is massless in this model, we will show that it can be generalized in such a way that the graviton gets a mass, while the problems mentioned above do not arise. One gets, therefore, a consistent theory which is well-defined below a certain scale and in which the Lorentz symmetry is spontaneously broken, the graviton is massive and obvious pathologies are absent. We will argue below that this theory is a perfect theoretical laboratory for studying modifications of gravity and may even be interesting from the phenomenological point of view.

The outline of these lectures is as follows. We start in Sect. 2 by discussing the generic obstructions to massive gravity. In Sect. 3 we outline the ways to overcome the difficulties and present a class of models where these ideas are realized and the graviton is made massive. We investigate this class of models at the liner level in Sect. 4. Sec. 5 deals with some phenomenological consequences of the graviton mass in a concrete model. In particular, the cosmological solutions are considered. Finally, in Sect. 6 we summarize the results and outline the open questions.

## 2. Theoretical obstructions to massive gravity

Let is discuss in more detail the origin of the problems which arise when one tries to modify the general relativity in such a way that the graviton gets a mass, namely, the appearance of ghosts, the vDVZ discontinuity and the strong coupling at too low energy.

*Instabilities and ghosts.* There may occur several types of instabilities. In the simplest case of a single variable $\phi$ the quadratic action has the form

$$\int d^4x \left( \alpha \dot{\phi}^2 - \beta (\partial_i \phi)^2 - m^2 \phi^2 \right),$$

where $\alpha$, $\beta$ and $m^2$ are some real coefficients which can take any value. We do not assume here the Lorentz invariance which would require $\alpha = \beta$. The equation of motion for the variable $\phi$ in the Fourier space reads

$$\alpha \ddot{\phi} + \beta k_i^2 \phi + m^2 \phi = 0. \tag{2.1}$$

One usually has $\alpha > 0$, $\beta > 0$, $m^2 > 0$ and the solution to eq. (2.1) is oscillatory. This is the "normal" case which corresponds to usual particles.

If $\alpha > 0$ but $\beta k^2 + m^2 < 0$ for some $k$, then the solutions to eq. (2.1) are exponentially decaying or growing, so there is an instability. Two cases should be distinguished. If $\beta > 0$ then for $k > \sqrt{|m^2|/\beta}$ the instability disappears. If this critical value of $k$ is very low, there may be not enough time for he instability to develop. Thus, this kind of instability is not necessarily a pathology. On the contrary, if $\beta < 0$, then the instability persists at an arbitrary large $k$ and is therefore arbitrarily rapid. Instabilities of this type are unacceptable.

Finally, if $\alpha < 0$ then the contribution of the kinetic term $\alpha\dot\phi^2$ into the energy of the system is negative and unbounded from below. This is physically unacceptable unless the field $\phi$ is completely decoupled from the rest of the system, which is an unrealistic situation. Note that if $\alpha < 0$ and $\beta k^2 + m^2 < 0$ at the same time, the solutions to eq. (2.1) oscillate, so the pathology does not show up in the equations of motion. However, both the kinetic and the potential term have negative energy. The field of this type is referred to as ghost.

The requirement of the absence of ghosts and instabilities is routinely used in field theory. It allows, for instance, to fix uniquely the conventional gauge-invariant form of the Lagrangian of the massless vector field, $-1/4F_{\mu\nu}F^{\mu\nu}$. This latter observation may be used to see very easily that there is only one Lorentz-invariant graviton mass term which gives a theory free from instabilities and ghosts. Indeed, consider a quadratic Lagrangian for the metric perturbation $h_{\mu\nu}$ around the flat Minkowski space and add all possible Lorentz-invariant mass terms,

$$\int d^4x \left\{ L^{(2)}(h_{\mu\nu}) + \alpha h_{\mu\nu}^2 + \beta\left(h_\mu^\mu\right)^2 \right\}. \tag{2.2}$$

The first term in eq. (2.2) is just the standard kinetic term which comes from the Einstein action $\int \sqrt{g}R$ and describes massless gravitational waves. Its precise form is irrelevant for our purposes, as will be clear shortly. The second and third terms are the only possible Lorentz-invariant combinations which are quadratic in $h_{\mu\nu}$ and do not contain derivatives.

Let us see that only at $\beta = -\alpha$ the action (2.2) describes a non-pathological theory. To this end consider a *particular* metric perturbation,

$$h_{\mu\nu} = \partial_\mu\xi_\nu + \partial_\nu\xi_\mu. \tag{2.3}$$

If the action (2.2) is non-pathological, it should lead to a non-pathological action for $\xi_\mu$. The latter action is straightforward to calculate. Indeed, the perturbation (2.3) has the form of the gauge transformation, so the first term in eq. (2.2) produces no contribution. Substituting eq. (2.3) into the two last terms of eq. (2.2) we find

$$2\alpha(\partial_\mu\xi_\nu)^2 + (2\alpha + 4\beta)\left(\partial_\mu\xi^\mu\right)^2.$$

This is a general Lorentz-invariant action for the vector field $\xi_\mu$. As is known from field theory this action describes a consistent model only when it is proportional to $(\partial_\mu \xi_\nu - \partial_\nu \xi_\mu)^2$, which leads to the condition $\beta + \alpha = 0$. Setting $\alpha = -\beta = -\mu^2 < 0$ one arrives at the Fiertz-Pauli model of the massive graviton.

*vDVZ discontinuity.* Let us now see that the Fierz-Pauli model predicts the bending of light by massive bodies which is different from GR even in the limit of zero mass. We need to compare the interactions of a given mass with a massive test particle and with the photon in GR and in the Fierz-Pauli model. In the language of quantum field theory these interactions are proportional to the amplitude of the one-graviton exchange which in turn is proportional to the graviton propagator, to be more precise, to its contraction with the energy-momentum tensors of the source and the test particle. The graviton propagator in both massive and massless cases has the form

$$P_{\mu\nu\lambda\rho} \propto \frac{\sum_i e^i_{\mu\nu} e^i_{\lambda\rho}}{p^2 - m^2}, \tag{2.4}$$

where the sum runs over all "polarization tensors" $e^i_{\mu\nu}$. In the massive case there are 5 such tensors. In the rest frame of a graviton they have the following form:

$$\frac{\sqrt{2}}{3} \begin{pmatrix} 0 & 0 & 0 & 0 \\ 0 & 1/2 & 0 & 0 \\ 0 & 0 & 1/2 & 0 \\ 0 & 0 & 0 & -1 \end{pmatrix}, \quad \frac{1}{\sqrt{2}} \begin{pmatrix} 0 & 0 & 0 & 0 \\ 0 & 1 & 0 & 0 \\ 0 & 0 & -1 & 0 \\ 0 & 0 & 0 & 0 \end{pmatrix},$$

$$\frac{1}{\sqrt{2}} \begin{pmatrix} 0 & 0 & 0 & 0 \\ 0 & 0 & 1 & 0 \\ 0 & 1 & 0 & 0 \\ 0 & 0 & 0 & 0 \end{pmatrix}, \quad \frac{1}{\sqrt{2}} \begin{pmatrix} 0 & 0 & 0 & 0 \\ 0 & 0 & 0 & 1 \\ 0 & 0 & 0 & 0 \\ 0 & 1 & 0 & 0 \end{pmatrix}, \tag{2.5}$$

$$\frac{1}{\sqrt{2}} \begin{pmatrix} 0 & 0 & 0 & 0 \\ 0 & 0 & 0 & 0 \\ 0 & 0 & 0 & 1 \\ 0 & 0 & 1 & 0 \end{pmatrix},$$

while in the massless case there are only two polarization tensors which can be written as follows (in the frame where the vector $\vec{p}$ is parallel to the $z$ axis),

$$\frac{1}{\sqrt{2}} \begin{pmatrix} 0 & 0 & 0 & 0 \\ 0 & 1 & 0 & 0 \\ 0 & 0 & -1 & 0 \\ 0 & 0 & 0 & 0 \end{pmatrix}, \quad \frac{1}{\sqrt{2}} \begin{pmatrix} 0 & 0 & 0 & 0 \\ 0 & 0 & 1 & 0 \\ 0 & 1 & 0 & 0 \\ 0 & 0 & 0 & 0 \end{pmatrix}. \tag{2.6}$$

The substitution of these polarization tensors into eq. (2.4) leads to the following graviton propagators,

$$P_{\mu\nu\lambda\rho}^{m\neq0} = \frac{1}{p^2 - m^2}\left\{\frac{1}{2}\eta_{\mu\lambda}\eta_{\nu\rho} + \frac{1}{2}\eta_{\mu\rho}\eta_{\nu\lambda} - \frac{1}{3}\eta_{\mu\nu}\eta_{\lambda\rho}\right.$$
$$\left. +(p\text{-dependent terms})\right\},$$

$$P_{\mu\nu\lambda\rho}^{m=0} = \frac{1}{p^2}\left\{\frac{1}{2}\eta_{\mu\lambda}\eta_{\nu\rho} + \frac{1}{2}\eta_{\mu\rho}\eta_{\nu\lambda} - \frac{1}{2}\eta_{\mu\nu}\eta_{\lambda\rho}\right.$$
$$\left. +(p\text{-dependent terms})\right\}.$$

The terms containing $p$ are of no interest since the propagator is contracted with the conserved energy-momentum tensor and these terms give no contribution.

The crucial difference between the two cases is the coefficient in front of the term $\eta_{\mu\nu}\eta_{\lambda\rho}$ which couples to the trace of the energy-momentum tensor. This difference does not vanish in the zero mass limit. Clearly, it is due to a different number of graviton polarizations in the massive and massless cases, as can be seen by comparing eqs. (2.5) and (2.6).

Measuring the interaction with the test massive particle, one can predict the bending of light in both cases. It turns out that the predictions are different. To see this let us denote the gravitational interaction constant as $G$ and $\tilde{G}$ in the massless and massive cases, respectively. In the two cases the interaction between the non-relativistic masses is proportional to the following combinations,

massless case:     $G\, T_{\mu\nu}\, P_{\mu\nu\lambda\rho}^{m=0}\, T'_{\lambda\rho} = \frac{1}{2}G\, T_{00}T'_{00}\frac{1}{p^2}$,

massive case:     $\tilde{G}\, T_{\mu\nu}\, P_{\mu\nu\lambda\rho}^{m\neq0}\, T'_{\lambda\rho} = \frac{2}{3}\tilde{G}\, T_{00}T'_{00}\frac{1}{p^2 - m^2}$.

This implies in the small mass limit that

$$\tilde{G} = \frac{3}{4}G. \tag{2.7}$$

In the case of the light bending by a non-relativistic mass the third term in the propagator does not contribute because of the vanishing trace of the electromagnetic energy-momentum tensor, $T'^{\mu}_{\mu} = 0$, and the result is the same in both cases,

massless case:     $G\, T_{00}T'_{00}\frac{1}{p^2}$,

massive case:     $\tilde{G}\, T_{00}T'_{00}\frac{1}{p^2 - m^2}$.

In view of eq. (2.7) the light bending predicted in the massive theory in the limit of the vanishing graviton mass is 3/4 of that predicted in general relativity. Thus, there is a discontinuity.

*The low-scale strong coupling.* The calculation of the gravitational potential and of the light bending outlined above were performed in the linearized approximation where the self-interaction of the gravitational field is neglected. In general relativity this is a good approximation at distances which are much larger than the Schwarzschild radius of the mass producing the gravitational field. One may wonder whether this approximation is correct in the massive gravity as well. It was first noted by Vainstein [29] who considered the spherically-symmetric solutions in the Fierz-Pauli model that for these solutions the weak-field approximation actually breaks down much further from the source than the gravitational radius. Vainstein argued that the validity of the linear approximation is controlled by the parameter

$$\epsilon = \frac{R_g}{m_g^4 r^5}, \tag{2.8}$$

where $m_g$ is the graviton mass and $R_g = 2M/M_{Pl}$ is the Schwarzschild radius corresponding to the mass $M$. Eq. (2.8) is remarkable in that the graviton mass enters in the denominator, so that the expansion parameter always becomes large when the mass of the graviton goes to zero.

Let us see what are the numbers. Assuming the solar system planets feel the gravitational field of the Sun in the linear regime implies that the graviton mass should be smaller than the inverse radius of the Pluto orbit,

$$m_g < (40AU)^{-1} \sim 3 \times 10^{-20} \text{ eV}. \tag{2.9}$$

Then the parameter $\epsilon$ at the Mercury orbit $r_M \sim 0.3AU$ equals

$$\epsilon \sim 15,$$

so that the motion of Mercury should be strongly affected by non-linearities. However, even the estimate (2.9) is way too optimistic. From observations of star motion in galaxies the graviton mass should be smaller than at least kpc$^{-1}$. The parameter $\epsilon$ at the Earth orbit around the Sun would then be $\epsilon \sim 10^{25}$, so that the gravitational interaction would be deeply in the non-linear regime. Thus, the problem of discontinuity is replaced by the strong coupling problem.

By itself, the strong coupling does not mean that the theory is inconsistent with observations. Indeed, the arguments of Ref. [29] were reconsidered in Ref. [30] where it was argued, within the DGP model of modified gravity, that particular non-linear effects may make the transition to the zero graviton mass continuous and weaken the experimental limits on the mass of the graviton.

In the language of quantum field theory, the onset of the non-linear regime shows up in the strong coupling at high energies. When graviton is given a mass, its longitudinal polarizations cause the growth of the graviton-graviton cross section with energy, as has been checked by the direct calculations [31], thus leading to the strong coupling. In fact, these two phenomena are interconnected and are manifestations of the same problem [33].

## 3. Constructing the massive gravity models

Our goal now is to construct a model where the graviton is massive and which is compatible with observations despite the problems outlined above. To be mode precise, we will require that the flat Minkowski space is a solution to the equations of motion, that there are no instabilities and ghosts in perturbation theory around the flat space, there is no discontinuity at the point where the graviton mass goes to zero, and that the theory is weakly coupled below some sufficiently high scale $\Lambda$. We will see that all these requirements can be satisfied if one allows the Lorentz invariance to be spontaneously broken.

To see whether this is possible one may consider a generic graviton mass term which preserves rotational invariance but not necessarily the Lorentz invariance. Follwing Ref. [34], this term may be written as

$$\frac{1}{2}M_{\rm Pl}^2\{m_0^2 h_{00}^2 + 2m_1^2 h_{0i}^2 - m_2^2 h_{ij}^2 + m_3^2 h_{ii}^2 - 2m_4^2 h_{00}h_{ii}\}, \tag{3.1}$$

where $m_i$ are 5 mass parameters which are, in general, different. The graviton mass (the mass of the tensor perturbations) is given by $m_2$. Clearly, the Lorentz invariance requires that only two of the mass parameters be independent,

$$m_1^2 = m_2^2 = -\alpha^2, \quad m_3^2 = m_4^2 = -\beta^2, \quad m_0^2 = \alpha^2 + \beta^2,$$

where $\alpha$ and $\beta$ are arbitrary. The Fierz-Pauli model corresponds to the case $\alpha^2 = -\beta^2$. It has been shown in Ref. [34] that the choice $m_0 = 0$ leads to a model free of ghosts, vDVZ discontinuity and low strong coupling scale. Thus, all the above requirements may, in principle, be satisfied at the same time. We will discuss other possible values of the parameters shortly.

The mass term (3.1) can be considered as the quadratic part of a more general action depending on the metric components,

$$S = \int d^4x \sqrt{g}\{M_{\rm Pl}^2 R + \Lambda^4 F(g_{\mu\nu}) + {\rm matter}\}, \tag{3.2}$$

where the function $F(g_{\mu\nu}) = F(g_{00}, g_{0i}, g_{ij})$ is assumed to preserve rotations but not necessarily the Lorentz invariance. The scale $\Lambda$ which will play the role of the cutoff of the model is related to the graviton masses as

$$m_i^2 \sim \frac{\Lambda^4}{M_{\text{Pl}}^2}.$$

We will assume in what follows that there are no other scales in the function $F$. Our purpose now is to investigate the models with the action (3.2).

The analysis is greatly simplified by the so-called Stückelberg's trick which consists in restoring the gauge invariance by introducing auxiliary scalar fields, a sort of inverse Higgs mechanism. To illustrate how this works consider an example of the (Lorentz-invariant) massive electrodynamics with the action

$$S = \int d^4x \left\{ -\frac{1}{4} F_{\mu\nu}^2 + m^2 A_\mu^2 \right\}.$$

There are three degrees of freedom: two transverse polarizations of the photon, and the longitudinal polarization. The gauge invariance is broken by the photon mass term. Let us now add a scalar field $\phi$ in such a way as to restore the gauge invariance,

$$S = \int d^4x \left\{ -\frac{1}{4} F_{\mu\nu}^2 + m^2 (A_\mu - \partial_\mu \phi)^2 \right\}.$$

The new action is explicitly gauge-invariant under the transformations

$$A_\mu \rightarrow A_\mu + \partial_\mu \phi,$$
$$\phi \rightarrow \phi + \alpha,$$

and has the same three degrees of freedom as the original action. Moreover, one may use the gauge transformation to set $\phi = 0$. Then the original action is recovered. On the other hand, one may concentrate on the "Goldstone" part of the action $m^2 (\partial_\mu \phi)^2$ and recover the action for the phase part of the Higgs field in the standard $U(1)$ Higgs mechanism. This is this freedom of the gauge choice which simplifies the analysis.

This trick can be generalized [33, 35] to the case of the action (3.2). The symmetries of general relativity are four coordinate transformations

$$x^\mu \rightarrow x^\mu + \xi^\mu(x).$$

Thus one has to introduce 4 scalar "Goldstone" fields $\phi^0$ and $\phi^i$. We stress that these fields are scalars and thus must transform as scalars under the coordinate transformations. For this reason it is not straightforward to introduce these fields into the action (3.2) in such a way that they restore the general covariance. The job is done by the following combinations [35],

$$X = \frac{1}{\Lambda^4} g^{\mu\nu} \partial_\mu \phi^0 \partial_\mu \phi^0,$$

$$V^i = \frac{1}{\Lambda^4} g^{\mu\nu} \partial_\mu \phi^i \partial_\mu \phi^0,$$

$$Y^{ij} = \frac{1}{\Lambda^4} g^{\mu\nu} \partial_\mu \phi^i \partial_\mu \phi^j. \tag{3.3}$$

The factor $1/\Lambda^4$ is introduced to make these combinations dimensionless. Thus, the general action for the massive gravity becomes

$$S = \int d^4 x \sqrt{g} \left\{ M_{\text{Pl}}^2 R + \Lambda^4 F(X, V^i, Y^{ij}) + L_{\text{matter}} \right\}. \tag{3.4}$$

Several remarks are in order. First, the Goldstone fields only enter the action through the derivatives as they should. Second, they only couple to the metric and not to the matter fields directly, so that they do not introduce extra interactions except the modification of gravity. Finally, as in our toy example, one can choose the gauge (the reference frame) in such a way that the action (3.4) reduces to the action (3.2), as we will now show.

Before that we need to discuss one more important point, the vacuum solutions in the model (3.4). More precisely, we have to determine under which conditions the flat space is the vacuum solution. To this end let us assume there is no ordinary matter. Then the Einstein equations derived from the action (3.4) are

$$R_{\mu\nu} - \frac{1}{2} g_{\mu\nu} R = M_{\text{Pl}}^2 T_{\mu\nu}^G,$$

where $T_{\mu\nu}^G$ is the energy-momentum tensor of the Goldstone fields. Since the left hand side of this equation is zero in the flat space, the right hand side should also be zero. Thus, we have to determine conditions under which the Goldstone energy-momentum tensor vanishes in the Minkowski background.

As usual, the calculation of the energy-momentum tensor is done by the variation of the Goldstone action with respect to the metric. Denote

$$\delta F \equiv F_X \delta X + F_i \delta V^i + F_{ij} \delta Y^{ij},$$

where

$$\delta X = \frac{1}{\Lambda^4} \partial_\mu \phi^0 \partial_\nu \phi^0 \delta g^{\mu\nu},$$

$$\delta V^i = \frac{1}{\Lambda^4} \partial_\mu \phi^0 \partial_\nu \phi^i \delta g^{\mu\nu},$$

$$\delta Y^{ij} = \frac{1}{\Lambda^4} \partial_\mu \phi^i \partial_\nu \phi^j \delta g^{\mu\nu}.$$

One finds then

$$\begin{aligned}
T^G_{\mu\nu} &= 2 F_X \partial_\mu \phi^0 \partial_\nu \phi^0 + F_i \left( \partial_\mu \phi^0 \partial_\nu \phi^i + \partial_\mu \phi^i \partial_\nu \phi^0 \right) \\
&\quad + 2 F_{ij} \partial_\mu \phi^i \partial_\nu \phi^j - g_{\mu\nu} \Lambda^4 F.
\end{aligned}$$

The requirement $T^G_{\mu\nu} = 0$ should be considered as the set of equations for the Goldstone fields. For an arbitrary metric these equations are impossible to satisfy because there are 10 equations (which are in general different) for only 4 unknowns. However, for the flat space the solution is easy to guess. Consider the Goldstone fields of the following form,

$$\phi^0 = a\Lambda^2 t,$$
$$\phi^i = b\Lambda^2 x^i,$$

where $a$ and $b$ are two constants. For this ansatz the equations $T^G_{\mu\nu} = 0$ reduce to the following two *algebraic* equations,

$$\begin{aligned}
2a^2 F_X(a^2, b^2) - F(a^2, b^2) &= 0, \\
2b^2 F_Y(a^2, b^2) - F(a^2, b^2) &= 0,
\end{aligned} \tag{3.5}$$

where we have used the notation $F_{ij} = F_Y \delta_{ij}$. Since these are two equations for two variables, in general they have a solution. Without loss of generality we will assume in what follows that this solution is such that $a = b = 1$. Thus, the vacuum in our model is

$$\phi^0 = \Lambda^2 t,$$
$$\phi^i = \Lambda^2 x^i. \tag{3.6}$$

Note that the fields themselves do not enter the action, so there is nothing wrong with them growing at infinity. However, since the vacuum values of these fields are space-time dependent, they break the Lorentz symmetry. The rotational symmetry remains unbroken if the action preserves the global rotations of the fields $\phi^i$ with respect to the index $i$. Indeed, in this case the space rotations of the vacuum manifold (3.6) can be compensated by the global rotations of the fields, so that of the two rotation groups the diagonal part remains unbroken.

We can now see that the action (3.4) is equivalent to the original action (3.2). Indeed, for an arbitrary metric we can choose the gauge in which the Goldstone fields equal to their vacuum values (3.6). Then we have

$$X = g^{00}, \quad V^i = g^{0i}, \quad Y^{ij} = g^{ij}, \tag{3.7}$$

so that the function $F$ becomes a function of the metric components as in eq. (3.2).

## 4. Linear perturbations

Let us discuss the behavior of the linear perturbations in the flat background. We have to consider both perturbations of the metric $h_{\mu\nu}$ and perturbations of the Goldstone fields $\pi_\mu$,

$$\phi^\mu = \Lambda^2 x^\mu + \pi^\mu,$$
$$g_{\mu\nu} = \eta_{\mu\nu} + h_{\mu\nu}.$$

The purpose is to show that there are neither ghosts nor instabilities present for some values of the mass parameters.

First, let us clarify the relation between the masses entering eq. (3.1) and the function $F$ in the action (3.4). The mass term (3.1) is recovered from eq. (3.4) in the unitary gauge where the perturbation of the Goldstone fields are zero. To calculate the mass parameters one has to expand $\sqrt{g}F$ up to the second order in metric perturbations. The zeroth order term is an irrelevant constant. The linear terms must vanish; this is the condition that our background is a solution to the Einstein equations. If we start with an arbitrary function $F$, the vanishing of the linear terms will be equivalent to the conditions (3.5) which ensure that the energy-momentum tensor of the Goldstone fields is zero. Finally, the quadratic terms should be identified with the mass parameters. The overall mass scale is already clear: assuming the function $F$ does not contain other scales apart from $\Lambda$, the masses are of the order $m_i^2 \sim \Lambda^4/M_{\text{Pl}}^2$. Carrying out the expansion one finds, for instance

$$m_0^2 = \frac{\Lambda^4}{M_{\text{Pl}}^2}\left(\frac{1}{2}F_X + F_{XX}\right),$$
$$m_1^2 = 2\frac{\Lambda^4}{M_{\text{Pl}}^2}(F_Y + F_{VV}),$$
$$\cdots$$

where the subscript on $F$ denotes the derivative with respect to the corresponding variable. An important observation which will be useful in what follows is that if one takes the function $F$ which depends only on the two arguments $X$ and

$$W^{ij} = Y^{ij} - V^i V^j / X, \tag{4.1}$$

then one has $m_1^2 = 0$. As we will discuss shortly, this gives a particularly interesting class of models.

The next question to discuss is the stability and ghosts. The problem comes from the longitudinal polarizations of the graviton, i.e., from the purely Goldstone sector. This part of the quadratic action may be obtained by substituting

$$h_{\mu\nu} = \partial_\mu \pi_\nu + \partial_\nu \pi_\mu \tag{4.2}$$

into the full quadratic action for perturbations. Since in general relativity the perturbation (4.2) is a pure gauge, the Einstein part of the action does not contribute, and the only contribution comes from the mass term which takes the form

$$\frac{1}{2} M_{\text{Pl}}^2 \left\{ 2m_0^2 (\partial_0 \pi_0)^2 + m_1^2 (\partial_0 \pi_i)^2 + m_1^2 (\partial_i \pi_0)^2 + \left(4m_4^2 - 2m_1^2\right) \pi_0 \partial_0 \partial_i \pi_i \right.$$
$$\left. - m_2^2 (\partial_i \pi_j)^2 - \left(m_2^2 - 2m_3^2\right) (\partial_i \pi_i)^2 \right\}.$$

We need to determine under which constraints on the masses this Lagrangian defines a consistent model.

It is convenient to use the internal $O(3)$-symmetry and separate the vector and scalar representations. The field $\pi_0$ is a scalar under $O(3)$, while the vector $\pi_i$ can be decomposed in the transverse and longitudinal parts,

$$\pi_i = \pi_i^T + \pi_i^L,$$

where $\pi_i^T$ is transverse,

$$\partial_i \pi_i^T = 0,$$

and $\pi_i^L$ is a divergence of a scalar, $\pi_i^L = \partial_i \pi^L / \sqrt{-\partial_j^2}$. The vector and scalar sectors separate. The Lagrangian of the vector sector reads

$$L^{(\text{vec})} = \frac{1}{2} M_{\text{Pl}}^2 \left\{ m_1^2 \left(\partial_0 \pi_i^T\right)^2 - m_2^2 \left(\partial_i \pi_j^T\right)^2 \right\}.$$

For the absence of pathologies it is sufficient to require $m_1^2, m_2^2 > 0$.

In the scalar sector the analysis proceeds in the same way but is more involved as one has to deal with the coupled system of equations. Without going into details of the calculations which can be found in Ref. [35], let us summarize the results.

- At general values of the mass parameters there are 6 propagating degrees of freedom (two tensor, two vector and two scalar modes). One of them is necessarily either ghost or unstable. The consistent model arises only in special cases.
- In the case $m_0^2 = 0$, as was found in Ref. [34], one of the scalar modes does not propagate. Five other modes, 2 tensor, 2 vector and 1 scalar, form five polarizations of the massive graviton. Note, however, that the masses of these modes are, in general, different. This is the manifestation of the Lorentz symmetry breaking. There are no pathologies in this model provided the masses $m_1^2 \ldots m_4^2$ satisfy certain inequalities [34].
- In the case $m_2^2 = m_3^2$ one of the scalars does not propagate. As in the previous case, the remaining 5 modes can be viewed as 5 polarizations of the massive graviton.
- Finally, in the case $m_1^2 = 0$ none of the scalar and vector modes are dynamical, so the only propagating degrees of freedom are the two tensor modes. These modes are massive and have the mass $m_2$. No ghosts or instabilities are present provided the masses satisfy certain inequalities.

There is one more important issue which has to be discussed in the context of linear perturbations. The action (3.4) is no more than the low-energy effective action. One should expect the appearance of higher terms suppresses by the powers of the energy divided by the cutoff scale $\Lambda$. These terms may contain, in particular, higher derivatives of the fields $\phi_0$ and $\phi_i$. Usually these terms can be neglected at low energies. However, the absence of instabilities requires the fine-tuning relations as was explained above. The violation of these fine-tuning relations may result in instabilities even if it is tiny. For instance, if a dispersion relation $\omega^2 = 0$ which corresponds to a non-propagating mode acquires a correction and changes to $\omega^2 = -\alpha k^4$, this may lead to a rapid instability at sufficiently high momentum even if the coefficient $\alpha$ is small. So, one has to make sure that the fine-tuning relations needed for the stability of the model can be protected by symmetries. This is probably not the case for the phase $m_0^2 = 0$ [35]. On the contrary, the phase $m_1^2 = 0$ can be protected against higher-order corrections by the symmetry

$$x^i \to x^i + \xi^i(t),$$

which is a part of the group of coordinate transformations. In terms of the Goldstone fields this symmetry reads

$$\phi^i(x) \to \phi^i(x) + \xi^i\big(\phi^0(x)\big). \tag{4.3}$$

It is easy to check that the latter symmetry is satisfied if the action depends only on the variables $X$ and $W^{ij}$ defined by eq. (4.1).

This is this last case that we will consider in more detail in the remaining part of these lectures. We will see that it has a number of other attractive features apart from being stable against higher-order corrections.

## 5. Some phenomenological implications

### 5.1. Newton's potential

Consider from the phenomenological point of view a particular class of models with the function $F$ of the form

$$F = F(X, W^{ij}).$$

The first question which we have to address is whether — and how — the Newton's law is modified in these models. Thus, we have to calculate the linear response of the system to a point-like source of the gravitational field.

It is convenient to work in the "unitary gauge" where the Goldstone fields are set to their vacuum values (3.6). In this gauge the remaining perturbations are the perturbations of the metric $\delta g_{\mu\nu}$,

$$g_{\mu\nu} = \eta_{\mu\nu} + \delta g_{\mu\nu}. \tag{5.1}$$

In the notations of Ref. [36] they are parameterized as follows,

$$\delta g_{00} = 2\varphi;$$
$$\delta g_{0i} = S_i - \partial_i B;$$
$$\delta g_{ij} = -h_{ij} - \partial_i F_j - \partial_j F_i + 2(\psi \delta_{ij} - \partial_i \partial_j E),$$

where $h_{ij}$ are the transverse and traceless tensor perturbations, $S_i$ and $F_i$ are the transverse vector perturbations, while $\varphi$, $\psi$, $B$ and $E$ are the scalar perturbations.

The quadratic Lagrangian for perturbations consists of the Einstein-Hilbert term, the mass term and the source term,

$$L = L_{EH} + L_m + L_s. \tag{5.2}$$

Explicitly, the Einstein-Hilbert part reads

$$L_{EH} = M_{Pl}^2 \left\{ -\frac{1}{4} h_{ij} \left( \partial_0^2 - \partial_i^2 \right) h_{ij} - \frac{1}{2} (S_i + \partial_0 F_i) \partial_j^2 (S_i + \partial_0 F_i) \right.$$
$$\left. + 4 \left( \varphi + \partial_0 B - \partial_0^2 E \right) \partial_i^2 \psi + 6 \psi \partial_0^2 \psi - 2\psi \partial_i^2 \psi \right\}, \tag{5.3}$$

while for the mass term one finds

$$M_{Pl}^2 \left\{ -\frac{1}{4}m_2^2 h_{ij}^2 - \frac{1}{2}m_2^2(\partial_i F_j)^2 + m_0^2\varphi^2 + (m_3^2 - m_2^2)(\partial_i^2 E)^2 \right.$$
$$\left. - 2(3m_3^2 - m_2^2)\psi\,\partial_i^2 E + 3(3m_3^2 - m_2^2)\psi^2 + 2m_4^2\varphi\,\partial_i^2 E - 6m_4^2\varphi\psi \right\}.$$
$$(5.4)$$

Note that the tensor, vector and scalar sectors do not mix, so they can be considered separately.

We also add the external source $T_{\mu\nu}$ which is assumed to be conserved, $\partial^\mu T_{\mu\nu} = 0$. The corresponding contribution to the Lagrangian can be written as

$$L_s = -T_{00}\left(\varphi + \partial_0 B - \partial_0^2 E\right) - T_{ii}\psi + (S_i + \partial_0 F_i)T_{0i} + \frac{1}{2}h_{ij}T_{ij}.$$

All the combinations of metric perturbations which enter this equation are gauge-invariant. The one multiplying $T_{00}$,

$$\Phi \equiv \varphi + \partial_0 B - \partial_0^2 E,$$

plays the role of the Newtonian potential in the non-relativistic limit.

*Tensor sector.* In the tensor sector there is a single equation of the form

$$(-\partial_0^2 + \partial_i^2 - m_2^2)h_{ij} = 0. \tag{5.5}$$

This equation describes propagation of a massive gravitational wave. Note that this wave has only two polarizations. This is, of course, only possible because of the violation of the Lorentz invariance.

*Vector sector.* The field equations in the vector sector are

$$-\partial_j^2(S_i + \partial_0 F_i) = -T_{0i}, \tag{5.6}$$
$$\partial_0\partial_j^2(S_i + \partial_0 F_i) + m_2^2\partial_j^2 F_i = \partial_0 T_{0i}. \tag{5.7}$$

Taking the time derivative of eq. (5.6) and adding it to eq. (5.7) gives

$$F_i = 0,$$

provided that $m_2^2 \neq 0$. Thus, the vector sector of our model behaves in the same way as in the Einstein theory in the gauge $F_i = 0$. There are no propagating vector perturbations.

*Scalar sector.* The field equations for the scalar perturbations are

$$2\partial_i^2 \psi + m_0^2 \varphi + m_4^2 \partial_i^2 E - 3m_4^2 \psi = \frac{T_{00}}{2M_{Pl}^2}, \tag{5.8}$$

$$2\partial_i^2 \Phi - 2\partial_i^2 \psi + 6\partial_0^2 \psi - \left(3m_3^2 - m_2^2\right) \partial_i^2 E$$
$$+ 3\left(3m_3^2 - m_2^2\right)\psi - 3m_4^2 \varphi = \frac{T_{ii}}{2M_{Pl}^2}, \tag{5.9}$$

$$-2\partial_i^2 \partial_0^2 \psi + \left(m_3^2 - m_2^2\right)\partial_i^4 E$$
$$- \left(3m_3^2 - m_2^2\right)\partial_i^2 \psi + m_4^2 \partial_i^2 \varphi = -\frac{\partial_0^2 T_{00}}{2M_{Pl}^2}, \tag{5.10}$$

$$2\partial_i^2 \partial_0 \psi = \frac{\partial_0 T_{00}}{2M_{Pl}^2}. \tag{5.11}$$

Eq. (5.11) implies

$$\psi = \frac{1}{\partial_i^2}\frac{T_{00}}{4M_{Pl}^2} + \psi_0(x^i), \tag{5.12}$$

where $\psi_0(x^i)$ is some time-independent function. From Eqs. (5.8) and (5.10) one finds

$$\varphi = \frac{1}{\Delta}\left\{2m_2^2 m_4^2 \psi + 2\left(m_3^2 - m_2^2\right)\partial_i^2 \psi_0\right\}, \tag{5.13}$$

$$\partial_i^2 E = \frac{1}{\Delta}\left\{\left(3\Delta - 2m_0^2 m_2^2\right)\psi - 2m_4^2 \partial_i^2 \psi_0\right\}, \tag{5.14}$$

where

$$\Delta = m_4^4 - m_0^2\left(m_3^2 - m_2^2\right).$$

Finally, substituting eqs. (5.12), (5.13) and (5.14) into eq. (5.9) one finds the gauge-invariant potential $\Phi$,

$$\Phi = \frac{1}{\partial_i^2}\frac{T_{00} + T_{ii}}{4M_{Pl}^2} - 3\frac{\partial_0^2}{\partial_i^4}\frac{T_{00}}{4M_{Pl}^2} + \frac{m_2^2}{\Delta}\left(3\Delta - 2m_0^2 m_2^2\right)\frac{1}{\partial_i^4}\frac{T_{00}}{4M_{Pl}^2}$$
$$+ \frac{m_2^2}{\Delta}\left(3\Delta - 2m_0^2 m_2^2\right)\frac{1}{\partial_i^2}\psi_0 + \left(1 - \frac{2m_2^2 m_4^2}{\Delta}\right)\psi_0. \tag{5.15}$$

The first two terms on the r.h.s. of eq. (5.15) are the standard contributions of the Einstein theory, the first becoming the Newtonian potential in the nonrelativistic limit. The third term on the r.h.s. is the new contribution specific to massive gravity. We will discuss this term in more detail shortly. Finally, the terms on the second line of eq. (5.15) are also new; they are all proportional to the time-independent integration constant $\psi_0(x^i)$. The value of this constant is determined by the initial conditions. Since there are no absolutely static objects in the Universe, the value of $\psi_0$ is not related to the gravitational fields created by massive bodies. Thus, these terms are irrelevant for us here.

Therefore, the gauge-invariant potentials $\Phi$ and $\psi$ in differ from their analogs in the Einstein theory $\Phi_E$ and $\psi_E$ by the mass-dependent third term on the r.h.s of eq. (5.15),

$$\psi = \psi_E,$$

$$\Phi = \Phi_E + \left(3 - \frac{2m_0^2 m_2^2}{\Delta}\right) \frac{m_2^2}{\partial_i^4} \frac{T_{00}}{4M_{Pl}^2}. \tag{5.16}$$

This term vanishes if all masses uniformly go to zero, which implies the absence of the vDVZ discontinuity as expected. In the coordinate space this term leads to the extra contribution to the potential which has the "confining" form, so that the whole potential becomes

$$\Phi = G_N M \left(-\frac{1}{r} + \mu^2 r\right), \tag{5.17}$$

where

$$\mu^2 = -\frac{1}{2} m_2^2 \left(3 - \frac{2m_0^2 m_2^2}{\Delta}\right). \tag{5.18}$$

The growth of the second term indicates the breakdown of perturbation theory at distances $r \gtrsim 1/(G_N M \mu^2)$.

An interesting situation arises when there is no modification of the Newtonian potential. This may happen if

$$3\Delta - 2m_0^2 m_2^2 = 0 \tag{5.19}$$

and $\Delta \neq 0$. In this case the static interaction between two massive bodies is described by the standard Newtonian force proportional to $1/r^2$, so the deviations from the standard gravity would not be possible to detect in the Cavendish-type experiments. Note that eq. (5.19) does not require that the mass of the graviton $m_2$ is zero. Thus, in the case when the condition (5.19) is satisfied the non-zero graviton mass coexists with the long-range force. This is yet another manifestation of the violation of the Lorentz invariance in this model.

The condition (5.19) can be ensured by imposing the following dilatation symmetry,

$$t \rightarrow \lambda t,$$
$$x^i \rightarrow \lambda^\gamma x^i, \tag{5.20}$$

where $\lambda$ is the transformation parameter and $\gamma$ is a constant. Requiring the symmetry (5.19) is equivalent to a taking a particular dependence of the function $F$ on its arguments,

$$F = F(Z^{ij}) \tag{5.21}$$

where

$$Z^{ij} \equiv X^\gamma W^{ij}.$$

We will see in the next section that the model obtained in this way has a number of phenomenologically interesting features.

### 5.2. Cosmological evolution

There is one more reason to consider the models possessing the symmetry (5.19). To understand this reason we have to discuss the cosmological solutions in massive gravity. The action of our model is a full non-linear action of the low-energy effective theory. Thus, we can study the non-linear gravitational fields and, in particular, the cosmology, provided the relevant energy scale is below the cutoff scale $\Lambda$.

The homogeneous and isotropic ansatz in the spatially-flat case reads

$$ds^2 = dt^2 - a^2(t)dx_i^2, \tag{5.22}$$
$$\phi^0 = \phi(t),$$
$$\phi^i = \Lambda^2 x^i.$$

For this ansatz the variable $W^{ij}$ takes the form

$$W^{ij} = -\frac{1}{a^2}\delta_{ij},$$

so the function $F$ becomes a function of $X$ and the scale factor $a$ which one can consider as two independent variables. The equations which determine the cosmological evolution are the Friedmann equation

$$\left(\frac{\dot{a}}{a}\right)^2 = \frac{1}{6M_{Pl}^2}\{\rho_m + 2\Lambda^4 X F_X - \Lambda^4 F\} \equiv \frac{1}{6M_{Pl}^2}\{\rho_m + \rho_1 + \rho_2\}, \tag{5.23}$$

where $\rho_m$ is the energy density of the ordinary matter not including the Goldstone fields, and the field equation for $\phi^0$,

$$\partial_t \left( a^3 \sqrt{X} F_X \right) = 0. \tag{5.24}$$

It is straightforward to solve this system of equations for any given function $F(X, a)$. After the integration, Eq. (5.24) gives an algebraic relation between $X$ and the scale factor $a$. The dependence $X(a)$ as found from Eq. (5.24) determines the behavior of the Goldstone energy density $\rho_1 + \rho_2$ as a function of $a$. This makes Eq. (5.23) a closed equation for the scale factor $a(t)$.

Rather than solving eqs. (5.23) and (5.24) (see Ref. [37] for details) let us discuss general properties of the solutions. We are interested in solutions such that $a \to \infty$ at late times. Let us assume that the function $X(a)$ which results in the solution of eq. (5.24) asymptotes to some power of $a$. Then there exists a constant $\gamma$ such that the combination $X^\gamma / a^2$ goes to a non-zero value as $a \to \infty$. Then eq. (5.24) implies

$$\rho_1 = \text{const} \frac{1}{a^{3-1/\gamma}}, \tag{5.25}$$

i.e., one of the contributions of the Goldstone fields to the energy density behaves like a matter with the equation of state $w = -1/(3\gamma)$.

By construction, we have $Z^{ij} = X^\gamma W^{ij} \to Z_0 \delta_{ij}$, where $Z_0$ is some constant. If we assume that the function $F$ is regular in the limit $a \to \infty$, then it also has to go to some constant $F \to F_0 = F(Z_0)$. Thus at late times of the evolution the function $F$ depends on the combination $Z^{ij} = X^\gamma W^{ij}$. In other words, this point is an attractor of the cosmological evolution. This is another reason to consider the actions which depend on the Goldstone fields in the combination $Z^{ij} = X^\gamma W^{ij}$. Note that the second contribution to the energy density behaves at late times as a cosmological constant.

## 5.3. Experimental signatures

Let us first discuss the experimental constraints on the graviton mass. We will concentrate on the case of the functions $F$ satisfying the constraint (5.21). In this case the Newtonian potential remains unchanged, so the existing constraints from the Cavendish-type experiments and solar system tests [38, 39] do not apply. On the contrary, the presence of the mass affects emission of the gravitational waves. Precision observations of the slow down of the orbital motion in binary pulsar systems [40] imply that the mass of the gravitational waves cannot be larger than the frequency of the waves emitted by these systems. The latter is determined by the period of the orbital motion which is of order 10 hours, implying the following limit on the graviton mass,

$$m_2 \lesssim 2 \times 10^{-4} \text{Hz} \sim 10^{-19} \text{eV} \sim (10^{14} \text{cm})^{-1}.$$

This is a very large mass in cosmological standards. It would certainly be ruled out if it implied the Yukawa-type deviations from the Newtonian potential.

The non-zero mass of the graviton leads to interesting consequences for the primordial gravitational waves as well. The massive gravitons can be produced during the cosmological expansion. In the expanding Universe eq. (5.5) is modified in the following way,

$$\left(-\partial_0^2 - 3H\partial_0 + \partial_i^2 - m_2^2\right)h_{ij} = 0, \tag{5.26}$$

where $H = \dot{a}/a$ is the Hubble constant. This equation is identical to the one which governs the behavior of a massive scalar field such as axion. Thus, the massive gravitons will be efficiently produced during inflation (cf. Ref. [41,42]). One may estimate the amount of the gravitational waves produced. Assuming the inflationary scenario in which the Hubble parameter is constant (e.g., as in the hybrid models of inflation [43]), the spectrum for the massive gravitons is that for the minimally coupled massive scalar field in the de Sitter space [44–47],

$$\langle h_{ij}^2 \rangle \simeq \frac{1}{4\pi^2} \left(\frac{H_i}{M_{Pl}}\right)^2 \int \frac{dk}{k} \left(\frac{k}{H_i}\right)^{\frac{2m_2^2}{3H^2}}. \tag{5.27}$$

Superhorizon metric fluctuations remain frozen until the Hubble factor becomes smaller than the graviton mass, when they start to oscillate with the amplitude decreasing as $a^{-3/2}$. The energy density in massive gravitons at the beginning of oscillations is of order

$$\rho_o \sim M_{Pl}^2 m_2^2 \langle h_{ij}^2 \rangle \simeq \frac{3H_i^4}{8\pi^2}, \tag{5.28}$$

where we integrated in Eq. (5.27) over the modes longer than the horizon. Today the fraction of the energy density in the massive gravitational waves is, therefore,

$$\Omega_g = \frac{\rho_o}{z_o^3 \rho_c} = \frac{\rho_o}{z_e^3 \rho_c} \left(\frac{H_e}{H_o}\right)^{3/2}, \tag{5.29}$$

where $z_o$ is the redshift at the start of oscillations, $H_o \sim m_2$ is the Hubble parameter at that time, $H_e \approx 0.4 \cdot 10^{-12}$ s$^{-1}$ is the Hubble parameter at the matter/radiation equality, and $z_e \approx 3200$ is the corresponding redshift. Combining all the factors together one gets

$$\Omega_g \sim \cdot 10^4 \left(m_2 \cdot 10^{14} \text{cm}\right)^{1/2} \left(\frac{H_i}{\Lambda}\right)^4 . \tag{5.30}$$

This estimate assumes that the number of e-foldings during inflation is large, $\ln N_e > H^2/m^2$, which is quite natural in the model of inflation assumed. Thus, the amount of the gravitons produced at inflation may be enough to constitute the dark matter of the Universe.

Let us estimate the amplitude of the gravitational waves assuming that they comprise all of the dark matter in the halo of our Galaxy. The energy density in non-relativistic gravitational waves is of order $M_{Pl}^2 m_2^2 h_{ij}^2$. Equating this to the local halo density one gets

$$\langle h_{ij} \rangle \sim 10^{-10} \left(\frac{2 \cdot 10^{-4} \text{Hz}}{m_2}\right) . \tag{5.31}$$

At frequencies $10^{-6} \div 10^{-5}$ Hz this value is well above the expected sensitivity of the LISA detector [48]. Note that in the close frequency range $10^{-9} \div 10^{-7}$ Hz there exists a much lower bound [49] on the stochastic background of the gravitational waves coming from the timing of the millisecond pulsars [50,51], which is at the level of $\Omega_g < 10^{-9}$. Thus, it is possible that the massive graviton as a candidate for the dark matter can be ruled out by the reanalysis of the already existing data on the pulsar timing.

The relic graviton abundance depends on both the specific inflationary model and the details of the (unknown) UV completion of massive gravity. Therefore, in general, massive gravitons may not comprise the whole of the dark matter in the Universe. In that case the exclusion of the graviton as the dark matter candidate does not necessarily rile out the model of massive gravity and massive gravitational waves may still be present at a lower level. These gravitational waves differ from the conventional stochastic gravitational wave background in that they are monochromatic with the frequency equal to the graviton mass. It is important that the expected LISA sensitivity allows to detect the presence of such gravitational waves at a significantly lower level than in Eq. (5.31).

## 6. Summary and outlook

Let us summarize the conclusions.

– Existing attempts to give graviton a mass in a Lorentz-invariant way suffer from severe problems of which the strong coupling is the most harmless one. It is not clear at the moment whether one can deal with the strong coupling efficiently and construct a phenomenologically viable Lorentz-invariant model of massive graviton.

– The breaking of Lorentz invariance introduces enough freedom to circumvent these problems. Namely, one arrives at a variety of models which at the linear level can be parameterized by 5 graviton mass parameters. Some regions in the space of these parameters lead to a consistent low-energy effective theory with massive gravitons.

– A particular class of models, the ones possessing the residual symmetry (4.3), has a number of attractive features. The two transverse traceless polarizations of the graviton are the only propagating degrees of freedom in these models. The cosmological evolution has an attractor point possessing an additional dilatation symmetry (5.20). In this point the two contributions of the Goldstone fields into the energy-momentum tensor have the form of the cosmological constant and of matter with the equation of state which depends on the parameters of the model. The graviton masses go to finite constants during the expansion of the Universe.

– In the models possessing symmetries (4.3) and (5.20) the non-zero mass of the graviton coexists with the unmodified Newtonian potential – the possibility which is due to the violation of the Lorentz invariance. This allows for relatively large masses of the graviton $m_2 \lesssim (10^{14}\text{cm})^{-1}$ to be phenomenologically acceptable. Massive gravitons can be produced in sufficient amount in the Early Universe and are a new candidate for the dark matter.

– The relic massive gravitons produce an easily identifiable monochromatic signal in the gravitational wave detectors. Among those LISA has a large potential to probe the presence of massive gravitational waves and to rule out graviton as a dark matter candidate.

At the same time, there remain quite a number of open questions which require further study. Here are some of them:

– Modern cosmological observations are becoming more and more precise. In order to be in accord with these observations any alternative theory of gravity has to successfully address several issues, one of which is the structure formation. The first stage of this process, the linear growth of perturbations, is straightforward to study in the massive gravity model.

– When solving for the linear perturbations in sect. 5 we have seen that there appears an integration constant $\psi_0(x^i)$. We have set this constant to zero since we were interested in the gravitational fields of the massive bodies which are not related to this constant. In the cosmological context this constant is also present. Presumably it is determined dynamically and is driven to zero at the inflationary stage, but this remains to be demonstrated.

– An interesting special case $\gamma = 1/3$ deserves attention from another perspective. In this case both contributions $\rho_1$ and $\rho_2$ have the vacuum equation of state $w = -1$. As a result, the acceleration rate of the late de Sitter phase is a dynamical quantity, determined by the initial conditions in the Goldstone sector rather than by parameters of the action. This is similar to the situation in the unimodular gravity [52] where the cosmological constant is also a constant of integration. Thus, the massive gravity models may shed light on the cosmological problem.

– It is worth studying in more detail the theories which do not possess the dilatation symmetry (5.20). In these models the gravitational potential is modified at large distances from the source. It is possible in principle that these models may provide an explanation of the galactic flat rotation curves alternative to the dark matter.

– Finally, having the full non-linear action one can study non-linear solutions of the model, in particular, black holes. The latter are of particular interest in view of the expected progress in their experimental observations [53].

## References

[1]  C. Brans and R.H. Dicke, Phys. Rev. **124**, 925 (1961).

[2]  D.N. Spergel *et al.*, astro-ph/0603449.

[3]  M. Milgrom, Astrophys. J. **270**, 371 (1983).

[4]  J. Bekenstein and M. Milgrom, Astrophys. J. **286**, 7 (1984).

[5]  R.H. Sanders and S.S. McGaugh,  Ann. Rev. Astron. Astrophys. **40**, 263 (2002), [astro-ph/0204521].

[6]  J.D. Bekenstein, Phys. Rev. **D70**, 083509 (2004), [astro-ph/0403694].

[7]  R. Gregory, V.A. Rubakov and S.M. Sibiryakov,  Phys. Rev. Lett. **84**, 5928 (2000), [hep-th/0002072].

[8]  G.R. Dvali, G. Gabadadze and M. Porrati, Phys. Lett. **B484**, 129 (2000), [hep-th/0003054].

[9]  L. Pilo, R. Rattazzi and A. Zaffaroni, JHEP **07**, 056 (2000), [hep-th/0004028].

[10]  G.R. Dvali, G. Gabadadze and M. Porrati, Phys. Lett. **B485**, 208 (2000), [hep-th/0005016].

[11]  C. Deffayet, Phys. Lett. **B502**, 199 (2001), [hep-th/0010186].

[12]  C. Deffayet, G.R. Dvali and G. Gabadadze,  Phys. Rev. **D65**, 044023 (2002), [astro-ph/0105068].

[13]  M.A. Luty, M. Porrati and R. Rattazzi, JHEP **09**, 029 (2003), [hep-th/0303116].

[14]  A. Nicolis and R. Rattazzi, JHEP **06**, 059 (2004), [hep-th/0404159].

[15]  D. Gorbunov, K. Koyama and S. Sibiryakov, Phys. Rev. **D73**, 044016 (2006), [hep-th/0512097].

[16]  S.M. Carroll, V. Duvvuri, M. Trodden and M.S. Turner, Phys. Rev. **D70**, 043528 (2004), [astro-ph/0306438].

[17]  S.M. Carroll *et al.*, Phys. Rev. **D71**, 063513 (2005), [astro-ph/0410031].

[18]  S. Capozziello, V.F. Cardone and M. Francaviglia,  Gen. Rel. Grav. **38**, 711 (2006), [astro-ph/0410135].

[19]  M. Amarzguioui, O. Elgaroy, D.F. Mota and T. Multamaki, Astron. Astrophys. **454**, 707 (2006), [astro-ph/0510519].

[20]  A. Nunez and S. Solganik, hep-th/0403159.

[21]  A.E. Dominguez and D.E. Barraco, Phys. Rev. **D70**, 043505 (2004), [gr-qc/0408069].

[22] T.P. Sotiriou, Class. Quant. Grav. **23**, 5117 (2006), [gr-qc/0604028].

[23] I.I. Kogan, S. Mouslopoulos and A. Papazoglou, Phys. Lett. **B501**, 140 (2001), [hep-th/0011141].

[24] T. Damour and I.I. Kogan, Phys. Rev. **D66**, 104024 (2002), [hep-th/0206042].

[25] D. Blas, C. Deffayet and J. Garriga, Class. Quant. Grav. **23**, 1697 (2006), [hep-th/0508163].

[26] M. Fierz and W. Pauli, Proc. Roy. Soc. Lond. **A173**, 211 (1939).

[27] H. van Dam and M.J.G. Veltman, Nucl. Phys. **B22**, 397 (1970).

[28] V.I. Zakharov, JETP Lett. **12**, 312 (1970).

[29] A.I. Vainshtein, Phys. Lett. **B39**, 393 (1972).

[30] A. Gruzinov, New Astron. **10**, 311 (2005), [astro-ph/0112246].

[31] A. Aubert, Phys. Rev. **D69**, 087502 (2004), [hep-th/0312246].

[32] N. Arkani-Hamed, H.-C. Cheng, M.A. Luty and S. Mukohyama, JHEP **05**, 074 (2004), [hep-th/0312099].

[33] N. Arkani-Hamed, H. Georgi and M.D. Schwartz, Ann. Phys. **305**, 96 (2003), [hep-th/0210184].

[34] V.A. Rubakov, hep-th/0407104.

[35] S.L. Dubovsky, JHEP **10**, 076 (2004), [hep-th/0409124].

[36] V.F. Mukhanov, H.A. Feldman and R.H. Brandenberger, Phys. Rept. **215**, 203 (1992).

[37] S.L. Dubovsky, P.G. Tinyakov and I.I. Tkachev, Phys. Rev. **D72**, 084011 (2005), [hep-th/0504067].

[38] B. Bertotti, L. Iess and P. Tortora, Nature **425**, 374 (2003).

[39] G. Esposito-Farese and D. Polarski, Phys. Rev. **D63**, 063504 (2001), [gr-qc/0009034].

[40] J.H. Taylor, Rev. Mod. Phys. **66**, 711 (1994).

[41] A.A. Starobinsky, JETP Lett. **30**, 682 (1979).

[42] V.A. Rubakov, M.V. Sazhin and A.V. Veryaskin, Phys. Lett. **B115**, 189 (1982).

[43] A.D. Linde, Phys. Rev. **D49**, 748 (1994), [astro-ph/9307002].

[44] T.S. Bunch and P.C.W. Davies, Proc. Roy. Soc. Lond. **A360**, 117 (1978).

[45] A. Vilenkin and L.H. Ford, Phys. Rev. **D26**, 1231 (1982).

[46] A.D. Linde, Phys. Lett. **B116**, 335 (1982).

[47] A.A. Starobinsky, Phys. Lett. **B117**, 175 (1982).

[48] P.L. Bender, Class. Quant. Grav. **20**, S301 (2003).

[49] A.N. Lommen, astro-ph/0208572.

[50] M.V. Sazhin, Soviet Astronomy **22**, 36 (1978).

[51] S. Detweiler, Astrophys. J. **234**, 1100 (1979).

[52] S. Weinberg, Rev. Mod. Phys. **61**, 1 (1989).

[53] C.M. Will, gr-qc/0510072.

# WORKING GROUP REPORTS

## Dark Matter on the Smallest Scales

*E.R. Siegel, G. D'Amico, E. Di Napoli, L. Fu, M.P. Hertzberg,*
*N.T.T. Huong, F. Palorini and A. Sellerholm*

## Dark Matter in Minimal Trinification

*E. Di Napoli, D. George, M. Hertzberg, F. Metzler and E. Siegel*

## Scalar Gravity: Post-Newtonian Corrections via an Effective Field Theory Approach

*R.A. Porto and R. Sturani*

## A Pedagogical Primer on Preheating

*M.A. Amin, A. Brown, A.D. Chambers, S.E. Jorás,*
*A. Kumar, F. Metzler, J. Pritchard and N. Sivanandam*

## Trans-Planckian Physics from Cosmology

*S.E. Jorás, G. Marozzi and R. Ansari*

## Braneworld Graviton Interactions in Early Universe Phase Transitions

*R.U.H. Ansari, C. Delaunay, R. Gwyn, A. Knauf,*
*A. Sellerholm, N.R. Shah and F.R. Urban*

501

# DARK MATTER ON THE SMALLEST SCALES

E.R. Siegel[1], G. D'Amico[2], E. Di Napoli[3], L. Fu[4],
M.P. Hertzberg[5], N.T.T. Huong[6], F. Palorini[7] and A. Sellerholm[8]

[1] Department of Physics, University of Wisconsin, 1150 University Avenue, Madison, WI 53706, USA
[2] Department of Physics, Scuola Normale Superiore, Pisa 56100, Italy
[3] Department of Physics and Astronomy, University of North Carolina, CB#3255 Phillips Hall, Chapel Hill, NC 27599, USA
[4] Joint Center of Astrophysics, Shanghai Normal University, 200234 Shanghai, China
[5] Center for Theoretical Physics, MIT, Cambridge, MA 02139, USA
[6] Centre for High Energy Physics, Vietnam National University – Hanoi, 334 Nguyen Trai Street, Hanoi, Vietnam
[7] Université C. Bernard, 69662 Villeurbanne cedex, France
[8] Department of Physics, Stockholm University, 10691 Stockholm, Sweden

## Abstract

This work investigates the dark matters structures that form on the smallest cosmological scales. We find that the types and abundances of structures which form at approximately Earth-mass scales are very sensitive to the nature of dark matter. We explore various candidates for dark matter and determine the corresponding properties of small-scale structure. In particular, we discuss possibilities for indirect detection of dark matter through small-scale structure, and comment on the potential of these methods for discriminating between dark matter candidates.

## 1. Introduction

Perhaps the most fundamental goal of cosmological physics is to determine the composition and evolution of the universe. Cosmology has recently reached a state where the vast majority of measurements point towards the same result; the universe is composed of $\sim 5$ per cent baryons, $\sim 20$ per cent dark matter, and $\sim 75$ per cent dark energy. This result is quite puzzling, as dark matter and dark energy have only been detected by their gravitational effects on cosmological scales. While much is known about baryonic structure as well as the particle

503

properties of baryons, much less is known about dark matter structure and its particle properties, and even less is known about dark energy. As a result, the nature of these two latter components of the universe, which compose 95 per cent of its total energy density, are presently the greatest unsolved puzzles in cosmology.

This work investigates the nature of dark matter by examining its effects on cosmologically small scales. In particular, constraining the properties of small-scale structure will constrain the nature of dark matter. The layout of this paper is as follows: section 2 presents an overview of the dark matter problem, focusing on astrophysical evidence. Section 3 identifies the leading particle candidates for dark matter, with a view towards their distinguishing properties. Section 4 describes and contrasts the properties of small scale structure which forms in various dark matter models. Finally, section 5 provides a discussion on possible observable signals which could point towards the nature of dark matter.

## 2. The dark matter problem

The discrepancy between the amount of mass inferred from observations of light and the amount of mass inferred from gravitation on cosmological scales is known as the dark matter problem. In our solar system, nearly all of the mass is contained in the sun, and so it seems reasonable to assume that light traces mass. Since properties of stars are well known, it is possible to infer the amount of mass present in luminous sources ($M_\star$), such as stars. From measurements of $M_\star$, it is possible to infer the fractional energy density in stars, $\Omega_\star$, where $\Omega_i$ is defined as the energy density in species $i$ over the critical energy density, $\rho_c$. Observations indicate that $\Omega_\star \simeq 0.005$ [1], or that luminous matter composes about 0.5 per cent of the total energy density in the universe.

On the other hand, it is also possible to reconstruct the matter density of the universe from its gravitational mass (yielding $\Omega_m$) instead of its luminous mass ($\Omega_\star$). Many methods exist for determining $\Omega_m$, including extrapolating the matter density from peculiar velocities within galaxy clusters [2], rotational and virial properties of individual galaxies [3], gravitational lensing [4], and from large scale structure [5]. All of these methods yield a range that $0.15 < \Omega_m < 0.35$, consistent with one another at the 2-$\sigma$ level. The best measurement of matter density comes from a combination of the latest microwave background data [6] with type Ia supernova data [7], yielding $\Omega_m = 0.26 \pm 0.02$. This discrepancy between $\Omega_m$ and $\Omega_\star$ is the dark matter problem; the amount of mass in stars is only 2 per cent of the total matter density! The remaining 98 per cent, the non-luminous matter, is known as dark matter.

## 3. Dark matter candidates

Once the existence of dark matter has been established, it becomes a fundamental question to inquire what its nature is. Many candidates have been proposed (see [8] for a review), but a confirmed experimental detection has been thus far elusive. This section therefore investigates the properties and viability of four generic models of dark matter: baryons, neutrinos, thermal relics, and nonthermal relics.

Baryons are the most obvious candidate for dark matter, as baryonic dark matter is observed in the form of planets, low luminosity stars, and diffuse gas clouds. One consequence is that baryonic dark matter will fragment and collapse to form MAssive Compact Halo Objects (MACHOs). Unfortunately, the quantitative effect of Silk damping in large-scale structure [9], the insufficient number of MACHOs [10], and big bang nucleosynthesis [11] all indicate that $\Omega_m \simeq 0.04$. Finally, direct observations of X-rays from interacting galaxies show that dark matter does not interact in the same fashion as baryons [12], which means that although baryons compose a significant fraction (about 15 per cent) of the total matter, they cannot be responsible for all of the dark matter.

It is therefore unavoidable that the majority of dark matter be non-baryonic. The standard model contains a candidate for non-baryonic dark matter: the neutrino. However, neutrinos are produced thermally and have very low masses, conflicting with constraints from the Lyman-$\alpha$ forest that limit the mass of thermal dark matter to be at least $2 \, \mathrm{keV}$ [13]. These observations, combined with the constraints on neutrino masses, rule out standard model neutrinos as a significant component of dark matter.

Heavier thermal relics are allowed experimentally, however, and particle physics beyond the standard model provides us with many well-motivated dark matter candidates of this type, including the lightest supersymmetric particle and the lightest Kaluza-Klein particle. These particles do not interact either strongly or electromagnetically, and are therefore classified as Weakly Interacting Massive Particles (WIMPs). Thermal relics are defined by the fact that, at some point in the past history of the universe, the particles were in thermal equilibrium with the primordial plasma. At some point, their abundance freezes out (a process known as chemical decoupling), and the number of dark matter particles remains constant thereafter. At a later point, the dark matter particles cease to scatter off of the plasma (kinetic decoupling), and thereafter evolve solely gravitationally.

Nonthermal relics, on the other hand, were never in thermal equilibrium with the primordial plasma. Instead, these dark matter candidate particles can be produced during phase transitions in the early universe. Candidates include axions [14] and massive gravitons [15]. Unlike thermal relics, nonthermal relics evolve only gravitationally from the moment of their creation, receiving only a gravitational imprint from the primordial plasma. As will be discussed in sections 4 and 5, present-day gravitational effects of dark matter have the potential to shed light on its nature, and may be able to distinguish between thermal and nonthermal relics.

## 4. Small scale structure

Cosmological structure formation is the key process that allows an understanding of the types and abundances of structures which will form in the universe given a set of initial conditions. While large-scale structure is well understood, small-scale structure is more of a challenge due to the uncertainty of the nature of dark matter and the problems associated with gravitational collapse in the deeply nonlinear regime. The remainder of the work in this paper represents the initial stages of an ambitious research project to probe the nature of dark matter via astrophysical observations on small (i.e., sub-galactic) scales.

Just as baryonic structures are suppressed on mass scales below $\sim 10^6$ solar masses ($M_\odot$) due to Silk damping prior to decoupling, WIMPy structures will be suppressed on scales smaller than the horizon at the time of kinetic decoupling. By contrast, nonthermal relic dark matter will have no such suppression, owing to the fact that they were always decoupled from the primeval plasma. The fluctuations at the time of decoupling are frozen in, and the nonlinear structures on small scales which eventually form from the collapse of dark matter will be very sensitive to the epoch of kinetic decoupling. Hence, the exact mass and abundance of these cosmic microstructures depend very sensitively on the dark matter particle properties. In particular, they are sensitive to dark matter mass, which is directly tied to the times of kinetic and chemical decoupling, and to whether the dark matter was produced thermally or nonthermally.

One challenge of understanding small scale structure is to understand the dark matter's evolution through kinetic decoupling. This difficulty is evidenced by the varied results obtained by authors using different approximations [16, 17]. We assume that the dark matter is composed of neutralino WIMPs, and follow the approach given by E. Bertschinger [18]. By calculating the transfer functions for cold dark matter fluctuations beginning with the full Boltzmann equations describing scattering between WIMPs and the plasma, the uncertainty created by the aforementioned approximations is eliminated. We begin by treating the dark matter particles as an unperturbed fluid. Analysis allows us to obtain an expression for the temperature at which kinetic decoupling occurs,

$$T_d = 0.2528 \, g_{\text{eff}}^{1/8} \left( \frac{m_{\tilde{L}}^2 - m_\chi^2}{G_F m_W^2 m_\chi^2 \tan \theta_W} \right)^{1/2} \left( \frac{m_\chi^5}{m_{\text{Pl}}} \right)^{1/4}, \tag{4.1}$$

where $g_{\mathrm{eff}}$ is the number of thermal degrees of freedom, $G_F$ is the Fermi constant, $\theta_W$ is the Weinberg angle, and $m_\chi$, $m_{\tilde{l}}$, and $m_W$ are the masses of the neutralino, slepton, and $W$ boson. There is an associated time for kinetic decoupling, $t_d$. For a slepton mass of 200 GeV, this yields a temperature for kinetic decoupling of

$$T_d \simeq 15.7 \left( \frac{m_\chi}{100\,\mathrm{GeV}} \right)^{5/4} \mathrm{MeV}. \tag{4.2}$$

By including perturbations in the gravitational field, one can derive the density transfer function for cold dark matter through the epoch of kinetic decoupling. We find that, on scales outside the horizon at kinetic decoupling ($k/a < 1/t_d$), acoustic oscillations average out, resulting in a logarithmic growth of cold dark matter fluctuations. However, on scales inside the horizon ($k/a > 1/t_d$), density fluctuations exhibit damped acoustic oscillations, which suppress the formation of structure.

The density transfer function can then be evolved through electron-positron pair annihilation and matter-radiation equality, yielding predictions for small scale cold dark matter structure. The Press-Schechter mass fraction can be derived, and indicates a suppression in the mass of collapsed structure per mass interval on mass scales below $\approx 2.3 M_d$, where $M_d$ is the mass contained in a typical density fluctuation at kinetic decoupling. For a neutralino of mass $m_\chi = 100\,\mathrm{GeV}$, this indicates that the number of cold dark matter structures which form on mass scales below approximately 20 Earth masses will be suppressed. The root-mean-squared mass density perturbation containing a mass $M$, $\sigma(M)$, is suppressed as

$$\frac{d\sigma(M)}{d\ln M} \propto \left( \frac{M}{M_d} \right)^{2/3} \tag{4.3}$$

for masses $M \ll M_d$, which then results in a suppression of WIMP microhalos of equivalent mass when converted into the nonlinear regime. By comparison, nonthermal dark matter such as axions does not have a suppression in its Press-Schechter mass fraction, as it was always kinetically decoupled [19]. For axion-like dark matter, this would indicate a comparatively much larger number of collapsed structures below about 20 Earth masses, resulting in nonlinear structures we term Nonthermal Axionic Collapsed HalOs (NACHOs).

## 5. Detectability and future work

It is unknown whether these dark matter microhalos, regardless of whether they are WIMPs or NACHOs, will be able to survive hierarchical mergers and galactic infall, and thus be found intact within our own galaxy [20]. Recent work [21], however, indicates that this is a quantitative question and not a qualitative one, as tidal stripping is not sufficient to completely destroy all of these collapsed structures. One extremely important question to ask is, "what does small scale structure look like today?" $N$−body simulations yield varied results, and there is no consensus as to the density profiles and core concentrations of these objects. Additionally, the Press-Schechter approach may be invalid, as monolithic collapse may be more important than hierarchical mergers on small scales. Ideally, therefore, detection methods which probe the density profiles and abundances of these microhalos will be able to not only constrain the nature of dark matter, but will provide information about the types of structures which form through the deeply nonlinear regime.

If these microhalos survive intact to the present day, a large number should be present within our own galaxy. We therefore seek to uncover methods to detect these microhalos and investigate their properties. For neutralino dark matter, there will be a significant annihilation cross-section, which could result in an observable gamma-ray signal. For a 100 GeV WIMP forming an Earth mass microhalo with an NFW profile, we calculate a gamma-ray flux of $\sim 10^{22}$ photons sec$^{-1}$. As abundance estimates indicate that the nearest microhalo should be nearer than the nearest star [20], this flux has the potential to outshine even gamma-rays emitted from the galactic center. Gravitational lensing due to a microhalo transit is also a possible effect. While WIMP microhalos are too diffuse, NACHOs may form much more dense structures [19], which may leave observable lensing signals. One very interesting possibility currently being investigated by two of the authors is that a dark matter microhalo transiting across the line-of-sight from an observer to a pulsar could cause a shift in the pulse arrival time due to the gravitational time delay [22]. Finally, interactions between dark matter microhalos and stars or gas clouds may be important. We note that if a dark matter microhalos of a few Earth masses were gravitationally captured by our sun, it would cause an anomalous acceleration towards the sun at significantly large distances.

Cosmological structures on small scales have the potential to shed light on the nature of dark matter, as many of the methods of detecting small-scale dark matter structure are sensitive to its mass and/or method of production. Success of any of the above methods would provide the first definitive confirmation of the presence of dark matter within our own galaxy. Although a tremendous amount of nonlinear processing has occurred since the creation of dark matter microhalos, present and future searches for small-scale cosmological structures may hold the key to determining the nature of the non-baryonic dark matter in our universe.

## Acknowledgments

We would like to thank the organizers of the 86th Les Houches summer school, Francis Bernardeau and Christophe Grojean, as well as the staff and our colleagues for providing a stimulating and productive working environment. We also acknowledge Sergio Joras, Laura Lopez-Honorez and Gilles Vertongen for their contributions to our group. E.R.S. acknowledges Hai-Ping Cheng and Pierre Ramond for support.

## References

[1] M. Fukugita and P.J.E. Peebles, Astrophys. J. **616**, 643 (2004).

[2] R.P. van der Marel, J. Magorrian, R.G. Carlberg, H.K.C. Yee and E. Ellingson, Astron. J. **119**, 2038 (2000).

[3] T.G. Brainerd and M.A. Specian, Astrophys. J. **593**, L7 (2003).

[4] C. Heymans *et al.*, Mon. Not. Roy. Astron. Soc. Lett. **371**, L60 (2006).

[5] J.A. Peacock *et al.*, Nature **410**, 169 (2001).

[6] D.N. Spergel *et al.*, arXiv:astro-ph/0603449.

[7] A.G. Riess *et al.*, arXiv:astro-ph/0611572.

[8] G. Bertone, D. Hooper and J. Silk, Phys. Rept. **405**, 279 (2005).

[9] C.J. Miller, R.C. Nichol and X.l. Chen, Astrophys. J. **579**, 483 (2002).

[10] C. Alcock *et al.* [MACHO Collaboration], Astrophys. J. **499**, L9 (1998).

[11] B.D. Fields and K.A. Olive, Nuc. Phys. A **777**, 208 (2006).

[12] D. Clowe, M. Bradac, A.H. Gonzalez, M. Markevitch, S.W. Randall, C. Jones and D. Zaritsky, arXiv:astro-ph/0608407.

[13] M. Viel, J. Lesgourgues, M.G. Haehnelt, S. Matarrese and A. Riotto, Phys. Rev. D **71**, 063534 (2005).

[14] R.D. Peccei and H.R. Quinn, Phys. Rev. Lett. **38**, 1440 (1977).

[15] S.L. Dubovsky, P.G. Tinyakov and I.I. Tkachev, Phys. Rev. Lett. **94**, 181102 (2005).

[16] A.M. Green, S. Hofmann and D.J. Schwarz, JCAP **0508**, 003 (2005).

[17] A. Loeb and M. Zaldarriaga, Phys. Rev. D **71**, 103520 (2005).

[18] E. Bertschinger, Phys. Rev. D **74**, 063509 (2006).

[19] K.M. Zurek, C.J. Hogan and T.R. Quinn, arXiv:astro-ph/0607341.

[20] J. Diemand, B. Moore and J. Stadel, Nature **433**, 389 (2005).

[21] T. Goerdt, O.Y. Gnedin, B. Moore, J. Diemand and J. Stadel, arXiv:astro-ph/0608495.

[22] E.R. Siegel, J.N. Fry and M.P. Hertzberg, *in preparation*.

# SCALAR GRAVITY: POST-NEWTONIAN CORRECTIONS VIA AN EFFECTIVE FIELD THEORY APPROACH

Rafael A. Porto[1] and Riccardo Sturani[2]

[1] *Department of Physics, Carnegie Mellon University, Pittsburgh, PA 15213*
[2] *Physics Department, Univ. Geneva, 24 Quai E. Ansermet, 1211 Geneve, Switzerland and INFN, Presidenza dell'INFN, Roma, Italy*

## 1. Introduction

The problem of motion in GR has lost its academic status and become an active research area since the next generation of gravity wave detectors will rely upon its solution. In this short contribution we will show, within scalar gravity, how an EFT approach can be used to solve the problem of motion in a systematic fashion. We will calculate the Einstein-Infeld-Hoffmann (EIH) action for the case of two *scalar-gravitating* bodies, accurate up to 1PN. As we shall see a systematic perturbative expansion puts strong constraints on the couplings of non-derivative interactions in the theory.

## 2. Scalar gravity

The starting point of the EFT approach consists of a theory of point particles coupled to a real scalar field $\phi$ we shall call the "s-graviton". For simplicity we will consider here a massive $\phi^3$ theory in a Minkowski background, though we will discuss other type of models later on. The action is $S = S_g + S_{pp}$, with

$$S_g = \int d^4x \left( \partial_\mu \phi \partial^\mu \phi - \mu^2 \phi^2 - \lambda \phi^3 \right), \quad S_{pp} = -\sum_a m_a \int d\tau_a \sqrt{1 + \frac{\phi}{M}}$$

describing the s-graviton dynamics and motion of the binary system ($a = 1, 2$). In this equation $M$ sets the coupling to matter, and $\lambda, \mu$, the self-interaction and

511

s-graviton mass respectively.[1] One important aspect of a $\phi^3$ theory is its dimensionful coupling, and the fact that the perturbative approach breaks down at distances of order $1/\lambda$. This is connected to IR divergences (in the massless limit) which appear in the perturbative expansion due to factors of $\lambda/E$, with $E$ the energy of the s-graviton. To regularize these divergences we adopted a small s-graviton mass. Notice that a mass term can be produced by a tadpole mechanism. We will see in what follows how a well defined perturbation theory puts strong constraints in the self-interaction coupling of the theory.

## 3. NRGR

The power of the EFT formalism resides in a manifest power counting in the expansion parameter of the theory, in this case the relative velocity $v$. We will pinpoint the necessary steps and refer to Goldberger's contribution for further details. The expansion of the worldline Lagrangian leads to

$$L_{pp} = \sum_a \frac{m_a}{2}\left[\mathbf{v}_a^2 - \left(1 - \frac{\mathbf{v}_a^2}{2}\right)\frac{\phi}{M} + \frac{1}{4}\mathbf{v}_a^4 + \frac{1}{4}\frac{\phi^2}{M^2}\right] + \dots, \qquad (3.1)$$

where we have chosen $x^0$ as the worldline parameter. The propagator for the field $\phi$ appearing in $L_{pp}$ is still fully relativistic, and therefore a small velocity expansion has yet to be performed. To deal with this problem it is convenient to decompose the s-graviton field into potential modes ($\bar{\phi}$) with momentum scaling $k^\mu \sim (v/r, 1/r)$ (notice they can never go on shell), and radiation modes ($\Phi$) whose momentum scales as $k^\mu \sim (v/r, v/r)$. In the EFT approach one computes the effective action perturbatively, based on systematic power counting rules. For convenience one first introduces $\Phi_\mathbf{k}$, where the large momentum piece of the potential s-graviton is factored out. The scaling laws are ($\mu r < v$)

$$\bar{\phi} \sim v/r, \quad \Phi_\mathbf{k} \sim \sqrt{v}r^2 \to \Phi \sim M\frac{v^2}{\sqrt{L}}, \qquad (3.2)$$

where $L = mvr$. The last arrow follows from the assumption that the leading order potential is given by $1/r$ and hence the virial theorem, $v^2 \sim \frac{m}{M^2 r}$, applies. However, $\lambda \neq 0$ can easily jeopardize our power counting due to the introduction

---

[1]The choice of matter coupling is meant to resemble Einstein case, at least for the $h_{00}$ mode, with $M$ playing the role of the Planck Mass. The normalization of the kinetic piece is also chosen to mimic the graviton propagator.

of a dimensionful coupling. Let us power count the first correction due to $\lambda$. The diagram is shown in fig. 2a and it scales as

$$\text{fig. 2a} \sim \left[ dx^0 \frac{m}{M} \Phi \right]^3 \left[ dx^0 \delta^3(\mathbf{k}) \lambda \Phi^3 \right] \sim L v^2 (\lambda M r^2), \tag{3.3}$$

$v^2(\lambda M r^2)$ times the leading order term which scales as $L$. It is easy to see higher order terms in $\lambda$ follow the same pattern. For $\lambda = g^2 M$, with $g$ a dimensionless coupling, we end up with $r < \frac{1}{gM}$ for the validity of the perturbative approximation and power counting. If we demand our leading order potential to match the Newtonian case we will set $M \equiv m_{Pl}$ and therefore the perturbative expansion is valid for $r < l_{Pl}/g$, with $l_{Pl} = 10^{-33}$ cm, the Planck length. To avoid entering the quantum realm we will have to fine tune $g$ to an extremely small number of the order of $10^{-40}$ for typical binary systems in the solar mass range in the inspiral regime. This obviously defies naturalness arguments and puts a flag on the phenomenological viability of such theory. Notice that the problem does not lie in the self-coupling itself but in the strength of the worldline coupling which determines the leading order scaling laws. The condition $\mu \leq v/r$ also implies a stunningly tiny s-graviton mass of the order of $10^{-30} ev$. These are consistent, and somehow equivalent, to solar system constraints. We will hereon assume $\mu r \sim v$, $\lambda r \ll 1$, and proceed with this theory as a playground.

## 4. Einstein-Infeld-Hoffmann

The leading order one s-graviton exchange can be shown to reproduce Newtonian gravity. Here we will calculate the 1PN correction to the gravitationa potential. The diagrams are shown in figures 1a, 1b and 1c for the one s-graviton exchange, and in figures 2a, 2c the non-linear effects. We will concentrate in detail in fig. 2a,

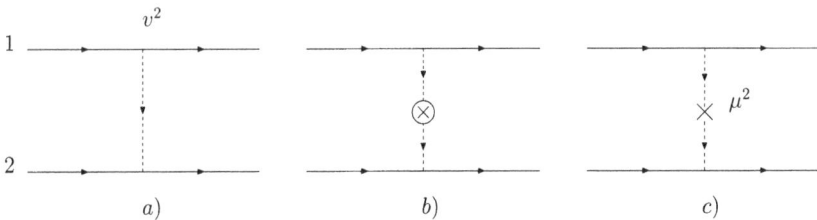

Fig. 1. One s-graviton exchange contribution at 1PN. The $\otimes$ represents a correction to the propagator, and $\times$ a mass insertion.

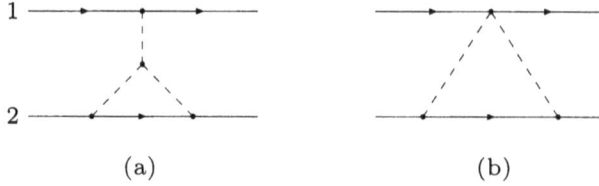

Fig. 2. Non linear contributions at 1PN.

we will display the full result later on. For the three s-graviton diagram we will have

$$\text{fig. 2a} = \frac{1}{2}\left(\frac{-im_2}{2M}\right)^2 \frac{-im_1}{2M}\int dt_1 dt_2 dt_{2'}\langle T\big(\Phi(x_1)\Phi(x_2)\Phi(x_{2'})\big)\rangle. \tag{4.1}$$

The next step would be to compute the three-point function and plug it back into (4.1), get a finite result which we will have to further expand in powers of $\mu r \sim v$ and keep the leading order piece, already at 1PN for $\lambda M r^2 \sim 1$. In the EFT spirit a better way to proceed is to treat $\mu$ as a perturbation, in the same way time derivatives are treated, by expanding the propagators in powers of $\mu/|\mathbf{k}|$. For the one s-graviton exchange this represents no harm. For higher orders one faces the problem that IR divergences will only cancel out after all the terms are included. If we are willing to accept that, one can calculate the 1PN correction by taking the massless limit and keep the (non-constant) finite piece. Therefore, introducing $d = 3 + \epsilon$ and taking $\epsilon \to 0$ one gets.[2]

$$\begin{aligned}
\text{fig. 2a} &= i\lambda\frac{3m_1 m_2^2}{64 M^3}\int dt \frac{d^3 k_2}{(2\pi)^3}\frac{d^3 k_1}{(2\pi)^3}\frac{1}{\mathbf{k}_1^2 \mathbf{k}_2^2 (\mathbf{k}_1 + \mathbf{k}_2)^2}e^{-\mathbf{k}_1\cdot(\mathbf{x}_1 - \mathbf{x}_2)}\\
&= i3\lambda\frac{G_N m_1 m_2^2}{64\pi M}\Gamma(\epsilon/2)\int dt\left(\frac{|\mathbf{x}_1 - \mathbf{x}_2|^2}{4}\right)^{-\epsilon/2}\\
&\to -i3\lambda M G_N^2 m_2^2 m_1 \int dt \log(\mu|\mathbf{x}_1 - \mathbf{x}_2|)
\end{aligned} \tag{4.2}$$

with $G_N = \frac{1}{32\pi M^2}$. Our final task consists in collecting the other few pieces. We refer to Golberger's lectures for details since the calculations are almost identical. Let us however compute the result for the new term in fig. 1c due to the s-graviton mass insertion,

$$-i\frac{m_1 m_2}{8M^2}\int dt \frac{d^3 k}{(2\pi)^3}\frac{\mu^2}{\mathbf{k}^4}e^{-i\mathbf{k}\cdot(\mathbf{x}_1 - \mathbf{x}_2)} = \frac{i}{2}\int dt\, G_N m_1 m_2 \mu^2 |\mathbf{x}_1 - \mathbf{x}_2|,$$

---

[2]The full result is a Bessel function, $K_0(\mu r)$, whose leading order piece in $\mu r$ reproduces the logarithmic potential.

which is nothing but the $\mathcal{O}(v^2)$ piece in the expansion of the Yukawa potential, $-\frac{e^{-\mu r}}{r} \sim \mu - \frac{1}{r}(1 + \frac{\mu^2 r^2}{2} + \ldots)$. Putting everything together, including mirror images, we finally obtain

$$
\begin{aligned}
L_{EIH} \; = \; & \frac{1}{8} \sum_a m_a v_a^4 - 3\lambda G_N^2 M m_2 m_1 (m_1 + m_2) \, \log\!\big(\mu|\mathbf{x}_1 - \mathbf{x}_2|\big) \\
& + \frac{G_N^2 m_1 m_2 (m_1 + m_2)}{2|\mathbf{x}_1 - \mathbf{x}_2|^2} + \frac{1}{2} \, G_N \, m_1 m_2 \mu^2 |\mathbf{x}_1 - \mathbf{x}_2| \\
& + \frac{G_N m_1 m_2}{2|\mathbf{x}_1 - \mathbf{x}_2|} \left[ v_1^2 + v_2^2 + (\mathbf{v}_1 \cdot \mathbf{v}_2) - \frac{(\mathbf{v}_1 \cdot \mathbf{x}_{12})(\mathbf{v}_2 \cdot \mathbf{x}_{12})}{|\mathbf{x}_1 - \mathbf{x}_2|^2} \right]
\end{aligned}
\tag{4.3}
$$

The logarithmic potential introduces a very interesting feature, namely a $1/r$ force into the equations of motion which resembles a *dark matter* type of effect. This is however by no means a serious candidate and we mention it only as a curiosity.

## 5. Discussion – Conclusions

The EFT approach is a powerful tool within the PN framework. Using no more than dimensional analysis many conclusions can be already drawn before dwelling into the details of the calculations. We applied the techniques in the case of a massive $\phi^3$ theory as a playground but the ideas can be easily extended to more complicated scenarios. From the NRGR power counting rules we learned that in order to produce a well defined perturbative expansion, $\lambda$ had to be fine tuned to a ridiculously small (compared with $m_{Pl}$) scale. One might however ask whether this is a feature of a $\phi^3$ theory or will this be faced in other scenarios. Let us consider a more general case,

$$
S_\phi = \int d^4 x \big( \gamma(\phi) \partial_\mu \phi \partial^\mu \phi + B(\phi) \big)
\tag{5.1}
$$

In what follows we will consider two distinct possibilities. Let us start with $B = 0$. If we expand $\gamma(\phi) \sim 1 + \phi/M + \ldots$, we will get a kinetic piece plus a potential $V(\phi)$ with terms like $(\phi/M)^n \phi \, \partial^2 \phi \; n \geq 1$. This theory is not renormalizable, and it is easy to show it resembles Einstein case. We can also show that the perturbative approach is under control by power counting the contribution from a generic term in $V(\phi)$. For an $(n + 2)$ s-graviton diagram we will get,

$$
\sqrt{L}^{n+2} \left( \frac{v^2}{\sqrt{L}} \right)^{n+2} M^2 r^2 / v \sim L v^{2n}.
\tag{5.2}
$$

For instance the first term in the expansion, $\phi^2 \partial^2 \phi$, resembles the three graviton coupling in Einstein theory (up to tensor structure). Had we chosen this interaction we would have ended up with a similar 1PN correction as in the original Einstein-Infeld-Hoffmann action.

The $B \neq 0$ case is substantially different. Let us study a generic term, $g M^4 (\phi/M)^n$, with $g$ a dimensionless coupling and $n \geq 4$. The $n$ s-graviton diagram will scale as $g M^4 r^4 v^{2n-1} \sim g L^2 v^{2n-7} \sim L \, g(m/M)^2 v^{2n-8}$. We would have then to impose $g v^{2n-8}(m/M)^2 < 1$ to get a controlled perturbative expansion. For the marginal case, $n = 4$, setting $M = m_{Pl}$ one needs $g < \left(\frac{m_{Pl}}{m}\right)^2 \sim 10^{-70}$! for solar mass binary constituents. Notice that this problem arises at the *classical* level since the coupling to elementary particles is already too small to represent any trouble. Is only in the superposition of terms, which build up the massive lump of the star, that the perturbative expansion breaks down. From this analysis we conclude that in pure scalar gravity non-derivative self-interactions are extremely constrained.

One could then wonder about more general models including scalar fields, like tensor-scalar gravity. In the latter in addition to the graviton field a scalar interaction is added with an action similar to (5.1) in a curved spacetime background. Within this type of scenarios the problems we encountered here can be cured by modifying the power counting. For instance, a large mass can be added to the scalar field (larger than the inverse of the solar system distance), which will render the field a negligible short range interaction. Another possibility would be to keep it nearly massless but weaken the coupling to matter to a much feeble strength $M \gg m_{Pl}$. In this case the 3-scalar diagram (fig. 2a) will now scale as $L \frac{\lambda}{mv^2} \left(\frac{m}{M}\right)^3$. For $\lambda \sim gM$ one needs $g(m/M)^2 < v^2$ in order to be treated as a perturbation. By naturalness argument one would expect $g \sim 1$, and we will then have a very tiny coupling to elementary particles. For instance, the coupling to a proton will be of the order of $m_{proton}/M \sim 10^{-60}$! For a $\phi^4$ theory the 4-scalar diagram would scale as $\tilde{\lambda}(m/M)^3 (Mrv^2)^{-1}$ with respect to the Newtonian potential. Again, for $\tilde{\lambda} \sim 1$ one needs $M \sim m$ and the coupling to elementary particles is also highly suppressed. Both solutions will keep the theory consistent with experimental data, both rely however in the introduction of a high mass scale into the theory, much higher than the Planck scale. Perhaps this is an indication that non-derivative self-interactions are not present in nature.

## Acknowledgments

We would like to thank Walter Goldberger and Ira Rothstein for helpful comments and discussions. Special thanks to Christophe Grojean and Francis Bernardeau for organizing such a great summer school and give us the chance to contribute to this volume.

# DARK MATTER IN MINIMAL TRINIFICATION

E. Di Napoli[1], D. George[2],
M. Hertzberg[3], F. Metzler[4] and E. Siegel[5]

[1] *University of North Carolina, Department of Physics and Astronomy, CB 3255 Phillips Hall,*
*Chapel Hill NC 27599-3255, USA*
[2] *The School of Physics, University of Melbourne, Victoria, 3010 Australia*
[3] *MIT, Center for Theoretical Physics, 77, Massachusetts Av., Cambridge MA 02139, USA*
[4] *Institut für Theoretische Teilchenphysik, Universität Karlsruhe, 76128 Karlsruhe, Germany*
[5] *University of Wisconsin, Department of Physics, 1150 University Avenue, Madison WI 53706, USA*

## 1. Introduction

In this letter we would like to present the preliminary and suggestive results of a more ambitious and extensive research project. We study an example of Grand Unified Theory (GUT) in the context of certain requirements dictated by cosmology. In other words we require the model in exam to address questions like "Is there a Dark Matter (DM) candidate? How abundant is this at present?" or "Can we find successful mechanisms for Baryogenesis and Reheating?". These questions arise from the more general program of using present cosmological data to constrain the enormous proliferation of phenomenological works describing physics beyond the Standard Model (SM).

We are going to concentrate on one of such models, known as "trinification", introduced for the first time in 1984 by S. Glashow [1], successively studied in detail by Babu et al. [2] and more recently by Willenbrock et al. [3, 4]. In particular we consider the minimal formulation of the model (respect to how the SM is embedded in it), and focus on the question of Dark Matter. We show that the model does *not* have a stable DM candidate compatible with $\Omega_{DM} \leq 0.24$. We find that if we adjust the parameters such that we have a stable candidate, there is far too much DM. On the other hand, making the candidate unstable conflicts with Big Bang Nucleosynthesis (BBN) constraints. In section 2 we are going to briefly present the model, list its salient features and focus on its advantages and downsides. We present our results in section 3. We conclude in

517

section 4 discussing possible improvements of the model that would circumvent its difficulty in providing a viable DM candidate.

## 2. Trinification in a nutshell

The name "trinification" comes from its gauge symmetry: a triple replica of $SU(3)$ conventionally written as $SU_C(3) \times SU_L(3) \times SU_R(3) \times \mathbb{Z}_3$. The discrete group $\mathbb{Z}_3$ guarantees the gauge couplings of the single $SU(3)$ factors are the same at the GUT scale. The SM embedding is obtained by identifying the $SU_C(3)$ with the QCD gauge symmetry while the electroweak gauge group emerges as a result of breaking the other two $SU(3)$ factors. Each SM fermion generation is embedded in a $\mathbf{27} = (1, 3, \bar{3}) \oplus (\bar{3}, 1, 3) \oplus (3, \bar{3}, 1)$ representation of the gauge group. In order to better understand the field content we can re-express this representation in terms of SM quantum numbers

$$E \equiv \left(1, 2, -\frac{1}{2}\right) \qquad N_1 \equiv (1, 1, 0) \qquad E^c \equiv \left(1, 2, \frac{1}{2}\right)$$

$$L \equiv \left(1, 2, -\frac{1}{2}\right) \qquad N_2 \equiv (1, 1, 0) \qquad e^c \equiv (1, 1, 1)$$

$$Q \equiv \left(3, 2, \frac{1}{6}\right) \qquad u^c \equiv \left(\bar{3}, 1, \frac{2}{3}\right) \qquad d^c \equiv \left(\bar{3}, 1, -\frac{1}{3}\right)$$

$$B \equiv \left(3, 1, -\frac{1}{3}\right) \qquad B^c \equiv \left(\bar{3}, 1, -\frac{1}{3}\right).$$

We immediately notice that each generation includes the usual lepton and quark doublets and singlets plus some additional fields: we get two additional lepton doublets $E$ and $E^c$, two neutral singlets $N_1$ and $N_2$ and two quark singlets $B$ and $B^c$.

In order to have unification of the SM couplings at the GUT scale we also need two copies of scalars both in the same $\Phi^a(\mathbf{27}) \equiv \Phi_\ell^a(1, 3, \bar{3}) \oplus \Phi_{q^c}^a(\bar{3}, 1, 3) \oplus \Phi_q^a(3, \bar{3}, 1)$ $a = 1, 2$ representation the fermions are in. Let's concentrate only on the $\Phi_\ell^a$ and write its field content in terms of SM quantum numbers

$$\phi_{1\ell}^a \equiv \left(1, 2, -\frac{1}{2}\right) \qquad \phi_{2\ell}^a \equiv \left(1, 2, \frac{1}{2}\right) \qquad \phi_{3\ell}^a \equiv \left(1, 2, -\frac{1}{2}\right)$$

$$S_{1\ell}^a \equiv (1, 1, 0) \qquad S_{2\ell}^a \equiv (1, 1, 1) \qquad S_{3\ell}^a \equiv (1, 1, 0)$$

The breaking $[SU(3)]^3 \rightarrow SU_C(3) \times SU_L(\_) \times U_Y(1)$ is obtained giving vevs to some of the singlets $S_{i\ell}^a$. The most general choice being ($S_{2i}^a$ are electrically charged so they cannot assume a vev)

$$\langle S^1_{3\ell}\rangle = v_1 \qquad \langle S^2_{1\ell}\rangle = v_2 \qquad \langle S^2_{3\ell}\rangle = v_3, \tag{2.1}$$

with $v$'s $\sim O(M_{GUT} = 10^{14}\,\text{GeV})$ [3]. In the same fashion the electroweak symmetry is broken giving vevs to the electrically neutral components of the doublets $\phi^a_i$ charged under the $SU_L(2)$. A very general choice is

$$\langle(\phi^2_{1\ell})^0\rangle = n_1 \qquad \langle(\phi^2_{2\ell})^0\rangle = n_2 \qquad \langle(\phi^2_{3\ell})^0\rangle = n_3$$
$$\langle(\phi^1_{1\ell})^0\rangle = u_1 \qquad \langle(\phi^1_{2\ell})^0\rangle = u_2, \tag{2.2}$$

with $u_i, n_i \sim O(M_{EW})$. The other scalar fields $\Phi^a_{q^c}$ and $\Phi^a_q$ do not acquire any vev since they carry color charge and would break $SU_C(3)$. They are generically very heavy due to radiative corrections to their mass terms and do not show up in the low energy spectrum of the theory. From now on we will assume their masses to be of the order of $M_{GUT}$.

## 2.1. A simple case

In a simplified version of the the model, we set $n_1 = n_2 = n_3 = v_3 \equiv 0$ but keep all other vevs ($u_1, u_2, v_1, v_2$) non-zero. The qualitative results will be exactly equal to the more general case and suffice to illustrate our point. A linear combination of the $\phi^a_{i\ell}$ and four of the $S^a_{i\ell}$ are eaten by twelve of the gauge bosons[1] that become heavy with masses proportional to the $v_i$. Fine-tuning the quartic couplings, it is possible to obtain at most 5 light $(\sim M_{EW})$ Higgs doublets. At the same time Yukawa terms give masses for the fermions at the tree level. In general, such terms can be built pairing two fermion doublets and a scalar singlet or one scalar doublet contracting indices with one fermion doublet plus one one fermion singlet. In the first case we end up with a heavy (mass $\sim M_{GUT}$) Dirac fermion meanwhile in the second case the fermions are light (mass $\sim M_{EW}$). Limiting our analysis to one fermion generation we obtain at tree level

$$m_B \simeq \sqrt{g_1^2 v_1^2 + g_2^2 v_2^2} \qquad\qquad m_E \simeq \sqrt{h_1^2 v_1^2 + h_2^2 v_2^2}$$
$$m_u = g_1 u_2 \qquad m_d \simeq g_1 u_1 \qquad m_e \simeq h_1 u_1 \tag{2.3}$$
$$m_v = h_1 u_2 \qquad m_{N_1} = h_1 u_2 \qquad m_{N_2} \simeq \frac{h_1^2 u_1 u_2}{m_E},$$

where $h_a$ and $g_a$ are couplings associated with the Yukawa terms proportional to $\phi^a_{i\ell}$. This spectrum has some positive qualitative features and a negative one.

---

[1] The gauge bosons are in the $\mathbf{24} = (1, 1, 8) \oplus (1, 8, 1) \oplus (8, 1, 1)$ adjoint representation of $[SU(3)]^3$.

$B$, $B^c$ and $E$, $E^c$ pair up to become very heavy Dirac fermions. The up and down quarks as well as the electron get different light masses. Unfortunately $N_2$, $N_1$ and $\nu$ are also light, with the additional inconvenience that the last two seem to pair up to form a light Dirac fermion. This is highly undesirable and can be fixed invoking radiative correction induced by cubic scalar couplings [3, 4]. Calculating the one loop contribution to mass terms for light fermions and using a seesaw mechanism the spectrum for these light neutral fields become

$$m_{N_{1,2}} \sim g_q^2 F \qquad m_\nu \sim \frac{h_1^2 u_2^2}{g_q^2 F} \tag{2.4}$$

where $F \leq M_{GUT}$ is a factor of pure one-loop origin. In the presence of other two fermion generations there are additional $N_i$'s and they can mix up and appear in a sort of hierarchy where the lightest of all can be pushed as far down as $10^5$ GeV.

## 2.2. Light and darkness of trinification

We end this section by briefly reviewing the positive and negative features of this GUT model. Thanks to the two sets of scalar fields (six weak doublets $\phi_{i\ell}^a$) this model achieves unifications of the SM gauge couplings at a scale around $10^{14}$ GeV. The mechanism is similar to that of generic SUSY SU(5) GUT (although this unifies at $10^{16}$ GeV): the six Higgs contribution to the $\beta$-function is equivalent to the contribution of the two SUSY Higgs doublets and their fermion superpartners.

SM masses for fermions can be arbitrarily adjusted through Yukawa couplings to fit the experimental values. From this point of view Trinification is not any more predicting than the SM itself. Masses for the scalars are not protected and receive one loop quadratic corrections requiring fine tuning for at least one Higgs light doublet in $\Phi_\ell^a$. Trinification does not provide any mechanism to solve the so called hierarchy problem.

Unlike SU(5), in trinification gauge bosons conserve Baryon number and so do not mediate proton decay: the proton can only decay through Yukawa interactions. An acceptable value for its lifetime is recovered without the need of fine-tuning the involved Yukawa couplings.

Due to abundant number of scalar fields, it is possible that baryogenesis is achieved at GUT scales [6] or at electroweak scales through a first order phase transition [7]. Heavy scalars may be important in some Inflationary scenario. Moreover light neutral singlets may play the role of sterile neutrinos and be used to invoke a Leptogenesis mechanism [5]. The punch line being that this model offers a wide variety of possible developments in the context of Cosmology. In the next section we will address one of these issues. We investigate the possibility that the lightest of the neutral singlets can function as a viable candidate for Dark Matter and try to give an indicative answer.

## 3. A Dark Matter candidate

Let us indicate the lightest neutral singlet as $N_\chi$. There are two main requirements that $N_\chi$ needs to satisfy in order to be a possible Dark Matter candidate. First, its relative abundance $\Omega_{N_\chi}$ has to be less than or equal to 0.24. Second, its decay time needs to be much longer than the age of the Universe.

### 3.1. Relative abundance

The relative abundance at present (assuming the particle is stable) is calculated using the Boltzmann equation. The form of its solution depends on the regime (relativistic or not) at which it is approximated. To check if $N_\chi$ was non-relativistic at freeze out we need to verify under which condition the inequality $x_f = \frac{m}{T} \geq 3$ is valid. If we assume that $N_\chi$ is non-relativistic the value of $x_f$ depends logarithmically on the annihilation cross section and the mass of $N_\chi$ [8].

$$x_f \simeq \ln\left(0.038\frac{g}{\sqrt{g_*}}M_{PL}m_{N_\chi}\langle\sigma v\rangle\right) \tag{3.1}$$

Since we are interested only in estimating this expression we consider the dominant contribution to the annihilation channel. This is given by the following tree level Feynman graph[2]

$$(3.2)$$

Assuming $\phi_{i\ell}^a$ and $\phi_{i\ell}^{*a}$ are light Higgs we are lead to the s-wave expression for the cross-section

$$\langle\sigma v\rangle \sim \frac{h^4}{(4\pi)^2 m_E^2}. \tag{3.3}$$

---

[2] The contribution coming from two $N_\chi$ singlets annihilating in a virtual heavy $X$ boson is even more suppressed since it is proportional to $\frac{\alpha_{gauge}^2}{m_\chi^2}$ and $m_\chi^2 \sim v_i^2 g_{gauge}$.

Here $h$ is a generic Yukawa coupling. Plugging this expression in (3.1) and assuming the reasonable range $10^5 \, \text{GeV} \leq m_{N_\chi} \leq 10^{10} \, \text{GeV}$ we arrive at the conclusion that $\frac{v_i}{h} \leq 10^9 \, \text{GeV}$ in order to have $x_f \geq 3$. Since $v_i \sim M_{GUT}$ the inequality cannot be satisfied implying that $N_\chi$ is highly relativistic when it freezes out.

We then equate the annihilation rate of $N_\chi$ to the Hubble rate $H$ to obtain the freeze-out temperature

$$T_f \approx 9.1 \frac{\sqrt{g_*(T_f)}}{M_{Pl}\langle \sigma v \rangle} \sim 10^{12} \sqrt{g_*} h^{-2} \, \text{GeV}. \tag{3.4}$$

This gives a present-day abundance that is rather insensitive to $\langle \sigma v \rangle$ and much too large, namely

$$\Omega_{N_\chi} \sim 2 \times 10^8 g_{*S}^{-1} \frac{m_{N_\chi}}{\text{GeV}}. \tag{3.5}$$

### 3.2. Decay time

The previous estimate is based on the assumption that $N_\chi$ is stable over the course of the universe requiring that $\tau_{N_\chi} \gg 10^{10} \, yr$. However, this is not correct for most reasonable choices of parameters in this model. We find that it decays not too long after it freezes out, and tends to destroy Big Bang Nucleosynthesis (BBN) unless we adjust some parameters appropriately. In order to estimate the lifetime of $N_\chi$ we will consider only the dominant contribution, as we did earlier for the cross section. The most favorable decay channel is given by the following

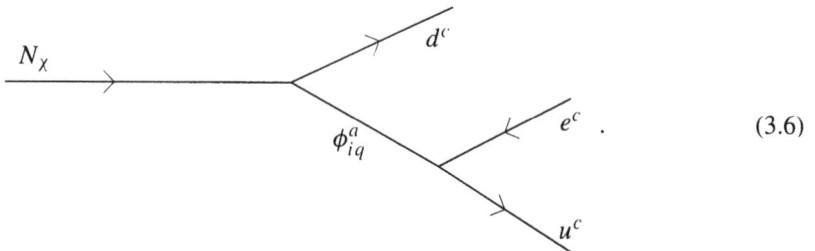

$$\tag{3.6}$$

In this graph $\phi_{iq}^a$ is one of the scalars carrying color charge with mass of the order of $M_{GUT}$. In calculating the decay rate it is a very good approximation to consider the decay products massless since they are highly relativistic. In such an approximation the decay rate is

$$\Gamma_{N_\chi} \sim \frac{128 g_q^4}{(2\pi)^2} \frac{m_{N_\chi}^5}{m_{\phi_{iq}^a}^4} \tag{3.7}$$

with $g_q$ being the Yukawa coupling associated with the heavy colored scalar. For $m_{\phi^\alpha_{t_q}} \sim 10^{14}$ GeV and $m_{N_y} \sim 10^5$ GeV we get the following estimate for the decay time

$$\tau_{N_\chi} \sim 10^{30} g_q^{-4} \, \text{GeV}^{-1} \sim 10^6 g_q^{-4} \, s. \tag{3.8}$$

For a typical value of $g_q = 0.1$ the decay time is about a 600 years making $N_\chi$ rather unstable and completely absent at present time in the universe. Moreover its decay products are so highly relativistic and get produced in such abundance that they destroy any product of BBN. We can reverse the reasoning and put an upper bound on the Yukawa coupling that stabilizes $N_\chi$

$$\tau_{N_\chi} = 6.3 \times 10^{-2} g_q^{-4} \, yr \gg 10^{10} \, yr \quad \Longrightarrow \quad g_q \ll 2 \times 10^{-3}. \tag{3.9}$$

This fine-tuning of $g_q$ is not so unreasonable, but it may negatively influence the one-loop corrections to the mass of the neutral singlets compromising the efficiency of the radiative seesaw mechanism.

## 4. Conclusion and discussion

We have seen that the lightest of the neutral singlets in the minimal trinification model is greatly overabundant at freeze out with the consequence that it is completely ruled out by cosmological data. This may be overcome if the particle is not very stable but then it ruins BBN and creates another big problem. As it stands this GUT model does not seem to be a complete model of particle physics since it does not withstand one of the most needed cosmological requirements: the existence of Dark Matter.

There is room for improvement though. On one side we may introduce a $\mathbb{Z}_2$ symmetry like the one introduced in SUSY models and make $N_\chi$ more stable without the need of adjusting the Yukawas opportunely. Such a discrete symmetry would forbid the Yukawa terms responsible for the dominant decay channel and stabilize the singlet. On the other hand we have to admit that our calculation of the cross-section for the relative abundance is a little "naive" since it doesn't take into account possible mixing of $N_i$'s with the other neutral Weyl fermions of the model. For example within just one generations of leptons we have $E_0 \in E$ and $E_0^c \in E^c$ plus the two neutral singlets, $N_1$ and $N_2$, and the neutrino $\nu$. The mass matrix of these five neutral fields is far from diagonal and can provide some mixing between them. The net result is that the lightest among them may have some non-zero coupling with the $Z^0$ boson greatly enhancing the cross section amplitude. This mechanism may help in reducing the abundance to more accepted values.

*E. Di Napoli et al.*

## Acknowledgments

We would like to thank the organizers of the LXXXVI "Les Houches" summer school, the staff and all our colleagues at the school that made our stay enjoyable and this work possible.

## References

[1] S.L. Glashow, in *Fifth Workshop on Grand Unification* edited by K. Kang, H. Fried and F. Frampton (World Scientific, Singapore, 1984). p. 88.
[2] K.S. Babu, X.G. He, and S. Pakvasa, Phys. Rev. D**33**, 763 (1986).
[3] S. Willenbrock, Phys. Lett. B**561**, 10 (2003).
[4] J. Sayre, S. Wiesenfeldt, and S. Willenbrock, Phys. Rev. D**73**, 035013 (2006).
[5] M. Fukugita and T. Yanagida, Phys. Lett. B**174**, 45 (1986).
[6] D.V. Nanopoulos and S. Weinberg, Phys. Rev. D**20**, 2484 (1979).
[7] J.M. Cline and P.A. Lamieux, Phys. Rev. D **55**, 3873 (1997).
[8] E.W. Kolb and M.S. Turner, *"The Early Universe"* Addison and Wesley (1990).

# TRANS-PLANCKIAN PHYSICS FROM COSMOLOGY

S.E. Jorás[1], G. Marozzi[2] and R. Ansari[3]

[1] *Institute of Physics, Federal University of Rio de Janeiro – IF-UFRJ, P.O. Box 68528, 21941-972, Rio de Janeiro, RJ – Brazil*
[2] *Dipartimento di Fisica, Università degli Studi di Bologna and INFN via Irnerio 46, I-40126 Bologna, Italy*
[3] *School of Physics, University of Hyderabad, Hyd-46, India*

## 1. Introduction

Inflationary models provide answers to many problems in standard Big Bang cosmology, in particular the origin of density fluctuations and the spectrum of cosmic microwave background anisotropies. The basis of the whole mechanism is the stretching of quantum fluctuations generated at sub-Hubble scales due to the exponential expansion of the spacetime during inflation. This model, however, has a serious "problem": if we consider a plain scalar-field driven inflationary model — say, chaotic inflation — the period of inflation lasts so long that the wavelengths of the fluctuations which at present correspond to cosmological scales were sub-Planckian at the begining of the inflationary phase. Therefore, the evolution of fluctuations at such scales is supposed to follow different rules from those provided by the standard theory of cosmological perturbations — which is based on quantum field theory and general relativity. The set of rules expected to hold above this energy scale is the so-called Trans-Planckian Physics (TPP from now on). The TPP could lead to deviations from standard predictions on the cosmic microwave background radiation (CMBR), which probes the scales we mentioned before. The question to be asked is "whether the predictions of the standard cosmology are insensitive to effects of TPP". This is the precise statement of Trans-Planckian problem[1] [2,3].

In Sec. 2 we present the standard approach for a test scalar field in a FRW universe and we introduce the WKB approximation. We conclude in Sec. 3 with a particular way to approach TPP: the modification of the dispersion relation.

---

[1] See [1] and references therein for the role of TPP in the black hole thermal radiation.

## 2. Standard approach

The FRW line element with flat spatial section without cosmological fluctuations is given by:

$$ds^2 = g_{\mu\nu}dx^\mu dx^\nu = -dt^2 + a^2(t)d\vec{x}^2 = a^2(\eta)\left[-d\eta^2 + d\vec{x}^2\right] \tag{2.1}$$

where $\eta$ is the conformal time. In the following, we restrict ourselves to the de Sitter case $a(\eta) = -1/(H\eta)$.

If we expand a test scalar field $\phi(\eta, \vec{x})$ in Fourier modes

$$\phi(\eta, \vec{x}) = \frac{1}{a(\eta)}\frac{1}{(2\pi)^{3/2}}\int d^3k\, \mu_k(\eta)\, e^{i\vec{k}\cdot\vec{x}}, \tag{2.2}$$

then the equation of motion for each mode is given by[2]

$$\frac{d^2\mu_k}{d\eta^2} + \left(k^2 - \frac{1}{a}\frac{d^2a}{d\eta^2}\right) = 0, \qquad \text{or} \tag{2.3}$$

$$\frac{d^2\mu_k}{dk_p^2} + \left(\frac{1}{H^2} - \frac{2}{k_p^2}\right)\mu_k = 0, \tag{2.4}$$

where we use the physical wavenumber $k_p \equiv k/a(\eta)$ as the independent variable instead of $\eta$ in Eq. (2.4). The equations above can be exactly solved in a de Sitter background.[3] The two-point correlation function and the related power spectrum are given by

$$\langle 0|\phi(\eta, \vec{x})\phi(\eta, \vec{x} + \vec{r})|0\rangle = \int_0^{+\infty}\frac{dk}{k}\frac{\sin kr}{kr}k^3 P_\phi(k) \tag{2.5}$$

$$k^3 P_\phi(k) = \frac{k^3}{2\pi^2}\left|\frac{\mu_k}{a}\right|^2. \tag{2.6}$$

For super-horizon modes ($k_p \ll H$) the spectrum is time-independent and scale-invariant, since in this limit $\mu_k \propto a$. Eq. (2.4) can also be interpreted as a Schrödinger equation for a stationary wave function corresponding to a particle with energy $E = H^{-2}$ in an effective potential $V(k_p) = 2k_p^{-2}$ which is a function of the "spatial" variable $k_p$. Whenever WKB approximation [5] holds, i.e., when

$$WKB \equiv \left|\frac{1}{\Omega^2}\left[\frac{3}{4}\left(\frac{\Omega'}{\Omega}\right)^2 - \frac{1}{2}\frac{\Omega''}{\Omega}\right]\right| \ll 1, \tag{2.7}$$

---

[2]This is the same equation of motion for tensor perturbations. Scalar perturbations are absent if the background is exactly de Sitter [4]. They are, however, perfectly well-defined in a slow-roll approximation and in other power-law models.

[3]It can also be solved at first order in slow-roll approximation.

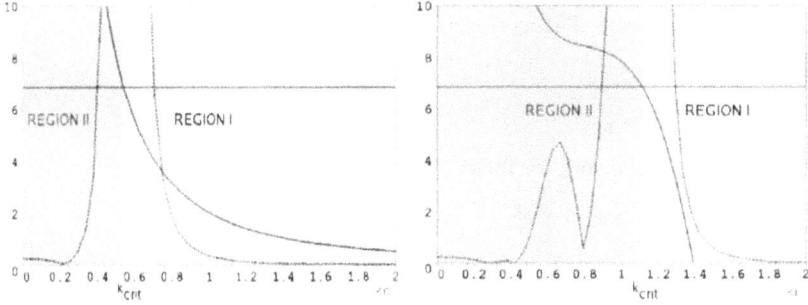

Fig. 1. Total energies (full straight lines), effective potentials (full lines), and WKB parameters (dashed lines) as a function of the physical wavenumber $k_p$ (in Planck-mass units). In both panels the WKB parameter has been multiplied by 2 and $H = 0.38$ (for the sake of comparison with Ref. [4]). **Left panel:** for the linear d.r.. **Right panel:** for the d.r. given in Eq. (3.2).

where ( $'$ ) $\equiv d/dk_p$ and $\Omega \equiv \sqrt{E - V(k_p)}$, the solution of Eq. (2.4) is given by a linear combination of plane waves with frequencies $\pm\Omega$. Figure 1 (left panel) shows the behavior of the WKB parameter (2.7) as a function of $k_p$. According to this plot (left panel), the WKB approximation holds for $k_p \to \infty$ but not for values close to zero, where it equals 0.125 (i.e, *not* much smaller than 1). The WKB approximation fails exactly where it is supposed to: at a classical turning point and where the effective potential is too steep. The critical value is $k_{crit} \equiv \sqrt{2}H$. In region I, where $E \gg V_{eff}$, the solution of Eq. (2.4) is a plane wave with comoving frequency $\omega = k$ in the conformal time, as expected.

## 3. Trans-Planckian behavior: non-linear dispersion relation

Unruh [6] proposed a modification of the dispersion relation (d.r. from now on) for describing high-energy Physics for studying the black-hole radiation emission. He was inspired by sonic holes, since the equations of motion for the sound waves cease to be valid when the wavelength gets closer to and smaller than the lattice spacing. Jacobson and Corley [1, 3] also proposed nonlinear terms in the d.r. that could be justified by the inclusion of higher-order derivatives in the Lagrangian. Other changes can also be introduced by arguing that the spacetime symmetries migh not survive at high energies [7]. In this approach one substitutes the comoving wavenumber $k$ in Eq. (2.3) using the physical frequency $\omega_p$ with $a(\eta)\omega_p(k_p)$ and assumes that $\omega_p$ is a nonlinear function of $k_p$ which differs from the standard (linear) one only for physical wavelengths closer to and smaller than the Planck scale. That corresponds to the substitution

$$\frac{1}{H^2} \rightarrow \frac{1}{H^2} \frac{\omega_p^2(k_p)}{k_p^2} \tag{3.1}$$

in Eq. (2.4). In this work we focus on the d.r.

$$\omega_p^2(k_p) = k_p^2 - \alpha k_p^4 + \beta k_p^6, \tag{3.2}$$

with $\alpha > 0$ and $\beta > 0$, proposed in [8]. The above expression can be seen as a mere series expansion, but it is also found in solid state physics [9] and describes the *rotons*. Following the discussion in Section 2, we plot in fig. 1 (right panel) the analogous quantities to the total energy $E = H^{-2}$, the WKB parameter (2.7) and the effective potential, with the latter given by[4]

$$V_{\text{eff}}(k_p) \equiv \frac{\alpha}{H^2} k_p^2 - \frac{\beta}{H^2} k_p^4 + \frac{2}{k_p^2} \tag{3.3}$$

As it is clearly seen, the WKB approximation is valid only when[5] $k_p \rightarrow \infty$. The paper [10] provides a convenient change of variables to circumvent this problem. Upon the changes of coordinates $x \equiv \ln(H/k_p)$ and $u \equiv \mu \exp(x/2)$, Eq. (2.4) is written as

$$\frac{d^2 u}{dx^2} + \tilde{\Omega}^2(x) u = 0, \quad \tilde{\Omega}^2(x) \equiv -\frac{9}{4} + e^{-2x} \left(1 - \alpha H^2 e^{-2x} + \beta H^4 e^{-4x}\right)$$

One can promptly verify that the WKB approximation is valid for all values of $x$ except in the vicinity of the turning point $\tilde{\Omega}^2(x_{crit}) = 0$. The exact value of $x_{crit}$, where WKB parameter diverges, depends on $\alpha$ and $\beta$. Since the behavior of the WKB parameter is qualitatively different from the linear case (see fig. 1) and even the divergence point depends on the values of the parameters, it is expected that the outcome (namely, fluctuations on the CMBR) will be different as well. We are currently calculating [11] the spectrum for different sets of $\{\alpha, \beta\}$.

## Acknowledgments

We would like to thank A. Sarkar and M. Schulze for discussions during the Inflation Working Group, R.H. Brandenberger for useful correspondence and the organizers and the staff of the 86th Session of Les Houches Summer School, who made such a nice working environment for all of us. S.E.J. acknowledges financial support from the School and from CNPq.

---

[4]We used $\alpha = \sqrt{24/7}$ and $\beta = 1$, picked in the range suggested by [4].
[5]To be confronted with [4].

# References

[1] S. Corley and T. Jacobson, Phys. Rev. **D 54**, 1568 (1996).

[2] R.H. Brandenberger, hep-ph/9910410.

[3] J. Martin and R.H. Brandenberger, Phys. Rev. **D 63**, 123501 (2001) and R.H. Brandenberger and J. Martin, Mod. Phys. Lett. A **16**, 999 (2001).

[4] J. Martin and R. Brandenberger, Phys. Rev. **D 68**, 063513 (2003). R.H. Brandenberger and J. Martin, Phys. Rev. **D 71**, 023504 (2005).

[5] Eugen Merzbacher, "Quantum Mechanics" (John Wiley & Sons), 3$^{rd}$ edition.

[6] W. Unruh, Phys. Rev **D 51**, 2827 (1995).

[7] S. Cremonini, Phys. Rev. **D 68**, 063514 (2003); J. Kowalski-Glikman, Phys. Lett. **B 499**, 1 (2001) and H.-C. Kim *et al.*, Phys. Rev. **D 72**, 103523 (2005).

[8] M. Lemoine, M. Lubo, J. Martin and J.-P. Uzan, Phys. Rev. **D 65**, 023510 (2002).

[9] A.A. Abrikosov, "Methods of Quantum Field Theory in Statistical Physics" (Dover Pub.). T. Jacobson, Phys. Rev. **D 44**, 1731 (1991).

[10] J. Martin and D.J. Schwarz, Phys. Rev. **D 67**, 083512 (2003).

[11] S.E. Jorás and G. Marozzi, in preparation.

# A PEDAGOGICAL PRIMER ON PREHEATING

M.A. Amin[1], A. Brown[2], A.D. Chambers[3], S.E. Jorás[4],
A. Kumar[5], F. Metzler[6], J. Pritchard[7] and N. Sivanandam[1]

[1] *Stanford University, USA*
[2] *Columbia University, USA*
[3] *Imperial College London, UK*
[4] *Federal University of Rio de Janeiro, Brazil*
[5] *New York University, USA*
[6] *Universitat Karlsruhe, Germany*
[7] *California Institute of Technology, USA*

## 1. Introduction

It is becoming increasingly difficult to deny that the inflationary paradigm [1] is a central part of our cosmology. Moreover, whilst slow roll inflation can be described in a seductively simple fashion – an (almost) uniform, slowly decaying scalar field drives an exponential expansion period, which ends with a decay into a hot mix of particles that "reheat" the universe – the details are subtle, complex and rich. With this in mind we seek, in this brief primer, to outline some of the details of the final stages of inflation. In particular, we shall describe the process of preheating [2, 3], a period where the energy of the inflaton field is rapidly transferred to another coupled, scalar field via the phenomenon of parametric resonance.

We will begin with an overview of reheating, describing both the old perturbative scenario, and its extension through the preheating mechanism. With respect to the latter we shall outline the salient features of parametric resonance and explain how the phenomenon is implemented for the inflaton. Following this overview, there will be a somewhat sharpened focus on the behavior of gravitational fluctuations during the era of preheating – we wish to address both how they evolve through preheating and also how they can be generated during that period [4].

Throughout this paper details will be sacrificed for clarity and brevity – the reader is urged to delve into the references to fill in the minutiae.

## 2. Inflation and its end

The dynamics of single-field, slow-roll inflation are controlled by the following equations:

$$\ddot{\phi} + 3H\dot{\phi} + V'(\phi) = 0 \tag{2.1}$$

$$H^2 = \frac{8\pi}{3m_{pl}^2}\left(\frac{1}{2}\dot{\phi}^2 + V(\phi)\right) \tag{2.2}$$

The former is, of course, the equation of motion for a scalar field in an FRW background, and the latter is the Friedman equation. Whilst the complete solution to this equation is involved, it can be shown that in the slow-roll regime the solution (for $V(\phi) = m^2\phi^2/2$) is given by $\phi \approx \phi_0 - \frac{m_{pl}m}{2\sqrt{3\pi}}t$, and in the reheating regime (i.e. the oscillatory one) by:

$$\phi \approx \sqrt{\frac{2}{3}}\frac{2}{mt}\sin mt \tag{2.3}$$

The naive way to extract the energy from this field would be to couple it weakly to some set of elementary particles, let it decay perturbatively and then allow the newly created particles to thermalize. If we do this, we find that observations of the CMB power spectrum constrain our coupling in such a way as to give a low reheat temperature. To see this, note that in this scenario the reheat temperature will be set by the energy density at which the inflaton decays. This, in turn, occurs when the decay rate is equal to the Hubble constant [2]:

$$\Gamma \sim H \sim \frac{g_* T_r^2}{m_{pl}} \Rightarrow T_r \sim 0.1\sqrt{\Gamma m_{pl}}, \quad \text{using} \quad g_* \sim 100 \tag{2.4}$$

For a simple coupling $g$ to some other field, we have $\Gamma \sim m_\phi g^2$. The microwave background constrains $m_\phi \sim 10^{-6}m_{pl}$; protecting this mass from quantum corrections ($\delta m_\phi^2 \sim g^2 m_{pl}^2$) keeps $g$ small, $g \sim 10^{-6}$. Putting all of this together gives a reheat temperature of $T_r \sim 10^9\text{GeV}$.

## 3. A simple model of preheating

As neat as the above scenario is, a closer examination of the decay process reveals that in significantly-sized regions of parameter space it is not the appropriate description. Rather, following Kofman et al. in [2], we need to consider preheating.

Here the coupling of the coherently oscillating inflaton to another scalar field results in decay through parametric resonance. The modes of this field self interact and decay into elementary particles which interact and thermalize.

To understand the preheating phenomenon, let us consider a simple model with the action:

$$S = \int d^4x \sqrt{-g} \left\{ \frac{m_{pl}^2}{16\pi} R - \frac{1}{2}(\nabla\phi)^2 - \frac{1}{2}(\nabla\chi)^2 - V(\phi, \chi) \right\} \tag{3.1}$$

$$V(\phi, \chi) = \frac{1}{2}m^2\phi^2 + \frac{1}{2}g^2\phi^2\chi^2 \tag{3.2}$$

Assuming the usual FRW metric and an oscillatory form for $\phi$ ($\phi(t) = \Phi(t)\sin mt$) we get the following mode equation for $\chi$:

$$\ddot{\chi}_k + 3H\dot{\chi}_k + \left[\frac{k^2}{a^2} + g^2\Phi^2\sin^2 mt\right]\chi_k = 0 \tag{3.3}$$

A cursory examination of this equation reveals that the effective mass of the $\chi$ field depends on $\phi$. Essentially, as the $\phi$ field goes though its minimum, the mass of the effective $\chi$ field becomes small, leading to an explosive production of $\chi$ particles. In fact, the above expression can be easily massaged into the form of the well-known Mathieu equation:

$$\chi_k'' + 3H\chi_k + [A - q\cos 2z]\chi_k = 0, \quad A = 2q + \frac{k^2}{m^2a^2}, \quad q = \frac{g^2\Phi^2}{4m^2} \tag{3.4}$$

where $z = mt$. For particular values of $A$ and $q$, this equation has unstable solutions, where the amplitude of a particular mode grows exponentially, $\chi_k \propto e^{\mu z}$. Ignoring the expansion, we find that for $q \ll 1$ exponentially-growing solutions are restricted in $A$ (narrow resonance); for $q \gg 1$ the unstable bands dominate (broad resonance). There are sharp qualitative differences between these modes – most notably that although for narrow resonance occupation of a particular mode grows smoothly, for broad resonance the effect takes place in jumps. The details of this distinction with illuminating figures (and our source for the interpretation) can be found in [3]. Including expansion complicates the above analysis. Specifically, the phase relationship between the two oscillators is destroyed. This leads to "stochastic preheating", where though the energy from the inflaton is still transferred efficiently, neither the occupation number of a particular mode nor its amplitude are monotonically increasing.

Preheating ends when $\chi$ becomes comparable to $\phi$; here non-linearities restrict the exponential growth ([5] and references therein). Typically parametric resonance lasts for less than a Hubble time. As a result, energy can be rapidly transferred from the inflaton field, even when we require its couplings to be small. After the preheating stage ends, the $\chi$ modes dissipate energy via self-interactions and decay into elementary particles (without a strong restriction on the coupling). These particles then thermalize and yield a reheating temperature. Since the path the to a hot universe can now be traveled more rapidly, substantially higher reheat temperatures are possible when preheating takes place.

## 4. Gravitational fluctuations

In our discussions so far, we have assumed a homogenous, isotropic background. Now we consider the possible effects of allowing linear perturbations to the metric:

$$ds^2 = \left(g_{\mu\nu}^{(b)} + \delta g_{\mu\nu}\right)dx^\mu dx^\nu \tag{4.1}$$

With this refined situation in mind, there are several questions that one could ask with respect to the mechanisms described above.

Perhaps the most obvious of these questions is whether or not the power spectrum of metric fluctuations from inflation evolves during preheating and reheating. More succinctly we wish to know:

$$P_{\delta g_{\mu\nu}}^{(i)}(k) \xrightarrow{\ ?\ } P_{\delta g_{\mu\nu}}^{(f)}(k) \tag{4.2}$$

In the standard scenario the answer is: "No". Causality prevents metric fluctuations with wavelengths much larger than the Hubble horizon at the time of reheating from being influenced; whereas perturbations with wavelengths smaller than the Hubble horizon are washed out during the radiation era. In the standard inflationary scenario, the ratio of tensor to scalar fluctuations is set by the scale of inflation. Preheating can change this significantly. In particular, tensor fluctuations can be generated at the end of preheating, sourced by the spatial gradients in the fields. The amplitude of these fluctuations can be independent of the energy scale of inflation and the peak of the spectrum is inversely related to the energy scale of inflation. For details regarding tensor fluctuations and their possible observational consequences we refer the reader to [4]. There is a possibility of the back-reaction of metric fluctuations influencing the dynamics of reheating [6] which we shall not discuss here.

To summarize, we have provided a brief overview of (p)reheating and the evolution and generation of metric fluctuations during (p)reheating in a simple scenario.

## Acknowledgments

It is our pleasure to acknowledge Lev Kofman for teaching us about (p)reheating and writing readable papers on the subject. We would also like to thank the participants of the summer school for stimulating discussions and the organizers and staff for providing an excellent working environment. SEJ acknowledges financial support from the School and from CNPq.

## References

[1] A.H. Guth, Phys. Rev. D **23**, 347 (1981), A.D. Linde, Phys. Lett. B **108**, 389 (1982).

[2] L. Kofman, A.D. Linde and A.A. Starobinsky, Phys. Rev. Lett. **73**, 3195 (1994) [arXiv:hep-th/9405187]. Y. Shtanov, J.H. Traschen and R.H. Brandenberger, Phys. Rev. D **51**, 5438 (1995) [arXiv:hep-ph/9407247].

[3] L. Kofman, A.D. Linde and A.A. Starobinsky, Phys. Rev. D **56**, 3258 (1997) [arXiv:hep-ph/9704452].

[4] S.Y. Khlebnikov and I.I. Tkachev, Phys. Rev. D **56**, 653 (1997) [arXiv:hep-ph/9701423], R. Easther and E.A. Lim, JCAP **0604**, 010 (2006) [arXiv:astro-ph/0601617], R. Easther, J.T. Giblin and E.A. Lim, [arXiv:astro-ph/0612294].

[5] G.N. Felder and L. Kofman, [arXiv:hep-ph/0606256].

[6] J.P. Zibin, R.H. Brandenberger and D. Scott, Phys. Rev. D **63**, 043511 (2001) [arXiv:hep-ph/0007219].

# BRANEWORLD GRAVITON INTERACTIONS IN EARLY UNIVERSE PHASE TRANSITIONS

R.U.H. Ansari[1], C. Delaunay[2], R. Gwyn[3], A. Knauf[3],
A. Sellerholm[4], N.R. Shah[5,6] and F.R. Urban[7,8]

[1] *School of Physics, University of Hyderabad, Hyderabad-500046, India*
[2] *Service de Physique Théorique, CEA Saclay, F91191 Gif-sur-Yvette, France*
[3] *Department of Physics, McGill University, Montréal, QC, H3A 2T8, Canada*
[4] *Department of Physics, Stockholm University, Albanova, SE-106 91 Stockholm, Sweden*
[5] *Enrico Fermi Institute, University of Chicago, Chicago IL 60637, USA*
[6] *HEP Division, Argonne National Laboratory, Argonne, IL 60439, USA*
[7] *INFN, via Saragat 1, I-44100, Ferrara, Italy*
[8] *Department of Physics, University of Ferrara, via Saragat 1, I-44100, Ferrara, Italy*

## 1. Braneworld cosmology [1]

Braneworld models have been suggested to solve the hierarchy problem and have received renewed attention in string–inspired cosmology. One assumes the brane to be a $(3 + 1)$–dimensional hypersurface embedded in a higher–dimensional bulk with the standard model fields confined to the brane (and gravity propagating in the bulk). Since string theory predicts extra dimensions and contains higher–dimensional objects such as D–branes, it seems natural to assume some underlying string theory motivation, although most braneworld models leave the features of the brane itself quite generic.

We will focus on Randall–Sundrum (RS) type models, which embed one or two branes into a five dimensional AdS bulk. In RS1 a static solution of the 5d Einstein equations was derived with two branes of opposite tension at the orbifold fixed points of the finite fifth dimension. The curvature in the AdS bulk leads to a "warping down" of physical scales and thus provides a hierarchy between the positive tension "Planck brane" and the negative tension "TeV brane". Unfortunately, in this setup general relativity is not recovered on the TeV brane (containing our observable universe). What one finds instead is a Brans Dicke theory with negative BD parameter, which is ruled out by observations. There are two ways to circumvent this problem: one can either consider the RS2 model, in which the negative tension brane is pushed to infinity and our universe is confined to the Planck brane;[1] or one could take the stabilisation of the radion (the scalar field associated with the brane distance) into account, which removes the constraint that the brane tensions have to be opposite, but leads to additional, model–dependent terms in the Friedmann equation.

In the first case standard cosmology is recovered at late times. For simplicity we only consider models with infinite, static fifth dimension (and only gravity in the bulk), but we do not restrict ourselves to the RS fine–tuning condition for the brane tension (i.e. we study an FRW brane with a non–vanishing 4d cosmological constant). The Friedmann equation for this case is:

$$H^2 = \frac{8\pi G}{3} \rho \left(1 + \frac{\rho}{2\lambda}\right) + \frac{\Lambda_4}{3} - \frac{k_4}{a^2}. \tag{1.1}$$

Here, $\rho$ is the energy density on the brane, $\lambda$ is its tension and $G$ is the four-dimensional gravitational constant; $k_4$ is the curvature of the 4d space, and the 4d cosmological constant is given by $\Lambda_4$. This shows that there is a high–energy regime where the dominant contribution to $H^2$ arises from $\rho^2$, whereas the low-energy regime is governed by the usual Friedmann equation $H^2 \sim \rho$.

Under the assumption that the 4d metric does not mix with the static fifth dimension, one finds the following relation between 4d and 5d quantities: $M_4^2 = 6M_5^6/\lambda$, and $2\Lambda_4 = \Lambda_5 + \lambda/M_4^2$. Only by setting $\Lambda_4 = 0$ does one recover the RS constraint, which is equivalent to $M_5^3 = kM_4^2$, where $k$ is the AdS mass scale and related to the bulk cosmological constant by $\Lambda_5 = -6k^2$.

Standard cosmology can also be recovered in RS1 models, at least in the low–energy limit, once we stabilise the radion. It is then no longer necessary to balance the bulk cosmological constant with appropriate brane tension. However, the high–energy behaviour is unclear and depends on the precise radion potential. We will therefore not elaborate further in this direction, but work with the infinite extra–dimensional model with FRW Planck–brane (or a TeV probe–brane).

## 2. Brane inflation [1] and the Electroweak phase transition

As previously noted, (1.1) has a high–energy regime in which $H \sim \rho/M_5^3$. As an immediate consequence, the dynamics of scalar field driven inflation is modified.

---

[1] In this scenario one faces again the hierarchy problem, but one can add a "probe brane" in the infinite AdS bulk at precisely such a distance that the effective scale on this probe brane is again the electroweak scale. The only requirement for the probe approximation to be valid is that the probe brane tension be much smaller than the Planck brane tension. This is actually an appealing scenario, because it would allow us to consider cosmology on a TeV brane with arbitrary (albeit small) tension.

We want to constrain a particular inflationary model using its predictions for the shape and amplitude of the temperature fluctuations (at the time of the last scattering surface) and comparing them to the WMAP data. We will not show the detailed procedure but only quote the main result. In order to have successful 4d inflation on the brane (in the case of chaotic inflation with a potential of the form $V \sim m_\phi^2 \phi^2$, where $\phi$ is the usual 4d inflaton field), one requires:

$$m_\phi \approx 10^{-4} M_5. \tag{2.1}$$

For simplicity, we make a naïve estimate of the reheating temperature. This is not precise, but nonetheless the scenario we describe below holds. Furthermore, a complete theory of (p)reheating is still lacking, and we are thus led to extract the reheating temperature by means of the relation $\Gamma_\phi \sim H$, where $\Gamma_\phi = \alpha m_\phi$ is a constant decay rate for the inflaton; $\alpha$ being its coupling constant with light degrees of freedom. Using this relation and assuming that at the end of the reheating process we are still in the high–energy regime (an assumption readily confirmed for low values of $M_5$), we obtain the approximate reheating temperature

$$T_{RH} \sim 10^3 \alpha^{1/4} m_\phi \sim 0.1 \alpha^{1/4} M_5, \tag{2.2}$$

where the result (2.1) was used in the second step. This shows that low–scale gravity also leads to low reheating temperatures, and it is possible to obtain temperatures lower than the Electroweak scale. This implies that the Electroweak phase transition can take place before the completion of reheating.[2] Consequently, in this scenario we should take into account the entropy dilution after the transition (coming from the inflaton decay), the faster expansion rate during the transition, and the impact of these effects on, for example, the generation of the baryon asymmetry via the Electroweak phase transition.

## 3. Gravitons [2]

In this section we will discuss how gravitons couple to matter fields on the brane. Since we work in an infinite extra–dimensional model, we have a continuous spectrum of KK graviton states, including the conventional 4d graviton as the zero mode. We will work in conformally flat space, keeping in mind that at high curvatures (e.g. in the inflationary epoch) this approximation must be corrected.

---

[2] Several issues arise here, and are mainly related to our lack of confidence with the (p)reheating stage. For instance, since the actual thermalisation mechanism is still unclear we don't know whether we have thermal equilibrium and thus whether it makes sense to speak of an equilibrium plasma, or what its temperature is.

Matter couples to gravitational fluctuations $h_{AB}$ via the energy–momentum tensor $T_{AB}$,

$$S_{\text{int}} \sim \int d^4x \, dy \, h_{AB} \, T^{AB}, \quad A, B = 0 \ldots 4, \tag{3.1}$$

where the 5d energy–momentum tensor of matter confined to the brane becomes (in the weak gravitational field limit) $T_{AB} = \eta_A^\mu \eta_B^\nu T_{\mu\nu}(x)\delta(y_c)$, and $h_{AB}(x, y)$ are the perturbations specified below in (3.2). This means, that after integrating out the fifth dimension $y$ in (3.1), the effective 4d coupling will be determined by the amplitude of $h_{AB}(x, y)$ at the (Planck or probe) brane position $y_c$. We parametrise the 4d perturbations[3] as

$$ds^2 = \left(e^{-2k|y|} \eta_{\mu\nu} + h_{\mu\nu}(x, y)\right)dx^\mu dx^\nu - dy^2, \quad \mu, \nu = 0 \ldots 3, \tag{3.2}$$

where we will introduce the factorisation $h_{\mu\nu}(x, y) = \hat{h}_{\mu\nu}(x) \, \psi(y)$. Plugging this ansatz into the 5d Einstein equations, one finds that $\psi$ obeys a Schrödinger-like equation. Its solutions show a strong localisation at the Planck brane for the zero mode, whereas the massive graviton modes are suppressed there and asymptotically approach a plane wave as we move far away from the brane. This means that the zero mode coupling at the Planck brane, given by $\psi(0)$, will be much larger than that of individual massive KK states (their contribution might be significant nevertheless, as we sum over a large number of them). On a TeV probe brane, on the other hand, we find the opposite coupling hierarchy.

We would like to study the impact of gravitational corrections on phase transitions. We will therefore compute the effective potential for a scalar field that undergoes spontaneous symmetry breaking, like the Higgs. For the Electroweak phase transition (which would be interesting e.g. for baryogenesis models) we would have to couple all standard model fields to the KK gravitons. As a slightly less ambitious step, let us consider the one–loop effective potential of a single scalar field coupled to gravity, which could be relevant for the inflationary era.

The one–loop effective potential is given as a sum over all 1PI diagrams with a single scalar/graviton loop. This can be re–summed into a one–loop vacuum diagram (a loop with all external legs removed). Our task is therefore to calculate this diagram for the massless graviton as well as the continuum of massive KK states. If we are located on the Planck brane, we should integrate over the full KK mass spectrum (up to a cutoff of the order of Planck mass); if we live instead on a probe brane somewhere in the infinite fifth dimension we only need to take masses $10^{-4}$ eV $\leq m \leq 1$ TeV into account (assuming a hierarchy such that the fundamental scale on the probe brane is precisely 1 TeV). In both cases we have

---

[3]Here we already decomposed the 5d KK modes into 4d tensor, vector and scalar perturbations, and considered only the tensor modes.

to impose a UV cutoff on the loop integral. This will not yield a normalisable result: even at one loop we will encounter an explicit cutoff dependence. This is due to the well–known fact that gravity cannot be renormalised as a QFT.

The KK decomposition, graviton propagators and their coupling to matter and gauge fields have been considered for toroidal extra dimensions; we will have to repeat the analysis for the infinite AdS bulk. We hope to see some non–negligible contribution to the symmetry–breaking potential that could result in interesting consequences for the (order of the) phase transition, inflationary dynamics and baryogenesis.

## Acknowledgments

We would like to thank the organisers and staff of the school for creating a very enjoyable and stimulating atmosphere. We are also grateful to Jim Cline and Osamu Seto for helpful discussions. F.U. is grateful to the HEP group at McGill University, where part of this work was carried out. F.U. is supported by INFN under grant n.10793/05. The work of A.K. is funded by an NSERC grant. A.S. is grateful for support from the Carl–Erik Levin foundation. R.G. is supported in part by a Chalk–Rowles fellowship.

## References

[1] R. Maartens, Living Rev. Rel. **7** (2004) 7, [arXiv:gr-qc/0312059], and references therein.
[2] V.A. Rubakov, Phys. Usp. **44** (2001) 871, Usp. Fiz. Nauk **171** (2001) 913, [arXiv:hep-ph/0104152], and references therein.

We apologize for the incomplete list of references, which is due to space limitations.

# INDIVIDUAL SEMINARS

Abstracts

# COSMOKINEMATICS

Mustafa Amin

*Stanford University, USA*

We present a kinematical approach for understanding cosmic acceleration and characterizing departures from the standard concordance model of cosmology (flat $\Lambda$CDM model). In our approach, the expansion rate is parameterized by the second and third derivatives of the scale factor $a(t)$ with respect to cosmic time, namely the present value of the deceleration parameter $q_0$ and the cosmic jerk parameter $j(t) = \dddot{a}/(aH^3)$. Elegant features of this approach include: (i) $j(t) = 1$ in the $\Lambda$CDM model which facilitates the characterization of departures from this model. (ii) A kinematical analysis is by definition independent of the underlying theories of gravity and hence comes with minimum theoretical prejudice. In comparison to the standard analysis using the equation of state parameter $w$ and $\Omega_m$, the kinematical approach uses a minimum amount of prior information. (iii) $j(t)$ naturally enters the equations of motion in modified theories of gravity (for example $f(R)$ theories). When analyzed using our kinematical framework, the distance-redshift measurements from type Ia supernovae and X-ray cluster gas mass fractions $f_{gas}(z)$ provide a clear statistical evidence for a late time transition from a decelerating to an accelerating phase. For models with a constant jerk parameter $j(t) = j$, we obtain $q_0 = -0.81^{0.14}_{0.14}$ and $j = 2.16^{0.81}_{0.75}$. A standard Analysis using $w$ and $\Omega_m$ yields $\Omega_m = 0.306^{0.042}_{0.040}$ and $w = -1.15^{0.14}_{0.18}$. Both approaches yield results that are consistent with $\Lambda$CDM at about $1\sigma$ confidence level [D. Rapetti et al., arXiv:astro-ph/0605683; R.D. Blandford et al., NOAO/Tucson proceedings 2004, arXiv:astro-ph/0408279]. We argue that both kinematical and dynamical techniques should be employed wherever possible.

# ELECTRIC CHARGE ESTIMATION
# OF A NEWBORN BLACK HOLE

## Anton Baushev

*LAPTH, BP 110 F-74941, Annecy-le-Vieux Cedex, France*

We estimate the electric charge of stellar mass astrophysical black holes. They are born most likely as a result of supernova explosions of massive stars. On late stages of evolution such a star consists of a compact and dense core and an extensive rare envelope. Finally the core loses stability and collapses into a black hole. The envelope is erupted out and observed as a supernova phenomenon. During all the process the collapsing core is surrounded by some plasma, which is very good conductor. Usually it is considered as a critical argument that the core, and, consequently, the newborn black hole cannot have any significant charge. However, it disregards that the core rotates and possesses strong magnetic field. If the resistance of the substance is negligible, electric field in the comoving frame of reference must be zero. In the static frame of reference electric field is defined by the Lorentz transform; generally speaking, the charge density $\rho \sim \operatorname{div} \vec{E}$ is not zero.

In order to estimate the charge of the core we calculated the magnetic field around the collapsing core, determined the electric field from it, and calculated the charge density and the integral core charge from the Maxwell equations. We have shown that the charge of a newborn black hole can be significant ($\sim 10^{13}$ Coulombs). Though the obtained value is much bigger than the usual estimations, the charge to mass ratio is small $Q/(\sqrt{G}M) \sim 10^{-7}$, and it is not enough to effect significantly neither on the gravitational field of the star nor on the dynamics of its collapse.

# SUPER ACCELERATION FROM DARK-DARK INTERACTION

## Subinoy Das

*Center for Cosmology and Particle Physics, New York University, USA*

An exciting possibility for dark energy is that its effective equation of state might well be $w < -1$ as various independent analyses of the "Gold" SNIa dataset already suggest. In [S. Das, P.S. Corasaniti and J. Khoury, Phys. Rev. D**73** (2006) 083509, arXiv:astro-ph/0510628], we show that an interaction between dark matter and dark energy generically results in an effective dark energy equation of state of $w < -1$. This can be obtained with a quintessence field $\phi$ which couples to the dark matter via, *e.g.*, a Yukawa-like interaction $f(\phi/M_{Pl})\bar{\psi}\psi$ where is $f$ is an arbitrary function of $\phi$ and $\psi$ is a dark matter Dirac spinor. In the presence of this dark-sector interaction, the dark matter energy density no longer redshifts as $a^{-3}$ but instead scales as $\rho_{DM} \sim f(\phi/M_{Pl})/a^3$.

For a run-away potential, the dynamics of $\phi$ is such that dark matter density redshifts slower than that of $\Lambda CDM$ during recent past. For a fixed matter density today it implies less matter in the past compared to $\Lambda CDM$. An observer unaware of this interaction and fitting the data assuming a $\Lambda CDM$ universe will ascribe this energy deficit in DM to the dark energy density. As a result, the effective dark energy density has two components and its density can be found to increase in the recent past, resulting in an effective equation of state $w < -1$.

We also note that this interaction can lead to other important cosmological signatures. With such a fifth force structure formation gets modified. The fact that baryons do not experience the fifth force, however, may provide us with an explanation for the recent bullet cluster observation.

# NEUTRINO SPIN OSCILLATIONS IN EXTERNAL GRAVITATIONAL FIELDS

## Maxim Dvornikov

*University of Jyväskylä, Finland and IZMIRAN, Troitsk, Russia*

The quasi-classical approach is used to describe the neutrino spin evolution in arbitrary gravitational fields [M. Dvornikov, Int. J. Mod. Phys. D **15** (2006) 1017, hep-ph/0601095]. First we examine neutrino propagation and oscillations in the weak gravitational field of a massive rotating object. The effective Hamiltonian for the description of neutrino spin oscillations and the transition probability are obtained. We also develop the theory of the neutrino spin light in an external gravitational field [M. Dvornikov *et al.*, Int. J. Mod. Phys. D **14** (2005) 309, arXiv:hep-ph/0406114]. Then we study neutrino spin oscillations in the strong gravitational field described by the Schwarzschild metric. This case is important for the neutrino motion in the vicinity of a massive black hole. The neutrino spin evolution is described and effective Hamiltonian as well as the transition probability are derived. We consider neutrino spin oscillations when a particle moves on various orbits around a black hole. It is demonstrated by means of the explicit calculations that spin transitions can occur if a neutrino propagates in the Schwarzschild metric.

# THE DYNAMICS OF SCALAR-TENSOR COSMOLOGY FROM RS TWO-BRANE MODEL

Laur Järv

*University of Tartu, Estonia*

We consider a Randall-Sundrum two-brane cosmological model in the low energy gradient expansion approximation by Kanno and Soda. It is a scalar-tensor theory with a specific coupling function and a specific potential. Upon introducing the FLRW metric and perfect fluid matter on both branes in the Jordan frame, the effective dynamical equation for the the A-brane (our Universe) scale factor decouples from the scalar field and B-brane matter leaving only a non-vanishing integration constant (the dark radiation term). We find exact solutions for the A-brane scale factor for four types of matter: cosmological constant, radiation, dust, and cosmological constant plus radiation. We perform a complementary analysis of the dynamics of the scalar field (radion) using phase space methods and examine convergence towards the limit of general relativity. In particular, we find that radion stabilizes at a certain finite value for suitable negative matter densities on the B-brane. Observational constraints from Solar system experiments (PPN) and primordial nucleosynthesis (BBN) are also briefly discussed. This presentation is based on [L. Järv. P. Kuusk and M. Saal, Phys. Rev. D **75** (2007) 023505, arXiv:gr-qc/0608109].

# MODULAR THERMAL INFLATION WITHOUT SLOW-ROLL APPROXIMATION

## Jinn-Ouk Gong

*International Center for Astrophysics, Korea Astronomy and Space Science Institute,
Dajeon, Republic of Korea*

We study an inflationary scenario where thermal inflation is followed by fast-roll inflation. This is a rather generic possibility based on the effective potentials of spontaneous symmetry breaking in the context of particle physics models. We show that a large enough expansion could be achieved to solve cosmological problems. However, the power spectrum of primordial density perturbations from the quantum fluctuations in the inflaton field is not scale invariant and thus inconsistent with observations. Using the curvaton mechanism instead, we can obtain a nearly scale invariant spectrum, provided that the inflationary energy scale is sufficiently low to have long enough fast-roll inflation to dilute the perturbations produced by the inflaton fluctuations.

# SECOND ORDER GAUGE-INVARIANT PERTURBATIONS DURING INFLATION

## Giovanni Marozzi

*University of Bologna and INFN Section of Bologna, Italy*

We evaluate perturbatively a second order curvature perturbation and a second order gauge invariant scalar field fluctuation during the slow-roll stage of a massive chaotic inflationary scenario, taking into account the deviation from a pure de Sitter evolution and considering only the contribution of super-Hubble perturbations in mode-mode coupling. The spectra resulting from their contribution to the second order quantum correlation function are nearly scale-invariant, with additional logarithmic corrections with respect to the first order spectrum. For all scales of interest the amplitude of those spectra depends on the total number of e-folds. We find, on comparing first and second order perturbation results, an upper limit to the total number of e-folds beyond which the two orders are comparable. This presentation is based on [F. Finelli et al. Phys. Rev. D **74** (2006) 083522, arXiv:gr-qc/0604081].

# FRACTAL ANALYSIS OF COSMIC MICROWAVE BACKGROUND RADIATION

## Sadegh Movahed

*Department of Physics, Sharif University of technology, Tehran, Iran*

Statistical properties of the Cosmic Microwave Background radiation (CMB) data is a unique tool to explore the physics of inflation. In particular, the Statistical Isotropy and Gaussianity of the CMB anisotropy have attracted considerable attention. Since the observed CMB sky is a single realization of the underlying correlation, the detection of Statistical Isotropy violation or correlation patterns pose however a great observational challenge.

In [M.S. Movahed et al., arXiv:astro-ph/0602461] we investigate the Statistical Isotropy and Gaussianity of temperature fluctuations of CMB data from *Wilkinson Microwave Anisotropy Probe* survey, using the Multifractal Detrended Fluctuation Analysis (MF-DFA), Rescaled Range (R/S) and Scaled Windowed Variance (SWV) methods. These methods verify that there is no evidence for violation of Statistical Isotropy in CMB data. The MF-DFA shows that CMB fluctuations has a long range correlation function with a multifractal behavior. By comparing our analysis with the artificial shuffled and surrogate series of CMB data, we conclude that the multifractality nature of temperature fluctuation of CMB is mainly due to the long-range correlations and the map is consistent with a Gaussian distribution.

# EFFECTIVE FIELD THEORY APPROACH TO GRAVITY

## Rafael Porto

*Carnegie Mellon University, Pittsburgh, PA, USA*

In recent years a new theoretical framework has been introduced to describe black holes and neutron stars dynamics. The approach relies on Effective Field Theory (EFT) techniques to deal with the proliferation of scales involved in the problem (see Goldberger's lectures). An EFT approach allows for a systematic account of the internal structure of the bodies through the inclusion of a set of Wilson coefficients into the effective action. Dissipative effects can be also accommodated by including new degrees of freedom in the worldline. Although we do not know the precise nature of these new operators it is possible to express physical observables in terms of correlators. Matching with known results in the full theory is therefore necessary to extract the relevant green functions which allows us to predict effects of absorption in more complicated scenarios. In this talk I will discuss this matching procedure for spinning compact bodies and calculate the energy loss due to absorption for spinning black hole binaries.

# PROBING THE COSMIC DARK AGES
# USING THE 21 CM LINE

## Jonathan Pritchard

*California Institute of Technology, Pasadena, CA, USA*

Observations of the redshifted 21 cm line offer a promising probe of the intergalactic medium (IGM) before reionization is completed. Anisotropies in the 21 cm signal arise from fluctuations in the IGM temperature, density, and neutral fraction and through the Lyman alpha flux. These fluctuations contain a wealth of information about astrophysics and cosmology. At lower redshifts ($z < 30$), the state of the IGM is determined by the nature and number of luminous sources. In this regime, 21 cm observations should shed light on the thermal history of the IGM and how reionization proceeds. At higher redshifts ($30 < z < 200$), it may be possible to probe truly primordial gas and make precision measurements of the D-H ratio. Additionally, if the matter power spectrum can be extracted from the signal, measurements of cosmological parameters can be made. In this talk, I discuss the theory underlying 21 cm observations and explore many of these possible applications.

# COSMOLOGY OF
# LEFT–RIGHT SUPERSYMMETRIC MODEL

## Anjishnu Sarkar

*Indian Institute of Technology, Bombay, Powai, Mumbai, India*

Supersymmetric unification in SO(10) presents several possibilities and many subtleties. Here we consider a particular realisation of supersymmetric SO(10) unification which remains renormalisable upto Planck scale. The cosmology of such a model passes through a Left-Right symmetric phase. When this symmetry breaks domain walls form. We show that a brief period of domination by the domain walls can ensure dilution of gravitinos and hence evade the related problem to cosmology. The requirement that domain walls disappear constrains certain soft parameters of the Higgs potential.

# ELECTRIC CHARGES AND CURRENTS FROM COSMOLOGICAL PERTURBATIONS

## Ethan Siegel

*University of Wisconsin, Madison, WI, USA*

Although the universe is isotropic and homogeneous on the largest scales, it is the miniscule inhomogeneities that are vital for seeding the large- and small-scale structure in the universe. In this talk, I examine the role these cosmological perturbations play in the evolution of charged particles in the universe. I illustrate that the combination of the gravitational force, the Coulomb force, and photon scattering leads to the generation of seed magnetic fields. These same interactions also eliminate any relic charge or currents left over from the early universe, which frees restrictions on models admitting the production of a net electric charge, and severely disfavors many exotic scenarios for the production of magnetic fields in the Early Universe. Based on [E.R. Siegel and J.N. Fry, astro-ph/0604526].

# BARYOGENESIS BY R-PARITY VIOLATING TOP QUARK DECAYS

Federico Urban

*University of Ferrara and INFN, Italy*

Generation of the cosmological baryon asymmetry in SUSY based model with broken R-parity in the baryon sector and low scale gravity is considered. We find that a GUT-like model of baryogenesis is viable at energies below the Electro-Weak phase transition. Moreover the model allows for a long-life time or even stable proton and observable neutron-antineutron oscillations.

# FASTER THAN LIGHT?

Alexander Vikman

*Ludwig Maximillians Universität München, Germany*

In this talk I discuss the nonlinear scalar field theories with nontrivial causal structure. Namely, these manifestly generally covariant and Lorentz invariant theories allow, due to the nonlinearity, the propagation of signals faster than light. These signals are small perturbations on the nontrivial dynamical solutions which break the Lorentz symmetry spontaneously and represent in this way a new kind of "effective ether". This new ether provides a preferred reference frame. There is no superluminal propagation of signals in true classical vacuum where the field and field derivatives vanish. The equation of motion for the field is hyperbolic and the resulting emergent geometry (metric), where the small perturbations propagate, has well defined past and future cones. In the cases studied in our papers [V.F. Mukhanov and A. Vikman, JCAP **0602** (2006) 004, arXiv:astro-ph/0512066; E. Babichev, V.F. Mukhanov and A. Vikman, JHEP **0609** (2006) 061; arXiv:hep-th/0604075, A. Vikman, arXiv:astro-ph/0606033], the effective metric describes stable causal spacetimes. Thus there is no violation of causality for these theories on these backgrounds. I briefly discussed a mechanism of enhancing the gravitational waves in inflationary models of this type. Finally, I concisely presented the results of our second paper where we have shown that it is possible to send information from the inside of the Schwarzschild horizon.

www.ingramcontent.com/pod-product-compliance
Lightning Source LLC
Chambersburg PA
CBHW050519190326
41458CB00005B/1596